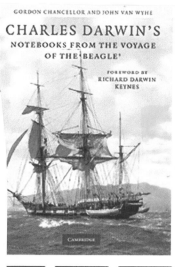

GORDON CHANCELLOR AND JOHN VAN WYHE

CHARLES DARWIN'S

NOTEBOOKS FROM THE VOYAGE
OF THE 'BEAGLE'

FOREWORD BY
RICHARD DARWIN
KEYNES

D0753420

...arles Darwin's Notebooks from the Voyage of the Beagle

1809-1882.

...ted by Gordon Chancellor
...versity of Essex

...n van Wyhe
...iversity of Cambridge

...reword by Richard Darwin Keynes

...sisted by Kees Rookmaaker

...rdback (ISBN-13: 9780521517577)

...lished July 2009

...stock (Stock level updated: 17:01 GMT, 30 September 2009)

...5.00

...iew

...: until now has it been possible to read in book form the immediate notes that
...win himself had written in the little field notebooks that he carried with him ...
...ch takes us all the way to what a young man born two hundred years ago once
...v when he was for some years very far from home.' Richard Darwin Keynes, editor
...Charles Darwin's Beagle Diary (Cambridge University Press) and great-grandson of
...rwin

...ails

...b/w illus. 10 maps 3 tables
   extent: 650 pages
   ?47 x 174 mm
   : 1.41 kg

...mbridge University Press 2009.

£85.- = $132.60

Charles Darwin's Notebooks from the Voyage of the *Beagle*

Transcribed, edited and introduced by
GORDON CHANCELLOR AND JOHN VAN WYHE

Darwin's *Beagle* notebooks are the most direct sources we have for his experiences on his epic voyage, and they now survive as some of the most precious documents in the history of science and exploration, written by the man who later used these notes to develop one of the greatest scientific theories of all time.

The book contains complete transcriptions of the 15 notebooks that Darwin used over the 5 years of the voyage to record his 'on the spot' geological and general observations. Unlike the many other documents that he also created, the field notebooks are not confined to any one subject or genre. Instead, they record the full range of his interests and activities during the voyage, with notes and observations on the rocks, fossils, plants and animals that he saw and collected. They also record his encounters with peoples of the various countries he travelled to, alongside maps, drawings, shopping lists, memoranda, theoretical essays and personal diary entries.

Some of Darwin's critical discoveries and experiences, made famous through his own publications, are recorded in their most immediate form in the notebooks, and published in their entirety here for the very first time. The notebook texts are fully edited and accompanied by introductions that explain in detail Darwin's adventures at each stage of the voyage, and focus on discoveries which were pivotal to convincing him that life on Earth had evolved.

GORDON CHANCELLOR is currently Business Manager at the UK Data Archive, University of Essex. He was formerly based at the Museums, Libraries and Archives Council, and before that, managed several regional museums in the UK. He graduated in geology from the University of Wales in 1976, and completed his PhD in palaeontology at the University of Aberdeen, with periods of research at the Universities of Uppsala and Texas at Austin. He carried out post-doctoral research at the University of Oxford in the early 1980s, and began work on the initial transcriptions of the notebooks. He has published works on Cretaceous palaeontology as well as several scholarly papers on Darwin and the *Beagle*, and is now writing introductions to Darwin's geological publications as Associate Editor of Darwin Online.

JOHN VAN WYHE is a historian of science currently based at the University of Cambridge. He has edited *Charles Darwin's Shorter Publications, 1829–1883*, also with Cambridge University Press. In 2002 he launched *Darwin Online*, the aim of which is to make available online all of Darwin's publications, unpublished manuscripts and associated materials. *Darwin Online* is the largest publication on Darwin ever created, and is used by millions of readers around the world. Van Wyhe lectures internationally, and appears frequently on TV, radio and in the press, to discuss the life and work of Darwin.

The *Beagle* notebooks arranged in order of first use.

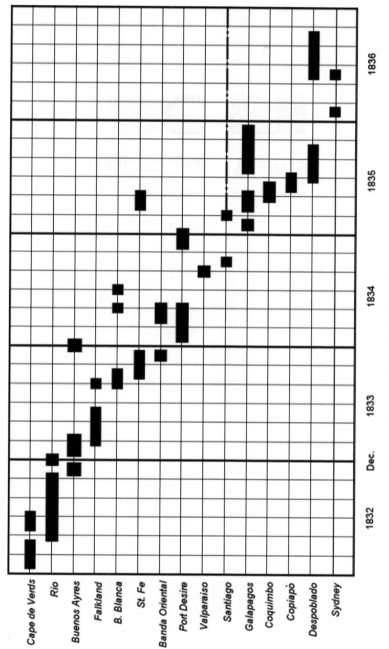

Diagram showing when the *Beagle* notebooks were in use.

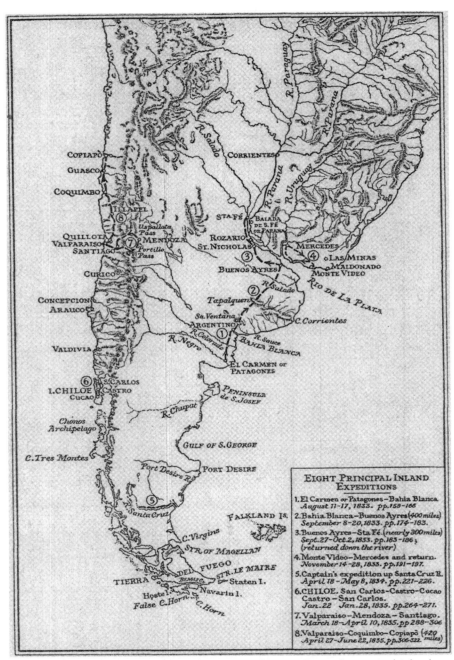

Map of southern portion of South America, showing Darwin's principal inland expeditions, from Barlow 1933.

Map of Rio de Janeiro area, extract from the chart 'Southern portion of South America' from *Journal of researches*.

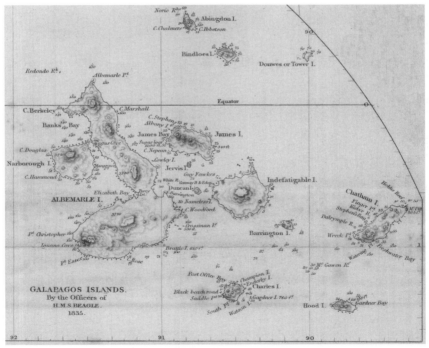

Chart of the 'Galapagos Islands by the Officers of H. M. S. *Beagle* 1835' from the 'Map of South America' in *Narrative* 1.

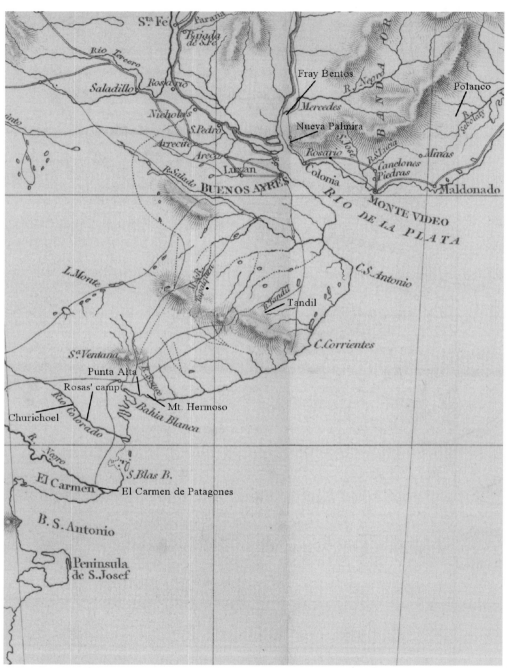

Map of Rio de la Plata, Buenos Ayres area, extract from the chart 'Southern portion of South America' from *Journal of researches*.

Map of Patagonia, extract from the chart 'Southern portion of South America'
from *Journal of researches*.

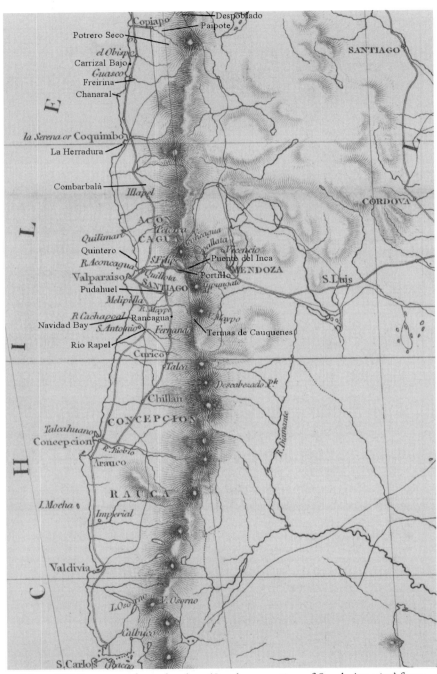

Map of Chile, extract from the chart 'Southern portion of South America' from *Journal of researchers*.

Charles Darwin's Notebooks from the Voyage of the *Beagle*

Transcribed, edited and introduced by

GORDON CHANCELLOR and JOHN VAN WYHE

With the assistance of Kees Rookmaaker

Foreword by Richard Darwin Keynes

CAMBRIDGE
UNIVERSITY PRESS

CAMBRIDGE UNIVERSITY PRESS

Cambridge, New York, Melbourne, Madrid, Cape Town, Singapore, São Paulo, Delhi

Cambridge University Press
The Edinburgh Building, Cambridge CB2 8RU, UK

Published in the United States of America by Cambridge University Press, New York

www.cambridge.org
Information on this title: www.cambridge.org/9780521517577

First published 2009

Printed in the United Kingdom at the University Press, Cambridge

A catalogue record for this publication is available from the British Library

ISBN 978-0-521-51757-7 hardback

CONTENTS

⌒

ILLUSTRATIONS

FOREWORD

~

Much has been written about the 1831–6 voyage around the world of H. M. S. *Beagle*, earning it a reputation as one of the most momentous of all voyages of scientific discovery. It was during its five years that the enthusiastic young Charles Darwin (1809–82) emerged as one of Britain's most promising scientists. He amassed large collections of rocks and fossils, plants, corals, fish, birds and mammals, and his far seeing observations on the geology and biology of many parts of South America visited by the little ship would by themselves have secured him an honoured place in the history of exploration.

It is not, however, for this reason that Darwin is often regarded today as one of the greatest of all naturalists, who has undoubtedly changed much of our understanding of our place in the universe.

The voyage of the *Beagle* was in Darwin's own estimation the most important thing that ever happened to him, and we now know that within a few years of his return to England in 1836 he had become convinced that all life had evolved by some natural process. His prime contribution was that he discovered how such a process might actually work, and then set about collecting the evidence to convince his fellow scientists – many of whom initially held deeply sceptical views – that he was right. In 1859, he at last published his great theory in *On the Origin of Species*, and although it must be admitted that there are still unbelievers, I am fully convinced that any true scientist will agree that Darwin's description of life on Earth as a great tree is correct in all its essentials.

I myself have fortunately been able in my several trips to South America and the Galapagos to visit quite a few of the places seen by Darwin. I was also very happy to find in the Cambridge University Library the surprisingly well preserved manuscript, on the blue note paper that he always used on board ship, of his *Zoology notes and specimen lists* from the *Beagle*, still unpublished in 1994 and now available in my transcription (*Zoology notes*). Not until now, however, has it been possible to read in book form the immediate notes that Darwin himself had written in the little field notebooks that he carried with him, today kept at his old home at Down House. Darwin's grand-daughter Nora Barlow – my mother's first cousin and my own god-mother, the finest of Darwin scholars – was the first to transcribe and publish

Darwin's original journal of the voyage (1933). She followed this with some of his letters home, and extracts from the notebooks (omitting the purely geological material) in her *Charles Darwin and the Voyage of the Beagle* (1945), and then by her invaluable edition of his *Beagle* ornithology notes in 1963.

So Gordon Chancellor's dream, to which I was first introduced about twenty-five years ago, of producing a complete edition of the field notebooks, has, with John van Wyhe's collaboration, come to fruition. We thus have another link which takes us all the way to what a young man born two hundred years ago once saw when he was for some years very far from home.

Richard Darwin Keynes
Cambridge
June 2008

INTRODUCTION

～

In their pages his impressions pour forth with an almost devotional enthusiasm; that they are hastily scribbled and intended for no eye but his own is obvious. But the lapse of more than a hundred years, with all that was to ensue from these fragmentary records, has given them a value like that of the first and imperfect impression of a precious etching.

Nora Barlow, *Charles Darwin and the voyage of the Beagle.* 1945, p. 2.

Charles Darwin created a vast amount of notes and records during the 1831–6 voyage of the *Beagle*. Probably the best known of these is the *Beagle diary*, which formed the basis of his classic work *Journal of researches* (1839). It is still in print. Yet Darwin spent only thirty-three per cent of the voyage at sea in the *Beagle*. Most of his time was spent on inland expeditions. While on shore Darwin usually carried one of fifteen small pocket field notebooks in which he entered his immediate impressions. The notebooks are thus the most direct source for Darwin's *Beagle* experiences. The revolutionary importance and fame of Darwin's work on the voyage of the *Beagle* makes these notebooks some of the most precious documents in the history of science. They are printed in their entirety here for the first time. They offer a rich new vein of material for further study.

As the editors of the *Correspondence* noted, Darwin's *Beagle* records formed five broad kinds: field notebooks, personal diary, geological and zoological diaries and specimen catalogues. The latter category comprised three specimen notebooks for dried plants and animals, three notebooks for specimens in spirits of wine and four geological specimen notebooks for rocks and fossils. Unlike the many other documents Darwin created during the voyage, the field notebooks are not confined to any one subject or genre. Instead they record the full range of his interests and activities during the voyage. They contain notes and observations on geology, zoology, botany, ecology, weather notes, barometer and thermometer readings, depth soundings, ethnography, anthropology, archaeology and linguistics as well as maps, drawings, financial records, shopping lists, reading notes, memoranda, theoretical essays and personal diary entries. They contain a wealth of new and untapped material. For example, dentists are mentioned eight times in the notebooks throughout the voyage and there is a note of a 'denture mended' suggesting previously unknown insights into the state of Darwin's teeth. The notebooks also contain Darwin's first known mention of the Galapagos islands and of their now famous finches. The notebooks also provide evidence of literature consulted or cited by Darwin that is not known to have been in the *Beagle* library. (See the reconstructed list of the *Beagle* library in CCD1 Appendix IV.)[1]

It is often remarked that Darwin had a poor grasp of foreign languages. Yet the notebooks attest the fact that Darwin experienced much of the voyage in Spanish. On St Jago in the Cape de Verds in 1832, at the start of the voyage, Darwin recorded using a 'Spanish interpreter' (*Beagle diary*, p. 30); in Bahia, Brazil, in March 1832 Darwin 'procured an Irish boy as an interpreter' (*Beagle diary*, p. 45) but there are no other references to interpreters until his visit to Tahiti in November 1835. Unlike the *Beagle diary* and correspondence, which were written retrospectively and for others to read, the notebooks were usually written concurrently with or on the same day as the events they record, and for himself alone. They thus preserve traces of Darwin's social experience amongst local peoples. In addition to using local names and expressions, Darwin sometimes began to write in Spanish such as the entry: 'Dom[ingo] Sunday 21 [June 1835]' (*Copiapò*, p. 81). Darwin began to write the day's name in Spanish, corrected himself, and proceeded in English. Another entry: '**Domingo 15**[th] [September 1833]' (*B. Blanca*, p. 55a) was even written over a second time by Darwin in ink, and not altered.

In a very few cases Darwin quoted directly from the notebooks in his publications as in *Journal of researches*, p. 24, 'I see by my note-book, "wonderful and beautiful, flowering parasites," invariably struck me as the most novel object in these grand scenes' (*Rio notebook*, p. 9b). During the Darwin centenary celebrations in 1909 three of the *Beagle* notebooks were displayed at Christ's College, Cambridge, and two at the Natural History Museum (as it is now known).[2] The notebooks were first described by Nora Barlow (1885–1989) in the preface to her edition of the *Beagle diary* in 1933. She later published fairly detailed descriptions of the notebooks, together with extensive extracts from the non-geological parts in her *Charles Darwin and the voyage of the Beagle* (1945), the only book length study of the field notebooks. However, Barlow omitted 'as much as nine tenths' of the complex geological content. Historians now appreciate that Darwin saw himself during, and long after, the voyage as a geologist (see Herbert 2005). Today the notebooks are preserved

---

1 The works clearly consulted by Darwin in the notebooks, but not known to be in the *Beagle* library, are: Anon. 1833, Boué 1830, Cleaveland 1816, Dillon 1829, Febres 1765, Mariner 1817, Péron 1807, Seale 1834, Thunberg 1795–6 and Waterton 1833. There was also a map, apparently by d'Albe from 1819, which has not been identified. Other works referred to, though not necessarily consulted, include Azara 1802–5, Azara 1809, Davy 1830, Fitzinger 1826, Funes 1816–7, Hacq 1826, Helms 1807, Luccock 1820, MacCulloch 1820, Pennant 1771, Scoresby 1820, Stevenson 1825, Vargas y Ponce 1788, Wafer 1699 and probably a report of Murray 1826.
2 [Shipley and Simpson] 1909, p. 37; Harmer and Ridewood 1910, p. 11.

by English Heritage at Down House.[3] The history of the notebooks, and how they came to be preserved at Down House in 1942, is told in Barrett *et al.* 1987, p. 2.

The present edition began in the 1980s when Chancellor prepared working transcriptions of the notebooks in longhand with the intention of publishing them as a book. Owing to unforeseen circumstances the work was never completed. In 2004 van Wyhe invited Chancellor to contribute his transcriptions to *The Complete Work of Charles Darwin Online* (http://darwin-online.org.uk/ [hereafter *Darwin Online*]). Chancellor's transcriptions were keyed into a computer and checked against digitized microfilm images of the notebooks by Kees Rookmaaker in 2006–7. These were then checked by Chancellor, Rookmaaker and van Wyhe against the microfilm and later against the notebooks themselves at Down House. Working transcriptions of the field notebooks were published for the first time on *Darwin Online*. For the present edition Chancellor provided definitions of the geological and palaeontological terms and van Wyhe revised the transcriptions and added editorial and textual notes.

The *Beagle* field notebooks are usually referred to by their former Down House catalogue numbers. Sometimes some of the text on their labels is used. However as the numbers are arbitrary, were not given by Darwin and convey no useful meaning and Darwin's labels are often rather long, we have assigned unique short names to each notebook. These are taken verbatim from the notebook labels written by Darwin. The following table collates these short names with the former Down House catalogue numbers and their English Heritage numbers.

Short name	Down House	English Heritage
Cape de Verds	1.4	88202324
Rio	1.10	88202330
Buenos Ayres	1.12	88202332
Falkland	1.14	88202334
B. Blanca	1.11	88202331
St. Fe	1.13	88202333
Banda Oriental	1.9	88202329
Port Desire	1.8	88202328
Valparaiso	1.15	88202335
Santiago	1.18	88202338

3 The notebook labelled 'Galapagos. Otaheite Lima' (*Galapagos notebook*), although fortunately micro-filmed with the others in 1969, had disappeared from Down House by the early 1980s. Its current whereabouts are unknown. There are three other notebooks at Down House which were not used in the field during the *Beagle* voyage and therefore are not included in this edition. These are *R. N.* or the *Red notebook* (published by Sandra Herbert 1980 and in Barrett *et al.* 1987), the *St. Helena Model notebook* (Chancellor 1990), which is entirely post-voyage, and a very fragmentary notebook labelled 1.1, which, to judge from the London address of CD's brother on the cover and inside back cover, and apparent 1870s dates on some of the many excised page stubs, is also entirely post-voyage.

Galapagos	1.17	88202337
Coquimbo	1.16	88202336
Copiapò	1.7	88202327
Despoblado	1.6	88202326
Sydney	1.3	88202323

The notebooks are here arranged in the chronological order of their first entries. This is approximately the order in which they were originally presented by Barlow and corresponds to the small circled numbers written on the inside covers by Barlow.[4] The present order differs slightly from Barlow's in the position of the *Galapagos notebook* which she placed between the *Despoblado* and *Sydney notebooks*. Our order agrees with the list adopted by the editors of the *Correspondence* who consulted Chancellor when preparing the list in CCD1.[5]

Some periods in the voyage are not covered in any of the field notebooks. Barlow believed that there were probably other, now lost, notebooks. We agree with Armstrong 1985 that this is unlikely, given the care with which Darwin preserved his *Beagle* notes and the existence of loose notes which cover some of the gaps in the notebooks. Loose sheets were used as field notes for Chiloe (January 1835), Hobart Town (February 1836), King George's Sound (March 1836), Keeling (April 1836), Ascension (July 1836) and Bahia (August 1836).

The field notebooks' role in the recording of Darwin's experiences during the voyage has been described by a number of authors from Barlow onwards (e.g. Armstrong 1985). Darwin used them to record in pencil his observations, often, but not exclusively, while he was on long inland expeditions hundreds of kilometres from the *Beagle*, perhaps with no other paper available. A notable exception to this generalization is the latter part of the *Santiago notebook*, which is effectively the first of Darwin's theoretical notebooks. In a letter to his Cambridge mentor John Stevens Henslow (1796–1861), Darwin remarked that he was keeping his diary and scientific notes separate. The field notebooks are documents prior to this distinction because they fed into both types of later manuscripts as well as correspondence.

Darwin later wrote about making notes in the field in *Journal of researches*, p. 598: 'Let the collector's motto be, "Trust nothing to the memory;" for the memory becomes a fickle guardian when one interesting object is succeeded by another still more interesting.' In 1849 he wrote in his chapter on geology for the *Admiralty manual*, p. 163:

---

4 Chancellor 1990, p. 206.
5 The list of notebook names given in CCD1: 545–6 is not always verbatim from CD's labels, e.g. 'Santiago' and 'Tahiti' are not on the *Galapagos notebook* and 'Bathurst' is not on the *Sydney notebook*.

[A naturalist] ought to acquire the habit of writing very copious notes, not all for publication, but as a guide for himself. He ought to remember Bacon's aphorism, that 'Reading maketh a full man, conference a ready man, and *writing an exact man;* and no follower of science has greater need of taking precautions to attain accuracy; for the imagination is apt to run riot when dealing with masses of vast dimensions and with time during almost infinity.

The notebooks were also part of Darwin's expedition equipment. In one of his tips for travellers, Darwin noted that 'by placing a note-book on [a flat piece of rock], the measurement can be made very accurately' (Darwin 1849, p. 161). A re-constructed list of the equipment Darwin carried on his expeditions is provided on p. 583.

The diagram of the *Beagle* notebooks at the front of the volume attempts to show when the notebooks were in use. Most of the notebooks were used on separate occasions, and sometimes more than one notebook was in use at any one time. Therefore the relationships between them are often complex. It should be stressed that this diagram indicates only that a notebook was used in any particular month. The text of the notebooks or the Chronological Register at the end of the volume should be consulted to see whether this was many pages of continuous use, or a few jottings. The Chronological Register, in addition to being the most complete itinerary of the voyage of the *Beagle* yet published, allows the dates and places Darwin recorded in the notebooks to be easily found, despite the fact that he often changed notebooks and therefore they themselves are often not chronological.

Four of the maps at the front of the volume are extracts from the map of the southern portion of South America included with the first edition of Darwin's *Journal of researches* (and in *South America*). For clarity, extraneous place names have been removed and others have been added.

Probably the main reason Darwin did not use one notebook until it was full before taking up another was to protect information not yet transferred to other notes. Once he was back on board the *Beagle* after an excursion he would use the information from the notebook just used as the basis for his lengthy geological, zoological and personal diaries written in ink. Since this process might take weeks, and therefore was often not completed before his next excursion, Darwin took a notebook with him ashore which could be spared, rather than risk losing field notes which had not yet been processed. In this way he had a conveyor belt of field notebooks in various states of use. One can only guess why he sometimes ended up with very incompletely used notebooks such as *Banda Oriental*. Perhaps he preferred using some notebook types rather than others.

There are six manufactured types of notebook which were used almost chronologically, perhaps reflecting successive purchases. This can be seen by the frontispiece which depicts the field notebooks arranged in order of first use.

Type 1: *Cape de Verds, Rio, Buenos Ayres* and *B. Blanca*

All four notebooks have red leather covers with blind embossed edges and are of a long rectangular shape (*c.* 130 × 80 mm) with integral leather pencil holding sleeve and brass clasps. All of the original pencils, if they were included with the notebooks when purchased, are missing. This is true for all of the *Beagle* notebooks; no pencils used are known to survive. The notebooks are between 104 and 112 pages long. Some pages bear the watermark 'J. Whatman 1830'.

Type 2: *Falkland* and *Red notebook*

These two notebooks are long and rectangular (164 × 100 mm) and have brown leather covers with embossed floral borders and brass clasps. The notebooks contained 184 pages, some bearing the watermark 'T. Warren 1830'. Although used at the most widely varying times for any of Darwin's notebooks, they are twins. Similarly, transmutation and expression *Notebooks D* and *M* and *Notebooks N* and *E* are manufactured pairs. The *Red notebook*, which is labelled on both sides 'R. N.', was first published by Sandra Herbert in 1980. Herbert referred to the notebook in 1980 'as the name suggests, red in colour, although the original brilliance has faded'.[6] Both notebooks are now brown though there are very slight traces of red on the front cover of *Falkland* which appear to be part of now faded colouring. Darwin at least twice referred to the former as the 'Red Note Book'.[7] On other occasions he referred to it as 'R N' (e.g. DAR 36: 466a, DAR 118: 103v). It is a curious name given that, so far as is known, Darwin did not name any other notebooks by their colour or appearance. Given that eight of the *Beagle* notebooks are still bright red, it seems unthinkable that the *Red notebook* would be named after its colour in their presence. It was probably named after the voyage when the *Beagle* notebooks were no longer in use. Perhaps *Santiago*, which is black, was the only other notebook in use at the time the *Red notebook* was labelled.

Type 3: *St. Fe* and *Banda Oriental*

These two notebooks (155 × 100 mm) are bound in brown leather with brass clasps. Unlike the preceding types they open along the long side like a book, rather than lengthways like a pocket book. Only *Santiago* opens in the same manner. The notebooks were 244 pages long. The end papers and page edges are marbled. Some pages bear the watermark 'W. Brookman 1828'.

---

6 *Red notebook*, p. 5. In Barrett *et al.* 1987, p. 17, the notebook is described as 'bound in red leather'.
7 See for example DAR 29.3.9, DAR 36.436–7 and the *Red notebook*.

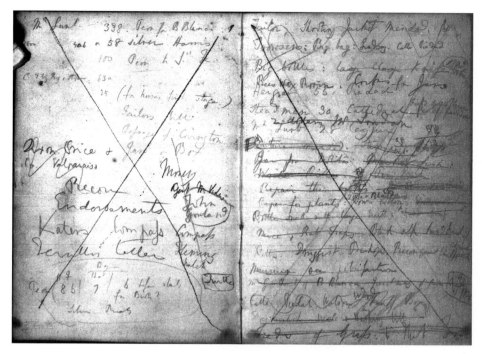

Pages 4a–5a of the *St. Fe notebook*.

Type 4: *Port Desire*

This long and rectangular (170 × 130 mm) notebook is bound in brown leather with floral embossed borders and brass clasp. Its original back cover was missing (probably the one referred to by Barlow 1945, p. 154) but has since been carefully restored with a new one. The last page of the notebook, p. 137, is heavily soiled where it was thus exposed. There were originally 146 pages, some of which bear incomplete watermarks which seem to read 'John Morbey 1830'.

Type 5: *Valparaiso, Galapagos, Coquimbo, Copiapò, Despoblado* and *Sydney*

These six 'Velvet paper memorandum' notebooks are bound in red or black leather (the first and last are black) with the borders blind embossed and with brass clasps. Integral pencil holding sleeves, extensions of the cover leather as in Type 1, are placed on the left inside of the front cover. The paper is yellow edged except for *Sydney* (*Galapagos* is unknown). The notebooks are of an almost square shape varying from 90 × 75 to 120 × 100 mm and were between 100 and 140 pages

Label from a 'Velvet paper memorandum book'.

long. This makes Type 5 the most variable of the *Beagle* notebooks. The inside front covers bear printed labels surmounted by an engraved lion and unicorn.[8]

The pages of these notebooks were treated or coated to react with the metallic pencils, now lost. The paper remains bright white and has a silky or velvety feel. Although the writing in these books looks at first like graphite pencil it is in fact a reaction between the metal of the pencil tip and the chemicals with which the paper was treated. This rendered the writing indelible.[9] Occasionally some notebooks of other types (with untreated paper) have very faint writing which are almost uncoloured scratches which may be from the metallic pencils.

Pages 36–7 of the *Copiapò notebook*.

8 The printed labels are all identical. The label depicted throughout this volume is from the *Coquimbo notebook*.
9 We are grateful to Louise Foster (personal communication) for useful insights on metallic pencils and for supplying us with various types of treated paper and metallic pencils.

Type 6: *Santiago*

This 100 × 165 mm notebook is bound in black paper with black leather spine and was originally 138 pages long. Four pencil holder loops, dovetailed along the opposite cover edges, held the notebook closed when a pencil was inserted. This is not seen on any other *Beagle* notebook. The inside covers are green paper. Inside the front cover there is a collapsing pocket. The manufacturer's label shows that the notebook was made in France, unlike all of the other *Beagle* field notebooks. As will be argued below, *Santiago* was used after the voyage and is labelled identically on both sides as is the *Red notebook* and the transmutation and expression notebooks, a post-voyage notebook labelling practice.

We are not aware of any other Darwin notebooks which match any of these field notebook types. Notebooks 1.1, *St. Helena Model*, the specimen notebooks and all other post voyage notebooks are different types. Some of the *Beagle* notebooks, such as *St. Fe* and *Valparaiso*, have long fine, almost parallel, knife cuts on their covers. These may be where Darwin sharpened his pen or otherwise cut something on the notebooks. Possibly he used them in excising pages from other notebooks. The cuts were made before the notebooks were labelled. It is unknown when Darwin labelled the *Beagle* notebooks, but it was clearly after they were completed, and not all at once as he used different versions of place names, such as 'Isle of France' on the label of the *Despoblado notebook* but 'Mauritius' on the label of the *Sydney notebook*.

The appearance of the notebooks today is less battered and frayed than they were when Nora Barlow first described them in 1933. The 1969 microfilm images reveal flaps of torn leather on their covers and in one case (*Port Desire*) a back cover torn off. The edges of some of the leather covers were worn away. The notebooks have since been carefully conserved so that they appear in rather better condition than when they returned form the *Beagle* voyage. The *Falkland, St. Fe, Banda Oriental* and *Port Desire* notebooks, for example, have had missing pieces of their leather bindings restored.

Editorial policy

The *Beagle* field notebooks are arguably the most complex and difficult of all of Darwin's manuscripts. They are for the most part written in pencil which is often faint or smeared. They were generally not written while sitting at a desk but held in one hand, on mule or horseback or on the deck of the *Beagle*. Furthermore the lines are very short and much is not written in complete sentences. Added to this they are full of Darwin's chaotic spelling of foreign names and cover an enormous range of subjects. Therefore the handwriting is sometimes particularly difficult to decipher. Alternative readings are often possible. Some illegible words are transcribed as well as possible, even when they are obviously not the correct word, when this seems more informative than just listing the word as illegible.

In the transcriptions we have strictly followed Darwin's spelling and punctuation in so far as these could be determined. The transcribed text follows as closely as possible the layout of the notebooks, although no attempt is made to produce a type-facsimile of the manuscript; word-spacing and line-division in the running text are not reproduced. Editorial interpolations in the text are enclosed in square brackets. The page numbers assigned to the notebooks are in square brackets in the margin at the start of the page to which they refer. Italic square brackets enclose conjectured readings and descriptions of illegible passages. Darwin's use of the ſ or long s (appearing as the first 's' of a double 's'), has been silently modernized. Darwin used an unusual backwards question mark (ʕ) which might be based on the Spanish convention of preceding a question with an inverted question mark (¿).[10]

Textual notes are given at the end of each notebook. The notebooks are almost entirely written in pencil. Where ink was used instead this is indicated in the textual notes. Brown ink was used except where otherwise indicated. Pencil text that was later overwritten with ink is represented in bold font.

The length, complexity and need to refer to the textual notes has been minimized by representing some of the features of the original manuscripts typographically.[11] Words underlined by Darwin are printed underlined rather than given in italics. Text that is underlined more than three times is double underlined and bold. (There are no instances of such entries also overwritten in ink.) Text that was circled or boxed by Darwin is printed as boxed. Also text that appears to have been struck through at the time of writing is printed as struck through text. Darwin's insertions and interlineations have been silently inserted where he indicated or where we have judged appropriate.

Paragraphs are problematic; often Darwin ran all of his entries together across the page to save space. We have made a new paragraph when there was sufficient space at the end of the preceding line to have continued there. We have silently added a paragraph break wherever Darwin made a line across the page, apparently at the time of writing, or short double scores between lines separating blocks of text. When long strings of notes are separated with stops or colons and dashes we have left these as written by Darwin.

We have ignored all later scoring through of lines, paragraphs and pages in the interest of readability. Virtually every page and paragraph is scored through, often several times, indicating that Darwin had made use of the material.[12] It has been our aim to make Darwin's notebooks widely accessible and readable, as well as a scholarly edition.

---

10 These marks seem to be present in his *c.* 1827 notes (DAR 91.115v) on reading John Bird Sumner's, *The evidence of Christianity*, see van Wyhe 2008b.

11 Sandra Herbert adopted a similar approach in the *Red notebook*.

12 As photographs of the manuscripts are due to be made available online by English Heritage, this should provide all the information on scoring that any scholar might need.

Finally, and perhaps most importantly, the field notebooks are quite different from all of Darwin's other notebooks, except the Glen Roy field notebook of 1838 (DAR 118), in that they all contain diagrams or sketches. With the exception of a small number of apparently meaningless cross-hatches and doodles, we include photographs of all of these drawings in the transcriptions.[13] To assist the reader the writing contained in these drawings is also transcribed and the sketches turned to the horizontal when necessary. We have provided captions to those that have been identified and which are not already captioned by Darwin or which do not appear to be self explanatory.

Some of the notebooks, notably *St. Fe*, are abundantly illustrated with geological sections which Darwin was able later to 'stitch together', via various intermediate copies, now preserved in the Darwin Archive at Cambridge University Library, into the versions he published in the three works which comprised his *Geology of the Beagle*.[14] Of at least equal interest are the far fewer but equally informative sketches of animals, apparently in some cases having been dissected, and the occasional crude diagram of human interest, such as the floor plan of a house, a tiny drawing of the *Beagle*, and a self-portrait of Darwin as a 'stick man' on the cliffs of St Helena.

With one partial exception, Darwin did not number the pages of the notebooks, and often wrote in them at different times from opposite ends. This edition therefore uses an 'a, b' page numbering system. When a notebook was used starting from opposite ends the pages written from the front cover are labelled 'a', and pages written from the back cover are labelled 'b'. In order to make the transcriptions readable the second sequence, starting with the back cover, is placed immediately after the end of the first sequence. With the original manuscript it is necessary to turn the notebook around and begin reading from the other end. Most of the notebooks have brass clasps and we use the convention of referring to the cover with the hinge attached as the back cover.

Many persons, places and publications are recorded in the notebooks which appear in no other Darwin manuscripts. To make the notebooks more accessible explanatory footnotes are provided. The notes identify persons referred to in the text and references to publications as well as technical terms or particular specimens when these could be readily identified in Darwin's other *Beagle* records. Technical terms are usually defined at their first occurrence or where necessary to clarify Darwin's meaning. The definitions are mostly intended for the general reader. Darwin used some rather outdated geological terms (for example as used by Alexander von Humboldt (1769–1859)) but for definitions he probably started to use those in Lyell's *Principles of geology*, vol. 3 (1833), which Darwin had with him on the voyage

---

13 Only some of the photographs could be supplied by English Heritage in time, therefore other images are taken, with permission, from reference photographs and microfilm.
14 This aspect of CD's 'visual language of geology' has been analysed by Stoddart 1995 and Herbert 2005.

after mid 1834. For more detailed discussion of Darwin's petrological (i.e. rock) terms see Pearson 1996. For zoological names and terms see *Zoology notes*, for ornithology Steinheimer 2004, for insects *Darwin's insects* and for botany *Beagle plants*. As these works have shown, Darwin used the European Latin names for animals and plants he encountered which reminded him of European species. He also used local names. Publications are given as 'author date', but given in full in the bibliography. Short titles are used for references to Darwin's books and articles and some of the standard works on Darwin, as cited in the *Correspondence*. Persons are identified fully at first occurrence and only subsequently if clarity is required. More detailed information about most of these individuals can be found in the Biographical Register in the *Correspondence*, now helpfully made available as an online database.

Content

Summarizing the content of the notebooks in a few paragraphs is not only impossible, but also rather pointless since Barlow 1945 provided an unsurpassably engaging précis. Even Barlow at times had to admit defeat in trying to convey an impression of the hundreds of pages of geological descriptions, diagrams and speculations which fill great swathes of the notebooks, especially those used during 1834 and 1835. A series of quotations, in more or less chronological order, suggests the potential of the notebooks:

> 'Solitude on board enervating heat comfort: hard to look forward pleasures in prospect: do not wish for cold night delicious sea calm sky not blue' *Cape de Verds notebook*, p. 46b.
>
> 'Lofty trees white holes the pleasure of eating my lunch on one of the rotten trees — — so gloomy that only shean of higher enters the profound.' *Cape de Verds notebook*, p. 85b.
>
> 'twiners entwining twiners. tresses like hair beautiful lepidoptera. silence hosannah' *Rio notebook*, p. 27b.
>
> 'View at first leaving Rio sublime, picturesque intense colours blue prevailing tint — large plantations of sugar & rustling coffee: Mimosa natural veil' *Rio notebook*, p. 2a
>
> 'always think of home' *Buenos Ayres notebook*, p. 4b.
>
> 'The gauchos … look as if they would cut your throat & make a bow at same time' *Falkland notebook*, p. 34a.
>
> **'nobody knows pleasure of reading till a few days of such indolence'** *B. Blanca notebook*, p. 12a.
>
> 'Most magnificently splendid the view of the mountains' *Valparaiso notebook*, p. 78a.
>
> 'Rode down to Port — miserable rocky desert' *Copiapò notebook*, p. 14a.
>
> 'But every thing exceeded by ladies, like mermaids, could not keep eyes away from them' *Galapagos notebook*, p. 18a.

The notebooks are of variable length. There are a small number of entries (mainly notes and drawings) which, although contemporary, are not in Darwin's

handwriting.[15] The fine sketch of the Baobab tree in *Cape de Verds*, p. 5a, is almost
certainly by Robert FitzRoy (1805–65), the *Beagle*'s Commander. A small number
of faint coastline sketches may also be by others. The last line of the inside back
cover and p. 1b of the *B. Blanca notebook* are in FitzRoy's handwriting. A few faint
numbers and letters on the labels could be later additions, but it seems impossible to
determine when they were written.

	Pages	Words	/per page	Sketches	Blank	Excised
Cape de Verds	112	3,660	33	20	0	2
Rio	106	4,340	41	13	0	16
Buenos Ayres	106	5,600	53	12	3	2
Falkland	180	11,150	62	10	25	2
B. Blanca	104	6,250	60	12	10	1
St. Fe	234	22,130	95	47	2	11
Banda Oriental	242	9,250	38	8	130	2
Port Desire	180	7,670	43	44	49	2
Valparaiso	104	5,250	51	5	0	2
Santiago	140	8,370	60	6	2	12
Galapagos	100	3,610	36	4	0	2
Coquimbo	136	11,330	83	20	0	0
Copiapò	102	8,040	79	12	39	4
Despoblado	136	6,300	46	20	19	14
Sydney	88	3,130	36	10	10	2
			Mean			
Totals:	2,070	116,080	53	241	289	74

Darwin's use of the notebooks gradually changed throughout the voyage. There
is a symmetry to the density of entries as they gradually became lengthier during the
first year of the voyage, reached a plateau in the middle years, then tailed off in the
last year. The first three notebooks gradually get longer, then there is a large increase
in the *Falkland notebook* which is not only almost twice as long as its predecessors but
for the first time is routinely used for lengthy descriptions. Darwin maintained this
Falkland style of use through the three South American years of the voyage, but it
'spiked' quite extraordinarily in the *St. Fe notebook* of which the majority dates from
early 1835. *St. Fe* is seven times longer than the two shortest notebooks, which are
those used at the beginning and end of the voyage. The daily rate of notebook entries
dropped after Darwin left South America.

The *Santiago notebook* was used at the same time as *St. Fe* and seems to mark a new
development in Darwin's note-taking. In *Santiago* for the first time Darwin started
to separate his theoretical notes from his more observational notes and kept *Santiago*
for theory, also using the exclusively theoretical *Red notebook* from May 1836.

15 Modern additions such as institutional accession numbers are given in the textual notes.

Previous scholars have assumed not only that the use of *Santiago* ceased when the *Red notebook* began, but that the transition from field notes to theory notes is to be seen in the *Red notebook* whereas we believe it is in *Santiago*.

Each notebook is provided with an individual introduction which is intended to assist the general reader to understand what Darwin was doing during the parts of the voyage when the notebook was in use. To facilitate comparing the notebooks with other Darwin manuscripts, such as the *Beagle diary* or *Correspondence*, place names used by Darwin are followed. When this differs from the present-day name, the latter is provided in square brackets on first mention.

The *Beagle diary*, *Correspondence* and other *Beagle* manuscripts, such as the *Zoology notes*, overlap with the notebooks. Citing them on every date in the notebooks would be cumbersome. It is essential to consult these works frequently when reading the *Beagle* notebooks.

Although there can be no substitute for reading the notebooks themselves, the introductions provide an overview of Darwin's scientific development during the voyage. The earlier introductions set the scene and introduce key scientific issues in Darwin's scientific context. As the voyage progresses the introductions become more detailed as Darwin climbed metaphorically and physically into higher and higher realms of geology.

A very important contribution to the literature on Darwin's work in Argentina, a special edition of *Revista de la Asociacion Geologica Argentina* 64, No. 1 (February 2009), appeared after this volume was going to press. Therefore it was unfortunately not possible to utilise this new research (helpfully published in English) in this book.

While the notebooks are overwhelmingly geological they also record Darwin's field work in botany and zoology. Particular attention is given in the introductions to Darwin's gradual accumulation of evidence that something was wrong with current views concerning the 'death' and 'birth' of species, even when this evidence is only faintly recorded in the notebooks. Our discussion culminates with Darwin's realization, recorded in the *Galapagos notebook*, that the land birds in the Galapagos were American types, implying an historical origin on the mainland rather than a special local creation to suit volcanic island conditions.

Darwin published his *Journal of researches* from the voyage in 1839, and until 1846, ten years after his return home, continued to publish his results from the *Beagle* voyage. Much of his later scientific career was an extension of the work he carried out during the voyage. Nora Barlow published *Charles Darwin's diary of the voyage of H. M. S. Beagle* in 1933 and many of his voyage letters with extracts from his field notebooks in 1945. It is our hope that, by now presenting his field notebooks in their entirety, it will be possible for the reader to see, for the first time, the full range of Darwin's activities during the voyage of the *Beagle*. The notebooks, read alongside Darwin's other published records from the voyage, provide an unparalleled opportunity to study the intellectual development of arguably the most influential naturalist who ever lived.

ACKNOWLEDGEMENTS

This work would never have begun if Nora Barlow had not dedicated much of her later life to researching and publishing her grandfather's *Beagle* manuscripts. Barlow's partial publication of the field notebooks is the true inspiration and foundation for our edition and we offer it as a tribute to her memory.

We owe a special debt to Richard Darwin Keynes, Darwin's great-grandson, for his gracious foreword, for his irreplaceable works on the voyage, for much advice based on his deep knowledge of the *Beagle* voyage and for many delightful discussions. We would also like to thank Anne Keynes for her patient hosting of several of these discussions. We also wish to thank Richard's sons, Simon and Randal, for their contributions of knowledge and support. We are very grateful to William Huxley Darwin and English Heritage for permission to publish Darwin's unpublished notebooks.

In an edition such as this, spanning thirty years of work, it is impossible to acknowledge every librarian, curator, archivist, publisher or other individual who has contributed in some way. It is a privilege, however, to acknowledge the contributions of Kees Rookmaaker, without whose checking and typing of the transcriptions the entire project would have been impossible. Kees's enthusiasm in tracking down countless obscure references and his dogged deciphering of hundreds of Darwin's most challenging scrawls, and his ever cheerful determination have made working with him a genuine pleasure and we are truly grateful to him.

Looking back over the years we wish also to record our especial thanks to Philip Titheradge, Peter Gautrey, Janet Browne, the late Frederick Burkhardt, Godfrey Waller, Duncan Porter, Patrick Armstrong, Jim Secord, Nicholaas Rupke, Frank and Jan Nicholas and Sandra Herbert for their untiring help and friendship on numerous occasions. We are also grateful for the help and support of Paul Pearson, Andrew Hart, Paddy Pollak, Frank Sulloway, Tom Glick, Jonathan Hodge, Jeff Ollerton, John Woram, Sergio Zagier, Patrick Puigmal, James Taylor, Adam Perkins, Holger Pedersen, Maria Beatriz Aguirre-Urreta, Louise Foster, Wesley Collins, George Beccaloni, Angus Carroll, Ludmilla Jordanova, Peter, Rikke, Johan, Victoria and Daniel Kjærgaard, Candace Guite, Colin Higgins, Ann Keith, Judith Magee, Joe Cain, Mike Benton, Tim Eggington, David Clifford, Martin

Rudwick, Geoffrey Martin, Antranig Basman and Joanne Poole. Chancellor wishes to acknowledge the help he received from the Royal College of Surgeons in the days when they managed the Darwin collection at Down House. More recently Tori Reeve and Cathy Power, successive curators of Down House, provided access to the notebooks. Duncan Brown provided digital images of some of the notebook pages. Simon Thurley provided much needed assistance.

Our families have given incalculable support over many years. Chancellor wishes especially to thank Karen and Eve for their unconditional support, patience and encouragement. He wishes also to acknowledge the profound contribution of his parents, John and Rita, to furthering his Darwin researches. Sadly neither lived long enough to see this work completed.

Van Wyhe wishes to thank his parents, Richard and Donna, for a lifetime of support and encouragement and Cordula van Wyhe for great understanding and countless cups of tea at the fireside where much of the final editing was carried out. Van Wyhe is deeply grateful to the late Malcolm Bowie and to the Master and Fellows of Christ's College, Cambridge, for welcoming a young Darwinian to Darwin's own College. Christopher Lewis, Suzanne Collins and Ann Mitchell provided relaxing and welcoming escapes from Cambridge. Van Wyhe is also grateful to the late Christopher Mitchell, an encouraging friend who is sorely missed.

Jacqueline Garget and, more recently, Martin Griffiths, our editors at Cambridge University Press, have smoothed many a bumpy path for us through the complex issues and negotiations for publishing an edition of this kind during the Darwin bicentenary. We wish also to thank Beverley Lawrence for her fastidious copy-editing and Margot Levy for her magnificent index.

We are grateful to the Arts and Humanities Research Council which funded *Darwin Online* between September 2005–September 2008, during which time the working transcriptions were digitized and initially edited and made freely available online. We also wish to thank the staff at the Centre for Research in the Arts, Social Sciences and Humanities for providing a base for the project in Cambridge. We are enormously grateful to an anonymous donor for generously providing funds to support *Darwin Online* from September 2008 to September 2009, when no other help was forthcoming.

Above all we are grateful to Charles Darwin for writing and preserving the notebooks, Robert FitzRoy for deciding to take a gentleman naturalist with him on his second trip to South America, and Josiah Wedgwood II for convincing Darwin's father, Robert Darwin, that such a trip would not be disreputable to his second youngest child's 'character as a Clergyman'.

Gordon Chancellor, UK Data Archive, University of Essex
John van Wyhe, Christ's College, University of Cambridge

NOTE ON THE TEXT

⮑

Transcription conventions

[some text] 'some text' is an editorial insertion

⌈some text⌋ 'some text' is the conjectured reading of an ambiguous word or passage

[some text] 'some text' is a description of a word or passage that cannot be
   transcribed, e.g. *[3 words illeg]*

< > word(s) destroyed

<some text> 'some text' is a description of a destroyed word or passage,
   e.g. *<3 lines excised>*

Editors' abbreviations

BC: back cover

CD: Charles Darwin

CUL: Cambridge University Library

DAR: Darwin Archive, Cambridge University Library

EH: English Heritage (Down House Collection)

FC: front cover

IBC: inside back cover

IFC: inside front cover

illeg: one word illegible

Symbols used by Darwin

∴   therefore

∠   angle

∠^{r} angular

Abbreviations used by Darwin

do   ditto

P.B.   Porphyritic Breccia

V   vide

mem   memorandum

THE NOTEBOOKS

Nothing of him that doth fade
But doth suffer a sea-change
Into something rich and strange

William Shakespeare, *The Tempest*

THE *CAPE DE VERDS NOTEBOOK*

~

The *Cape de Verds notebook* takes its name from the archipelago of the same name in the North Atlantic Ocean, off the western coast of Africa. It is bound in red leather with the border blind embossed: the brass clasp is missing. The front of the notebook has a label of cream-coloured paper (68 × 40 mm) with 'Cape de Verds Fernando Noronha Bahia Abrolhos Rio de Janeiro City' written in ink. The notebook has 56 leaves or 112 pages. The text is written in two sequences, pp. 1a–27a and pp. 1b–85b. The entries can be dated between 18 January to 29 March 1832 and 9 May to 10 June 1832. The notebook covers more or less the first half year of the voyage, although it overlaps, in April 1832, with the second notebook, *Rio*. The first dated entry in the *Cape de Verds notebook* is 18 January 1832, three weeks after the *Beagle* left England, and the last is 10 June 1832. The notebook therefore covers Darwin's birthday in February when a very seasick young man turned twenty-three.

Holes were through pp. 1a–16a (coast of Brazil, March 1832) and pp. 17a–26a (Rio de Janeiro, May–June); pp. 77b–85b (Rio de Janeiro, May–June) were bound together by string through single pierced holes and pp. 21b–back cover (Rio de Janeiro, April) were bound together via a larger hole punched through the centre of the pages. These allowed Darwin to fasten sections together, probably using string or thread so that he could easily open the notebook to make new entries on blank pages.

The notebook was the first of Darwin's *Beagle* field notebooks to be used and therefore contains some of his very first recorded observations during the voyage of the *Beagle*. The notebook begins with a rapidly written torrent of calculations, geological sections, measurements of angles, temperatures, barometer readings, compass bearings, diagrams and sketches. As such it is an extraordinary document, which vividly records how in the initial weeks and months of the voyage Darwin was maturing from a mere trainee into an accomplished geologist who had a powerful grasp of contemporary geological knowledge.

Cape de Verds, January 1832

The notebook was first used on the volcanic Cape de Verds Islands, [Republica de Cabo Verde]. The *Beagle* arrived on 16 January 1832, after twenty-one days at sea, and spent twenty-one days in the islands. In his *Autobiography*, p. 77, written for his children and grandchildren forty-four years later, Darwin recalled how 'the very first

3

place which I examined, namely St Jago [Sao Tiago] in the Cape de Verds Islands, showed me clearly the wonderful superiority of Lyell's manner of treating geology, compared with that of any other author, whose work I had with me or ever afterwards read.' Darwin immediately plunged himself into geological theorizing, devouring the first volume of Charles Lyell's (1797–1875) highly important book *Principles of geology* (1830), which was given to Darwin by the *Beagle*'s Commander, Robert FitzRoy.

Much has been written about Darwin's geologizing at St Jago, most recently and in most detail by Pearson and Nicholas 2007. Herbert 2005 devotes a whole section to Darwin's collecting and recording of sections and specimens and provides photographs of his specimens. These can be seen today at the Sedgwick Museum of Earth Sciences in Cambridge. Herbert explains how the young Darwin appropriated Lyell's method of geological interpretation, which was based on causes of change currently occurring (erosion, earthquakes, volcanoes etc.) a method known as actualism. Lyell's particular version of actualism, which later became known as uniformitarianism, argued that in reconstructing the geological past one should assume that the intensity of such causes of change was more or less uniform throughout time.[16] Darwin applied Lyell's method to unravelling St Jago's recent volcanic past, especially in relation to Quail Island [Santa Maria] and Flag Staff [Signal Post] Hill. There a band of limestone exposed in a cliff and obviously composed of marine fossil material and now elevated well above sea level, provided clear evidence of past subsidence and uplift.

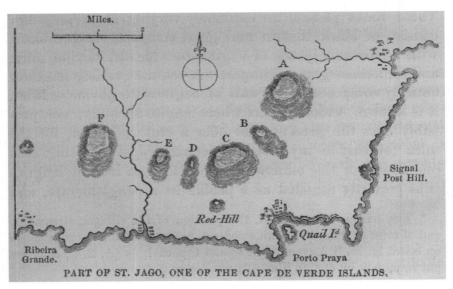

Fig. 1 from *Volcanic islands*.
Part of St Jago, one of the Cape de Verds.

16 CD's response to Lyell is brilliantly summarized in Rudwick 2008.

The *Cape de Verds notebook* records Darwin's on-the-spot geological section on the very first page (p. 1b). After his disappointment at not being able to go ashore at the Canaries, to see for himself the wonderful sights described there by Humboldt and so long dreamed of in Cambridge, Darwin was ecstatic to find that his geological skills were good enough to understand almost immediately the geological history of the Cape de Verds.

Darwin's field training the previous summer in North Wales with Adam Sedgwick (1785–1873), combined with his reading of Lyell and other works and what he was seeing with his own eyes intoxicated him with the realization that he too could become a geologist and explain the past events that had shaped the natural structures before him. In his *Autobiography*, p. 81, he recalled the moment when he first imagined his own geological book: 'That was a memorable hour to me, and how distinctly I can call to mind the low cliff of lava beneath which I rested, with the sun glaring hot, a few strange desert plants growing near and with living corals in the tidal pools at my feet.'

In the long run Darwin's Lyellian conversion was perhaps unfortunate for John Stevens Henslow (1796–1861), who recommended Darwin to read Lyell's book, and who acted as his scientific champion before and during the voyage. Henslow might have taken FitzRoy's offer of a place onboard himself if domestic commitments had not prevented him, and may in some ways have regretted that Darwin's experiences on the voyage eventually turned him away from a view of the Earth's history as a series of 'revolutions'. Henslow had, after all, asked Darwin to read Lyell's book, but 'on no account to accept the views therein advocated'.[17] Which of Lyell's views were not to be accepted is not entirely clear.

When Darwin revisited the Cape de Verds in September 1836, he struck through parts of his foolscap 1832 geology notes from the islands (now in DAR 32), especially their discussion of 'the long disputed Diluvium' i.e. deposits which might be evidence of a widespread catastrophe. This suggests the extent of his increased commitment to Lyell's actualistic methodology. Lyell, who adopted the term 'diluvium' from William Buckland (1784–1856), used it in the sense of a certain type of local superficial deposits. Lyell's methodology was to remain a cornerstone of Darwin's own for the rest of his life. He saw (and experienced) enough evidence of the forces at work on the Earth's surface to accept Lyell's explanations.[18]

Although Darwin accepted Lyell's actualistic methodology, he did not accept Lyell's uncompromising belief in a steady state or non-directional (uniformitarian) Earth history. In the second volume of *Principles of geology*, which Darwin read at the end of 1832, Lyell argued, applying actualism, that since no one had ever seen a

---

17 *Autobiography*, p. 101.
18 For further discussion of Diluvium see the introduction to the *B. Blanca notebook* and Herbert 2005, p. 397 note 59.

new species appear by natural means the origin of species must be some sort of supernatural process. Since new species seemed to appear regularly in the strata to replace those which became extinct (to maintain his steady state Earth), there must be 'centres of creation' to explain the appearance of new species in new environments.

Cape de Verds were previously uninhabited islands when they were discovered and colonized by the Portuguese in the fifteenth century. Darwin later realised that the animals and plants exhibited a close relationship to those of the neighbouring continent of Africa. When visiting the volcanic Galápagos Islands [Islas Galápagos] three and a half years later, he recognized how the animals and plants there were different from those in the Cape de Verds, yet obviously bore an identical relationship to those on the neighbouring continent of South America. The significance of these relationships became clearer to Darwin in the months leading up to his return visit to the Cape de Verds in September 1836, as he pondered why these otherwise similar volcanic archipelagos were populated by such different animals and plants.

By 1836 Darwin realised that the fact that the plants and animals on the Cape de Verds were of an African cast was because that is where they had originated. Any Lyellian Cape de Verds 'centre of creation' was a fiction.

The diagram on the inside front cover shows how FitzRoy used his sextant to help Darwin calculate the height of a Baobab tree. A very similar diagram occurs in the *Beagle diary*, p. 29, and the accomplished sketch on p. 5a is almost certainly the one Darwin mentioned in the *Beagle diary*: 'Cap FitzRoy made a sketch which gave a good idea of its proportion'. There are also references on pp. 10a and 40b to Robert McCormick, the *Beagle*'s surgeon (the role normally doubling with naturalist). The first known use by Darwin of 'entangled' occurs on p. 17b.[19] This word is evocatively used in the last paragraph of the *Origin of species*, p. 489.

St Paul's Rocks, February 1832

From the Cape de Verds the *Beagle* crossed the equator and called at St Paul's Rocks [Penados de Sao Pedro e Sao Paulo] on 16 February. There Darwin could immediately see that the Rocks were not volcanic.[20] The Rocks are in fact still of great geological interest since unlike almost all oceanic islands they are indeed non-volcanic and for this reason Darwin's specimens are still valued today.[21] On p. 77b Darwin compared a gneiss rock he saw on St Paul's to one from Bahia, and on p. 49b he noted the layer of St Paul's 'dung' [guano] which coats the rocks there, later described in *Volcanic islands* p. 45.

19 See Herbert 2005, p. 397 note 7.
20 See Armstrong 2004.
21 See Barlow 1967, p. 54, and Herbert 2005, p. 116.

Fernando de Noronha, February 1832

The *Beagle* remained almost entirely in the southern hemisphere for the next four and a half years. She next called briefly at the islands of Fernando de Noronha, which appeared to be based around a plug of volcanic rock, on 20 February (see p. 44b). The lichen mentioned on p. 45b is recorded as Darwin's dry specimen 309 'from the highest peak of Fernando Noronha'.[22] Darwin jotted a note on p. 53b wondering if the lack of active volcanoes on the Atlantic side of South America indicated 'no volcanic influence East of Andes!'. The wonderfully evocative line from p. 46b is from the sea passage south after this stop, and is dated 25 February. The following ten pages of the notebook seem to be reflections on geological features previously seen, with Darwin exercising his strengthening theoretical muscles while cruising at sea.

Bahia, March 1832

The *Beagle* was next stationed for nineteen days at Bahia [Salvador] in Brazil where, at the end of February, Darwin experienced the euphoria of his first visit to a tropical rain forest. There are many traces of his exhilaration at the time of writing and which were developed in classic pages in the *Beagle diary*. Many of these he published to great acclaim in *Journal of researches* in 1839, or have since been published in the letters he wrote home (see CCD1). In Bahia Darwin had his first taste of the New World proper. Here he first encountered the 'Primitive' [i.e. ancient] igneous and metamorphic rocks of the Brazilian continent (see for example p. 56b), albeit covered with the more recent 'Diluvium'.[23] On p. 72b he switched to 'gen[eral] obs[ervations]', i.e. not geology, describing a delightful interaction with some apparently aggressive ants, and attempting a sketch of a spider's web.

Abrolhos Islets, April 1832

The *Beagle* left Bahia on 18 March for a cruise down to Rio de Janeiro where she arrived, via an examination of the Abrolhos Islets, on 4 April. There is a dramatic demonstration of the enduring value to Darwin of his field notebooks in the 1 November 1839 letter he wrote to Humboldt in which he copied out the depths and temperatures of the sea off the Abrolhos from p. 13a of the notebook.[24] Darwin expanded on this in his section on Rio de Janeiro in *South America*, pp. 142–4.

Rio de Janeiro, April–June 1832

There is a letter from Darwin to his second cousin William Darwin Fox (1805–80), dated May 1832 and today preserved at their old Cambridge College, Christ's. In it Darwin tried to convey his excitement at geologizing: 'it is like the pleasure of

---

22 See *Zoology notes*, p. 372, and Armstrong 2004.
23 See Pearson 1996.
24 CCD2: 239.

gambling, speculating on first arriving what the rocks may be; I often mentally cry out 3 to one Tertiary against primitive, but the latter have hitherto won all the bets.'[25]

Henslow received a letter from Darwin written in Rio de Janeiro on 18 May 1832 from which he read extracts (together with extracts from other letters from Darwin) to the Cambridge Philosophical Society on 16 November 1835. These were printed as a pamphlet in December 1835 which could be counted as Darwin's first true scientific publication.[26] Henslow sent copies to Darwin's father and these had the affect of convincing Dr Darwin that his son would one day make a handsome return on his paternal investment in the voyage. The pamphlet also ensured that Darwin was being discussed in scientific circles back home. Darwin first learnt of the pamphlet in a letter from his sister Catherine (1810–66) which he received in June 1836. He was at first horrified by the news as he would have preferred the chance to check the extracts for accuracy, but quickly realized that this was a minor issue: 'after reading this letter I clambered over the mountains of Ascension with a bounding step and made the volcanic rocks resound under my geological hammer!'[27]

The pamphlet was important in preparing the scientific community to receive Darwin as an accredited geologist and naturalist when he returned to Britain in 1836. Sedgwick also read the letters to The Geological Society two days after Henslow's reading, and this event was singled out by Charles Lyell in his Presidential Address to that august body in February 1836. Lyell realized that Darwin was going to be a great asset to him as a rare early convert to his own gradualist principles.

In Rio de Janeiro Darwin was finally able to settle down for a period of three months, partly as an unplanned bonus from the *Beagle*'s needing to return north to Bahia to check a longitude discrepancy.[28] It was while the *Beagle* was away that three of Darwin's ship-mates, including the young Charles Musters, died of malaria. The last entries in the *Cape de Verds notebook*, from about p. 76b, were made in Rio de Janeiro as Darwin started to collect new species of animals and plants as well as rock samples.[29] These Rio de Janeiro entries overlap with the *Rio notebook* which deals with Darwin's famous excursion to the plantation where he encountered the realities of slavery.

On p. 84b there is a vague but very tantalizing entry that seems reminiscent of some of Darwin's later speculations about species: 'Scale in nature amongst spiders kept up by hymenopt[era] in absence of Carabid [beetles] supplied by the Ants. — may after been less of insects & caterpillars'. The last lines in the notebook, on p. 85b and dated 10 June 1832, are among of the most poetic and beautiful Darwin ever wrote.

25 CCD1: 232.
26 Darwin 1835; *Shorter publications*, pp. 2–15.
27 *Autobiography*, p. 82.
28 See CCD1: 227.
29 See the excellent summaries of his zoological and botanical collecting from Rio de Janeiro provided by Barlow, Keynes, Porter, Smith, Steinheimer *et al.* and the largest collection of published scientific descriptions and identifications of CD's specimens on *Darwin Online*.

[FRONT COVER]

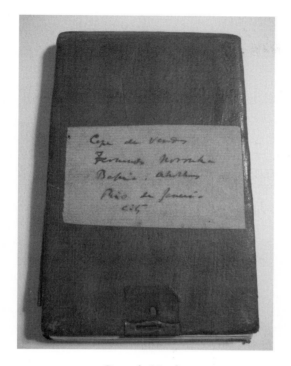

Cape de Verds
Fernando Noronha
Bahia Abrolhos
Rio de Janeiro City

[INSIDE FRONT COVER]

C. Darwin

13/4
  35
20
26 — 20
63 - 20   60 — 30
26 - 30
= + + -
BC : C A : : *[illeg]* : + ∠ B
11, 95 4 2 4
10. 30 2 5
1. 65 19 0 1
  45 39 14 28

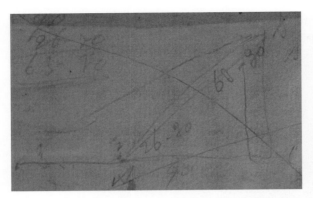

[right angled triangle for calculating height of Baobab tree]

[1a] length of *[illeg]*

[possible sketch of a coastline or mountain range]

       2.8   5.4
                  6
      36.3   32
       2.8

[2a] Tops of prickles scarlet red & other parts tile red[30]
  Don Manuel Rodriguez y da Silva Ponseca

[3a]

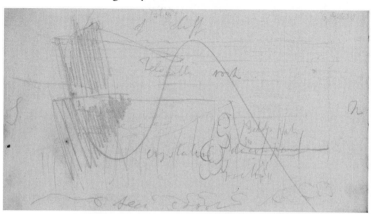

[geological sketch section of cliff] S top of cliff Feldspathic[31] rock Feldspar[32]
Crystalline decompose rock sea coral[33] N

[4a] Baobab[34] 6 ¾
  high ½ strap
  Strap 5" 4
  shadow 9 & per 0 ½ 12
  base 90 feet 26° ½ 12 ½
  90 from trees first position

---

30 *Asterias*, specimen not in spirits 140 in *Zoology notes*, p. 370.
31 Containing noticeably abundant quantities of feldspar minerals.
32 Name given to a group of very common silicate minerals usually whitish in colour occurring in crystals
  or crystalline masses.
33 Marine invertebrate animal with radial symmetry, often in colonies called reefs.
34 The *Adansonia* tree of Africa.

[5a]

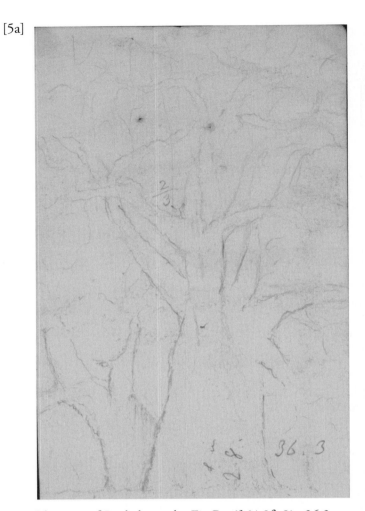

[drawing of Baobab tree by FitzRoy?] ⅔ 2ft 8in 36.3

[6a]   Cliff 42 feet Sand 28.3

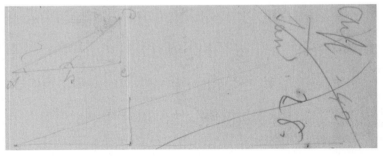

[mathematical sketch of triangle] A B C D

[7a]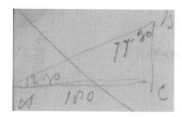

A B C 180 12 30 77 - 30

= + + -

BC : CA : : r : + ∠ B

n 45

Diam 13

Circum 2 feet from the ground 35

[feet]

13| 45.0 (3.4

    39

    ‾‾‾‾

    60

    52

12, 2 5 5 2 7

10, 6 5 4 2 4

‾‾‾‾‾‾‾‾‾‾‾‾

2, 6 0 1 0 3

29

[8a] 45 feet high

Lobularia[35] coating 15 & intermediate 15

Brick in modern Breccia[36]

Fistularia[37] squirt white threads from anus.

Cavolina long horns[38]

[9a] =   4th

Hill 1 E of North concentric crystalline white marble[39] &c

[illeg] conglomerate[40] above trap[41] hills

[Corundum] hills

F R overlies *[Trap]* from hill

[10a] Brick in modern Breccia

Kingfisher eating lizard Mac C

milk cream for *[illeg]*

Shells action under blowpipe try the old ones

Crabs that run & leap

---

35 *Lobularia* or dead man's fingers is a soft coral of order Alcyonacea. See *Zoology notes*, p. xvi.

36 A composite rock consisting of angular fragments of stone, etc., cemented together by some matrix such as lime: sometimes opposed to conglomerate in which the fragments are rounded or water-worn.

37 An echinoderm of order Apodida, specimen 61 in *Zoology notes*, p. 12.

38 An aeolidacean nudibranch, specimen 56 in *Zoology notes*, p. 11.

39 A metamorphic rock derived from limestone.

40 A coarse sediment in which the fragments are rounded (often opposite of breccia).

41 Trappean: hard, splintery lava. Derived from 'trappa', meaning steps in Swedish and referring to the step-like scenery associated with this rock.

Milk from goat on 7th [February 1832] examined on 11th composed of small globules about .0001 in diameter

Dineutes of MacLeay[42]

[11a] N P[t] 29–97

Capa 1/33   35/1

Temp 66°

(Blue colour)

Sounding 23/1 fathoms

26 Lat

Sea only in patches R in diameter

microscope showed small globules floating & transparent irregular shaped /fluids/ in the water.[43]

[12a] patches uniting & joining — showing irridescent colours

26th [March 1832] 10. am 82° 41 M 82° 101 M 81

2° 21' S E ~~Bahia~~.

West 60 ½ 200/1 between 230/1 & 30 bottom

30/1 ~~230/1~~ 250/1

18° 6' S

2° 21' E of /~~Bahia~~/

38 30 ₂

4̶0̶°̶ ̶5̶4̶'̶

36° 6' W

30/1 230/1 250/1

¾ & 7 = 21 [divided by] 4 [=] 5

About 20 ½

[13a] 27th [March 1832]

½ past 8 am	180/1	81°	⅔
— 9 am	150/1	81	⅓
— 10 am	200/1	81	¼
½ pa 11 am	250/1	81	½
¼ pa 1 pm	250/1	81	⅔
¼ pa 2 pm	30/1	81	⅔
3 pm	bottom 20	81	⅔
4 pm	bottom 22	81	½
5 pm	bottom 21	81	½
6 pm	bottom 21	81	½
7	27	81	

42 See *Darwin's insects*, p. 46: '213. 214. Gyrinus allied to Dineutes MacLeay (?) Hab Do. [Cape Verde Islands]' William Sharp Macleay (1792–1865), naturalist and diplomat who devised the Quinary System of classification in which the five main animal groups are represented by 'circles of affinity'.

43 See 'Oily matter on sea', *Zoology notes*, p. 31.

[14a] 27th [March 1832]

10 am Sea coated with irridescent oil.

ship going 5 K & 2 at ½ ~~hal~~ after 11 water yet oily having run 2 & ½ knots <u>containing Confervae[44]</u>

[15a] 27th [March 1832]

Lat 17° 43' S 45'

Long 1° 7' E 19'

Bahia 38° 30' W

[continate] of temp of sea 8 pm — 25 F. 81 ½

(sun set few minutes after 6 o'clock)

[16a] 28th [March 1832]

8 am bottom 28 79° ⅔

no sun. greenish blueish back

sky pale ultra marine & near horizon pale Berlin blue.

10 am   ~~80~~ 79 ¾

4 pm   78 ½

Colour very variable of sea during the day — in the evening very green

9 pm [Both] 10 . 20 F. 66° ½

[17a] Friday [April or May 1832]

NE by E. dip[45] S decomposing strata

Veins[46] of Quartz[47] W ½ S

Greenstone[48] pap.[49] long axis. NW by W massive with <u>garnets.</u>[50] surrounded decomposing gneiss[51]—

SW by W ([&] ½?)

Hills running NE by E ½ E

[18a] 27th [May? 1832] Porphyritic[52] gneiss — cleavage[53] scarcely perceptible on the large scale, but breaks easier in that direction. — Feldspar flattened in same direction

---

44 Microscopic marine algae. See *Zoology notes*, p. 30, and *Journal of researches*, pp. 14–20.

45 The angle by which a bed or stratum has been tipped up.

46 A sheet-like body of a metal ore.

47 The commonest mineral in the Earth's crust, composed of silicon dioxide (silica). Since it is very hard, it tends to remain, for example on beaches, when softer minerals have been destroyed.

48 Low-grade metamorphic rock usually derived from a fine-grained basic (containing relatively low amounts of silica) igneous rock.

49 A small hillock.

50 Silicate minerals generally found in <u>high-grade metamorphic rocks</u>.

51 High-grade metamorphic rock distinguished from granite by its foliated or laminated structure.

52 An obsolete and often miss-used term for a hypabyssal rock with phenocrysts: a crystal which stands out as being larger than the crystals of the matrix containing it.

53 In metamorphic rocks the splitting planes caused by re-alignment of minerals such as mica due to pressure, for example during mountain-building.

good mica[54] slate,[55] balls of diabase

Concentric cleavage has no connection with stratif: ?

Corcovado certainly not.—

What sort of gneiss decomposes?

[19a] 30th [May 1832] —

/Spring/ 74°

do   73 best observ.

Fern 8 ¼ circumference ~~18~~ 19 feet high, to the leaves

Plate on an average 3 foot /thick/ projecting

apparent diam 7 " 3

[20a] big tree circum 7[ft], another 9 [ft] 7'[56]

Running water 68.5°

do 64.5° good observ

beds of fractured quartz & decomposing gneiss dipping small angle S by N. —

[21a–22a excised]

[23a] ferruginous spring

2 sorts of trap

quartz occurs on a break below the hills.—

Traps vary in fracture

[24a] June 5th [1832]

Mica N E ½ E (?)

Mica slate above porphy gneiss?

bed[57] of /porph./ gneiss irregular both /crossed/ & /nearly/ by irregular veins of quartz

/ferrous/ mica slate little talc

Concentric rectangle

[25a] /slate visible/ melted domes /illeg/ they not /illeg/ in /cooling/ than acquire this fracture.

shell /meaning/ conformably[58]

Each hill run not same direction as Cleavage

Dip ever E

---

54 A silicate mineral which splits into thin sheets.

55 Metamorphosed shale in which cleavage is more obvious than bedding.

56 Trees measured during an ascent of Corcovado mountain, Rio de Janeiro. See *Beagle plants*, p. 161.

57 Layer of sedimentary rock.

58 Describes sediment apparently laid down on earlier strata with no evidence of tilting of the earlier strata.

[26a] 7th [June 1832] — Gneiss running NE (some not about) cut through by &
       superimposed by Trap — concent & partly column
       [gap] full of quartz & granitic veins
       Gavia subtending to the coast 42°. —
[27a] Lithomarge[59] Maximillian
       Spider[60]
       Humming Birds
       Planes
       Sea seen through the wood of shady trees
       Annales des Sciences Naturelles[61]
       Then W 66° (?)

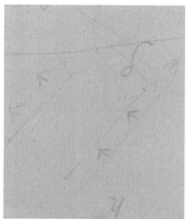

[dip and strike of beds] S N

59 Obsolete term for soft clay-like mineral such as kaolin (OED).
60 Possibly one of the spiders listed in *Zoology notes*, pp. 44–6.
61 Possibly Gay 1833.

[BACK COVER]

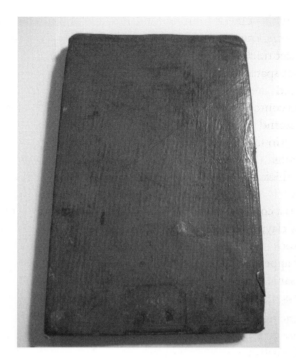

[INSIDE BACK COVER]
Bohn Coqu

                              C. Darwin

~~70~~ .74 N W
30.397[62]
70.5
beginning at 10 am 1.25 in ½ hour
in 6' .38 of rain fell +12" 35 in 3h[r] 1.6
9–11°   1° :8   3 .54

[1b]   30. 368
                    <u>28 58</u>

centre West 1.788
Quail Island
/Mr Farrar Lesson/
Above Breccia from /time/ the dislocation of parent strata

a Greyish white sand with Turbo like Odontoma
4 feet thick

62 Barometer reading.

b a /coarser/ with numerous shells

c Beneath, a conglomerate of black feldspathic rocks with numerous shells & corals

(a) about 4 feet thick

b about a foot spatangus[63] perfect quietly deposited —

[2b] N.W. of Island

(a) ~~heaved up~~ contorted together with inferior conglomerate: allowing a vein of superior breccia to descend

b & c much /trouble/ beneath a /bright/ red aluminous[64] soft rock

Hard feldspaths

[3b] rock 12 feet thick prismatic 5 sides having appearance with curved columns of stratification

Beneath (a) is a confused mass of decomposing rocks in which crystals are visible either red or yellow clay; upon which /rest/ about a foot thick of tufa[65] or carious[66] rock Then feldspathic rock

[4b] all the latter appears to have undergone an upheaving viz after deposition;

probably at same time that (a) was upheaved certainly N by E end of island white sand made up of shells upon which rests carious rock[67]

[5b] then prismatic feldspar

between these & former ones hard white rock with yellow spots

heap of white balls beneath white sand.—

[6b]   ~~carious~~

1 Feldspathic Lava

2 Overlying the white sand

3 beneath white sand

4 between Feldspathic & lower crystalline rocks /on/ white sand stone tufa

[7b] by a well marked vein in lower rock prismatic

crab in white sand

same process now going on shore living Iron found in it.

Upper breccia composed of lower rocks

63 A heart-shaped sea urchin (echinoderm).

64 Meaning silicate minerals containing aluminium.

65 Limestone formed chemically.

66 Decayed.

67 CD recalled in his *Autobiography*, p. 81, 'The geology of St. Jago is very striking yet simple: a stream of lava formerly flowed over the bed of the sea, formed of triturated recent shells and corals, which it has baked into a hard white rock. Since then the whole island has been upheaved. But the line of white rock revealed to me a new and important fact, namely that there had been afterwards subsidence round the craters, which had since been in action, and had poured forth lava. It then first dawned on me that I might perhaps write a book on the geology of the various countries visited, and this made me thrill with delight. That was a memorable hour to me, and how distinctly I can call to mind the low cliff of lava beneath which I rested, with the sun glaring hot, a few strange desert plants growing near, and with living corals in the tidal pools at my feet.'

[8b] new breccia with both upper & lower feldspathic rock
containing iron some inches deep is it cemented by decomposition of neighbouring rocks
standing from
/without wacke/
Some contemporaneous ~~of~~ Dykes[68] ~~standing from~~ *[2 words illeg]* running like trap dikes; upper divisions of prism filled with indurated[69] sand withstanding weathering

[9b] I should think this coast one of short duration —
sand white from decomposition of feldspar (?)
In places every rock is covered with pisiform concretions[70]

[10b] in places impossible to tell whether it is breccia of ~~moden~~ modern or older days.
Going for a hundred yards more South lower beds of white sand become filled with large boulders of lower rocks

[11b] beneath this comes a line of another stratum; more soily & contains large & more numerous shells
a regular bed of oyster remains attached to the rocks on which

[12b] they grew
alternation of sand boulders
/sandy & soily/ sand
The lower carious part of Feldspar perhaps owing to expansive power

[13b] of stream.—
Although in parts this old sea coast is 30 or 40 feet above present level of ocean yet in others the present breccia is again covering it

[14b] owing to it having sunk again most likely as it agrees with that which has been raised
What confusion for geologists.
coast crosses Island NE = SW.
On the E of island only appearing */proof/* is is E on W. in W.

[15b] Jan 18th. [1832]
Lower sorted beds full of Turritella[71] much better seen in S of island
Here the sand is only few inches thick, then the white concretion —
then about 6 feet of rubbish

[16b] which looks as if it were cemented by mortar.[72]
At S end of island sand becomes finer & thicker not so full of organic remains
Feldspathic 18 ft thick
Upper balls */taking/* a concretion

---

68 A sheet of igneous rock which has been injected into the country rock from beneath.
69 Hardened. Sediments are often said to be indurated if they are cemented, for example by lime.
70 A hard lump of rock within a softer matrix. These are often caused by chemicals migrating through the rock and cementing the existing sediment.
71 A genus of marine snail with high-spired shell.
72 Lime and sand cemented by water.

[17b] form balls even 3 feet in diameter

~~Enl~~ The upper rock in places has entangled the lower pyroxene[73] rocks

When does Pyroxene occur ??

[18b] procure more concrete <u>masses</u>

Aplysia hurts[74]

opposite Quail Island

beds higher than island. Carious & amygdaloid[75]

Augitic[76] conglomerate amygdaloid with white matrix

confused mass of Porphy Augit Alum.

[19b] Amygdaloid.—

Signs of sand bank running ~~the~~ N E. but dipping in *[wrong]*. —

concretions

long Dyke 18 inches wide, tortuous. Olivine[77] porphyritic: running some 100 yards

[20b] found the line of coast, about 2 miles — beneath it are olivine or dyke &c above

describ[ed] —

Mortar like

running NW

Sand 2 30 ∠

curiously furrowed like sea coast

[21b] something like top of Quail

Concentric amygdaloid ball more amygdaloid

Behind coast & beneath <u>Feldspathic</u> are the augite & <u>olivine rocks</u>

[22b] Agency of water

Indentation in mountain

Pebbles not rounded

2 system of table lands

Quail Island

[the] *[illeg]* another succession of such

[23b] rocks

Or Quail Island has sunk

Of course near the old coast land should be lower

[24b] 20th [January 1832] N of Quail Island —

Inferior table land

Diluvium[78]

---

73 Silicate mineral such as augite often found in basic igneous rocks.

74 A species of sea hare of order Anaspidea. See *Zoology notes*, p. 18, and *Journal of researches*, p. 6.

75 Of almond-shaped cavities in volcanic rocks caused by gasses that were subsequently infilled by mineral precipitates.

76 As augite, a pyroxene mineral, occurring mostly in volcanic rocks.

77 A silicate mineral generally associated with basic igneous rocks such as basalt.

78 Deposits which have the appearance of having been laid down by floods.

crossed by another <u>long</u> valley: NW in the <u>Baobob</u>
Perpendicular of grey rock of feldspar & augite: new soil (1)

[25b] $\boxed{\text{N 1 ?? . 104}^{79}}$

Sides of hill show great torrents
on side crystalline rock (2) amygdaloid porphyritic
prismatic [trap] of feldspathic rock (3)

[26b] Differences of augite & hornblende[80] ??
4. White rock in stream
5. crystalline altered rock by side of this the higher table land is composed
Dykes in upper crystalline rocks
Red hill
red lapilli[81]

[27b] & concretion balls
NE & by SW above 2 miles comes the chain of augitic hills like base of Quail Island
White magnesia[82] quantity contains fragment of lapilli
origin of concretion

[28b] line of former coast
// to the present
from NW. by W to S: ~ to E
NE. & extent of lava from Red
Hill

[29b] $\boxed{\text{Go to E beyond Praya to Hill}}$
Coast shells & concretions & appear in line between Red Hill & F upon Augitic ~~augiti~~
rock & covered by felspathic rocks
signs of destruction & veins in [lower] rock

[30b] Is the very centre of Island Augitic & the highest table land ???
Some rocks in Quail island have undergone a change [order] locality.
Upper Feldspathic

[31b] 23rd [January 1832]
Lumps of white conglomerate on [table] land of Feldspathic are [the] caught up
// Top Flag Hill & (2) beds of rubbish — dipping alternation red carious rock & breccia
with white matrix

79 Added rock specimen number.
80 An amphibole mineral.
81 Small fragments of rock blasted from a volcano (literally 'little stones').
82 Magnesium-rich limestone.

54
8
35
38
<u>25</u>
160

[32b] near here the covering of former coast 60 /feet/ is very thick

Under flag hill the line of former coast is lower.

Must have been a previous /fern/ hill

[33b] cliff 200.

remarkable ravine

sea coast rest on conglomerate

/1/ central augitic chain

great valley N & S augitic carious & covering conglomerate containing scoriae[83]

—

[34b] (2)

central chain with mica

/great/ valley

pebbles of <u>scoriae</u>

Feldspathic rock

Jan. 26.th [1832] —

higher land between St Martin (1)

[35b] Pap of scoriae. most regular cone surrounding country covered with blocks of this rock

rock beneath Red Hill much contorted & covering of Feldspathic rock

[36b] This cone situated in broad valley (2 & 3) of augitic rocks

2nd [February 1832]

Fuentes. Pap like hills, Feldspathic.

what relation to covering?

Hill N of Praya ~~red~~ like red hill

[37b] /northern white/

large hill

Fuentes — alternate red rubble & <u>Feldspathic</u> dipps NE & the /seven/

[38b] 3^d [February 1832] E coast

Cellular rock exudes at the fi very first formation of coast

lower parts boulders decomposing augitic rock, with dykes

like under flag

has been cut through by instantaneous stream of water

---

83 A solidified ash which is generally denser than water and has many gas holes.

[39b] leaving bed of red lapilli &c

over which thin column curved & Feldspathic rock twisted.

in these parts the rotten rocks beneath FR. more preponderate

[40b] wild cats. kingfisher

Lizard Mac Cormick[84]

Lava perhaps first stream & more melted

Concentric Balls before lava red lapilli

[41b] appear to have come in direction of Flag staff hill bringing with it white stuff & red

/lapilli/ /rubble/ red

Coast 90 feet thick.

Rock that

[42b] overlies carious full of lime crystallised by heat in diverging rays

Shells action under Blowpipe???

in /illeg/ coast

Conus[85] Isocardia[86] & /many/ Ostrea[87]

[43b] /Bottom most/

Milk for animalcule

Coast covered by running & leaping Crabs

[44b] Ornithomya Salient Feronia Leach[88]

20th [February 1832]

Lower rock vitreous feldspar no. 1. prismatic with acicular[89] Hornblende. distant peaks

the same. —

No monocotyledon[90] plants

[45b] ants nearby 3 feet high 2 thick with a tube at bottom

Lichen,[91] mosses

Terns & noddy on trees alight

beautiful pink flowers on the top of mountain on trees

[46b] vitreous feldspar & tufa alternate & dykes — white soil form bed

25th [February 1832]

Solitude on board enervating heat

comfort: hard to look forward pleasures in prospect: do not wish for cold

night delicious

sea calm sky not blue

---

84 Robert McCormick (1800–1900), Surgeon on the *Beagle* 1831–April 1832.

85 A marine turret snail.

86 A genus of bivalve mollusc.

87 The oyster *Ostrea* sp., a bivalve.

88 See specimens 225–7 in *Darwin's insects*, p. 46, collected on St. Paul's Rock.

89 Needle-like.

90 A plant with one seed leaf, with parallel-sided veins and three-part flowers etc., such as grasses and orchids.

91 Specimen not in spirits 309 in *Zoology notes*, p. 372, 'from the highest peak of Fernando Noronha'.

[47b] Geology —
Abrupt valley. Barancas[92]
poor specimen of upper strata. melt. —
interesting parts NB effects of /posterior/ current (red line)
Upper strata
Trap-form dendritic[93]

[48b] Reason for not think the sea has fallen
Pillars could stand on table-land
Lower rock very magnetic & all so
big hill show some appear of table

[49b] land elevated
Tufa angular
~~During~~
Quail Island composed of more compact perhaps from being a mountain upper
part of current
Amygalous with oblong cavities

[50b] St Paul's Lava shore —
outer rock curious weather —
Mistake for dung
Vitreous feldspar crystal remains /outside/ ?
Phonolitic /Fernand/ Noronha

[51b] Trap in do
small crystals of vitreous feldspar /crossing/ ~~each oth~~ not compact
Flag Staff Hill doubly submarine
Does not the cemented bed or diluvium prov water remain on it

[52b] Perhaps the sea really formed the larger valleys.
line of elevation N of Quail Island not nearly so violent
oysters first appear

[53b] on bed of boulders
Fernando
No volcanic influence East of Andes
Connexion of Trachyte[94] with basalt[95] interstratif ∴ of dyke & phonolite[96] on each side
Dyke /travers/ phonolite

92 A deep break or hole made by mountain floods, or heavy rains. Commonly used for a steep bank or
    ravine.
93 'Tree-shaped', such as the layout of streams into a river or finely branching minerals in a rock.
94 A fine-grained, acid igneous rock.
95 A fine-grained, basic igneous rock. The predominant lava of oceanic islands.
96 A rock similar to trachyte that rings when hit with a hammer (sometimes also called clinkstone).

[54b] & Trachyte hills most evidently appear of different origin from Pap-form hillocks
       Effects of P̶r̶a̶ table land to scenery at Praya
       Vegetation on Fernando

[55b] shows age
       /W or R./
       Formation of dyke /or mine/ from cooling & then cracking

[56b] 29th [February 1832]
       Coast crystalline rock dip NWW
       interstratif with bed of granite[97] little mica constituent part arranged in lines.
       chlorite sometimes in veins
       These too pass into each other
       Hornblende in weathered surface || layer crystalised ||

[57b]

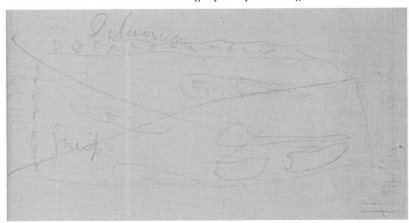

       [gneiss in hornblende rock, diluvium above][98] Diluvium E E Gneiss Hornblende rock

[58b] Diluvium angular gneiss in red clay above where small covering of globular diluvium
       Trap — stratified with Gneiss. — both crossed by a dyke of reddish granite —
       /Much/ conglomerate of the rock

[59b] rather above present water /mark/
       crossing trap there a several of the hornblendic & + dykes sending off branched
       ones. — —

       the junction in all these like knife   [branching dike]

---

97 Coarse-grained, acid igneous rock.
98 See the fair copy of this sketch in the geological diary DAR 32.41v.

[60b] Cary. 80.2
       Worthing 79
       Cary 79.5
       Jones 79.1
       Cary ~~79.5~~ 78.5
       Regulator 76.5
       Hygrom*[eter]*:

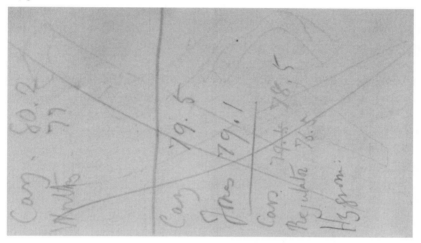

[61b] Gneiss with little mica. ∴ nearly *[Trap]* in neighbourhood large crystals of
       hornblende Gloves *[piece]* of fine hornblende rock imbedded
       Vide Plate
[62b] Will the clay effervesce? No
       Vertical shale[99] running SW by W *[decomps]* NW. by W
       March 3^d — [1832]
       Temp of Spring 80°
[63b] King[100] 3 & Stokes[101] 4
       5th [March 1832]
       About 15 feet of mica then larger mass red clay with nearly verticle *[streak]* of a rock
       resembling *[decompos]* gneiss — dip N by W hornblende crystal & on the top bed of
       quartz
[64b] pebbles // then bed from valleys must be a 2 hundred feet thick
       Spring 80
       red clay 101 ½
       in white sand 108 & spring 81
       Quartzy vein, not very regular in clay

99 A fine-grained sedimentary rock.
100 Philip Gidley King (1817–1904), Midshipman on the *Beagle*, 1831–6.
101 John Lort Stokes (1812–85), Mate and Assistant Surveyor on the *Beagle*, 1831–6.

[65b] blends with a /quartzose/ gneiss
       gneiss dipping regularly about 70° to NNW again
       beds of reddish & /variegate/ clay / carbonaceous matter / with hard veins crystaline
       veins no signs of deposition

[66b] Iron sandstone[102] dipping small 10 angle to the north from sea with beds of clay
       containing boulders of gneiss
       ~~I do not know if any differ from~~
       small formation or /if/ /noone/ go under the gneiss

[67b] The sea is now breaking down the strata
       I suppose has undergone changes as it must have been once horizontal

[68b] 13th [March 1832][103]
       Conglomeratic gneiss trap syenite[104] in a sandy (full of crystaline) base — alternation
       with ferruginous sand & green aluvium[105] slate dip N by E 5° —

[69b] line of contortion ? N by W
       Veins of coal covered by hard sandstone — in green clay

[70b] 14th [March 1832]

                                                                  Clay Slate
       20–30 fathoms extending 6 or 7 miles from the coast /bare. beach. copy/

---

102 A sedimentary rock composed of cemented sand.
103 '13th' is presumably a mistake for 14th as the *Beagle* was at sea on 13 March.
104 Coarse-grained, intermediate igneous rock.
105 Alluvium: associated with deposits laid down by rivers.

[71b]

[green sandstone with contortion?] S of green sandstone blue aluvium

[72b] rock like modern breccia dipping /to S/ by W

Gen: obser:

15th [March 1832]

Small black ant putting everything to flight,[106] — spiders & blattae[107] in great agitation

[73b] a brick stopped their course

Spider  regular web

14° " 19 20' South

38° " 8 West

21st March [1832]

23^d [March 1832]

[74b] 29 [March 1832]

4 o'clock

149/ no bottom

30/ at 4 bottom

N by W ∠ 12°

coarse /fine/ sandstone incrusted:

↑ ↑ 90

27

63

/1 Lava/ the dark

---

106 See specimens 357–8 in *Zoology notes*, p. 29. *Darwin's insects*, p. 48, reports that the specimens have not been found but the ants were "driver ants' (subfamily Dorylinae), probably of the genus *Eciton*.'

107 Cockroaches.

[75b] argillaceous[108] beds
      whiter sand stone *[illeg]* no in lines
      plates of iron
      hexagonal /meaning/ cleavage of trap 2 & ½ W 2 & ½ N

[76b] Basalt escarpement[109] W by ½ S overlying sandstone. & & lifted up with it — modern breccia
      columnar basalt with tufa, bears to other *[illeg]* SW by NE

[77b] Gneiss at Bahia & St Pauls
      Zoophyte[110]
      May 9th [1832] Lagoa[111]
      Red & white clay forming the rounded hill. /signs/ of beds D SSE
      Granite (?) Feldspar Porph concentric layers on the grand scale enormous block not /incumbent/ vegetation landscape in /momentum/

[78b] 3 salt water spiders one in W B: /one/ lengthened one /brown/ rounded hills running (Sugar loaf) W by S & N by E
      15th [May 1832]
      dip gneiss ESE ½ S Hills ENE
      19th [May 1832] — strike[112] SSW 59° D ESE gneiss fine grained beds. good observation.

[79b] beneath porphyritic syenite — garnets edge well marked. —
      /serpen/ soil decomposed like Bahia process visible. — boulders & greenstone lying about

[80b] 6. 6. ~~diameter~~ circum small
      height 4.1
      height 5.
      6.4 diameter *[illeg]*
      ~~height~~ 7 ¾ circum large
      height 6 ½
      if round. 1[f] " 2 ½
      if square. 1[f] " 4 ¼
      height 10 ¾

108 Very fine-grained sediment such as shale.
109 A ridge of landscape where a harder layer of tilted rock has resisted erosion more than the layers above or below.
110 A general term at the time for organisms which seemed to sit on the boundary between plants and animals, such as bryozoa, sponges, corals.
111 Lake or lagoon.
112 The compass bearing of the horizontal line perpendicular to the dip.

[81b] June 2^d. [1832] Inferior mica slate (Vide specimen) rock bed of about 6 inches of gneiss —

Superior gneiss abounding /patched/ with mica & garnets: both something inbedded in large crystals of Feldspar

Running NE by N

I think the water have levelled this valley. hard mud dyke

[82b] sky intense blue dead clear clouds at the Corcovado

fewness of bird swallows quiet. — general discussion on /rock/ bird /beneath/ &c

The /hardness/

gneiss decomposing & covered by a more compact part, is is if

where constituent parts

[83b] of the gneiss are more abundant, that this rock most easily decomposes

Clouds passing over Corcovado & yet not stopping. —

Leucauge ~~web~~ tissue above rock[113]

Tetragnathus Long Epeira[114] near the water web vertical

Travellers too much seek contrasts

[84b] shales in proportion so much darker

Scale in nature amongst spiders kept up by hymenopt[115] in absence of Carabid supplied by the Ants. —

may after been less of insects & caterpillars

June 10th [1832]

gneiss dipping SW

silence well exemplified

[85b] rippling of a brook —

Lofty trees white holes. the pleasure of eating my lunch on one of the rotten trees — — so gloomy

that only shean of higher enters the profound. — tops of the trees /enlumined/ cold damp feel

Textual notes to the *Cape de Verds notebook*

[IFC] *page written perpendicular to the spine.*

C. Darwin] *ink, written parallel to the spine.*

1.4] *Down House number, not transcribed.*

88202324] *English Heritage number, not transcribed.*

13/4…35] *written parallel to the spine.*

[1a] 2.8 … 32] *written upside down from other entries on the page. Page is heavily stained.*

---

113 An orb-weaving orchard spider. See *Zoology notes*, pp. 38–9.

114 A spider of the genus *Aranea*.

115 Hymenoptera: social insects (wasps, ants, bees, etc.).

[2a]   Don … Ponseca] *not in CD's handwriting.*

[6a]   *sketches drawn perpendicular to the spine, not in CD's handwriting.*

[7a]   *sketch drawn perpendicular to the spine.*
        13…52] *grey ink.*

[18a] Corcovado … not.—] *ink.*

[27a] *sketch written upside down from other entries on page.*
        Then …(?)] *written perpendicular to the spine.*

[IBC] *page written perpendicular to the spine.*
        'C Darwin' 'Bohn' and 'Coqu' *written parallel to the spine.*
        in 34' 1.6] *ink.*
        1] *Barlow number, not transcribed.*

[1b]   /Mr Farrar Lesson/] *written perpendicular to the spine, over other entries.*

[25b] 104] *added rock specimen number.*

[60b] *there are some marks that may be obscure sketches on this page.*

[70b] Clay Slate] *overwritten by* '20–30 … coast'. *There is possibly a faint sketch of concentric semicircular lines radiating downwards below the date.*
        20–30 … coast] *written perpendicular to the spine.*
        /bare … copy] *blue grey ink, written upside down from other entries on page.*

[71b] *sketch drawn perpendicular to the spine.*

[74b] ↑ ↑] *ink.*

[80b] *page in ink.*

THE *RIO NOTEBOOK*

⌒

The *Rio notebook* is named after Rio de Janeiro (River of January), Brazil. The notebook is bound in red leather with the border blind embossed. The brass clasp is intact. The front cover has a label of cream-coloured paper (53 × 44 mm) with 'Rio de Janeiro excursion city. M. Video Bahia Blanca' written in ink. The notebook had 53 leaves (or 106 pages) of which 49 leaves survive. The entries were written in two sequences, pp. 1a–20a and pp. 1b–86b. The entries relate to April 1832 (pp. 1b–41b), June–October 1832 (pp. 41b–86b) and December 1832 (pp. 15a–20a). The first entries date from the *Beagle*'s arrival in Rio de Janeiro in April. Almost half the notebook is devoted to Darwin's first inland expedition, after which the *Cape de Verds notebook* was used to cover the next six weeks of his stay in Rio de Janeiro, up to mid June.

The *Rio notebook* covers Darwin's first serious work on the geology of the eastern side of the South American continent and records many of his rapidly evolving interpretations of what he saw. The notebook also records Darwin's first excavations of fossil mammals, several of which were later identified as new species. Consideration of the links between these fossils and living species was one of the three main kinds of evidence that later convinced Darwin that species must change over time. It was this that Darwin referred to in the very first sentence of the *Origin of species*:

> When on board H. M. S. 'Beagle', as naturalist, I was much struck with certain facts in the distribution of the inhabitants of South America, and in the geological relations of the present to the past inhabitants of that continent. These facts seemed to me to throw some light on the origin of species — that mystery of mysteries, as it has been called by one of our greatest philosophers.

Darwin referred to the lasting impression of encountering fossil bones of extinct mammals which were clearly similar in certain characters, such as bony armour, to species known to live only in South America. Darwin was rapidly becoming familiar with South American creatures, such as the armadillo, which has bony armour, so the similarities between the fossil animals and the living ones were immediately apparent. It is remarkable that this realization dated from a very early stage in the voyage, in fact to September 1832 as this notebook reveals.

Darwin's fossils generated considerable excitement when they were unpacked in Cambridge and displayed at the British Association meeting in 1833. News of this reached Darwin, via Henslow, in March 1834. The fossils themselves were described by Richard Owen (1804–92) in *Fossil mammalia* (1838–40).

Rio de Janeiro, April–June 1832

The first forty-one pages used in the *Rio notebook* record Darwin's expedition to the estancia of an Irish merchant, Patrick Lennon, *c.* 250 km (155 miles) ENE of Rio de Janeiro. Darwin was in 'a perfect hurricane of delight'.[116] He described the expedition at length in the *Beagle diary*. The expedition started on 8 April 1832, p. 1b. Darwin had five travelling companions, including Lennon, who turned out, on reaching his estate, to be capable of great cruelty to his slaves. Darwin recorded the temperature at the start of the expedition as 104 °F in the shade at 3 p.m. There follow pages of delightful descriptions of the forest epitomised by 'wonderful, beautiful flowering parasites' which was quoted verbatim by Darwin from his 9 April entry, p. 9b in the *Beagle diary* and *Journal of researches*.

Darwin and his fellow travellers travelled from Praia Grande, now part of Niterói, and made their way through the lake country via Lagoa Marica and Mandetiba to Addea de St Pedro [now Sao Pedro D'Aldeia] on 11 April, having endured some tough travelling and 'waiting an hour for breakfast', p. 10b. They stayed the night of the 11th at a Venda about 5 km south of Marica at the entrance to the Rio Macaé and it was there that Darwin suddenly recorded the horrors of feeling seriously ill when a long way from home, p. 13b. Luckily, the next day, the 12th, 'Cinnamon and port wine cured me', p. 14b.

By the night of the 12th they arrived at the Sosego (or Socego) [Sossego] coffee plantation, some 50 km north of Macaé, where they stayed for two nights before proceeding to Lennon's estate. On p. 16b Darwin attempted a sketch of the layout of the rooms at the Fazenda, which he described in the *Beagle diary*. He seems to have been rather taken with a certain Donna Maria, p. 17b, the daughter of the Fazenda's owner Manoel Joaquem da Figuireda. The dangers of travelling so far from home are poignantly suggested by the way Darwin scratched out very emphatically the word 'villain' on p. 19b. This was presumably for fear of the entry being seen by 'our host' Senhor Figuireda, to whom the description clearly applies, immediately substituting the far more complimentary description of him as an 'enterprising character'.

Darwin was also impressed by the conditions in which the slaves were kept at Sosego, stating that 'I should think in this family the blacks were decidedly happy', p. 20b. In the *Beagle diary* he expanded on this, describing in some detail the management of the Fazenda and its more than one hundred slaves. Darwin felt that 'As long as the idea of slavery could be banished, there was something

---

116 CCD1: 232. The first part of this notebook was exceptionally well covered by Barlow 1945, pp. 158–65.

exceedingly fascinating in this simple & patriarchal style of living' and he believed that 'In such Fazendas as these I have no doubt the slaves pass contented & happy lives.' To Darwin a benevolent ruling gentry seemed synonymous with civilization. What mattered most to Darwin was cruelty, and the notebooks are strewn with references to his heartfelt sympathy for the suffering of other people and even more numerously of animals.

Darwin arrived at Lennon's estate on 15 April, p. 23b, and the notebook contains fascinating details of the crops being grown and some beautiful descriptions of the forest wildlife, such as the 'twiners entwining twiners. tresses like hair beautiful lepidoptera. silence. hosannah', p. 27b. He arrived back via an inland route through Madre de Dios [Rio Bonito] and Fregueria de Tabarai [Itaboraí] at Praia Grande in Rio de Janeiro on 23 April, but not before being 'plagued about our horses not having a passport', p. 38b. Many of Darwin's books, instruments and gun cases were damaged by a swamping in the surf when Darwin moved into a cottage in what is now Rio de Janeiro's South Zone at Botofogo [Botafogo] on 25 April, p. 39b. But he clearly enjoyed the company there of some English gentlemen and their families, such as the Astons, p. 41b.

At this point, Darwin switched to the *Cape de Verds notebook* for six weeks, until the *Rio notebook* came back into use in mid June. Considering the two notebooks together it is possible to see how Darwin became fascinated by the 'Primitive' meta-morphic rocks around Rio de Janeiro. He noted, *Cape de Verds notebook*, p. 77b; *Rio notebook*, p. 41b, the 'enormous blocks' of gneiss at Tajenka caught up within another type of gneiss, both types, however, foliated in the same direction and cut by a dyke of granite. This was described in *Volcanic islands*, pp. 132, 143. Unusually for Darwin, he made not attempt at an explanation for this complex field relationship.

This part of the *Rio notebook* covers the quite extraordinary occasion in Rio de Janeiro on 23 June when Darwin collected sixty-eight species of beetles in one day.[117] Unfortunately his field notes are very sketchy for that week, and are most notable for the sketch on p. 43b of the gneiss mentioned above which resembles, when turned to the side, two figures.

The *Beagle* left Rio de Janeiro for Monte Video on 5 July, p. 47b. Darwin was seasick much of the next twenty-one days of sailing when the weather was bad, such as the terrible gale of 15 July, p. 49b, and he made a sketch of an unusual halo in the sky on p. 48b. He also recorded a Grampus whale, porpoises and flying fish, and on quieter days he wrote many letters to his family, to old College friends and to Henslow.

The number of pages of the *Rio notebook* covering the period from June to December 1832 is approximately equal to those used only for the April expedition, and this reflects the long periods when the *Beagle* was cruising up and down the

---

117 *Journal of researches*, p. 38; see *Darwin's insects*, p. 58.

coast, as the cool southern winter gradually gave way to the spring and summer. These cruises took Darwin from Rio de Janeiro down to Monte Video [Montevideo], 5–26 July, from Monte Video down to Bahia Blanca, 19 August–6 September, Bahia Blanca back to Monte Video and Buenos Ayres, 19 October–2 November and finally from Monte Video down to Tierra del Fuego, 26 November–16 December. On this last cruise Darwin had the new second volume of Lyell's *Principles of geology* (1832) to read between bouts of seasickness.

Monte Video, July–August 1832

The *Beagle* arrived in Monte Video, in what was then the province of Banda Oriental [Uruguay] on 26 July, p. 52b, and stayed there until 19 August. In Monte Video Darwin immediately made notes on the mica slates and schists. According to the *Beagle diary* Darwin landed at The Mount on 28 July. On pp. 53b, 55b, 56b there are diagrams apparently made on The Mount, perhaps with the aid of surveying instruments from the officers of the *Beagle*. No other diagrams in the notebooks were made using a straight rule. His descriptions were eventually published in *South America*, pp. 145–7. On pp. 57b–59b he used the word 'entangled' or 'intangled' at least three times.

The *Beagle* crossed the Rio Plata to Buenos Ayres on 1 August but was prevented from landing by a quarantine for cholera, so Darwin spent more time collecting and geologizing in Monte Video. He was also caught up in the insurrection during the second week of August, but the notebook is silent on this. On p. 62b he mentioned the 'Capincha' or Capybara he shot on 15 August, after which there are no entries until 22 September.

Bahia Blanca, September–October 1832

The *Beagle* left Monte Video for Bahia Blanca in Patagonia (now Argentina) on 19 August arriving there around 6 September, staying until about 17 October. It was during this period that Darwin first experienced the thrill of unearthing fossil bones with his own hands. This feat took three hours of hard labour, but was only the beginning of a process of packing up and shipment to England, research and eventual publication in *Fossil mammalia*. On pp. 64b–65b he recorded the cliff section and on the very next page made a momentous entry:

> in the conglom[erate of pebbles and shells] teeth & thigh bone. Proceeding to the NW — there is a horizontal bed of <u>earth</u>, containing much fewer shells, but armadillo — this is horizontal but <u>widens</u> gradually hence I think conglomerate with broken shells was deposited by the action of tides <u>earth</u> quietly

This was a very significant discovery. Darwin thought that the conglomerate containing sea shells and bones was formed in an estuary environment and was *not* a relic of a sudden catastrophe.

When back on ship writing up his geological diary Darwin noted that it was 'impossible to behold [the conglomerate] without immediately saying that it is the mass of earth which a debacle tearing across the country would deposit' (DAR 32.53). He also remarked: 'Some geologist [sic] have been surprised that the extinction of land-animals, has not occurred, without the destroying the inhabitants of the sea; this would seem to be a case in point' (DAR 32.71-2). These important manuscripts need closer study. They are discussed in more detail in the introduction to the *Buenos Ayres notebook*.

Darwin went back to his original field notebook view of Punta Alta seven years later in his *Journal of researches*, p. 95, and then in much more detail in *South America*, p. 82 *et seq.*, where he published a woodcut version of the section drawn on notebook pp. 73b–75b.[118] Some of the South American mollusc species, plus a barnacle and two coral species, were identified by another great explorer, Alcide d'Orbigny (1802–57), as species still living on the same coast. By that time Richard Owen had described and named all the mammal fossils from the voyage in *Fossil mammalia* and had proved beyond doubt that most were extinct species, such as the *Megatherium cuvieri* referred to on p. 81a, closely related to the present day mammals of South America. Even today uncertainties remain as to why these species became extinct – although the arrival of North American predators and later of humans seems likely at least in some cases – but the point for Darwin was that a general catastrophe would surely have killed the marine species as well as the mammals.

In one of Darwin's first scientific papers, 'A sketch of the deposits containing extinct Mammalia in the neighbourhood of the Plata' read to the Geological Society on 3 May 1837, he referred to Bahia Blanca and indicated that he accepted a gradualistic Lyellian explanation for the extinction of the mammals. He also accepted Lyell's view that species have 'life spans' and that mollusc species have longer life spans than mammal species. Darwin also stressed the universality of this 'law' which applied to South America as well as to Europe.[119] On p. 79 there is a tiny drawing of a ship which occurs in a sketch map of Punta Alta. This ship is almost certainly the *Beagle*.

Monte Video, November 1832

The *Beagle* left Bahia Blanca for Monte Video on 19 October, arriving back at Monte Video around 26 October, then from Monte Video on 30 October to Buenos Ayres arriving there for the first time on 2 November, where the *Buenos*

---

118 The woodcut is reproduced in the introduction to the *B. Blanca notebook*. See also Herbert 2005, p. 100.

119 On Lyell's view of species life-spans see Rudwick 2005, p. 98 *et seq.*, and the introduction to the *Banda Oriental notebook*.

Ayres notebook was used for one week. This apparently trivial point gains importance when considering J. W. Judd's firm belief that Darwin first became aware of the potential relevance of his fossils to the species question in November 1832.[120]

It appears that the last pages of the 'b' sequence of the *Rio notebook*, i.e. pp. 85b–86b, relate to Monte Video around 14–26 November, as they contain discussion of cleavage. There is nothing in the *Rio notebook* which obviously relates to Buenos Ayres. The *Beagle* was back at Monte Video on 14 November and Darwin probably received his copy of the second volume of Lyell's *Principles of geology* there on 24 November.

In the first volume Lyell focused on the geological causes now in operation on the Earth's surface, but in the second volume he shifted focus to biological causes. So just at the time that Darwin was pondering on his extinct mammals, he was able to read Lyell's masterful review of all the evidence for the 'transmutation of species', which, if it occurred, would be a possible explanation for the fact that extinct species seemed to be replaced by new ones. Lyell examined the then current theories of transmutation, especially those of Jean Baptiste Lamarck (1744–1829), in order to reject them as untenable. Darwin might, therefore, have been disappointed that Lyell made no serious attempt to explain the origin of new species. Indeed it was a remarkable step in a work which argued for naturalistic causes to explain the history of the world, that the introduction of new species was allowed to seem so mysterious and beyond the power of naturalistic explanation.

Tierra del Fuego, December 1832

The last entries in the *Rio notebook*, pp. 19a–20a, which relate to Tierra del Fuego, tail off towards the end of 1832, although remarkably there is mention of three Australian places, Port Stephen, Jervis Bay and Bald Head, on p. 16a. The last dated entry is for 20 December, before the *Buenos Ayres notebook* was used instead. Many of the pages from the 'a' sequence were excised and have not been found.

120 Judd 1909, p. 353.

[FRONT COVER]

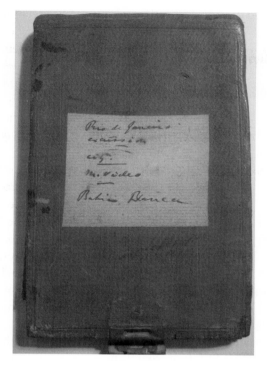

Rio de Janeiro
excursion
city.
M. Video
Bahia Blanca

[INSIDE FRONT COVER]
    R. NE. SW
    Has not M Video been denuded /of/ its Tosca??[121]
    Dip SSE
    /B R/ Point NNE
    Ship ½ D /more/ East
[1a–14a excised]
[15a] <u>de Azzara</u> (Don Felix)[122] Essai sur l'Histoire Naturelle des Quadrupedes du Paraguay.
    traduit sur le manuscrit par M. Moureau de St Mery 2 8 [v.] Paris 1801

---

121 A soft, dark brown limestone occurring embedded and sometimes stratified in the surface formation
    of the Pampas (OED). A local name from the Spanish 'tosco', rough, coarse. For further information
    about tosca see Herbert 2005, p. 98. Called caliche in North America.
122 Félix d' Azara (1746–1821), Spanish explorer and army officer who surveyed Spanish and
    Portuguese territories in South America. Azara 1802–5.

Voyage dans l'Amerique Meridionale; traduit par M. Walckenaer. Paris 1809. 4 Vols. 8vo. (the 2 last refer to Birds, & Sonnini)[123]

[landscape with a building on the left?]

[16a] Petrefactions or Preservation /island/ Coal at Port Stephen & Hat Hill. —
Volcanic rocks at Jervis Bay
Corals branches are yet on Bald Head: Where is it?
Sea almost entirely barren — from mouth of Rio Plata to entrance of San Blas

[17a–18a excised]

[19a] 19th December [1832]
Good Gneiss /Pg/
Slate dipping to SSE. (true) in two places & another
20th [December 1832] Snow lies on the ESE side of hill (true)
WNW ½ N (magnetic Policarpo Bay: 2[d] Hill same dip: altered layers of rock <u>ridge</u> at rt
∠ to strata
Water ∠ 25° bearing W. (mag)

[20a] <u>Started 20th</u> [December 1832]
Ought to go E by S or ESE (not true) 9 am
Dec 20th
Banks Hill. Slate cleaving: dipping to SSE true. ∠ 43°.
universal. hills not regular but topped with sharp ridge: (a)

123 Azara 1801 and Azara 1809.

[BACK COVER]

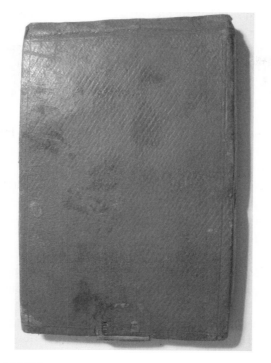

[INSIDE BACK COVER]

C. Darwin

<u>Sierra de Ventana</u>
Lat 38° 12' S
Long 61° 56' W
3^d Hill dip to South
20th
division of water N by W
S by E
SW by S Tent
[2 words illeg] Hill

[1b]   Hills near Rio runn *[ing]* SW ¼ S

[Again] SW by NW.

Sunday. —

Granite; porphyritic with Feldspar, crossed by veins of finer grained: decomposing like
Bahia: much quartz in veins: little dip or cleavage — bad observation
SW by S. dip. — hills generally rounded often bare flat alluvial valley

[2b] between them. village of Itho-caia. 12 miles from Rio. 3 PM. Temp: in white sand 104:
in shade
View at first leaving Rio sublime, picturesque intense colours blue prevailing tint —
large plantations of sugar & rustling coffee:
Mimosa natural veil Forest like but more glorious than those in the engraving; gleams
of sunshine: parasitical plants: lianas: large leaves: sun sultry all still. but large & brilliant
butterflies: much

[3b] water: surprised to see guinea fowls: our calvacade very quixotic: the banks most
teeming with wood & beautiful flowers; village of Itho regular like the Hottentots: the
poor blacks thus perhaps try to persuade themselves that they are in the land of their
fathers. The rock from which old woman threw herself. —

[4b] Temp of room 86° ½
Our dinner: eggs & rice: our host saying we could have anything. — About 4 o'clock &
arrived at our sleeping place about 9. — Sand & swampy plains & thickets alternating
passed through by a dim moonlight. — The cries of snipes: fire flies & a few noisy frogs
goat suckers. Solitary venda[124] slept on the table. could get little to eat. —
9.*th*Mon [April 1832]

[5b] Started about ½ after six. & passed over scorching plains cactuses & other succulent
plants (on the decayed & stunted trees beautiful parasitic Orchis with a deliceous smell)
glaring hot: therm: in pocket 96°. — inland brackish lakes with numerous birds. white
Egrets — Herons – whites & cormorants. lost our way
Geology: found a fragment on beach of sandstone with numerous Cardiums
Mactra.[125] — the whole line of country beach is composed of an extensive

[6b] flat or a lake. between which & sea are large sand hills. on which the surf roars (by night
fine effect) fresh land is gaining. — Sand emits a shrill sound Manatiba [Mandetiba]
dined Temp in shade 84°
our senses were refreshed by food & a more extended & prettier view: reflection very
clear in the lake. —

[7b] These Vendas consist of a house with a shed round it. in which are tables. — a court
yard in which the horses are turned into & & sleeping rooms in which straw reeds mats
are strewed. —
after dinner passed through a wilderness of lacoons. — some salt in which were Balani
others fresh in which

---

124 A poor inn on roads far from town or villages.
125 A marine bivalve mollusc.

[8b]   I caught a small turrited Lymnæa,[126] but in this the sea periodically flows perhaps at the
SW Gales: is not this fact curious? would not such circumstances produce tertiary[127]
strata, beds of sand full of Mactra easily cemented: then we went through impenetrable
forest. trees /eyle/ f very tall. stems

[9b]   white: wonderful beautiful flowering parasites & fire flies, /wren/ bird: large ant hills: we
have been nearly 10 hours on horseback:
slept at Ingetado; after a merry & sleepless night started at 5 oclock: the sky became
red & the stars died away & then the planets
(10th) [April 1832]

[10b]   all promised well.
but 15 miles before arriving at Addea de St Pedro nearly killed us together with waiting
an hour for breakfast, the road lay on the borders of lacoon. shore composed of an
infinite shells: at ½ after 12 started again; the road passed through sand with broken
shells, although

[11b]   some miles from the sea. & the trees attested how long things have thus remained: we
then entered the forest; beds of quartz boulders; —
after some miles came to Campos Novos good venda – in a open country. a pleasant
change: very cool on the turf only 74°. —

[12b]   went out collecting (having arrived at 4 o'clock) — & took a frog, & several
planorbis.[128] Helix[129] & Puccinea[130]
Saw more than 100 buzzards in a flock. —
11th [April 1832]
Passed through several leagues[131] of forest. very impervious trees not large: I here first
began to feel feverish shivering & sickness. much exhausted: could eat nothing at one
oclock

[13b]   which was the first time I got anything. — travelled till dark: miserably faint & trouble
with faintness. —
At night we slept 2 miles S of Marica: felt very ill in the course of day I thought I should
have dropt off the horse: horrors of illness in foreign

[14b]   country: during the morning C Frio appearing from refraction like
inverted tumblers. Gneiss dipping to the South (& then the north). —

126   Freshwater snail. See specimen not in spirits 138 in *Zoology notes*, p. 370.
127   The relatively recent period of earth history between the Quaternary (which includes the Ice Age and
the present) and the Mesozoic ('Secondary'). Characterized by the dominance of mammals over
reptiles and the absence of ammonites.
128   Freshwater snails.
129   A genus of land snail.
130   A small fungus; some rusts and mildews belong to this genus.
131   Distance of about three miles or five kilometres.

(12th). — [April 1832] Started in the morning & doubted whether I could proceed. — Cinnamon & port wine cured me

[15b] in a wonderful manner; passed through more swampy country & then entered a magnificent forest: sublime. trees lofty. well seen & contrasted in the cleared parts: palms, very thin stalks. beautiful in the forest cabbage palm. edible; /spinnach/

[16b] Arrived at the Fazenda of Sosego. situated in a Forest. square. coffee Mandioka.¹³² much game, number of horses cattle poultry & wild animals: patriarchal style of living

a long house. with a roof of reeds. [layout of the fazenda] at one end gay furniture a long dining room & bed rooms kitchen & large store houses: situated on a hillock. on the other side of the

[17b] square. sleeping rooms, & round the hill the house of more than one 100 blacks: children of these people stray into dining room. till driven away: daughter of our host Donna Maria, handsome & dignified married to Mr Laurie a scotchman & brother of our companion: on receiving a guest or the Signor a large bell is rung & cannon

[18b] fired. on leaving the house a crowd of blacks come to be blessed by the white man: one morning before daylight I was admiring the stillness of the forest. when it was broken by a Catholic morning raised by all the blacks. effect sublime:
our eating was sumptuous forced to taste everything:

[19b] this patriarchal style fascinating. but destroyed by our host being a ~~villain~~: enterprising character. has cut excellent roads through the wood: saw mill cutting up thick planks of Rose wood. something like a large-leaved acacia: dreadful the difficulty of procuring surgical aid: our host plenty of medicines: saw a canoe

[20b] building 70 feet long & 40 more left of solid & thick trunk: I should think in this family the blacks were decidedly happy. — The son in law coming only 2 days short journey found it necessary to bring 17 people with him: stayed at Sosego during the 13th [April 1832]

[21b] On the 14th [April 1832] started at midday for Mr Lennon's estate, after a beautiful ride stopped at a Facend. within a league of our end; blacks miserably worked long after dark: were received very hospitably by the only brazilian that I have yet seen with a pleasant expression
15th [April 1832] ~~Started early~~ saw some beautiful birds, Toucans & bee eaters all the rock is gneiss granite. Mica

132 Cassava: the source for tapioca.

[22b] dark coloured large plates: I should think rain had not much degrading effects: valley flat, well seen in the cleared parts;

15 [April 1832]

Started early for Mr Lennon's estate: it is the last cultivated piece till having passed over many miles of country: on our road saw some

[23b] bamboos. disappointed (day before saw a Papyrus?) & some small elegant tree ferns. — had a man to cut a road for us with a sword: when we arrived heard a disagreeable & most violent quarrel — between M^r L & Cowper his agent: they talked of pistols: so bad a character

[24b] that we were cautioned to recollect poison. — Blacks in a bad state; wet cold evening 75°. — threatened to sell his child as a punishment. yet most certainly a very humane man. —

I observe here & at Socego the clouds rest at a very low level not more than often at

[25b] 2 or 300 feet above the & only scarcely any above the adjoining country: Rio Macae is navigable the whole way & not /more/ 5 or 6 leagues in length & runs close by these places. — At here the air where passing over the forest on the level of the house becomes converted into cloud: rain has fallen

[26b] every day: a remarkable scarcity of rounded pebbles during all our course in the interior: 16th [April 1832] started early in the morning to the Juer do Paz. & proceeded to Sosego. pleasant ride & much enjoyed the glorious woods: Bamboo 12 inches in circumference. Several sorts of trees ferns:

[27b] 17th [April 1832] Sosego. twiners entwining twiners. tresses like hair beautiful lepidoptera.[133] silence hosannah. (Frog habits like toad. slow jumps. Iris copper coloured colours become fainter

Snake. Cobris de Corrall Fresh water fish. edible

Blaps musky shell. stain fingers red

one fish from salt Lagoa de Boacica. 2 from brook

one do. pricks the fingers

[28b] Manoel Joaquem da Figuireda

after clearing coffee & mandioka is planted. afterwards solely coffee. brother of Manuel has 95000 trees, producing 2 lb per tree (some produce 8 lb) rice on the swampy parts & some sugar cane 3 bags of rice produced

[29b] 320. —

Teijóa beans are cultivated. one bag bringing sometimes 80 bags: Mandioka stems & leaves eat by cattle. roots are ground. a slave holding them against the wheel. the pulp is then pressed dry & baken: excellent eating. —

March is the great season for planting. —

---

133 Butterflies and moths ('scaly wings').

[30b] the juice thus procured from the root is deadly poison: but the animals very fond of it. always dye: — from this Tapioka the made:

18th [April 1832]

Socego. mimosa exquisite foliage & ferns ditto. — trees average 3–4 feet circum in the bole.

A creeper circum. 1 ^{ft} " 4

[31b] spent the whole morning in thus rambling in the forests. sublime devotion the prevalent feeling: this days delay was owing to Mr Lennon going to visit his estate with Mr Cooper

19th [April 1832]

Left Socego & slept at Venda de Matto: took a most glorious walk on the beach high & magnificent surf. —

[32b] 20th. [April 1832] returned by the old route to Compos Novos. a tiresome ride all through a scorching & heavy sand. plain of rhododendrons had some difficulty in making our horses swim & in danger from a drunken man in canoe. — 21st [April 1832] time often by S † [Southern Cross] Started by daylight arrived after

[33b] a very long day almost without rest to Rio Comboata: miserable venda. — passed through an interesting cultivated country: this is the interior road, branching off at Paratra; Many of the fields from numbers of ants nest looked like Humboldt mud volcanoes

[34b] 22nd [April 1832] from our sleeping place to Fresqueria de Taboraa, torrents of rain during the whole day. destroying pleasure of a pleasant country: breakfasted at Madre de Dios. a nice village: on road beautiful & flowers passifloras & many birds: soil resulting from decomposing gneiss — generally

[35b] reddish clay producing sugar cane. little coffee: met several riding people & a few heavy carts dragged by 8 oxen. wheells almost a solid board. — no house in the whole country as good as a good farmers house. no road as good as a bad turnpike road in its worst parts. —

[36b] at the vendas seldom see a woman. not worth seeing distances most inaccurately known. — not above a score of murders or crosses. — In Heavens name in what are blacks better off than our English labourers? The national guardsman

[37b] is often a wretched looking mulatto with a sword strapped to his side. — Th. cold 62 ½ at Compos a supper of fowls & rice & biscuit & bottle of good wine, & coffee in evening & morning with fish in the morning

Indian corn & grass for our 3 selves & horses for 3 Mill reys[134] alltogether MR = 2s 6d. —

[38b] Acacia.

23rd [April 1832] Home a very pleasant day Acacias a much more /cultivated/ & pleasant.— some few pretty villages. Praia grande plagued about our horses not having a passport

134 The local currency in Brazil.

24th [April 1832] ~~Wednesd~~ Tuesday — Staid on board. I found a days rest so delightful. — they have turned the poop

[39b] cabin into eating room — change of officers. — riots in the town

~~Thursday~~ 25 [April 1832] Wednesday took my things to Botofogo shipwreck

26th [April 1832] ~~Friday~~ Thursday engaged in drying my things

27th [April 1832] ~~Saturday~~ Friday Worked my interior Bahia collection. went to Mr Astons[135]

pleasant like

[40b] Cambridge. not the Ambassador. —

28th [April 1832] Breakfasted with on board met Captains Talbot[136] & Hardy: Called on the Admiral[137] dined there pleasant evening. very gentlemanly the officers. —

29 [April 1832] Sunday quiet delightful day at Botofogo writing journal

[41b] 30th [April 1832] — Dined at Mr Astons.

31st [1 May 1832] — Worked at fresh water animals

16th [June 1832]. —

Tijeuka. — country much furrowed by water. enormous blocks. — chiefly a compact mica slate. — & quantities of trap. —

scenery always beautiful

Temp *[of 2 W]* 66°

Chain of *[hills]* NE & SW

[42b] Palm *[nearly]* 9 ft at top circum. 2 ⅔ at bottom 3 ⅓ — — 305 rings

184

131
___
59

90 (850)

Myrmecia[138] in corner of leaf Ctenus[139] in trees. — Spiders certainly see far —

[eyes of a spider?]

135 Arthur Ingram Aston (1798–1859), Secretary of Legation at Rio de Janeiro 1826–33.

136 Charles Talbot (b. 1801), commanded HMS *Warspite*, 1830–42, who rescued the Brazilian Royal Family from an insurrection on 6 April 1831.

137 Thomas Baker (d. 1845) Admiral, commanding the South American station, 1829–33. The South American station was the operational base for the Royal Navy in the South Atlantic.

138 An ant-mimicking spider, see *Zoology notes*, p. 46.

139 Hunting spiders that do not spin webs. See *Zoology notes*, pp. 51–2.

180	180
24	25
—	—
720	900
360.	3600
—	4500
4720	36
32	—
—	(77.)
795	

[43b]

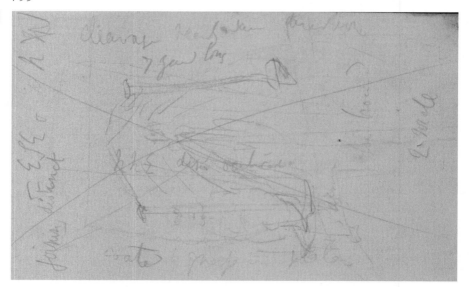

[very complex diagram which seems to show cleavage orientation and possibly two figures or animals] ENE & WNW joining distinct cleavage nearly same direction 7 yards long NNE dip vertical coated by gneiss plates [*illeg*] too broad 2 wide

[44b] Epeira pass through the web — —

Marvellous power of motion. generally full in aloes.

knew the diff of my finger & stick?

small ones.

centra mat-work

[45b] cleavage NE. (41 ∠) siliceous[140] mica slate

dyke N ½ W. — dyke of generally fine grained trap. much decomposed. angular lines of cleavage — decomposing into balls in places becoming [*slightly*] serpentine[141] in

140 Containing abundant silica (i.e. silicon dioxide) or quartz. Silicates are the most abundant rock-forming minerals in the Earth's crust.

141 A mineral derived from olivine.

[46b] balls very tough

X X

then another rock (& vein /partly/) with small veins /intangled/ into mica slate.

Dyke <u>at least</u> 100 yards long & 20 wide

/Mica/ 22 of iron

[47b] 5th [July 1832] Sailed

6th [July 1832]— Calm in sight of Sugar Loaf

7, 8th [July 1832] Very sick. not so bad Cape pidgeons whales

9 [July 1832] but little better.

Rio Corcovado.— Bones. *[illeg]*

[48b] (14th) [July 1832] ¼ 11 oclock PM

Cumili Fresh Breeze N Sky pale blue

few minutes duration

(Fine sight)

 [weather phenomenon, halo] reddish edge

luminous reddish greenish blue

~~whole~~ red diameter, 1° 45' whole diameter about double clouds passing larger colour indistinct ring. —

[49b] 14th [July 1832] Fine day & prosperous breeze.

Sunday 15 [July 1832] 160 miles since noon of yesterday; Grampus: uncertain weather gale: Morro de St. Martha: top gallant yards:

16th [July 1832] much sea sick

snake burrowing in the ground.

Pompilus¹⁴²

68 ½. 61 ½

59

---

142 Specimen 534 in *Darwin's insects*, p. 56.

[50b]   Barrell with spirits
        Bottles
        /pincer/ for insects & bottle
        Pix axe
        Buck shot
        18th [July 1832]
        Lat. 31° 37 S
                6 14 W
               6° 18 W
               <u>43   8</u>

        Long 49° 22 W
        Flying fish. Porpoises
[51b]   E by North — mice slate. — chloritic[143] schist.[144]
[52b]   <u>M Video</u>
        1'. spider tube in rock long. 1 & ½ inch long bag shaped
        ESE — slate (chloritic) or 5 yards thick <u>alternating</u> with mica slate. (<u>between</u> Rat
        Island & Mount strike various ESE E by N
        + chloritic slate cleavage running ESE. vertical smooth line of ~~cleavage~~ fracture
        running at ∠ and
[53b]   dipping to E 20°
        on the Mount — slate with crystals passes into trap. sonorous[145] (+ +)
        running E by N & in the line of Rat Island
        R Island

 [The Mount at Monte Video]
[illeg] by N E by N M ESE WSW E by N ½ NE ENE
quartzone mica slate & again slate chloritic

143 Containing conspicuous amounts of the mineral chlorite.
144 A metamorphic rock intermediate between slate and gneiss
145 Makes a ringing sound when struck.

[54b]   A. Hornblendic slate same site as last
(a point further on behind hill B paps of <u>compact</u> mica slate ~~containing feldspar~~ much fe
quartz vein of Quartz (Hyalomictite)[146]

[55b]

[survey map of Monte Video from The Mount, with Rat Island] River
M Bay R. Island

146 A glassy volcanic rock.

[56b]

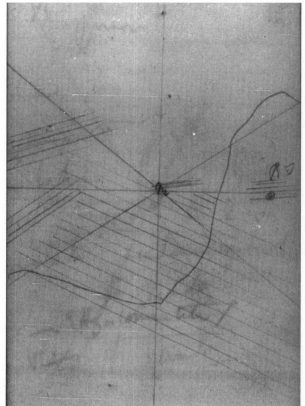

[larger scale survey map of Monte Video showing orientation of cleavage] R I

[57b]   Mica slate (fracture plains) dipping about ESE. at ∠ 12°

Chloritic slate (vide supra) entangled in mica slate & every where penetrated by curved veins of granitic crystalline quartz see specimen —

[58b]   in second or small piece of chloritic schist entangled: line of cleavage & fracture same as mica slate

planes of fracture even dipping SSW

Salt ~~springs~~ streams & bed of muscle shells under town show recent origin.

[59b]   August 15th [1832]

Chloritic E & W (G E by S)

Mica slate intangled with quartz veins (G E ~~by N~~) (cleavage vertical)

alternation ESE (G: E by N)

(G NE by E) compact slate specimen (A).—

(G) NE cleavage NE by E?

do (G NNE)

all these was ~~been~~ near the coast

[60b] near summit — (B) (G NNE) chloritic compact slate cleaving E by N

do (G NE by E)

(G E ½ S) cleav. E by N ½ N. slate

(G W NN) clear E by N D slate pushed over & contorted

[61b] right through centre of hill E by N [this] compact D <u>sonorous</u> slate conchoidal fracture[147]

(G SSE) clear ENE cleavage has a tendency to lean over & dip to NNW — on other side it leans or dips to opposite quarter of the compass

[62b] Capincha[148]

Girth 3' 2

length 3. 8 ½

Shoulder to toes 1" 9

weight 98 pounds

Linnaeus — Planaria[149]

Sept 22^d [1832] —

Entrance of creek. dark blue sandy clay much stratified dipping to NNW or N by W at about 6° ∠: —

[63b] on the beach a succession of thin strata dipping at 15° to W by S. & conglomerate quartz & jasper[150] pebbles with shells — vide specimens. —

On the coast about 12 feet high—

~~Proceeding to the~~

[64b] in the conglom teeth & thigh bone Proceeding to the NW. — there is a horizontal bed of <u>earth</u>, containing much fewer shells, but armadillo — this is horizontal but widens ~~as~~ gradually. hence

[65b] I think conglomerate with broken shells was deposited by the action of tides <u>earth</u> quietly

Is this above the clay which is seen a short ~~way~~ time previously?

covered by diluvium & sand hillocks

[66b] as earthy bank thickened & cropped out in direction NNW it probably overlies the clay. —

---

147 The way brittle materials such as glass fracture, not following any natural planes of separation.

148 The capybara, the world's largest rodent. See *Zoology notes*, p. 67; listed in *Mammalia*, p. 91, as *Hydrochœrus capybara*.

149 A genus of flatworm (Platyhelminthes). See *Zoology notes*, p. 66, and p. 67 note 1. Darwin 1844 published some descriptions of the specimens he collected; *Shorter publications*, pp. 179–87.

150 An amorphous silica.

[67b] <u>October 2^d [1832]</u>[151] /50/ feet high horizontal sand with pebbles
do
main bed of red earth:
lower parts containing limestone[152] in nodules & layers
[68b] or indurated marle[153] (1)
lower parts redder — all very compact
Vertebra of 2 animals
+ + /caught/ by the tide
14. feet red earth with layers of calcareous, about 12 of yellowish
[69b] softer earth. —
6 inches of indurated sandstone. blacker
20 feet of wavy sands ~~indu~~ cemented.
In NW by W section on coast — lower beds crop out & thicken:
[70b] Punta Alta. Octob 6th [1832]
The beds have been deposited in the sea. hence inequalities in the lower conglom is
filled up by the Tosca & then again by the upper bed which is sometimes conglom &
sometimes earthy.
circumstances were rather different during its deposit & present though: the
[71b] sea is now destroying it & has destroyed a bank of some miles leaving only lower
horizontal conglom: whilst on other places it is now heaping up bank. — believed the
cliffs in a bit of table land formerly bottom of sea — & the line of heaped
[72b] up bank & on this is nearly continuous with the rest of coast. which show there was not
such very great dissimilarity of circumstances: subsequent to the general elevation the
strata have been washed away but that now the sea has begun again to heap

---

151 There is a gap in the notebook between 23 September to 1 October 1832.
152 A sedimentary rock composed mainly of calcium carbonate, usually of biological origin (e.g. from
     corals).
153 Lime-rich clay.

[73b] them up:

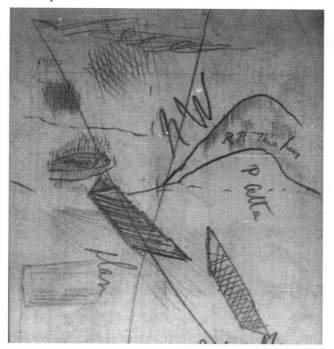

[map and section of Punta Alta] NW RR Thickness P Alta plan red earth resembles that at Mt. Hermoso

[74b] section

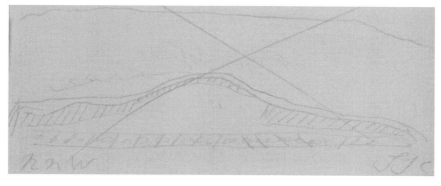

[section of Punta Alta] NNW SSE

[75b]   All the conglom contains quartz & fine sandstone pebbles like the Rio Negro tides flow
from the south

[section of Punta Alta] detritus red earth gravel red earth lower gravel argillaceous or
marly thin beds

[76b]   Look for Azzarras book

Edge of formation NNW, SSE.

Having dug three feet deep, it consisted of the same hard earthy matter as cliff. there was
a fragment of Ostrea. — The upper <u>foot</u> was stained black

[77b]   by the vegetation

I think it clear this belongs to the Punta Alta & M. Hermoso formation: but that from
some cause neither sandstone nor conglom were deposited; ~~perhaps this was further t~~
plain turf level — probably caused

[78b]   by sudden irruption of earthy matter from the shore — The sea probably flowed over it
for some time as sand hillocks show: <u>sand hillocks</u> running W by N. E by S.

(K^2) bearing to tent WSW ½ W ~~bearing to~~

[79b]   the line in which hole was dug.

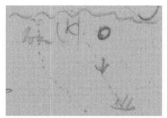

ship was thus   (K) O

[map of Punta Alta showing ship]

The pumice[154] pebbles are most remarkable lying in numbers on the surface. — were
they left when the Tosc**a** was

---

154 Solidified frothy lava, generally lighter than water.

[80b] deposited or by a subsequent inundation probably contemporaneous with detritus on Punta Alta —

~~Lava~~ Pumice pebble[s] probably washed up with the banks at first deposit: sand

[81b] hillock not connected between anchor chain & P. Alta

Tosca

Upper detritus or diluvium deposited in water on account of pebbles shells quartz sandstone greenstone marly limestone & pumice. Hills of do. probably formed when Megatherium[155] like armadillo case

plants like heath Salt places.

[82b] was lifted up: Tosca & upper gravel pass into each other. Tosca contains smaller pebbles & bits of shells:

Sand hill behind the plain strikes to the coast. from the /Alta/ on each side

[83b] 2 small chain run. the one by the creek perhaps partly modern: within of this /side/ the Tosca plain bounded by N & S Line extends

Tosca has been denuded: gravel beneath extends horizontal: —

~~Parrot~~ eggs

Hence Detritus is diluvium / green Phonolite

[84b] Examine diluvium; are there any ~~lava~~ pumice pebbles: Are there shells in Tosca: Has the deposit of sandstone cut up the Tosca; examine action of tide on the latter: Are there old shells on hillocks: examine general appearance of latter: Is the Tosca externally harder; (no)

(Lava pebbles) siliceous sandstone & the sandstone broken up by quartz pebbles. Swifts in flocks.

Curious habits of Lizard

NNW by W line in which the strata thicker —

remarkable gradual shoal to the anchor:

(5 to 7) **2** miles black clay formed by the action of water on the Tosca

[85b] ~~Phylla~~ Mica Slate with straigh[t] bands of siliceous when most frequent is altered into Phyllade[156] — — frequent alternation — crossed by irregular beds or dykes of (granitic?) rock which runs into it the quartz /oze/ bands, only differs in not

155 The extinct giant ground sloth (meaning 'big beast').
156 A finely splitting metamorphic rock, generally intermediate between slate and schist.

[86b] possessing cleavage ∴ — if mica slate was deposited in horizont layer this must
       have been a piece in middle projecting upward /without/ any—
       Rowlett[157] about trousers
       Sea eggs[158] 50 fathoms off the Straight of Magellan

Textual notes to the *Rio notebook*

[FC]   *a faint word or number in pencil is now illegible.*
[IFC] 1.10.] *Down House number, not transcribed.*
       88202330] *English Heritage number, not transcribed.*
[IBC] C. Darwin] *ink.*
       1A] *added by Nora Barlow, pencil, not transcribed.*
[1b]   NW] *ink, obscured by doodle in ink.*
[11b] Campos Novos] *added pencil.*
[48b] *page written parallel to the spine.*
[53b] R Island] *ink.*
[54b] Hyalomictite)] *ink.*
[81b] Megatherium … places.] *written perpendicular to the spine over* 'Upper'.

---

157 George Rowlett (d. 1834), Purser of the *Beagle*.
158 Sea urchins.

THE *BUENOS AYRES NOTEBOOK*

⌒

The *Buenos Ayres notebook* takes its name from the city of Buenos Aires (Buenos Ayres in Darwin's spelling, from 'Good Air' or 'Fair Winds'), on the southern shore of the Río de la Plata, on the coast of Argentina. The notebook is bound in red leather with the border blind embossed; the brass clasp is intact. The front cover has a label of cream-coloured paper (68 × 17 mm) with 'Buenos Ayres (city) *Beagle* Channel Ascent of P. Desire Creek.' written in ink. The notebook has 56 leaves or 106 pages. It was written in two sequences, pp. 1a–90a and pp. 1b–22b. The notebook records events in November 1832 (pp. 1a–12a), January and February 1833 (pp. 13a–79a), December 1833 (pp. 87a–88a) and November and December 1832 (pp. 1b–21b).

The notebook was essentially used in three chronological parts. The first part consists of the front thirteen and the back nineteen pages which seem to have been filled in more or less simultaneously; they both have pin holes indicating that Darwin pinned pages 1a–14a and 1b–17b together. This first part covers Darwin's exploration of Buenos Ayres in the first week of November 1832, thus filling a gap in the previous notebook. In a letter to his second eldest sister Caroline (1800–88), Darwin wrote 'I much enjoyed this long *cruize* onshore', which was delayed three months by cholera quarantine restrictions.[159]

The second chronological part is in the middle section of the notebook and picks up from the last pages of the *Rio notebook* in mid January 1833, with Darwin's first exploration of Tierra del Fuego.[160] The notebook was then used continuously until around 7 February, when the *Falkland notebook* was used instead. There is, however, a third chronological part of the notebook which is in the last couple of pages of this middle section of the notebook. These pages are dated to the last few days of an unspecified month but which is clearly December 1833. These pages relate to Port Desire [Puerto Deseado] in Patagonia, 1,600 km (1,000 miles) south of Buenos Ayres. The *Beagle* arrived there on 23 December 1833.

159 CCD1: 277.
160 See Armstrong 2004.

Buenos Ayres, November 1832

The Buenos Ayres descriptions seem to begin just before 1 November 1832 with a mixture of geological observations, zoological memoranda, 'Capinchas dung smells very sweet', p. 16b, and notes on interesting facts Darwin was told, 'it is said that Crocodile occur & small water turtles' p. 2b. Some of these were apparently local superstitions, e.g. 'the water has power of turning small bones into large ones', p. 11b. Darwin jotted down various other *aides mémoire* to himself, 'Museum open every 2nd Sunday', p. 1a; 'Is M. Video built on granite or the Gneiss?', p. 9b. There is a reference to the unusual 'Mendoza waggons' on p. 6a, which Darwin attempted to draw in the *Beagle diary*, p. 114, and a delightful allusion when riding the next day, five leagues west of the City, to botanizing in the Fens: 'very like Cambridgeshire from Poplars & Willows', p. 6a. Darwin's rather limited draughtsmanship is again exposed in some tiny sketches of tables on p. 15b.

There are long lists of names and addresses, things to buy and have repaired, 'Dentist', 'Watch mended' pp. 1a–2a, books to refer to 'Caldcleughs S. America' on p. 2a (Caldcleugh 1825) which was in the *Beagle*'s library; 'Spix' on p. 10b; (Spix and Martius 1824) also in the library with the inscription in volume 2 'Chas. Darwin Octob: 1832 Buenos Ayres';[161] and loans not to be forgotten: 'Mr Hamond owes me 27 paper dollars' p. 3a. It was Lieutenant Robert Hamond (1809–83) who shared Darwin's interest in the 'Spanish ladies beautiful dresses & walk' on p. 5b, an entry which continues with the wonderfully pithy 'went out riding bad roads good horses'. The name Charles Hughes also occurs several times; he was a great help to Darwin in Buenos Ayres and there is a memorandum of information from him in DAR 34.14. As the irreplaceable Biographical Register in the *Correspondence* indicates, Hughes attended Shrewsbury School in 1818–9, the year Darwin started at the school. Darwin was also impressed by Colonel Harcourt Vernon (1801–80), 'great traveller', p. 7a, probably one of the first European tourists to tackle an overland trek across South America. Jeff Ollerton has kindly explained (personal communication) the rather garbled and previously mis-transcribed reference to John Tweedie (1775–1862), a plant collector who had a botanic garden at Retiro. The fact that Tweedie is also mentioned on p. 5a of the *St. Fe notebook* and that there is a reference in DAR 34.13 to information derived from Tweedie, tends to suggest the possibility that the two men met.

There are many references to bones obviously reflecting Darwin's great interest in fossils since his dramatic discoveries at Punta Alta a few weeks previously; 'Oakleys fossil one scapula in true Tosca', p. 15b; 'Mr Flint an American Merchant has a tooth', p. 13a, indicating where they might be found or bought. Darwin probably met the English merchant Edward Lumb (1804–75), who played an important

---

161 See CCD1: 564.

role with the fossil mammals, on this excursion, pp. 5a, 7b. Oakley 'a joiner with red hair', p. 13b, was the agent of Woodbine Parish (1796–1882), the British consular representative.[162]

There is one entry in another hand, presumably John Meggett's; see p. 10b, and a wistful entry for 1 November on p. 4b: 'Very calm delightful days: quietness seems to shorten the distance. always think of home'. This entry became 'A calm delightful day. I know not the reason, why such days always lead the mind to think of England and home' in the *Beagle diary*, p. 113.

Tierra del Fuego, January–February 1833

After the above entries made in Buenos Ayres in November 1832 the notebook was not used again until Tierra del Fuego on 19 January 1833, p. 14a. Since the last dated entry in the *Rio notebook* is 20 December 1832, there is an unrecorded gap of about four weeks, obviously due to the appalling weather, which prevented any land excursions. The field notebooks seem only to have been used at sea for occasional musings, latitude and longitude positions, weather notes, depth soundings and so forth, but also for lists of equipment which Darwin jotted, presumably in preparation for going ashore. While on the *Beagle*, Darwin generally bypassed the notebooks and wrote directly into his *Beagle diary*, scientific notes or, less regularly, letters.

It is curious that there is no reference in the last pages of the *Rio notebook* to Darwin's first encounter with the native Fuegians, which is described in the *Beagle diary* entry for 18 December, but this could be explained by the fact that some of the *Rio notebook* pages have been excised. As Darwin declared in the *Beagle diary*, 'it was without exception the most curious and interesting spectacle I ever beheld', p. 122, and recorded in his *Autobiography*, p. 80, how 'the sight of a naked savage in his native land is an event which can never be forgotten'. Darwin was struck with the realization that his own ancestors lived in a similar way only a few thousand years before.

Because of the gap in the notebooks, there is also no mention of the terrible storm of 13 January which wrecked several ships in the seas around the southern tip of South America. This storm nearly brought the *Beagle* voyage, and the lives of all the men onboard, to a premature end. FitzRoy, in *Narrative* 2: 126, reported that even in the normally quiet harbour of Berkeley Sound in the Falklands, the whaler *Le Magellan* was totally wrecked that day. He also wrote 'Mr. Darwin's collections, in the poop and forecastle cabins on deck, were much injured' and the *Beagle* lost one of its beautiful whale-boats at 1.45 p.m., when 96 km (60 miles) WSW of Cape Horn. This latter event was chosen for its drama and historical moment by John Chancellor (1925–84) for his 1982 painting of the *Beagle Sorely Tried*.[163]

---

162 See Winslow 1975.
163 See Chancellor 2008.

Sorely Tried, HMS *Beagle* off Cape Horn, 13 January 1833 at 1.45 p.m.,
by John Chancellor.

The Tierra del Fuego entries start on p. 14a at the eastern end of the *Beagle*
Channel.[164] There are three pages of geology, then the Fuegians enter the account
on p. 17a as FitzRoy began his excursion in four boats to return the three Fuegians
to Jemmy Button's country in Ponsonby Sound, together with the young missionary
Richard Matthews (1811–93).

FitzRoy took four living Fuegians, and one dead one preserved for dissection
at the Royal College of Surgeons, to England after the *Beagle*'s first South
American commission.[165] Of the four living Fuegians, FitzRoy's favourite, who
he named Boat Memory (c. 1810–30), sadly died of small-pox in England. The
three who survived their sojourn in England and who were returned were the
faithful Orundellico, known as James 'Jemmy' Button, (?1816–63), the young
girl Yokcushlu, known as Fuegia Basket, (?1821–?1883) who had so enjoyed
being presented to Queen Adelaide, and the taciturn and by all accounts rather
sinister El'leparu, known as York Minster (?1804–c. 1871), an Alikhoolip man
named after an islet near Cape Horn Island. It is interesting to note that FitzRoy,
often described as an arch Tory, had the Fuegians examined by a phrenologist in
London in 1830. The phrenologist's reports were published in the Appendix to
FitzRoy's *Narrative*.[166]

Darwin filled several pages with detailed geology and some descriptions of the
scenery, then on p. 20a noted the natives' 'wild appearance on hill. naked long hair'

164 See Armstrong 2004.
165 See the first chapter of *Narrative* 2.
166 See van Wyhe 2004b. On phrenology see van Wyhe 2004a.

and on the next page 'innocent <u>naked</u> most miserable very wet'. On p. 22a he indicated that there was some 'fighting with savages'. Then there are more pages of geology before Darwin recorded on p. 29a camping on a starry night surrounded by 'J Button's quiet people' who although naked were perspiring by the large camp fire. Darwin noted that 'after breakfast' many more natives arrived having 'run so fast that their noses were bleeding', p. 30a. This dramatic incident eventually appeared in *Journal of researches*, p. 240.

There is a remarkable entry in Darwin's geological diary dated January–February 1833 which, quite apart from showing how much zoology Darwin mingled with his geology notes, also demonstrates that barely a year after leaving England he was noticing how animals are not always found in places for which they appear suited. In the entry Darwin wrote 'It is very remarkable that J. Button says there are no foxes or Guanacoes in Hoste island, which in every way appears equally well adapted for them [as Navarin Island]. I found however one mouse!' (DAR 32.101B).

At Woollya Bay [Wulaia] it was decided to build a settlement where Matthews would stay, so the ship's hands cleared the ground and built a hut. Jemmy's three brothers and his mother arrived and Darwin noted, p. 33a, how the natives were surprised by the white mens' skins and their habit of washing themselves. Word spread, and on p. 36a it is recorded that Jemmy's uncle arrived, with friends.[167]

It is remarkable that Darwin scarcely broke the flow of his geological observations to record these extraordinary encounters. He was clearly intrigued by the rather complex field relations of the metamorphic rocks (slates, greenstones, etc.) and was particularly concerned to disentangle original bedding from superimposed cleavage, as he was taught by Henslow and especially by Sedgwick in the rather similar rocks of North Wales.[168] It is very difficult to distinguish original bedding from later super-imposed cleavage in clay-slates like those Darwin was examining, and it is doubtful if he could have made much sense of the rocks without having visited Wales with Sedgwick.

'Beds' or 'bedding planes', are the original sedimentary layers, representing the sea bed where the clay accumulated, but they may later be found tipped up (that is 'dipping') or, in extreme cases, upside down. 'Cleavage' adds to the confusion as it results from microscopic re-orientation of certain minerals, such as mica, when subjected to great regional forces, such as the horizontal compression which occurs when continents collide. This was the case with the rocks Darwin saw in North Wales, where the cleavage planes are sometimes perfect enough to be made into billiard tables. A clay rock (generally called a shale), once subjected to cleavage becomes a slate. Sedgwick showed Darwin how, if one can be sure what is bedding and what is

---

167 There is a fine photograph of Woollya in Bartolomé and Glickman 2008, p. 76, a book which conveys the sheer emptiness, even today, of many of the places in South America visited by the *Beagle*.
168 See Secord 1991.

cleavage, it is possible to begin to work out the structures which the compression created, and hence to reveal where the pressure came from. This is why page after page of Darwin's notebook is filled with dips and compass bearings. Darwin was amazed to discover the immense area over which the cleavage revealed the same orientation.

On p. 38a, 26 January 1833, Darwin mentioned some heteromerous beetles he found and after a curious note, 'Bahia Blanca V[ide] De la Beche?' which may be a reference to De la Beche 1831, the 'arrival of women' and the fact that the men seemed to sit watching the women work. On p. 44a the odd events of the night of 27 January were recorded; strangers arrived and there was 'extreme treachery of character', p. 45a. On the following day FitzRoy, Darwin and party departed in two whale boats in an expedition to examine the islands along the western entrance to the *Beagle* channel.

Darwin's geological observations are interspersed with zoological notes, such as 'Fringilla in flocks' and 'whales blowing', then 'fearless barbarians' arrived, p. 50a. Darwin noted that Jemmy may have forgotten his native language but not his prejudices, as he would not eat land birds 'because they live on dead men' (*Beagle diary* , p. 137), and on p. 54a, Darwin described the beautiful beryl blue of the glaciers.[169] On the same page Darwin mentioned casually his contribution to saving one of the boats from being swept away by an iceberg-induced wave. FitzRoy, in *Narrative* 2: 217, left an account of this incident which could have ended in disaster. In that account we see plainly Darwin's natural modesty and athletic prowess.

Camping on the rocky shore near the boats on the night of the 30–31 January 1833 was clearly unpleasant and was singled out for a rare dash of ironic humour on p. 59a: 'Miserable sleeping place big stones putrefying sea weed & middle watch. not all pleasure'. On 1 February near Gordon Island Darwin saw '2 Whale within pistol shot enormous backs & tails', p. 62a.

It is difficult to do justice to the many pages of geology, which are partly the basis for Darwin's statement in *South America*, p. 151, that his notes on Tierra del Fuego were copious. There is reference to various rock formations such as greywacke, granite, gneiss, greenstone, trap, slate, serpentine, schist, hyalomictite, 'amphibolic formation' and so forth. Darwin's extensive mineralogical vocabulary bears witness to his childhood fascination for chemistry and crystallography. There are pages of what today would be called structural geology, following Sedgwick's interests. Darwin was always trying to make out the big picture: see for example his map of the entrance to Ponsonby Sound on p. 37a. It is perhaps a pity that he never published the geological map of southern South America which is preserved in DAR 44.13.[170]

169 See Herbert 1999.
170 It is published in colour in Herbert 2007, p. 315, and as a fold-out facsimile in van Wyhe 2008a
    p. 19. See also Stoddart 1995, p. 10.

This might have made his published descriptions of the whole region rather easier to follow.

Darwin also saw for the first time great boulders which were obviously dropped by glaciers, p. 64a. This first-hand experience of glaciation would be put to good use in 1841 when Darwin, under the influence of Louis Agassiz's (1807–73) glacial theory, re-interpreted the mountains of Snowdonia examined with Sedgwick in 1831, without seeing the now obvious signs of recent ice action.[171] The dramatic and inhospitable scenery surrounding the little *Beagle* left an indelible impression on the young Darwin who, on p. 76a, allowed himself a romantic flourish: 'Is not Tierra del the Ultima Thule', that is, a distant place located beyond the 'borders of the known world'.

Around 6 February an amusing evening was spent bartering for fish, then, on p. 74a, Darwin noted Matthews's disappointment at the Fuegians' behaviour. The next evening, on p. 77a, Darwin was 'followed by savages' but by firing over their heads he or someone in his party 'frightened them away'. This is the last reference to the Fuegians in this notebook.

On 7 February the *Beagle* was at Navarin Island [Isla Navarina] and Darwin made a rather obscure entry about a 'trench of diluvium', p. 78a. This entry is an interesting link to his 1842 paper on erratic boulders in which he wrote 'I cannot more accurately describe the appearance of the cliffs around Navarin Island, than by the remark which, at the time, I entered in my note-book, "that a vast debacle appeared to have been suddenly arrested in its course."'[172] The entry Darwin referred to is actually in his geological diary and is as follows: 'A Debacle sweeping along has been arrested in its course' (DAR 32.118). This is the first of two known cases of Darwin mis-quoting his *Beagle* notes; it consists of the addition of the adjective 'vast' and the adverb 'suddenly'. The second mis-quotation is in *South America* and is in connection with the Pampean Formation and appears to have been a deliberate attempt to bolster his interpretation of that Formation against that of Alcide d'Orbigny. The original statement is from the *St. Fe notebook* (see the introduction to that notebook) and consists in the substitution by 'tint and compactness' for the more prosaic 'colour and hardness'.

Port Desire, December 1833

The last few dated entries in the *Buenos Ayres notebook* relate to Port Desire, and were written just after Christmas 1833. This festival was celebrated by FitzRoy ordering almost all hands on shore for an 'Olympic Games' involving such curious sports as 'Slinging the monkey', as featured in Conrad Martens' (1801–78) watercolour

---

171 See Darwin 1842b; *Shorter publications*, pp. 140–7.
172 Darwin 1842a, p. 420; see Herbert 1999, p. 341.

sketched on the spot.[173] By this time the *Beagle* was accompanied by the schooner *Adventure* which FitzRoy purchased in March 1833, at his own expense, and without Admiralty approval.

There are some interesting zoological references on these pages, such as 'scorpion eating other scorpion', p. 87a; this is the 'cannibal scorpion' referred to in a footnote of *Journal of researches*, p. 194. 'Good eye sight of Guanaco', p. 88a, and the date of Christmas Eve given in *Zoology notes*, p. 182, for the guanaco shot by Darwin at Port Desire and described on pp. 80a–81a is especially valuable. Darwin used his handkerchief as an improvised ruler to measure various parts of the creature's anatomy.

Sadly there is no mention either here or in the continuing entries for Port Desire which occur at the start of the *Port Desire notebook* of the 'Avestruz Petise' (lesser rhea) shot by Martens at this time. This bird, closely related to the larger *Rhea americana* Darwin was familiar with from further north, was skinned and eaten before Darwin suddenly remembered gaucho reports of the smaller type. He rescued 'the head, neck, legs, wings, many of the larger feathers, and a large part of the skin' which were sufficient for John Gould's description in 1837 of what he thought was the new species *Rhea darwinii*.[174]

The Avestruz Petise was significant for Darwin because it appeared to be a different kind from the rhea he already knew but with a contiguous or overlapping geographical range. In other words, one replaced the other going south. This spatial relationship between two closely related bird species struck him as in some way connected to the temporal relationship he discovered between the fossil and living mammals of the same area. Darwin knew that the relationship between the two rheas was a signal of deeper meaning. It was one of the most crucial of the 'certain facts in the distribution of the inhabitants of South America' cited in the first sentence of the *Origin of species* which led to his theory of evolution.

As so often when on his own, in the last pages of the notebook Darwin waxed lyrical as he reflected, 'how many hundred years has been. how many will be … sublime view fine colour of rocks', p. 87a. This reflection was expanded in the *Beagle diary* to 'All is stillness & desolation. One reflects how many centuries it has thus been & how many more it will thus remain. — Yet in this scence [sic] without one bright object, there is high pleasure, which I can neither explain or comprehend', p. 209. Finally this became in *Journal of researches*, p. 198, 'One reflected how many ages the plain had thus lasted, and how many more it was doomed thus to continue.'

173 See Keynes 1979, p. 173.
174 *Journal of researches*, p. 108. See *Birds*, pp. 124–5. D'Orbigny had already described it as *Rhea pennata* in 1834. *Red notebook*, pp. 108–9 provides excellent photographs of both species.

Rhea darwinii, from eastern Patagonia. Plate 47 from *Birds*.

[FRONT COVER]

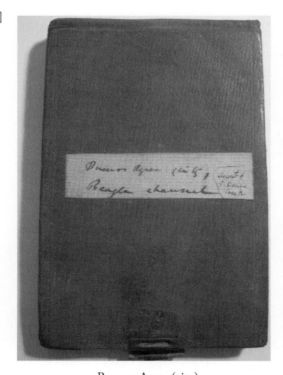

Buenos Ayres (city)
Beagle Channel
ascent of P. Desire creek.

[INSIDE FRONT COVER]

	C. Darwin	
Basker *[illeg]*	27	90
158 Calle[175] *[Victor]*	3	25
M[r] Hughes[176]	10	65
Hargrave	16	
Calle de la puedad	5	
Hill near Fouche Hotel	61	
Calle de la Cathedral		

[1a]   Hargrave
        Paper Bramah pens:[177] ┌Note book┐

---

175 Spanish for street.
176 Charles Hughes, a school fellow of CD's, attended Shrewsbury School, 1818–9, and resident in
     Buenos Ayres, 1832–3.
177 Joseph Bramah (1748–1814), patented a machine in 1809 to cut multiple nibs from a single quill.
     Bramah pens were the first widely used pens with separate nibs slipped into a holder.

Scissors
The /belt/
Watch key & glass
Dentist
Thermometer
Bramah Pens
Note book
Watch mended
Mr Chas Hughes Mr Rodgers /Bird/
~~Spurs~~
Museum open every 2nd Sunday
St Domingo
Mr Lumb[178] to collect shells in Limestone

[2a] Owe Mr Rowlett[179] one paper dollar
Pay for boys bringing to the Inn
Chaffers[180] one paper dollar
The very next door to Mr Griffiths[181] a French Dentist
Cigars Dentist
Mr Griffith Calle de Florida
Caldcleughs S America[182]

[3a] Mr Hammond[183] owes me 27 paper dollars
3 hooks
10 Williams
16 riding
50
Dr Browne Union Reading Rooms
Aperea;[184] animal without tail or Conejo[185] del Campo
Mr Flint,[186] owe him 1 dollar + 60

[4a] Calle de Victoria
158 No^s. Mr Breed R & Co

---

178 Edward Lumb (1804–75), English merchant in Buenos Ayres.
179 George Rowlett.
180 Edward Main Chaffers, master of the *Beagle*.
181 Charles Griffith, British Consul at Buenos Ayres, 1834.
182 Caldcleugh 1825. Alexander Caldcleugh (d. 1858), Private Secretary to British Ambassador to Chile, later trader and plant collector living in Santiago. Owner of copper mines at Panuncillo.
183 Robert Nicholas Hamond (1809–83), temporary mate on the *Beagle*, who often accompanied CD on shore excursions.
184 A guinea-pig, see specimen 1266 in *Zoology notes*, pp. 164–5; listed as *Cavia cobaia* in *Mammalia*, p. 89.
185 Spanish for rabbit.
186 Mr Flint, an American merchant at Buenos Ayres.

M[187] Stedman opposite theatre Calle de Cangallo: Bookseller

Famaterra (Rioja) Gold & Silver specimens

[5a] Tweedee[188] Retiro

J Tweedee

Tweedee

Public Garden

Retiro Retiro

M[r] Rowlett 30 dollar

M[r] Edward Lumb: 56 Calle de la [Paz]

[6a] Nov 4th [November 1832] — Sunday — Convents idolatry — gay appearance: Museum, civil manners; rode out in the evening along the beach: Mendoza waggons

5. [November 1832] Went about 6 leagues into the campo — & rode fresh horses back again: open flat country — very green tall thistles —

number of small owls. very like Cambridgeshire from Poplars & Willows

[7a] 6th [November 1832] Very busy in collecting informations & specimens shopping & Ladies

7 — [November 1832] Expecting to go off wasted the day dined with M[r] Gore[189] & met the Colonel Vernon,[190] great Traveller pleasant evening

[8a]

$$\begin{array}{ll} 14 & 14 \\ \underline{6} & \underline{4} \\ 84 & 56\ [\div] \\ \underline{7} & 8 \\ 91 & \end{array}$$

Bill 76 paper dollars

I paid 6 silver & 6 paper

I paid 15 silver dollars for horses: having had 2 rides. —

Captain owes me 21 dollars

[9a–12a excised]

[13a] M[r] Flint an American Merchant has a tooth: B. Ayres

Pay Rowlett for Boat

El Colonel [Ryuela].

[14a] January 19th [1833]

Entrance of *Beagle* channel all the mountains & ENE of it rounded — Slate —

187 Mr Steadman, a bookseller in Buenos Ayres. See CCD1: 319 note 2.

188 John Tweedie (1775–1862), Scottish gardener and plant collector based at Buenos Ayres, who owned a garden at Retiro. We are grateful to Jeff Ollerton for providing this information.

189 Philip Yorke Gore (1801–84), Chargé d'affaires in Buenos Ayres, 1832–4.

190 Colonel Francis Venables-Vernon-Harcourt (1801–80). See *Beagle diary*, pp. 115–6.

at the very Southern entrance there is a large bed, intervening, of a greywacke.[191] —

sonorous splintery fine grained siliceous scales of mica (A) &; therefore not covered

ferruginous, decomposing red oval balls of Iron by white lichen & with large angular cleavage like Trap: lines of cleavage dipping at about 80° to SSW ∴ running ESE & WNW

other lines not so well marked dipping at same

[15a] angle to SSE: in the former case almost jasper. the SSW dip is right, shown by the occurrence of rocks distant in channel with same bearing approaching character of slate some miles S of Entrance cliffs of diluvium, the same as formerly, <u>no</u> sign of deposition

SSW — sleeping place: slate (B) not very fissile. many lines of cleavage so as to render it impossible to be certain, perhaps

[16a] nearly vertical dipping to SSW. the upper beds at waters edge are banded for long distance with horizontal white lines from preponderance of silex.[192] — these stripes are sometimes curved & the planes /inclined/ in ~~contrary~~ 2 directions. but generally horizontal. with slight dip to S. — Cleavage is rather parallel to them: Hence I am unable to know any thing about stratification. the hills are not

[17a] remarkably parallel to B. Channel & on both sides rounded: started 4 boats, fair wind pretty spectacle. scenery very interesting. trees & verdur to waters edge: not very luxuriant & few trees. — Encampment tranquil, smoke, Cove of Islands: few inhabitants: Accident Robinson:[193]

[18a] 20th [January 1833] Southern bank rounded hills: /till/ Break jaw hills. but between them & channel yet slate hills. —

Northern side same only ~~crystalline rocks begin sooner (& at same point come down to water ?)~~ crystalline serrated ~~rocks~~ ridges begin sooner to appear ~~slate~~

from 1st sleeping place to 2^d ~~opposite~~ (NE of Break jaw) on each side similar rounded cliffs of white diluvium about 60 feet high generally very similar on opposite sides from observing almost island

---

191 'A conglomerate or grit rock consisting of rounded pebbles and sand firmly united together' (OED). The meaning of this term was changing during the *Beagle* voyage; Lyell for example implied that it was usually of 'early Secondary' age whereas other authors (e.g. CD's teacher Robert Jameson (1774–1854)) implied that it was of Transition age.

192 Amorphous silica, such as flint, chert, opal, agate or jasper.

193 'Cutfinger Cove. (This name was given because one of our party, Robinson by name, almost deprived himself of two fingers by an axe slipping with which he was cutting wood.)' *Narrative* 2: 202.

[19a]   in centre (at 2^d sleeping) it is evident these beds stretched across the channel (same colour on both sides (when above them were different rocks?): not so many boulders as Goree sound & whiter & at 2^d ~~break~~ sleeping finer grained & signs of deposition, therefore deposited from more tranquil water

[20a]   these cliffs remarkable in scenery: covered with coarse grass: Southern slate mountains parallel but no parallelism in sides of channell: long pull in boat. — scenery same, astonishment & following of savages wild appearance on hill: naked long hair: (give them many things. slings dinner time)

[21a]   attempt to drive them away by fires, innocent naked most miserable very wet (21st) [January 1833]
5 or 6 miles & close to 2nd sleeping place; North shore fine grained pale — rather fissile slate. (C): containing siliceous beds: cleavage dip about 60° SSW ∴ run ESE: & parallel fissures ~~parallel running NW by N & SE by E~~ N by E & S by W: also nearly horizontal

[22a]   lines are seen at ~~the~~ a distance — which /if/ there is stratification? the first is nearly parallel
▣ to the run of channel

▣ (the slate chain ends in the serrated jawbone ridge.) ∴   [sketch of 'jawbone ridge']
Cliff of alluvium on Southern shore extending to North by E of Break jaw & not so far on northern shore
wet night yet comfortable ~~at starting afraid of~~ fighting with savages women & children retreated signs of great fire:

[23a]   /on/ side of hill. B. Halls[194] Volcano.
▣ ▣ these horizontal planes dip at small angle to the SSW. in ~~in~~ the rock waving line of pale jaspery slate are parallel to them: (behind the slate ridge is quadrangular ~~hill~~ mountain of crystalline rocks?)
This must be strata
▣ Supplementary ridges about 150 feet high with pebbles
▣

[24a]   These supplementary and parallel ridges. most frequently by dip to about NE. directly contrary to general mountain. (Von Buch)[195]
The serrated rocks appeared nearer to shore on Northern bank than on the Southern when so the slate from a straight row of island (∴ these serrated mountains run in a NE direction?)
(D) Serpentine pebbles on the beach

194 Basil Hall (1788–1844), naval captain and explorer. Hall 1824.
195 Christian Leopold von Buch (1774–1853), pioneering German geologist and palaeontologist. Von Buch 1813.

[25a] Morning — 22^d [January 1833]
Last night comfortable & quiet. scenery begins to be beautiful: snow covered mountain: hot day
5 miles ~~W~~ E of Ponsonby Sound a large amphibolic[196] formation.
(E) upper beds most /horn stone/
(F) ~~variety~~ with Feldspar
(G) alternating with slate: on the coast 2 or 3 beds of (G) about one yard thick alternate

[26a] with the slate (H) slate (decomposing red) altered harsher generally ~~cleave~~ cleave dips $\angle$ 75° to SSW. — but near greenstone slightly altered.
sides /touching/ greenstone parallel — This amphibolic evidently protruded through & rests on slate: country with rounded paps desolate few trees
at this place on N shore serrated

[27a] mountains come down to water's edge with ~~exception of~~ chain of (slate?) islands at base height of trees curiously regular reaching nearly to the patches of snow
23^d [January 1833] — Morning entrance of Ponsonby Sound — Slate rather fine grained (K) dipping to the East $\angle$ 23° fissures very smooth planes NW by N $\angle$ 45° & others to E by S ½ S $\angle$ 87° cleave to S by W $\angle$ 50°
slate contains oval nodules of darker slate

[28a] crossed by many dykes ~~veins~~ of Trap (L) light coloured fine grained slate altered by junction (M) Sometimes where in contact dips ENE (K) common slate with white bands How much of this difference of dip from the SSW is to be attributed to the Trap: & how much to the serrated chain? (Trap)

[29a] W of Ponsonby Sound ~~after~~ (last night) after quiet delightful pull through the channel smooth water surrounded by peaked mountains between 2 & 300 feet high — the upper parts of which are brilliant with snow & lower dark with green wood found a snug cove: large fire. naked savages around it. Starlight

[30a] large fire: chorus of singers: savages perspired Tekenika J Buttons[197] quiet people:
after breakfast a large body came over the hill they had run so fast that their noses were bleeding. When we started to go J Buttons place within Ponsonby Sound. — 12 canoes accompanied & from the bright sun & hot day the scene resembled the drawing amongst Pacific isles:

196 Associated with the amphibole group of silicate minerals, such as hornblende, common in igneous and metamorphic rocks.
197 Orundellico known as James 'Jemmy' Button (?1816–63).

[31a] we out sailed them & ~~found~~ Jemmy guided us. none of Jemmy's immediate friends
were there, but doubtless the news wide spread:
within Entrance (East side) of Ponsonby Sound. slate strata generally dipping
southwards & curiously contorted almost anticlinal[198] line. perhaps continuation of
Break jaw mountains: North of island & South of W peninsula of Navarin I. the slate
~~generally~~ dips Southwards

[32a] The evening spent in cutting wood & clearing ground for Garden — Jemmy Buttons 3
brothers & Mother. J can talk but little: not much affection on Jemmy's part: excellent
spot: Guanaco[199] — everything favourable — 3 brothers came to meet Captain
favourably: Jemmy recognising voice: extraordinary strength of voice:
24th [January 1833] The hands busily employed in building a hut & going on with the
Garden:

[33a] Savages are quiet. sit in row. naked by the trench: will not much work watch everything:
our washing & white skins surprise them most: canoes slings spears fishing, manner
of life
~~mi~~ Guanaco in winter: very wet day uncomfortable walked up mountain to shoot
Guanaco could not get near to them: large & numerous trees. *[many]* decayed. —
summit a swampy dreary plain.

[34a] 25th [January 1833] Slate greyish blue rather fissile O (same as *[usual]*) dipping at ∠ 53°
to S by W ½ W. apparently beds; & coloured bands (not white stripes as before) &
cleavage all the same:
between the beds & ∴ parallel are beds of breccia small bits of slate in a Greenstone base
(N) very frequent & forming large masses in the mountains as

[35a] abundant as the slate: both greenstone & slate with flattened oval large 3 or 4 inches
cavities: perhaps contained some mineral now washed out
P̅ Greenstone with large fragments & passage Greenstone sonorous conchoidal
fracture & ~~in~~ natural fissure most parallel & smooth — so as to form very regular
~~quadrangular~~ rhomboidal pieces: alternations of slate & it, frequent well defined: does
the slate answer to greenstone as phyllade to mica slate

[36a] Many more Fuegians arrived in their canoes: uncle & friends of Jemmy. very civil
painted white. like millers sit quiet watching & begging for everything: never pass the
trench. — Is not the amphibolic formation at Eastern Extremity Navarin Is where strata
are contorted —
signs of anticlinal line. the subterranean agent by which Breakjaw Mountains were
elevated does not

198 Up-fold of strata (opposite of syncline, a down-fold).
199 A species of South American llama (*Lama guanicoe*).

[37a]   the bearing of chain countenance it? — NB. near anticlinal line strata dip at much greater angle

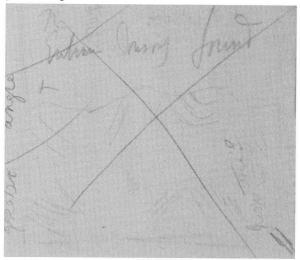

[geological section]²⁰⁰ N Entrance Ponsonby Sound Green Stone ?

[38a]   26ᵗʰ [January 1833] High up slate very fissile (Q) ferruginous cleavage dipping S by E ∠ 70°

Perhaps from the northerly chain in Hoste Island country. large elevated flattish track with these slate knolls between slate beds (R) trap like angular mass with fragments slate as before

N. B. Beetles on sandy

[39a]   plain Heteromerous.²⁰¹ Bahia Blanca V. De la Beche?²⁰²

Arrival of women. above 120 people. men sit watching women work

The specimen (P) is part of a mass with prismatic or trap-like cleavage. — & is curious as on one side looking like 2 pieces of slate entangled, on the other a gradual alteration

[40a]   if the slate & trap are the same in every thing but form, we must suppose the slate entangled & the greenstone by the contact loosing its crystalline power

27ᵗʰ [January 1833] Near the sea frequent about 50 or 60 alternations of narrow beds: varying in width of slate & greenstone sometimes one

---

200 See the fair copy of this sketch in the geological diary DAR 32.98a.
201 Coleoptera, see *Darwin's insects*, p. 71, and *Zoology notes*, p. 128.
202 Possibly a reference to De la Beche 1831.

[41a]

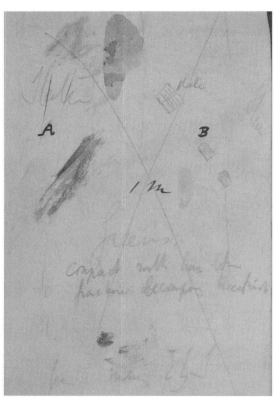

[section of slates][203]

Slates slate *[illeg]* A B Greens. compact with trap like fracture decomposing blackish
[lines] running E by S

[42a] broadest sometimes another slate indurated decomposing ferruginous greater spec.
gravity banded like jasper line of junctions <u>most</u> fine, parallel, straight, could not have
been protrusion or deposition: if one place had been only seen. trap appears to have
torn from beneath slate. & near it to have entangled & separated pieces so as to give
different cleavage lines Vide *[(P)]*

[43a] 9 inches of pure mould over old shells from wigwam
Temp of Salt water 55° in shade 70°
uncongenial climate glaciers & no crysomelidae[204]
▣ vide Plate (both fact not in one greenstone) the beds run E by S & W by N- & dip
nearly vertically to N by E ie exactly contrary

---

203 See the fair copy of this sketch in the geological diary DAR 32.93v.
204 Leaf beetles.

[44a] to general character of upper range of hills. Von Buch
Some beds not more than 2 feet thick

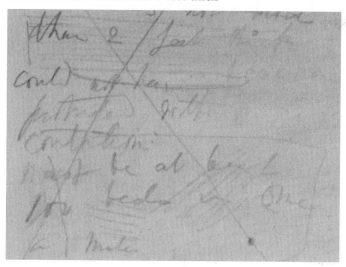

[section of strata]
could not have been protruded without contortion:
must be at least 100 beds in only ½ a mile
last night two men crawled near tent: return from walk found every
woman &
[45a] child & canoes & nearly all men gone: strangers arrived: watched from hill:
spears & spot of destination: change of manners: hurry all goods in tent & retreat
to little beautiful cove York &c not frightened impossible to judge what was the
matter: misfortune: extreme treachery of character. anecdote of Captain last voyage.
shark in Cove:
[46a] 28th [January 1833] Spider green with eggs enveloped in very strong web
brown on a bush
Bugs under Bark Settlement: Fringilla[205] in flocks.
Returned to Settlement, all quite quiet: Jemmy brother & mother returned:
sent away 2 bad men: aweful night for Matthews:[206] pretty scene: canoes
after fishes Entrance of Ponsonby Sound section of mountain curious contorted
to the W & N

---

205 A genus of birds of the finch family.
206 Richard Matthews (1811–93), missionary of the Church Missionary Society on board the *Beagle* to
establish a mission at Tierra del Fuego. This was abandoned in 1833 and Matthews rejoined the
Beagle until New Zealand in 1835.

[47a]

[sketch section] NNE SSW 3 or 400 yards

[48a]   the prevailing SSW dip appears to hold good but within the serrated ridges are out of
all order: — In *Beagle* Channell mountains not very parallel: about 18 miles northern
shore. W of Ponsonby Sound. Mica slate & few miles W of a big Bay & again Green
/talcose/ schist. — dipping to the South (SSE & SW by W) cleavage contorted
involving beds of quartz: all

[49a]   the pebbles do. ∴ ~~the~~ probably the neighbouring serrated ridges are of this formation
(J Button white limestone) Probably all serrated ridge are thus primitive[207] & that on
N: shore they come near to water opposite Ponsonby Sound
Snow & wood joining rivulets. whales blowing excessively hot

[50a]   Having put up tents (mica slate place) unfortunately a party of 7 Fuegians appeared:
perhaps had never seen Europeans. no way of frightening them, only rubbed theirs
heads when pistols were fired close to them. & laughed at flourish cutlass: obliged to
pack up from these fearless barbarians, fight like

[51a]   animals: found in the dark a quiet nook. kept watch. solemn scene till one oclock —
distant bark of dog:
— J. Button forget languages not prejudices: not eat land birds
only the lower trees change colour of leaf in autumn
(29th) [January 1833]
Just within N. Branch. Mica Slate. Phyllade. Hyalomictite dip SSW. — angular blocks
of Granite:

207 Coarse-grained rocks which did not show evidence of having been laid down under the sea. Often
incorrectly assumed to be of 'Primary' age. Re-named hypogene by Lyell (1833, pp. 13, 374). Closely
related to Protogine or Protogene. See p. 136 and notes 241 and 463.

[52a] about 5 miles within grand formation of Mica slate. small silvery scales (W̲) with largish dodecahedron garnets, cleavage ~~tortuous~~ serpentine dipping to the SW by S or SSW ∠ 65° great numbers of blocks of ~~garnets~~ granite /very/ small crystals /as/ (also Hornblende & Mica rocks) evidently the peaked summit of mountain (such forms

[53a] I have observed within the serrated ridge) channel /above/ 1 & ½ miles wide hills on both side above 2000 feet high & the Southern side the strata are evidently seen dipping to the SSW & formed of mica slate: sides very parallel

/chan/

[54a] entered N. branch of B channel Scenery very retired many glaciers uninhabited beryl blue <u>most beautiful</u> contrasted with snow: dinner great wave boat &c pack up. grand sight. — enormous block of granite 30 yards /around/ oblong in figure about 6 above ground & much below — bordering

[55a] ridge formed of such ~~blo~~ blocks- about ~~50~~ 60 feet high: glacier cliff to sea about 40 feet: blue by transmitted & reflected light: channell covered with small icebergs miniature Arctic Ocean:

30th [January 1833] (opposite /bite/ sleeping place. mica slate & narrow beds of green mica rock dipping to the SW by S

[56a] on S shore at the bend in channell there occurs slate fine grained (roofing slate /~~gre~~ more modern./

alternating with coarse beds & with beds of dark green greenstone small acicular crystals of Hornblend (X) & beds of white Feldspathic rock (Y) the dip nearly vertically to the SW by W. Is this from the Trap (Yes it is for further back

[57a]

all the cleaves) to SW by S   This clay slate formations extends on both sides the ~~be~~ leading bend. — The slate at point where ~~Islands~~ first lazy Bay Southern /(o)/ begin are anticlinal & contorted: N shore grand peaked granitic chain: ~~Begins~~ near here apparently ~~beyond~~ greenstone |?| Beyond it again talcose

[58a] schist & phyllade dipping to the SSW — & I suppose old formation — opposite the place (sleeping place). the northern hills not so serrated: ~~How:~~ (Grand views during day. — exceedingly jagged chain snowy clouds, blue sky — scenery generally spoilt by one chain & low point of

[59a] views. — vegetation like Mt Edgecombe.[208]

---

208 Mount Edgecombe, an estate in Cornwall overlooking Plymouth Sound. See *Beagle plants*, p. 166.

Miserable sleeping place big stones putrefying sea weed & middle watch. not all pleasure (31st) [January 1833]

N shore NW of last sleeping place Ph green phyllade — becoming Hornblendic (1). — characteristic /torture/ cleavage to SSW. like fire wood channell running /E/ ½ N & W ½ S the slate of course ESE & WNW the granite which has

[60a] upheaved the last slate it: Granite small black mica large crystals of quartz (2): hill globular quite barren

occurs at a projecting point on N. shore — also (from sight) evidently below the 2 ridges of slate opposite last nights sleeping place. it is seen thus in channell from the latters oblique /le/ direction: Again another alternation may

[61a] be seen. — These alternations extend N of the grand Southern entrance from N. Branch into Pacific

during heavy rain rather miserable we are now amongst the islands on the W coast called in chart Gordon Island Sleeping place. rock Granite with mica (3) passing into green mineral constituents so arranged as to show tendency to become Gneiss: covered with trees from feldspar

[62a] Feb 1st [1833] Gordon Island

very wet. squally true Tierra del. — weather. — 2 Whale within pistol shot enormous backs & tails: resting on /the/ the gneiss like granite in Mica slate (dining place) it extends for a considerable dista of the most N side (& I think perhaps central isles yes): the mica slate fine grained mica black in fine rows /lines/ dips to the SW !! Central Islands (the

[63a] main new island). Phyllade (4) (& mica slate (5) or gneiss ?) repeated by alternating & dipping to the SSW: the opposite coast perhaps again granite. extraordinarily barren; all hills on both sides not much elevated the great height ceased where the granite first was touched:

[64a] All country most desolate & quiet: never seen by any European; fur seals: (great accumulation of Boulder from glacier)

Feb 2^d [1833]

Few miles W of the slate chain of New Island range of granite, massive form. black mica. hornblende white feldspar & little quartz (6) small grained this as formerly would appear to intervene between two

[65a] slate ranges: The western extremity entering granite. The Eastern end slate. between 2 ranges of granite & dipping to NNE backwards: The East end of Stuart Isles greenstone the outer line of mountain in this & Londonderry look like granite. I should think this greenstone & porphyritic greenstone abounding with the nearest granite contains hornblende

[66a] Feldspar /was/ primitive. During the day been examining some large new islands the furthest point west & channels M Sarmiento. Cold day (Curious little green slate hill surrounded apparently on all sides by bare granite ones) In the granite were dyke & black mica rock

Feb 3^d [1833]

Miserable weather barren *[watery]* country: outside

[67a] passage rather lucky entered the southern arm of B channel. — the SE end of Londonderry Island. (Phyllade?) having become (7) greenstone slate: 9 miles within B channel both sides ~~pebbles of~~ *[near]* ~~true granite~~ Granitic greenstone occasional crystal of quartz external forms like Granite (8) has not the form cleavage sonorous sound of Hermit Island: —

There would <u>appear</u> to be remnants of slate on *[all]* islands perhaps washed away:

[68a] Feb 4th [1833]

On both sides S. arm of B channell. — granitic greenstone (9). In Feb 3^d place there were pebbles with acicular hornblende like Hermit Island. — per contra does not contain pyritic & above reasons. — NS 5 miles from bifurcation this granitic rock occurs mingled with micaceous[209] slate & above them

[69a] <u>appears</u> to be slate: on S side 2 miles from do evidently comes to water edge. with dykes traversing it — (this order of course depends on the run of channell) vegetation & form bespeaks slate at bifurcation true (10) (roofing) very fissile & coloured in bands // to cleavage. — some of the planes tortuous others

[70a] straight dipping (from the nearness of granitic rock) to SW by W — It is probable that this rock runs all along the S side of N arm as on the N side the peaks are not serrated but peaks. ∴ perhaps granitic. the lower ranges having been seen to be mica slate. all slate at bifurcation the tops of mountains had

[71a] coloured bands nearly horizontal. & on S side of N arm there were these bands to a great degree: —

very rainy. in afternoon passed bifurcation could see a great many miles through the channel both ways. nothing interrupted the view:

In the *[eel]* looking fish slime comes through white spots[210]

[72a] (5th) [February 1833] 5 miles W of big bay

The micaceous slate on the (28th) is ~~the~~ I think certainly the most Eastern which comes down to the channell & it cleaves to the SSE. there would appear to be roofing slate resting on it. — & from this place extends ~~W~~ Easter*[ly]* On the South bank the whole range is evidently roofing slate: the serrated ridges N of ~~Ponsonby~~ *Beagle* Sound & W of Ponsonby Sound

---

209 Containing conspicuous amounts of mica and so tending to reflect light and glisten.

210 Possibly a dogfish; see specimen in spirits 840 in *Zoology notes*, p. 347: '840 F X Dog Fish: Color pale "Lavender purple" with cupreous gloss. — sides silvery do. — above with regular quadruple chain of circular & oblong snow white spots.'

[73a]   — *[Dog jaw]* are evidently slate & perhaps only summit of roofing slate as in G Success Bay ? H *[Merely]* running over same ground in the evening amusing bartering scene for fish:
(6th) [February 1833] W Entrance of Ponsonby Sound. — abundance of Greenstone. dyke traversed by different *[qualities]*

[74a]   Surrounded by slate dipping South. — Trap forming large portion of mountain — cause

of irregularity of Mountain range. & as it exists on both sides of the  cause of it — Subsequent to the SSW dip —
for Disappointment with the Fuegians Matthews:
East centre of Ponsonby

[75a]   Sound. — Slate indurated narrow beds dipping to S. *[more]* folded & contorted traversed by vertical dyke of decomposed Trap-like clay: on one side beds upturned from the Southerly inclination to running N & S —
This same slate truly near *[bare]* mountain dip vertically to SSW. & has fissure dipping to NW by S ∠ 65°. flinty blue slate

[76a]   Is not Tierra del the Ultima Thule.[211] not volcanic occurrence of garnets in mice slate *[G]* SW end of Navarin Island Greenstone (perhaps altered slate). Greenstone every specimen different irregular. *[rubbly]* fracture. decomposing not highly crystalline containing pyrites[212] — some specimens very fine grained. other

[77a]   largely porphyritic with white Feldspar. —
The greater part of Hoste Island appears to be slate is this greenstone — (no Feldspathic rock?
Perhaps Pack saddle I. from its angular shape is greenstone of the Hermit *[Group]*
In the evening followed by Savages.

[78a]   fired over them & frightened them away found a quiet little cove on outside coast fire surrounded by trees:
7th [February 1833]. —
From E W of an Island of S. coast of Navarin Island: a *[trench]* of Ł of *[diluvium]* common & *[illeg]*

[79a]   & in an ENE direction *[illeg]* the Island
In Goree Sound the cleaving & coloured band — dipped vertically to SSW & believe with a degree of other side
distant about 34 miles:

211 Meaning a distant place located beyond the borders of the known world.
212 Usually iron pyrites, iron sulphide, common in slates, but can be other sulphide minerals.

[80a] length 1 handkerchief & ½ circumference handkerchief — 3 & ½ inches — Tail — AA beneath B

[leg of guanaco][213]
60 ½ extremity of claws to joint

[81a] most wide part of sole. 1 & 3/8 inch of foot 2. 7/8 $^{in.}$ — 1.70 minus inside *[7]* feet long tip to tip circumference of chest 4^{ft} 8^{in}. Tail
Ibis[214] iris pink & legs do. — stomach scropions cicadæ Lizard. Coleoptera[215]
ENE. 300

[82a] At ~~Fort~~ Sea 30.303. Temp 65.7
Shells   30.138. Tem 71.5
Plain of fort of 2 hills 30.030 Temp 70.7
shells *[at]* sea   30.283 T 72.7
One Carrancha[216]
Avecasina[217] on desert plain

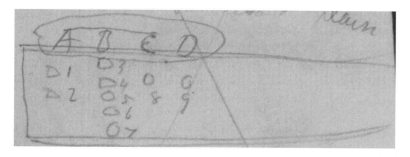

213 A footnote in CD's *Animal notes*, 30v reads: 'NB. I have dimensions of Specimen shot at Port Desire, especially about the foot, to compare with those killed in T. del Fuego, which are supposed by some to have broader feet.' See *Zoology notes*, pp. 182–3.

214 Specimen 1773 in *Zoology notes*, p. 186, 'Ornithological notes', pp. 229–30; listed as *Theristicus melanops* in *Birds*, pp. 128–9.

215 Beetles.

216 Possibly the crested caracara, a falconoid bird of prey, see *Zoology notes*, p. 212; listed as *Polyborus brasiliensis* in *Birds*, pp. 9–12.

217 A woodcock (*Scolopax paraguayae*). 'In la Plata the Spaniards call them 'Avecasina'', 'Ornithological notes', p. 212.

A	B	C	D
o 1	o 3		
o 2	o 4	o	o
	o 5	8	9
	o 6		
	o 7		

~~Foot~~: bed of free stone earthy looking rock yet with crystalline particles

[83a–84a excised]

[85a] yellow lines like water lines in Sandstone. (V2). These rocks instead of being modern beds are subordinate to great Porphyry (1 & 2) 2 is inferior to one — 4 5 6 7 are all beneath the porphyry (3) — These rocks pass into each are so soft as to *[resemble]* earth & lined with yellow lines. — The porphyry (3) is only bed 9 & 10 feet thick, a superincumbent

[86a] becomes 8. — then 9 both very sonorous, the latter with linear arrange[ment] of constituents about horizontal. — The bed of soft rock dip about ~~12~~ 10° somewhere near to S or SSE — & superincumbent porphyry do These rocks rest at foot of great porphyry range of hills lower than plain running

[87a] ENE WSW. —
origin evidently igneous[218] over aqueous
Scorpion eating other scorpion.[219] —
(~~27~~ 29th?) [December 1833] on the South side 8 miles up white calc: sands covered by thin bed of hard ferruginous sandstone lying ~~above~~ upon A & B varieties semicrystallized of porphyry: cliff land much subject to diluvium action, excessively solitary. delightful walk reflecting how many hundred years has been. how many will be

[88a] without tree sublime or animal excepting Guanaco which stands sentinel to its herd: bits of primitive rock scattered about. seen in the vallies: —
highest hill SW by W compass.
~~28~~ 29th [December 1833] mud banks good eye sight of Guanaco. —
river! horse! many varieties of porphyry fragments & crystals: high cliffs, almost ended vallies. sublime view fine colour of rocks (D) — [4] specimens

---

218 Rocks formed from the molten state (literally by fire), either 'intrusive' e.g. granite, or 'extrusive' e.g. basalt.

219 The 'cannibal scorpion' referred to in a footnote of *Journal of researches*, p. 194.

[89a–90a blank]
[BACK COVER]

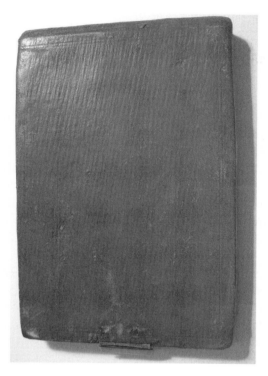

[INSIDE BACK COVER]
  For the pistol 6 & ½ Buck shot. — Captain
  8 for the Gun

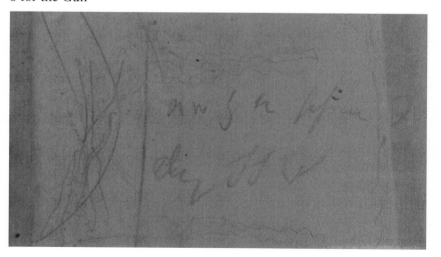

[geological section with fissure] NW by N fissure? dip SSW

/S/M. s/ /silty/ green /idem look/ schist

[1b] /M^r/ M

Rio Parana waters very black: /R/: m Rio Negro black

medicinal qualities excellent drinking sarsaparille[220]

M^r /Hujet/

Water pale (Both say) of Uruguay in latitude 32° — Palm Forest

[2b] it is said that Crocodile occur & small water turtles. —

silicified (/stink/ fire) wood: Lime said to be had at Aroya de la China: some shells

St Jago de Estero ~~felt was~~ was quite overthrown in an Earthquake not within in memory

of man

[3b] Tosca thick town built on it — cut into vallies — red compact — with ~~nearly~~

houses little inclined beds of (soft siliceous Limestone?): lines of fracture

separated & filled up by the Tosca: stops water wells dry to the surface or some

little way out it

[4b] November 1st [1832] Very calm delightful days: quietness seems to shorten the

distance. always think of home

Nove[mber] 2^d [1832] Anchored ~~at 3 am~~ before noon in the roads. We passed our

friend the Guards ship. who this time knew better than to fire at us. — Landed & were

carried in carts on shore — went out riding &

[5b] walk about the town. —

November 3^d [1832] B Ayres large city: /much/ regular streets (handsome plaza viceroys)

Quadras & square houses: — number of excellent shops: general European appearance

excepting a few Gauchos.[221] —

Spanish ladies beautiful dresses & walk

went out riding

[6b] bad roads, good horses; hedges agaves & fennel — — flat enclosed country — ditches.

most uninteresting. —

Saw Bullock killed: the feeling of being on shore very pleasant:

Rock which rings at Rio Negro. Trap? or thereabouts ie in B Ayres

Fresh water shells at Ensenada: — ?

[7b] Fresh water shells in the rock, above level & some distance from the present river burnt

for lime

Ensenada

Also on the coast opposite to Mr Lumb estancia Mya[222] &c living ones

---

220 Sarsaparilla, a species of *Smilax*, a perennial vine used for medicinal properties. Its name derives from
    the Spanish words zarza for 'shrub' and parrilla for 'little grape vine'.

221 A mixed European and Indian race of equestrian herdsmen of the South American Pampas.

222 A genus of soft-shell clams.

[8b] Stone from the mole comes from Martin Garcia & Colonia. Syenite mica slate.
Greenstone large crystals & slaty Hornblendic rock with crystals & Feldspar &c &c.
Spurs. Dentist Hughes Frenchman & Curiosities.
Mr Chaffers do. — /Botanical/
Captain The cliffs specimens.
Mr Lumbe (? direction & shells in Lime)

[9b] Mention about the animals being only skinned & snakes /Lizards/ &c &c
Is M. Video built on granite or the Gneiss?

[10b] Spix. Abrolhos. 120P.[223]
Will the gradual deposit of river explain the Tosca.
John Meggett
Messrs. Brittain & Co.
Buenos Ayres.
The Estancia de los Yuquerises
Entre Rios
Desert of Atacama
25 Leagues from the Port

[11b] of Cobija — alias La Paz great Iron Formation connection with Volcanic action
at San Antonio de [Arico] distant from Buenos Ayres 30 leagues: the water has power of
turning small bones into large ones.
Agraciado. Banda Oriental. 15 leagues from Martin Garcia. small shells are burnt
for lime: —

[12b] (a): Granitic undulating chain of hillocks. running North & South
Bramah pens: Stirrups. Note books
Calle de Porton
— Watch mended
— Great teeth
— Information about whale.
— Joiner
What M. Video built on ?

[13b] Oakley[224] a joiner. with red hair: Monte Video: Can be heard of at an Hotel formerly
kept by Browne:
Casks: length of red cloth: Lamb well & teeth
Mr Fitus Hotel

---

223 Spix and Martius 1824, 1: 120, discussed the Abrolhos Islets.
224 Oakley was the agent of Woodbine Parish (1796–1882), the British consular representative, see
Winslow 1975.

Calle de St Felipe

[Turtle]

[14b] South of Rio Negro cliffs of Tosca — [La] Paz?

15 miles above Pisandu Limestone near Arora de la China: Above Arora del [Palma]

near Segnor Manuel [Esquire] [Maga]

teeth belonging Mad [Barborza] Buenos Ayres

Teeth found at Salto between near head of Rio Arrecife &

[15b] [arrecos] ?)

Viscache[225] shit as dogs

There is a small Villa at Luxan near B Ayres (Oakley)

Gypsum[226]

Directions parchment & paint.

Custom House

Mr May[227] Hammer & [tools]

Oakleys fossil one scapula in true Tosca

 [sketches of tables]

[16b] Ask Henslow[228] open Pill boxes —

Seeds

Mr Stokes

Washing clothes

Capinchas dung smells very sweet —

On Friday Gossamer web

Conjuror

Hammer

[17b] Settle with Fuller[229]

<u>No</u> jars or paper

washing

---

225 A burrowing rodent related to the chinchilla which resembles a large rabbit, but with bigger gnawing teeth and a long tail and three toes behind like the agouti. It is common in the pampas in the neighbourhood of Buenos Ayres; see specimen 1442 in *Zoology notes*, pp. 180–1; listed as *Lagostomus trichodactylus* in *Mammalia*, p. 88, and spelled as 'Bizcacha' (elsewhere spelled 'Viscacha' or 'Biscatcha' by CD).

226 Hydrous calcium sulphate, often in the form of crystals (selenite). CD explained that in Chile gypsum was called 'Padre de la sal', while potassium sulphate ('potash') was called 'Padre de la sal'.

227 Jonathan May was the *Beagle*'s carpenter.

228 John Stevens Henslow (1796–1861), Cambridge botanist and mineralogist, CD's scientific mentor when at Cambridge, who received and preserved CD's *Beagle* specimens during the voyage.

229 Harry Fuller, Captain's steward on the *Beagle*.

Bull 90

Cooper 10

83:1

  3:4

  5.4

  <u>24</u>

116.0

  <u>48</u>

  68

[18b] Owe Hamond[230] 30 (1) /Had me 2/ silver dollars

Row[l]ett 4

Owe Hammon[d] as balance 21 Dollars.

Rowett 4   Black Duck

~~Plus Boro~~ ; Borr 1 & Dinner.   Covington[231]

Trousers Drawers & shirt.

Good Success. Bay. dip to West

[19b] at small angle.

Owe Peterson[232] ~~knife~~ net.

(John < >

[20b blank, partly excised]

[21b–22b excised]

Textual notes for the *Buenos Ayres notebook*

[IFC] 1.12] *Down House number, not transcribed.*

88202332] *English Heritage number, not transcribed.*

C. Darwin] *ink.*

  [6a] *eight marks in ink on this page may be the testing of a nib.*

[24a] *this and the facing page are very dirty.*

[38a] *an ink mark appears to be the testing of a nib.*

[40a] *there is a watercolour stain on this page.*

[41a] *there are watercolour stains and apparent ink nib tests on this page.*

A B] *ink.*

[IBC] 2] *added by Nora Barlow, pencil, not transcribed.*

---

230 Robert Nicholas Hamond (1809–83), mate on HMS *Druid*, loaned to the *Beagle* in November 1832, and returned to England in May 1833.

231 Syms Covington (1816?–61), 'Fiddler and boy to the poop cabin' on *Beagle*'s second voyage, he became CD's personal servant from 22 May 1833 until 25 February 1839.

232 John Peterson (1787–?), Quarter-Master on the *Beagle*.

[1b]   *3 ink marks near* 'Negro' *appear to be nib tests.*
[10b] John ... Post] *not in CD's handwriting.*
[11b] of Cobija ... Paz] *not in CD's handwriting.*
[15b] *sketch drawn perpendicular to the spine.*
       Gypsum] *ink.*
[19b] *rest of page excised.*

THE *FALKLAND NOTEBOOK*

↬

The *Falkland notebook* takes its name from the Falkland Islands, the archipelago in the South Atlantic Ocean, located 480 km (300 miles) from the coast of Argentina. The notebook is bound in brown leather with the border blind embossed: the brass clasp is intact. The front of the notebook has a label of cream-coloured paper (72 × 34 mm) with 'Falkland Maldonado (excursion) Rio Negro to Bahia Blanca' written in ink. Darwin created a pencil holder inside the front cover by pasting in a leather sleeve. The notebook has 90 leaves or 180 pages. The entries date from February to May 1833 (pp. 1a–86a; IBC–3b) and August to September 1833 (pp. 86a–142a; 4b–13b).

The earliest entries in the *Falkland notebook* date from around Darwin's twenty-fourth birthday in early February 1833. According to Sulloway 1985 this was in the middle of the time, from November 1832 to April 1833 when Darwin was at his lowest emotional ebb in the whole voyage, which seems understandable in view of the weather and seasickness he experienced around Tierra del Fuego and the Falklands, even though it was summertime. The prospect of the voyage lasting three years more than originally planned may have contributed. Using content analysis of Darwin's letters to Henslow during the voyage, Sulloway argued that Darwin's psychological development from a somewhat diffident tyro naturalist into a self-assured theoretician was the largest single intellectual legacy of the voyage. Sulloway showed, as far as the Henslow correspondence indicates, that Darwin was at his most anxious concerning his abilities and responsibilities at the time the *Falkland notebook* was in use. It was not until April 1834, almost half way through the voyage, that Darwin received any feedback about his collections from Henslow, by which time his confidence as a geologist was already on the rise. As Sulloway stressed, none of Darwin's observations alone was sufficient to establish the fact of evolution. He needed the expert identifications of his specimens that would be made only upon his return to England.

Darwin's use of the *Falkland notebook* was more or less continuous from early February until the end of August 1833, although there are significant gaps in June and July. The first continuous series of entries is from February to the end of May, pp. 1a–86a; IBC–3b. Use of the notebook was then sporadic until Darwin made his first major inland excursion in August which is recorded at length on pp. 86a–142a. The rest of the front pages of the notebook are blank. The *B. Blanca notebook* replaced the *Falkland notebook* around 29 August, although the last of the back pages of the *Falkland notebook*, c. pp. 4b–13b, seem to date from September. This dating seems confirmed by the reference on p. 85a to use of the *Falkland notebook* from 'end of Feb to Sept'.

Tierra del Fuego, February 1833

The first entries in the *Falkland notebook*, pp. 1a–7a, relate to Tierra del Fuego where the middle section of the *Buenos Ayres notebook* finished. They are mostly rather disjointed thermometer and barometer readings.

Falklands, February–April 1833

Entries then cover the *Beagle*'s first of two cruises to the Falkland Islands, 26 February to early April, pp. 7a–27a. The Falklands episodes are very well covered by Armstrong 1992 and Armstrong 2004. Keynes 1979, pp. 118–19, *Beagle diary*, p. 147 note, and Keynes 2003, pp. 137–8, also quoted extensively from the *Falkland notebook* to show that it was in the Falklands that Darwin first started to consider island endemism. The notebook is completely silent regarding the troubled political history of the islands.[233]

It is remarkable that in almost his first entry made on the eastern island (Darwin did not visit the western island) on 2 March 1833 he asked himself on p. 8a: 'To what animals did the dung beetles in S. America belong.' Then, apparently written after this is a second, more general question which seems to relate to the first: 'Is not the close connection of insects & plants as well as this fact point out closer connection than migration'. Grove 1985, p. 419, suggested that this was an indication that Darwin was intrigued by the similarity between island species and those on the nearest continent, an idea which Darwin took much further after he visited the Galapagos.[234]

There is no mention of the tragic death of Edward Hellyer, FitzRoy's Clerk, who drowned trying to recover a duck 'of a kind he had not before seen', on 4 March.[235] Darwin was in the party that found and recovered the body, which was buried the following day.

On 12 March Darwin noted the 'aberration of instinct' in the introduced horses which were 'fond of catching cattle'. He described what he called a 'mantle ridge' of the quartzite which he interpreted as a 'point of upheaval', p. 19a, running parallel to

233 See CCD1: 304 note 6.
234 See the introduction to the *Galapagos notebook*.
235 *Narrative* 2: 272.

Berkeley Sound, and on p. 21a he described watching a cormorant 'playing' with a fish, as an otter might, or a cat with a mouse. From the note on p. 26a he seems to have broken his watch.

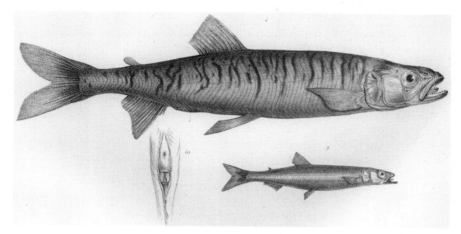

A freshwater fish (*Aplochiton zebra*) caught by Darwin in a freshwater lake on East Falkland in May 1833. Plate 24 from *Fish*.

Darwin's Falkland geological observations, which have been discussed in depth by Armstrong 1992, fitted comfortably neither in his book on the geology of *South America* nor in his book on *Volcanic islands*, so he published them separately in the *Proceedings of the Geological Society of London* (Darwin 1846). Darwin did, however, publish a short account of the Falklands geology in *South America*.

On p. 11a Darwin drew a simple sketch showing the peat, which is a prominent feature of the islands, underlain by clay and overlain by sand. On p. 23a he noted the small bones of mammals found in the peat causing him to speculate that these animals 'like rats' were the original inhabitants of the islands. It is unclear when Darwin first noticed the Falkland earthworms. He referred to them in his very last book, published the year before he died (*Earthworms*, p. 121).

Darwin noted the various slates and sandstones encountered, and was intrigued to 'ask Chaffers [i.e. Edwin Chaffers, Master of the *Beagle*] where gneiss came from', p. 13a. From a geological point of view, the fossils Darwin collected just south of Port Louis on 22 March were probably the most exciting treasures of this Falklands period, see p. 24a. The fossils were mainly brachiopods,[236] but also some crinoids.[237] The brachiopods were described by John Morris and Daniel Sharpe in a

---

236 'Lamp shells'; a group of fossils in mainly pre-Tertiary rocks. Bivalved, attached to seabed.
237 'Sea lilies'; a group of fossil and living echinoderm invertebrates.

paper (Morris and Sharpe 1846) immediately following Darwin's paper on the Falklands.

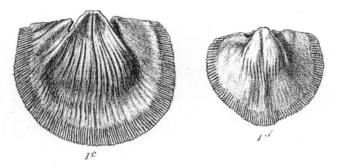

Devonian brachiopods collected by Darwin on East Falkland, from Morris and Sharpe 1846.

It is difficult to overstate the importance of these fossils. At the time of their discovery fossils like these were little known beyond Europe and were regarded as almost the oldest known life on Earth. These were the sorts of rocks and fossils which Darwin would have seen with Sedgwick in Wales in 1831 and at the time would have been called 'Transition', signalling their position somewhere between the oldest 'Primary' rocks, and the 'Secondary' rocks which underlay, for example, much of Southeast England.

There is a very interesting reference to the fossils in the *Copiapò notebook*, p. 72, '(NB. fossils of Falklands of hot country??)', demonstrating that Darwin was well aware of the value of fossils for reconstructing past environments. In 1846 Darwin dated the fossils as Silurian or Devonian (the term 'Devonian' having only been proposed by Roderick Murchison (1792–1871) and Adam Sedgwick in 1839 (see Rudwick 1985), but they are now dated firmly to the Devonian. The age of the specimens is now thought to be the same as those of the South African Bokkeveld, thus Emsian–Eifelian (Stages of the Devonian), very approximately 386 million years old. What is now the southern part of the African Plate was, in Devonian times, close to the now mainly submerged Falkland Plateau, which is an extension of the South American Plate. The similarity of the Devonian geology of South Africa and the Falklands is a clear demonstration of continental drift.

One suspects that Darwin was pleased when the *Beagle* left the Falklands and headed back to the mainland. As he wrote to his sister Caroline on 30 March he longed to see the tropics again: 'No disciple of Mohamet ever looked to his seventh heaven, with greater zeal, than I do to those regions.'[238]

238 CCD1: 303.

St Joseph's Bay, April 1833

The second week of April is not recorded in the *Falkland notebook*, as the *Beagle* was at sea. The notebook was used again briefly on p. 28a on 17 April 1833 in St Joseph's Bay [Golfo San José] in Patagonia. FitzRoy had specific instructions from the Admiralty to call there to explore the harbour (see *Narrative* 2: 27) giving Darwin a chance to explore what he called in the *Beagle diary*, p. 151, 'an El Dorado to a Geologist'. On p. 29a Darwin made a cross reference to the *Beagle diary* for around 19 April: 'if new paper is used it will be [p.] (313)'.

Maldonado, May–June 1833

The *Beagle* went next to Maldonado for the winter, p. 30a, via a stop in Monte Video. Darwin took up residence on shore in Maldonado on 29 April and it was while there that he made a long series of entries for the whole of May (this is mainly the 'Maldonado (excursion)', from 9–21 May, noted on the notebook label). A parallel series of entries for the middle of May occurs at the back of the notebook (IBC to c. p. 3b).

The descriptions in the notebook from Maldonado are classic: the gauchos 'would cut your throat & make a bow at the same time', p. 34a, and on p. 38a 'it is necessary to lounge all evening amongst drunken strangers.' On the night of 11 May Darwin stayed at the house of the 'very sick' Manuel Fuentes, but the house which was described more fully in the *Beagle diary* was 'thoroughily uncomfortable' and the food just as bad. Darwin's great-grandson Quentin Keynes visited the house in 1970 and found that it was still standing, but was 'partly tumbled down'.[239]

The notebook continues with pages of geology, interspersed with colourful notes of the travelling life: 'the people all look at me rather kindly but with much pity & wonder', pp. 44a–45a; 'Inglishman last night most hospitable: fresh horses', p. 47a; 'I am considered such a curiosity that I was sent to be shown to a sick woman', p. 48a; 'at night curious drunken scene; knives drawn', p. 54a; 'This days ride interesting it is an alpine country in miniature slept at the most hospitable house', p. 63a. At the end of this excursion Darwin listed a considerable number of books to read including 'Humboldt (of course)', p. 74a; almost the same list occurs in a letter to his younger sister Catherine begun on 22 May in Maldonado.[240]

The geology around Maldonado is described in *South America*, p. 1, on elevation, p. 90 on the Pampean Formation, p. 144 on the crystalline rocks. In the *Falkland notebook*, p. 49a, Darwin noted the 'trap' rocks (i.e. hard splintery lava rocks) which were of 'endless varieties: I have only selected a few: I could not bring any more'.

239 Keynes 2004, p. 172.
240 CCD1: 311.

On p. 43a Darwin stated 'I am inclined hence to believe this whole country to be of transition origin like so many primitive others:' and 'now no doubt that the whole country is Transition formation', p. 66a. Geologists of the time tended to regard the crystalline rocks (i.e. those such as granite and gneiss which did not show obvious signs of having been laid down under the sea) as the oldest, hence 'primitive'. Whilst it is generally true that the oldest rocks on Earth are crystalline, Darwin was already doubting that a granite, for example, had to be older than 'transition' and might in fact be even younger.[241]

There are extensive zoological notes around 31 May, pp. 76a–83a, pp. 1b–3b, reflecting Darwin's comments in letters that while in Maldonado he was very busy collecting. It is difficult to pick out any particular animal as more interesting than another. The slug *Limas* described on p. 2b, as the *Zoology notes*, p. 150, show was collected on 14 May and was described in unusual detail. On p. 77a Darwin seemed to wonder if the presence of beetles in horse dung argued for horses being original inhabitants. Around p. 80a there are many birds listed, sometimes with fascinating behavioural observations. On p. 78a, for example he mentioned a 'Picus [woodpecker] sits crossways on a branch like common bird', which formed the basis of a discussion in his 'Ornithological notes'. This bird, the campo flicker, was eventually referred to in the *Origin of species*, p. 184 (see the introduction to the *Banda Oriental notebook*).

Entries suddenly become difficult to date around p. 84a. June and July 1833 are not clearly registered in the notebook. The *Beagle diary*, p. 160, makes clear that up to June Darwin was extremely busy with collecting zoological specimens and on 29 June he 'arrived safely on board with all my Menagerie'. He spent several days on paperwork, then the *Beagle* sailed for Monte Video on 9 July, then back to Maldonado on 13 July. His collections went off with a letter to Henslow on the 20th. On 24 July the *Beagle* sailed for the Rio Negro.

Rio Negro to Bahia Blanca, August 1833

The next dated entry, which is in ink on p. 86a, was apparently just before 2 August when the *Beagle* arrived at the mouth of the Rio Negro.[242] Much of the notebook for early August is concerned with the economy of the region, an aspect of Darwin's interests during the voyage that requires more study. It is well known that his reading a tract by a political economist, Thomas Malthus (1766–1834), chanced to be one of the most important influences on the formulation of his theory of natural selection (see Schweber 1985), although Darwin's insight was derived from a biological argument about population growth rates.

241 See Pearson 1996 for detailed discussion. See notes 207 and 463.
242 See also *Beagle diary*, p. 164 note 1, and p. 166 note 2, for quotations from these pages.

There are extensive notes for early August, pp. 86a–124a, pp. 4b–8b, covering Darwin's first major inland excursion, from Patagones [Carmen dé Patagones], about 30 km (19 miles) up the Rio Negro and then northeast to Bahia Blanca. This excursion number 1 lasted from 11 August to 17 August. Darwin visited the 'Salina or great Salt Lake' and noted the sandstone which seemed to predominate and was covered by gravels with various types of pebbles including tosca. He was intrigued to know what relation these beds bore to those of Bahia Blanca which he saw the previous year and which he hoped soon to revisit. He concluded, p. 98a, that the gravel was deposited under the sea. The question to be answered was whether these vast spreads of sediment were the result of gradual deposition, or of catastrophic floods or mega-tsunamis. '**I am inclined to think the Pampas not diluvial although little above level of sea**', p. 113a. This is from a section of notebook which Darwin laboriously wrote over a second time in ink.

There are scattered zoological references, for example, flamingos, p. 101a; shrike, p. 106a; toco toco, pp. 106a–107a; guanacos, p. 111a; cuckoos, p. 117, and many references to the fruits and crops grown in the area. The reference to the 'toco toco' (tuco-tuco) is especially significant. At the time of making the notes on the habits of the animal from the Rio Negro on pp. 106a–107a, Darwin was unsure of its relationship to the tuco-tuco with which he was already familiar from his Maldonado explorations. These notes were expanded in the *Zoology notes*, p. 165, and by the time he wrote up his *Animal notes* sometime before mid July 1836, he had worked out that the Rio Negro species was separate from the tuco-tuco. Nevertheless the two species had many similarities, including the fact that they lived in burrows and suffered from inflammation of the nictitating membrane of their eyes, often leading to blindness, hence the note '**said to have no tail (?) & blind (?)**', p. 107a. Darwin was struck that an animal should possess eyes which were generally of no use and were in fact so easily damaged that the animal often went blind. Why would any animal created for living in burrows have such an imperfect organ? As Darwin wrote in his *Animal notes*, p. 8v:[243]

> Considering the subterranean habits of the tuco-tuco, the blindness, though so frequent, cannot be a very serious evil. Yet it appears odd that an animal should possess an organ constantly subject to injury. The mole, whose habits are so similar in every respect, excepting in the kind of food, has an extremely small protected eye, which although possessing a limited vision, seems at once adapted to its manner of life.

These observations were repeated in Darwin's *Journal of researches*, p. 60, and eventually became a cornerstone of his discussion of 'use and disuse' in the *Origin*

243 See also *Natural selection*, p. 295.

of species, p. 137. In 1859 he could explain that the tuco-tuco had eyes because it was descended from a surface-dwelling ancestor and he suggested that natural selection would eventually result in some protection for the animals' eyes. It is clear from the *Animal notes* that the case had already started him thinking some time before the end of the *Beagle* voyage, but is unclear how much this contributed to his growing doubts about the special creation of perfectly adapted organisms.

Darwin described the Indians as '**brown statues**', p. 106a, and his gaucho companions '**in a line, robes flowing**', p. 113a. On 12 August he recorded information from them about the most powerful man in the Buenos Ayres region: 'General Rosas Extraordinary man nearly 300 thousand cattle', p. 114a. Juan Manuel Rosas (1793–1877), Argentinean cattle rancher, was the Governor of Buenos Ayres, 1829–32 and 1835–52, who ruled as dictator of Argentina. Between 1833 and 1835 Rosas was overseeing a war of extermination against the Indians. Darwin's party travelled from Posta to Posta, the staging posts along the main routes across the Pampas. Darwin eventually met Rosas at the Posta by the Rio Colorado on 15 August.

Darwin must have realised that he was witnessing part of Rosas's systematic eradication of the Indians, usually excused as retaliation: '**Posta 5 men murdered**', p. 123a. Sometimes Darwin could not travel and was frustrated: '**no clean clothes, no books. I envied the very kittens playing on the mud floor**', p. 124a, and he was appalled by the treatment of the horses 'dreadful inhumanity riding such horses', p. 126a.

There is an interesting series of entries at the back of the notebook which relate to battles between the Spaniards and the Indians in the region around the Rio Negro, including what seems to have been a massacre in a church in San José in 1810 (pp. 7b, 10b). These entries begin on p. 7b where Darwin also recorded that 'All the Indians in the neighbourhood died of Small Pox & a sort of plague in throat'. The entries lead into the opening pages of the *B. Blanca notebook*, and tally with the fuller account in the *Beagle diary* for August and September 1833 where the extraordinary bravery of the Indians under the Cacique Pinchera is described.

On p. 10b Darwin noted that the Rio Negro 'is in a Box, very rapid' and in the *Beagle diary* entry for 6–7 August he estimated its width as four times that of the Severn at Shrewsbury. On p. 12b Darwin mentioned some information from an Englishman who told him about the Indians who 'plant little Wheat & Indian Corn' in the 'very bad country' 70 leagues (*c.* 330 km) up the Negro at 'Islands of Churchills'. What Darwin heard as 'Churchills' was actually the Island of Churichoel, which became 'Cholechel' in *Journal of researches*, p. 123, and is today known as Choele Choel, where there is a small neighbouring town called Darwin.

In the *Beagle diary* entry for 7 September Darwin added that Churichoel was a very important point at which horses could cross the river, and in the *B. Blanca notebook* entry for that date (p. 27a) he recorded seeing a large arrowhead, unlike any then in use, found 'at Churichol' where apparently they were common, giving some indication of the Pre-Columbian Indians' weaponry. This type of arrowhead became obsolete once the Indians obtained horses from the Europeans, strongly suggesting to Darwin that there was 'No horse' on the Pampas before 1535. Once he was certain that the fossil horse teeth he was finding were definitely Pre-Columbian, this begged the question why had the original South American horses become extinct in a country so suitable for them? This subject is discussed in more detail in the introduction to the *St. Fe notebook*.

Entries continue unabated to 22 August when, because the *Beagle* was not around for its rendezvous, Darwin returned to the site of his first fossil bone excavation at Punta Alta. Several important entries occur on pp. 131a–138a which show Darwin's altered view of the geology. These have been much discussed by historians since Barlow 1945, p. 194, first published them. The most celebrated of these passages, which refers to the beds in which Darwin found fossil rodents during a half hour stop at Monte Hermoso on 19 October 1832, p. 138a, is as follows:

> My alteration in view of geological nature of P. Alta is owing to more extended knowledge of country it is principally instructive in showing that the bones necessarily were not coexistent with present shells, though old shells: they exist at M: Hermosa pebbles from the beds of which occur in the gravel: Therefore such bones if same as those at M: Hermosa must be anterior to present shells: How much so. Quien Sabe?[244]

Unfortunately the visit to Monte Hermoso does not appear to be recorded in the *Rio notebook*. The *B. Blanca notebook* also deals with this second visit to Punta Alta at the end of August, and its geology is further discussed in the introduction to that notebook.

There is a delightful note at the end of the front pages of the *Falkland notebook* which conveys Darwin's intense satisfaction at being back on board the *Beagle* on 27 August and his close friendship with FitzRoy: 'Whole day consumed in relating my adventures & all anecdotes about Indians to the Captain', p. 142a. In the following note written the next day, Darwin was preparing for Buenos Ayres: 'Most delightful the feeling of excitement & activity after the indolence of many days in the last fortnight, spent in the Spanish settlements!'

These entries are followed by a sequence of blank pages as the *B. Blanca notebook* was then used for excursion number 2, the 600 km (370 miles) continuation of the overland trek to Buenos Ayres in September. The very last few pages of the 'b' sequence of the *Falkland notebook* appear to overlap with the *B. Blanca notebook*.

---

244 For more detail on this passage see for example Hodge 1983, p. 100, note 23 and *Beagle diary*, p. 176 note.

[FRONT COVER]

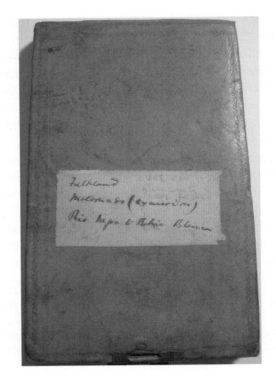

Falkland
Maldonado (excursion)
Rio Negro to
Bahia Blanca

[INSIDE FRONT COVER]
[1a–2a excised]
[3a] Stirrup at Whites
      Map —
      Poncho /Racket/ bay
      Charge /Ruone/
      Old Man R. Negro
      Barranca
      Salt Steam Quantity: 30.000
      Sand to the P. St Antonio (Yes)
      Salt petre
      Stirrups: Salt Petre: Girth: Poncho
      Bay. Blas. Spirits Salt Steam
      thirty thousand /fanego/ 1/16

[4a blank]

[5a] N: Orange Bay Feb: 16th [1833] —

(A) Sandy Slate. drusy[245] cavities. summit of hill with crystals

(B) (C) in close neighbour head. — angular conchoidal fracture. sonorous

(D) fine grained greenstone with pale concretions lower part of range

A a slight tendency to show a columnar structure.

Lime tolerable good slate before arriving at these Sandstone hills. —

~~They are~~ latter are nearly as lofty as the N slate range

[6a] A dips to the SW by S at above ∠ 25°

Southern ocean not phosphorescent:

At sea side

AM 6 ¾

Barometer 29. ~~27~~ 28 2

Attached Therm[246] 51.5

Therm 48.5

Banks Hill

9 AM

Barom 27.734

attached Thermometer 45°

Thermo 42

[7a] On return 11 ¾ AM

Barometer 29.348

Attach 57°

Therm 58°

Beech trees: lower down dwarf ones less snow on the hills

548	51.5	58
282	54.2	48.
.066	29.312	5
.033	27.734	53
57	1.578	.283
51.5		.033
5.5		.316
2		
2.7		

[8a]   Falkland Islands

~~51.25~~

March 2 [1833]

---

245 With cavities into which crystals grow.

246 An 'attached thermometer' is one affixed to another instrument, as a barometer, used to obtain its operating temperature. Throughout the notebooks CD abbreviated as 'A'.

to what animals did the dung beetles in S America belong — (Is not the close connection of insects & plants as well as this fact point out closer connection than migration) Scarcity of Aphodius[247] ?

Well not reversed cases of close

Vide Annales des Sciences for Rio Plata:

A crest of Quartz rock running E & W & dipping to the South ∠ 50°: hills occurring in same line at some <u>distance</u>. (A)

The rock in places is pure quartz (?) — dip

other crests S by W. & to the North of the former one

The peat not forming

[9a]   at present & but little of the Bog Plant of Tierra del F: no moss; perhaps decaying vegetables may slowly increase it. — beds varying from 10 to one foot thick.

Great scarcity in Tierra del of Corallines,[248] supplanted by Fuci[249]: — Clytra prevailing genus

Procure Trachaea of Upland Goose[250]

6th: [March 1833] South of the ship

running ENE &c dip SSE ∠ 46° —

Slate: (B) scales of mica black veins. Phyllade /prime/ South of the Ship

[10a] on the next crest pure quartz rock (?) sometimes arenaceous (C)

~~run E &~~ dip SSE

So that the slate here underlies the Quartz. whilst North of the ship. it is superior. — (according to the dip:

Saturday [9 March 1833]

Quartz rock becoming slaty & less Quartzose

A passes instantly into slate & more earthy particles.

Slate (B) & which has incurved — planes. dips to S by W ∠ 65° (& to the S)

247 A genus of dung beetles. See *Darwin's insects*, pp. 76–81 and 103.

248 Keynes noted in *Zoology notes*, p. xiii: 'The principal problem in classification encountered by CD in the 1830s lay in determining the true nature of some of the colonial plant-like invertebrates then still known colloquially as Zoophytes or Polypes, and nowadays separated into Cnidaria such as hydrozoa, anthozoa (including corals) and scyphozoa, Bryozoa and sponges. The smallest of these were the corallines, but thanks to the classical studies of John Ellis it had been accepted in many quarters by the end of the 18th century that like some of the coelenterates closely similar to them in appearance, they belonged to the animal kingdom…. At the beginning of the voyage, CD referred to all such animals indiscriminately as corallines or coralls, although some of them were in fact hydrozoa or hydrocorals, some bryozoans, and some coralline algae. When in the end he had concluded that his 'true corallinas' were indeed algae such as *Corallina* and *Amphiroa*, he listed this group as Nulliporae. The bryozoans were generally 'encrusting corallines' or Flustrae, and the reef-building hydrocorals were Madrepores. He had thus improved on the still prevailing confusion in the classification of the Zoophytes or Polypiferous Polypi in the accounts of Cuvier and Griffith that he had with him on the *Beagle*.'

249 *Fucus* is a genus of brown alga (seaweed).

250 See *Zoology notes*, p. 213; listed as *Chloephaga magellanica* in *Birds*, p. 134. The trachea of this bird, specimen in spirits 904 (*Zoology notes*, p. 349) was dissected and described in Eyton 1838.

[11a] but there are irregular planes like stratification dips NW by N $\angle$ 35°

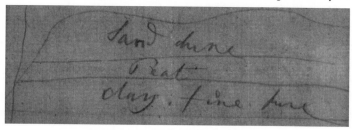

Sand dune Peat clay. fine pure
all now /consuming/ by the sea
The planes (if cleavage was absent) would be pronounced slate above one foot apart. It is evident the slate here according to dip overlies slate but the bed is not of great length: slate sometimes with many scales of mica: Transition insensible:

[12a] Tuesday 12th. — [March 1833]
The purest Quartz rock is of a granular variety, & contains always some very small proportion of the aluminous matter: the common rock in places has a breccia appearance. from the Quartz & aluminous matter being in larger quantities & masses Examine Balanus.[251] in fresh water
it is about 15 inches beneath high water mark:
Horses fond of catching
cattle aberration of instinct
~~snipes rather~~ /tame/
Examine pits for Peat

[13a] Specimen of do — Have there been any bones ever found &c. ~~or~~ Timber
Are there any reptiles? (no) or Limestone?
19th [March 1833]
Slate slippy fissile. S ~~of scale~~ Johnstone Creek dip to SSW or little
/more/ South
(ask Chaffers where gneiss came from)
Head of fishing place
Rock quartzose. ferruginous with partial slaty structure splintery in beds resembling strata dipping to SW ½ S $\angle$ 7°.
I think the point of compass difficult to be ascertained:

[14a] Within /Basin/
Slate slippy much laminated <u>fossiliferous</u> dipping to the N ½ E $\angle$ 65°; & from this to Vertical

251 Barnacle of the order Thyrostraca. See *Living Cirripedia* (1851).

22^d [March 1833] true observ

Slaty sandstone, abounding with fossils dipping ∠ 14° to W by S: layer of shells a seam
of hard /blue/ conchoidal slate. shells parallel to seam of slate & strata. covered above &
beneath by the vertical plates of slate dipping to N ½ E: ~~Fissil~~ cleavage of slate same
above & beneath nearly vertical: Sandstone bed about 4 feet thick

[15a]   This sandstone has not Mica the dip is not quite certain

It may be here remarked that the Sandstone, in many places has a decided tendency to
break vertically as if affected by the vertical cleavage: in the more perfect slaty Sandstone
the scales of Mica seem to determine splitting parallel to Stratification

Sandstone beds are in places 12 feet thick:

[16a]   20th [March 1833] St Salvador Bay

Siliceous ferruginous slate (passage) dip to S by E ∠ 30°, strata & cleavage also nearly
pure quartz rock head of great Bay containing small beds of micaceous slaty
sandstone & vertical cleavage of slate:

Bearing from ship (SSW Compass) Quartz with specks of mica slaty parallel to strata &
cleavage. dip to S ¼ W ∠ 19° parallel to hills higher up hill: bearing from last place ESE
ie more central part of range

[17a]   on north side strata dip to N ½ E ∠ 28°

Mantle ridge dips to W by N ½ N ∠ 16° (rocks all granules quartz)

on summit flattish strata

on S side dip to S ½ W ∠ 29°

On the east side of "Cone" dips ∠ 14° to ENE generally flattish & connected with a
range. which on South side has the usual Southerly dip

[18a]

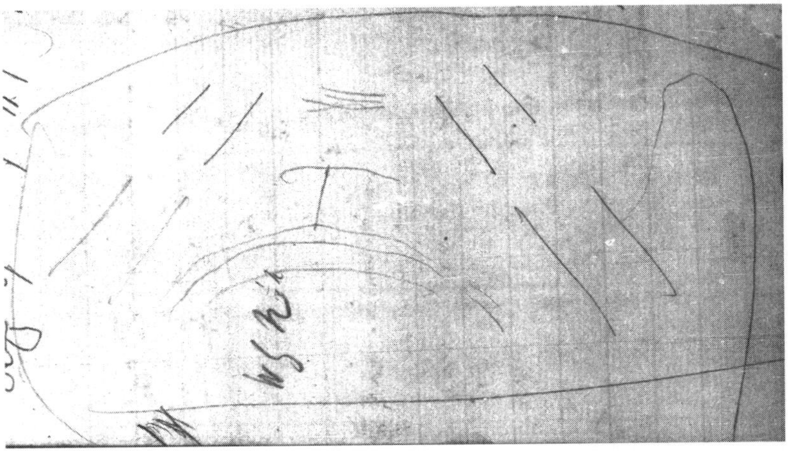

————[curved strata] W by N ½ N

The hills bear from Town

~~SE~~ S by E (true):

[19a] Desolation Island. said to be Volcanic with hot springs:
All the Shetland islands with very hot springs & vesicular lava
Capt Brisbane.[252] —
& South Orkney Volcanic products
It is clear that, in the crests, there has been in the mountain a <u>point</u> of upheaval, when strata have become mantle shaped instead of crests:

[20a] 21st [March 1833]—
Fossil seams dip SSE $\angle$ 20
————————— S by W
Then slightly curved horizontal or dipping to opposite sides S or Northerly
There would seem to be two principal ranges of rock in East Falkland. both of quartzose nature, both nearly parallel to Berkeley Sound & both with a Southerly dip: Slate occupies the lower districts & intermediate:[253] perhaps the lower from weathering so easily:

[21a] Saw a cormorant catch a fish & let it go 8 times successively like a cat does a mouse or otter a fish:[254] & extreme wildness of shags:
Then sandstone more contorted the site bears N by W (true) of the ~~conical~~ "cone"
Then <u>cleavage</u> of Slate Vertical running & E & W
Sandstone dips 15° to N

[22a] Read Bougainville[255]
In 1784, from return of Gov: Figueroa,[256] buildings amounted 34, population including 28 convicts 82 persons, & cattle of all kinds 7,774:
22[d] [March 1833]
Proceeding round Basin, ~~saw~~ observed before W by S $\angle$ 14°
Sandstone dips to SSW
————————— dip to S by W ½ W $\angle$ 52°
always /uprising/ on the highly inclined plates of slate, N ½ E dip
Sandstone SW by S

---

252 Matthew Brisbane (1787/8–1833), the first official British Resident in the Falkland Islands. He was murdered at Port Louis 26 August 1833.

253 Igneous rocks intermediate in composition between acid and basic.

254 See specimen not in spirits 1756 in *Zoology notes*, p. 396; listed as *Phalocrocorax carunculatus* in *Birds*, pp. 145–6.

255 Louis Antoine de Bougainville (1729–1811), French sea captain and navigator who circumnavigated the globe in *La Boudeuse*, 1767–8. CD refers to Bougainville 1772.

256 Augustín Figueroa, military administrator of the Spanish Settlement of Port Soledad, Falkland Islands 1784–6.

[23a]   cleavage of slate dip N ½ E

East of basin, peat above 12 feet thick resting on Clay & now eaten by the sea. — lower parts very compact but not so good to *[illeg]* as higher up; small

bones are found in it like Rats — argument for original inhabitants: from big bones must be forming at present but very slowly: Sandstone dips about N by East

[24a]   Fossils in Slate: opposite points of dip: & mistake of stratification: What has become of Lime?:

It will be interesting to observe difference of species & proportionate numbers. which always appear characteristic of different habitations:

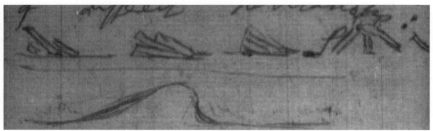

[strata]

Migration of Geese in Falkland Islands, as connected with Rio Negro?

[25a]   March 25th [1833]

Near the Ship

(x) <u>gradual</u> transition of the ferruginous quartzose rock into slate with vertical cleavage: it is within the latter (& to the North):

ferrug: quartz: rock has perhaps a NNE dip of Stratification: — also <u>passage of form</u>

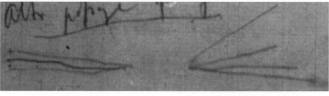

[diagram of strata]

[26a]   March 28th [1833] —

Crest of granular pure quartz — Crest running W by N & E by S. — (Z)

Emberiza[257] in flock

Send watch to be mended

All the country for 9 (?) miles N of the Ship: quartz rock ~~generally rather~~ pure: —

NB. This occurs to the N: of Alumino-Quartz Rock: Is the latter in that state from approaching to the grand slate formation:

---

257 A genus of birds of the bunting family. Possibly specimen 1232 in *Zoology notes*, p. 155; listed as *Crithagra? brevirostris* in *Birds*, pp. 88–9.

[27a] Compare Nebalia[258] & Desma[rest][259] with my Zoea[260]
Whether my Spec: Notopod is a Porcellina[261]
Enquire period of flooding of R. Negro & Plata
Is the cleavage at M: Video (an untroubled country) very generally vertical or what
is the dip? —
Tuesday 9 —
Saturday 6th [April 1833]

[28a] April ~~18~~ 17th. [1833]
Entrance of St Joseph Bay. section of cliffs running W by N & E by S. Shows numerous
nearly horizontal lines. which ~~leave~~ rise from the water towards the West: Thus
[diagram of strata]

E   W
nearly insensibly:
Covered by <u>stratified</u> blacker mass. which follows inequalities in the white: they both
seem to rest on a white rock at entrance of St Joseph bay: lines of

[29a] stratification <u>most parallel</u> in white mass: the line between black & white was seen
extending for about 20 miles in length:

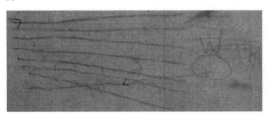 either beds of equal thickness dip somewhere near east: or they thicken
towards west & are intersected by water lines [*illeg*]
Private journal if new paper is used it will be (313)
It

W [diagram of strata]

---

258 Nebaliacea, a small, marine order of Crustacea.
259 Anselm-Gäetan Desmarest responsible for the Plates of Crustacea, Zoea, etc., in *Dictionnaire des sciences naturelles*. Paris, 1816–30.
260 The name for crustaceans during their larval stage. Keynes noted 'a knowledge of [them] was one of the most valuable outcomes of [CD's] dissections of marine invertebrates during the voyage.' *Zoology notes*, p. xiii.
261 Porcellana: a crab.

[30a] Capt Cook[262] Maldonado
Slaty Syenite ~~hills~~: running NE by N & SW by S with a SW dip?
NNE & (ESE dip)
Ampullaria[263] crawling on sand with sea shells
~~Near~~ on road Laguna Potrero siliceous slate running in usual direction: near head of
lake Limestone conchoidal semipellucid, veins of feldspar & crystals of Hornblende &

[31a] Mica: lines of partition running NNE & SSW. (dipping if anything Northerly)
accompanied lying in a soft decomposing ~~mica slate~~ /gneiss/ with much mica:
In a hill to the North there is usual gneiss, with red feldspar & imperfect mica: running
apparently in same direction as Marble: difficult to be ascertained

[32a] 356
Thursday: 9th. — [May 1833]
The chain of hills from Whale Point & East side of Laguna are the same & ~~may be~~ are
called Sierra larga: they are remarkable by the very straight white bands where the
cleavage edges project: the chain is composed of one nearly continuous lines of hills &
2 or 3 other imperfect ones: it is said to to reach to Brazil ? it is ~~nearly~~ the last chain
before the level country of Maldonado: Chaca

[33a] runs about NNE & SSW or more N: rock (as formerly described), dipping WNW:
Between this chain & that of Pan de Azucar the whole country is elevated: this level
country being seen beneath through the gaps: This rock passes into fine gneiss: & is
coarse gneiss at first sleeping place & near to limestone
Started in high spirits with troop of horses companions names, &c arms, describ/e/

[34a] Recon:[264] near limes kiln: dinner, wild sort of gauchos: white men run wild:
astonished at Compass & Promethians (before heard of murder of traveller & lives lost
in the rivers). hospitable place for night:
The gauchos, dissolute proud expression yet civil: moustaches, long black hair; great
spurs: pale faces: tall men. look as if they would cut your throat & make a bow at same
time: ostriches tame: made sail:

[35a] Friday 10th. — [May 1833]
A few miles beyond sleeping place; & crossing the Salco the country became more
elevated & hilly. hills 3- to 500 feet high & more stony. the rock generally more or less
perfect gneiss; & some quartz; also a little blue siliceous slate: our /course/ was about
NN by N. & in this journey we were generally upon high land: but it was impossible to
trace any very regular course of hills, after leaving the Sierra Larga:

[36a] The cleavage in many places was very striking, yet it so varied that it is hard to say what
was prevalent ~~after tea~~ near R Salco it dipped to the SW: then we had for some extent a
NNW dip ∴ & last near Minas ….. So that this irregularity agrees with characters of

---

262 James Cook (1728–79), naval captain and circumnavigator.
263 Tropical water-snail.
264 Spanish for saddle.

hills & is caused perhaps by the crossing of N & S & E & W chain of hills:
which also the course of the waters point out:

[37a] This ride had little interest except from novelty: the country is more stony
more hilly & possesses a very few trees: every where however there is high
green turf between the rocks which supports large flocks of cattle: our road was
mere track & quite unfrequented I ~~believe~~ did not meet one single
person: Arrived at night at Minas: small nice quiet town the only one in the
whole

[38a] country except Maldonado: The vendas miserably uncomfortable, as they have no
rooms & it is necessary to lounge all ~~day~~ evening amongst drunken
strangers. —
Saturday 11th. [May 1833] —
~~Near~~ At Minas there is a coarse blue, also pale slate: also much quartz. some pink &
granular dipping to NNE. — There would seem to be a W & E range of hills N of
Minas & N & S, E of Minas:

[39a] But the whole country is composed of irregular chains: at the Calera[265] E of
Minas there was a semitransparent limestone: & another reddish & earthy
penetrated by highly crystalline veins & containing breccia: at ~~the~~ 1st pit the
limestone occurred to the South of fine quartz range. & had some cleavage
running NE by E & SW by W: it was surrounded by imperfect gneiss in which
large red

[40a] crystals of feldspar were only visible: perhaps ranges of hills here were nearly the same as
the cleavage: Presently Mon there was ~~blackish~~ pellucid Monomenos beautiful
crystalline marble imbedded in rotten slate as at R Sauce: Presently there was a blackish
slaty limestone dipping at high angle to the SE. I think these rocks from the principal
chain in this part:

[41a] at sleeping place there was to my surprise a breccia or rather greywacke, with a
reddish base & resembling our old red Sandstone.[266] This nearly proves together
with characters of some of the limestone that they are of transition origin:
Our course was about East so that the next day we crossed the Marmaraya

[42a] Appearance of Minas: ostriches describe Don Manuel Fuentes[267] house &c &c
Minas, pretty, african looking from separate houses in fertile plain surrounded
by rocky hills: dung many ostriches flocks from 20–30. beautiful on the brow
of a hill: Don Manuel Fuentes: on first entering a house. after sitting for a

265 Spanish for quarry or lime kiln.
266 The red sandstones of Wales and the English west country thought to have been laid down in arid
conditions before the Carboniferous. Robert Jameson included the Old Red Sandstone in the
Secondary. In the late 1830s it was generally dated to the Devonian.
267 See *Beagle diary*, p. 156.

short time, ask as a matter of course: very rich number of cattle: horses, & killed & guests: house thoroughily uncomfortable: no furniture, no windows: only meat, savollas[268] & water for supper: Wretched room for sleeping, in which a very sick man slept:

[43a] Sunday 12th [May 1833]

The formation of red sandstone is very extensive & forms hills: it varies in character from a jaspery rock. to a sandstone & passes lastly into a coarse breccia: To the north of these hills there are others of coarse granite in which all the constituents are in large crystals: & not far distant: I am inclined hence to believe this whole country to be of transition origin like so many primitive others:

(our course to day has

[44a] been chiefly north) (NB. before Minas in the gneiss there was some siliceous blue slate) (NB. Examine peat at M. Video)

After then much Syenite in abrupt paps or cones: then chiefly quartz:

Our view to the East showed a much more level country: we travelling along hills the same as before:

The country is precisely the same as formerly the people all look at me rather kindly

[45a] but with much pity & wonder: Saw method of catching partridges:

Monday 13th [May 1833]

At sleeping place a reddish crystalline rock; together with basalt in small columns with green crystals: at head of R Tapas much Trap: rock with earthy Feldspar & much iron: Then much granite & gneiss & gneiss some imperfect the rest very perfect feldspar bright red: After crossing Rio Barrija large fine granite felspar red forming remarkable looking boulders

[46a] we then at the Calera of Don Juan Fuentes find limestone white beautiful; house built of it: from that across the Polanco all the rocks are blackish with white veins; or white limestone & pure quartz or red: — Limestone contains veins of quartz & other of Rhomboidal gypsum: a league to the north of the junction the whole

[47a] country is limestone:

Country more level: with more trees & rather different appearance more undulating. Furthest point North: Inglishman last night most hospitable: fresh horses: our guide a curious old Paraguay man: delight at meeting countryman

[48a] Tuesday 14th [May 1833] —

Found some limestone variegated in its colours & Found 1st limestone 2d granite 3d limestone 4 granite & in a NNW & SSE section of country granite forming ridges or low hills: Also saw some limestone S of the Barrija negra:

returned. I am considered such a curiosity that I was sent to be shown to a sick woman At our sleeping place between R Tapas & former

268 Spanish for pumpkins.

[49a]   sleeping place. There is a grand formation of these appearing Trap rocks: it is certain they extend for some distance & probably join those near former sleeping place a distance of two miles, /indeed/ here there was black basalt much resembling the former: These trap rocks are of <u>endless varieties</u>: I have only selected a few: I could not bring any more

[50a]   The amygdaloid is very abundant & very much more amygdaloid than the specimen:
There is are specimen of a remarkable reddish rock with large & abundant rock /drusy/ cavities of crystallised quartz. these are arrange in planes so as to give to the rock a decidedly lined cleaving appearance which dip to the East at angle about 45° — Now it is very remarkable

[51a]   this rock is situated between the above trap rocks & an imperfect gneiss which forms the whole hill to the West of it: This gneiss is only quartz & feldspar in planes not distinct & which quite fade away higher up: These planes nearly run in a N & S (within a point) direction: The whole of the 3 rocks are within pistol shot of each other

[52a]   & I must consider it the reddish rock as a passage one: It is certain these trap rocks form a large valley surrounded by the immense granitic formation, but from covering of turf I could not see junctions, nor do I believe they exist there being only transitions: (NB all specimens unnamed in knapsack handkerchief are from this site)

[53a]   (NB) I was told gypsum occurs some days riding to the North of Polanco)
(A) dubious Limestone   }?
(B) Limestone quarried; arranged in grains in lines slaty Cleavage
(D) Trap rock same as others in handkerchief
(E) Curious red transition rock; with druses of Quartz
From 2^d sleeping place R. Tapes

[54a]   At night curious drunken scene; knives drawn, evidently showing the usual manner of quarrelling: the instantaneous manner & striking & rushing out of the room:
Wednesday 15th [May 1833]
Our former fine weather has left us: & we are confined from bad weather: very stupid work (NB enquire whether there was much wind this morning at M Video

[55a]   & whether much lightning the night before).
Curious amusement of impromptu singing: general much politeness.
(NB) Ask whether Gold has ever been worked at Pan de Azucar
We only rode about a league to the place where we slept one night before: I here found with the Basalt & red crystallised rocks other varieties.
together with the pale variety (D) which is very abundant These rocks extend the whole

[56a] distance between the two sleeping places

Thursday ~~15~~ 16ᵗʰ [May 1833]

Returned to a house 4 leagues from Minas (in NE direction ? . Cutting off the angle at Don Fuentes house. — We thus crossed the range of red-sandstone rocks: I /think/ that it lies on both sides of a granite ridge. (ie S & N side) The rock is here accompanied by Basalt: In proceeding from the

[57a] last sleeping place /I &/ the Trap rocks extended about one mile to the /Southard/ & then came the granitic usual rocks & clearly separates them from the Sandstone ones: // Rivers rather too full, one of the inconveniences /fine/ of uncivilized country from one days rain:

[58a] I also here found ~~some~~ much amygdaloid with Limestone something like that from R. Tapes: taking into consideration having found the Sandstone on each side of the granite I think it probable that the whole formation is of one origin: whether contempareous with granite I cannot tell: (NB Basalt

[59a] not much (not primitive looking)

Friday 17ᵗʰ. [May 1833] —

About 2 miles ~~from~~ N Minas abundant white limestone dipping at high angle to the NNE or more N together with & in same direction as Clay Slate: To the South we crossed a broad mountain band or elevated ~~chain~~ tract of mountains

[60a] entirely composed of blue slate, generally very siliceous: ~~he~~ occasionally containing Lime; often pure quartz; often much iron. — ~~I~~ here ~~noticed~~ very generally was that extreme sort of contorted cleavage in which every possible curve was present many resembling small

[61a] Gothic windows, which it is difficult to imagine any force to have produced: There certainly is an extensive chain South of Minas as in the Maps

Here the cleavage was tolerably regularly ~~I once~~ in the first part E & W & afterward (by Gold

[62a] Mine) NNE & SSW: I should not have expected so much slate from what I saw before of the more Eastern section where perhaps the N Granite ridge /more/ interferes: about 10 miles to the South of Minas. (I do not think mass /or ridge/ Gold has been found: it is in a greenish sometimes talcose slate & chiefly in accompanying quartz & copper malachite beds. which contains drusy cavities with

[63a] much iron. & has a rotten appearanc: Gold obtained by washing. — // This days ride interesting it is an alpine country in miniature slept at a most hospitable house // beautiful boys

[64a] Saturday 18ᵗʰ [May 1833]

After leaving this place for some miles there was slate & much quartz, with some imperfect gneiss; We then skirted an irregular chains of hills running NNW & SSE. of several miles in length, entirely composed of breccia of various degrees of coarseness: the ₜincluded fragments were generally ~~quartz~~ coloured siliceous rock, granite & green slate (my specimen is

[65a] not characteristic) some of the fragments were large; the base is essentially crystallised feldspar: & in some specimens the rock contained so much well crystallized feldspar that it would be thought to be imperfect granite: rock extremely hard, it weathers into those round blocks heaped on each other & generally characteristic of Granite: I have

[66a] now do doubt that the whole country is Transition formation: It is very remarkable observing something like a transition ? between a Breccia & ~~Trap~~ Volcanic looking rocks with Granitic ones: This Breccia may be ~~cal~~ said to be formed ? in Granite:

At our breakfast place ~~when it~~, about 4 miles to the N of Pan de Azucar

[67a] There is greenish ~~chl~~ chlorite slate: (Saw lassooing) It will be safer to say there were so many crystals of Feldspar — that I am even yet & was then inclined to think these formed on the spot:

~~At Pan de Az~~ at Pimiento's[269] house plenty of well crystallised feldspar & quartz the rock so prevalent in this country no slaty structure:

[68a] Sunday 19th [May 1833]

Sierra las Animas (to the NW of Pan de Azucar ~~appear~~ & South of Betel) appears to be most lofty ~~part~~ hills in this country: To the West there is seen the great flat extensive plains, which is only broken by the Sierra de M: Video: To the East the endless hillocks which I have traversed: The hills run N & S: irregularly & must be same

[69a] with the Breccia ones already described: Las Animas is entirely composed of a more or less compact Feldspathic rocks with crystals of feldspar (sometimes quite Hornstone)[270] generally contains & sometimes specular; also occassionally crystals of Hornblende & quartz: the rocks is very hard, irregularly massive & of a pale red:

[70a] Pan de Azucar; which lies out of the chain, is chiefly composed of the mixture of largish crystals feldspar, & quartz, also some Syenite & a good deal of the same rock as Las Animas: The ~~plain~~ valley in which Pimento's house &c & the Arroyo de Pan de Azucar flows besides the green slate & felsic granite (already mentioned) chiefly is composed of

[71a] a greenish porphyry with crystals of feldspar: in other places amygdaloid with quartz (V specimen) the paps which surround the base of the larger hills is composed of the former: (NB At Las Animas there is a crystalline foliated brown rocks same as found with breccia near Juan Fuentes house; interesting as showing strong connection V Specimen)

269 Sebastian Pimiento, estate owner in Maldonado district. See *Beagle diary*, pp. 158–9.
270 A very fine-grained and hard rock derived from volcanic ash.

[72a] Spent the day at Sebastien Pimentos house; pretty daughter, beautifully dressed & hair;
       menial offices: view with numerous cattle & sheep. almost pretty with rising sun:
       piles of stones on Las Animas; small; said to be belong to the old Indians; not so great as
       in N Wales; curious the universal desire of man to show he

[73a] has ~~visited~~ frequented the highest points in his country from the sealed bottle of the
       traveller to these little piles
       Monday 20th. [May 1833]—
       Returned: granite of feldspar & quartz W of Lg [Laguna] Potrero ………..
       9–20

[74a] Palms: grow near Rio Marmaraga
       Azara's book[271]
       Fleming's philosophy of Zoology[272]
       Pennants Quadrupeds[273]
       Paul Scrope on Volcanoes[274]
       Scoresby arctic regions[275]
       Humboldt (of course)[276]
       Burchell's Travels[277]
       The occurrence of Palm trees: worth noting: & Maize

[75a] mentions rattle snakes: are not true rattle snakes peculiar to N. America?
       Davys consolation in Travels[278]
       Playfair Hutton[279]
       ~~Humboldt~~
       (/short/ measure)
       Radiator ([illeg] scalpel)
       Book of chemistry
       M: Mice traps: wadding & small Vasculums:[280]

[76a] Palms: Perhaps the earthy covering in this country is rather modern, accounting for the
       paucity: undulating surface: traversed in every direction by streams, beautiful climate,
       earth, no trees: Paucity of trees common to all the formations. Examine coarse
       Limestone pebbles at M Video. —
       (31st —) [May 1833]
       Female of white shrike with little grey on the

---

271 Possibly Azara 1801.
272 Fleming 1822.
273 Pennant 1771.
274 Scrope 1825.
275 Scoresby 1820.
276 Humboldt 1814–29.
277 Burchell 1822–4.
278 Davy 1830.
279 Playfair 1802.
280 Containers used to keep specimens cool and humid.

[77a] back ⟨General scarcity of coprophagous⟩[281] Exception ⟨1181 & 1225⟩[282]
Is not abundance of ~~dung~~ beetles in Horse dung — an argument for original habitation
of these animals
V Humming Bird Amer
Fly catcher with red wings iris yellow: eyelid blackbird do: base of lower mandible
especially yellow:
Long billed Certhia.[283] tongue shouldered *[slightly]* *[illeg]* & bristle projecting but not
recurved; moderately long: tail used
[78a] Furnarius[284] walks
Aperea trots
female of small Icterus brown
Black Icterus bubbling noise[285]
Connection between note of B B bird & Furnarius
Picus[286] sits crossways on a branch like common bird
Flycatcher[287] red wings running
F B B no do
[79a] Capincha tame pig
rabbit. curious *[profile]*
[Same hoots ones]
Kingfisher hovers
L Tailed
shrike   } sings
Red
Icterus[288]

Toco *[rough]*. flash without top or bottom & big scissors
[80a] Kingfisher continually elevates its tail:
Thrush with note like English.

281 Feeding upon dung. Used by CD to indicate dung beetles.
282 Specimens not in spirits 1181, 1225, see *Darwin's insects*, pp. 72–3.
283 Genus of birds of the tree-creeper family, see *Zoology notes*, pp. 157–8.
284 Specimen 1200 in *Zoology notes*, p. 151, listed as *Furnarius rufus* in *Birds*, p. 64. See 'Ornithological notes', p. 214.
285 Specimen 1211 in *Zoology notes*, p. 152, listed as *Molothrus niger* in *Birds*, pp. 107–8. See *Journal of researches*, pp. 60–2.
286 *Picus*: Latin name for woodpeckers. Specimen 1238 in *Zoology notes*, p. 155, listed as *Chrysoptilus campestris* in *Birds*, pp. 113–4.
287 Possibly specimen not in spirits 2191 in *Zoology notes*, p. 405: '2191 B Certhia (red wing) female'; which Keynes suggests may be listed as *Synallaxis humicola* in *Birds*, p. 75.
288 Possibly specimen 1244 in *Zoology notes*, p. 156, listed as *Amblyramphus ruber* in *Birds*, pp. 109–10.

Rat with upper lip, from ~~centre of no~~ *[illeg]* the two nostrils to bifurcation the retrecisement of the upper lip 3/12 which gives the upper jaw a peculiarly lengthened appearance

BB Bird[289] dust itself active in the evening: tame

Comadreja chico.[290] intestine full of remains of insects: chiefly ants & some hemipterous insect —

[81a]   active in the e

Scolopax — Perdrix different colour breast

Comadreja grande,[291] weigh flask with water, without bottom & with ⅔ of bullet:

Mouse[292] (?) Gerbilla weighs two turnscrew: has long hair in eye brown: very large eyes: tail found injured: caught with cheese

Alecturus[293] in stomach large Lycosa[294] & Coleoptera: appears very curious in flight first feather in wing curiously excised.

[82a]   The white & grey shrike fly in circle & alight again more so than the long tailed one which feeds more amongst the bushes

Big Rat weighs flask with water without bottom; 2 bullets 4 pellets

Ampullaria[295] length of time they live burries itself in the sand by revolving motion, lying on its under surface — returning towards edge of shells it acts like a centre bit

[83a]   & gradually sink*/s/* very much lower

shot at big lake

bird with long tail much on the ground not in thickets

Buy */strong/* oils

Tow? Paper: Essential oil — Jars

[84a]   As the Quartz at Falklands is aqueous origin it would be interesting to examine if any ruins exist

de origine Venarum

13

$\frac{6}{78}$

$\frac{22}{3}$

---

289 Specimen 1222 in *Zoology notes*, p. 154; listed as *Furnarius cunicularius* in *Birds*, pp. 65–6. See also 'Ornithological notes', pp. 217–8.

290 A weasel, specimen not in spirits 1283 in *Zoology notes*, p. 387; listed as *Didelphis brachyura* in *Mammalia*, p. 97 and plate 32. Comadreja = weasel.

291 Specimen not in spirits 1281 in *Zoology notes*, p. 387; listed as *Didelphis azarae* in *Mammalia*, p. 93.

292 Specimen not in spirits 1284 in *Zoology notes*, p. 387; listed as *Reithrodon typicus* in *Mammalia*, pp. 71–2.

293 A South American species of tyrant, specimen 1275 in *Zoology notes*, p. 160; listed as *Alecturus guirayetupa* in *Birds*, p. 51.

294 A genus of spider.

295 See *Zoology notes*, p. 57 note (c).

Compare Art. Climate in *[illeg]* with Tierra de Fuego

124

125

Reed Equivalents

Lightning? Maclean? Scientific *[Friar]* | Gran Bestia |

Another Parus: Cassiatus: & Reed Bird

Capt Fitz Roy — Reed bird give me

La struthious[296] breed at Port **Malaspina**

Ordinary $1^£.6^s$

∴ 16..18

Able 1.14

∴ $22^£..2^s$

[85a] How far do the cliffs extend?

How many young has Aperea?

What animals are there?

What is export of Salt: are bones found beneath deep beds?

When do the Geese arrive? end of Feb to Sept:

Fresh water fish[297]

Great bones in cliff?

Gum for cleaning Teeth

Springs of water? sections of wells?

Hard rock near town?

Falk[ner][298] about S. Pelagic birds a R. Grande.

Does Peat occur: — Bones. No

Examine Diluvium. Pumice?

Mineral springs: Gypsum. Nitre ?

Snow: Earthquakes: Thunder Storms

[86a]   What Fruits.

Grapes. Peaches. Nectarines. Quinces Standain?

Apples. Pumpkin. Wheat. barley. Indian Corn. Water. Musk Melon. Cherries

Patacas *[dulies]* sweet potatoes Cactus not differ species plentiful. — Potatoes

Olive? | Fig. ? | Palm oranges

---

296 The Greater or American Rhea (*Struthio rhea*), listed as *Rhea americana* in *Birds*, pp. 120–3. See also
     'Ornithological notes', pp. 268–71.
297 Specimens in spirits 553–5 in *Zoology notes*, p. 337; listed as *Aplochiton zebra* in *Fish*, pp. 131–2.
298 Falkner 1774.

Dessalines d'Orbigny[299]

Up to the — 2ᵈ [August 1833] —

Light contrary winds interrupted by a few gales:

3ᵈ [August 1833] arrived off the Mouth of R Negro — pleasant evening in the comfortable Schooner & slept at the Pilots house

[87a] **4ᵗʰ [August 1833] Walked to South Barranca general appearance of cliff land & great valley:**

5ᵗʰ [August 1833] Rode to the town pleasant ride banks of river. very unpicturesque country: Indians attacked a /house/:

Ornithology different: only small Icterus; not so very tame: some pidgeons; different parrots different partridge BBB Birds common rose starling Finch with black Sparrow Pteru Pteru[300] one days shooting many new birds.

[88a] The Gypsum in Tosca but /may obscuring/ affects the horizontal layers of St Josephs bay: shells are only organic remain are merely obscure cavities: & one shells Gypsum not abundant

[89a] **4ᵗʰ [August 1833]**

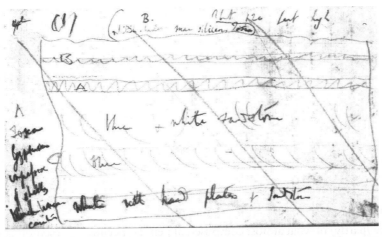

[section]

(1) B. About 120 feet high (whitish harder more siliceous Tosca) B A blue & white sandstone C blue **white with hard plates of sandstone**
A Tosca **Gypsum impressiones of shells black linear cavities**

299 Alcide Charles Victor Dessalines d'Orbigny (1802–57), a palaeontologist sent out by the French government to South America. His findings were published in Orbigny [1834]–47.

300 The pied lapwing (*Hoploxypterus cayanus*), with wings armed with sharp spurs. See specimen 1602 in *Zoology notes*, p. 163; listed as *Philomachus cayanus* in *Birds*, p. 127.

[90a]

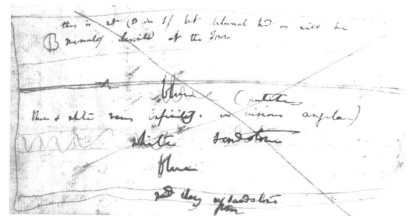

[section]

B this is not (B in 1) but diluvial bed as would be minutely described at the Town
A **blue** (patches blue & white veins infinitely. in curious angular)
white sandstone **blue red clay upon sandstone**[301]

[91a] **General** ~~stuff~~ **formation**
**(1) sandstone varying much in hardness: it is generally blackish, with stalactitic
particles harder. Black** ~~true~~ **cavitities. separation lines sometimes resembling beds
in Buenos Ayres. parts with cleavage from currents many lines Greenish blue with
little** *[illeg]* **also white beds: Above (A) thin bed of Argillo-Calcareous rock with
dendritic margins vide** */Lovett/* **specimens: has a alliance to B Blanca:**
In the section further (to SW)

[92a] **to the SW beneath Sandstone; bed of about 2 feet of rosy clay resting on
ferruginous & black Sandstone: in this section (A) has thinned away almost to
nothing: But a bed** ~~in~~ **(Fig. 1) which is called (B) (a)** */Diluvium/* **there is a gritty
Tosca seen** */partly/* **to** */become/ /ex panded/* **with a rubbly bed with calcareous
matter which extends along whole length of**

[93a] cliff & may */partly/* of what almost composed of pebbly */peculiarly/* resembling those at
St */Josephs/* Bay: pebble all small
The impression or Cavities in the Tosca more like Isocardium
(~~could not~~
The pebbles were mixed with pieces of calcareous matter resembling those at
Maldonado. — I think this old bed followed inequalities

---

301 There is a change in the date sequence at this point, from May 1833 to August 1833 on p. 91a.

[94a]   **6th & 7th** [August 1833] **The whole country round the town sandstone, with waving cleavage & beds. nearly horizontal. — Houses built on & in it. overlooking great flat valley. about 8 or 9 miles across. with projecting headlands on each side: — The sandstone is covered with bed of pebbles amongst which is very much of stuff resembling mortar with few** /minute/ **pebbles. —** The basis of whole land **is so much of this nature, that I think this mortar stone has been formed at the spot in lumps. which have partly been disintegrated**

[95a]   **subsequently: — It is the same which occurs at S. Barrancas — now this & the** ~~modern shell~~ **gravel bed at St Joseph I should certainly think contemporaneous: perhaps** /also/ **Alluvium of Banda Oriental from occurrence of Calcareous Matter: what relation it bears to B. Blanca is doubtful & will require examination of pebbles** (~~whe~~ /does/ **Pumice occur at B. Blanca**)
I should think the Tosca bed at S. Barranca — alluded to St Josephs from Gypsum & few shells: In the vegetable

[96a]   **mould near to the river Pumice pebbles which floated on water: —**
Flat rich ground on opposite banks with <u>lakes</u>**: plain behind town like Port Praya all bushes with prickles & dust between. — River fine stream 3 4 times Severn:** /Toldas/ **miserable race fed by Governor: Like Fuegians: much more ingenious: — 7 8th** [August 1833] **Went to the Salina, or great Salt Lake: The export of this one lake is at present & will be still** ~~more~~ **the main wealth of the Rio Negro: our**

[97a]   salt not very valuable
road lay along the Barrancas of the river: The bank at about 5 leagues up are covered with willow trees & the diluvial lands being cultivated afford a pleasant prospect: All the good land of this country is diluvial, produces corn every year from 16–18 fold: **about 30 leagues; there is much fine timber:**
On the road all rock sandstone which in places contains calcareous bed 4 or 5 inches thick: Vide Specimen light porous: perhaps much to the calcareous puddingstone also a Tosca bed inferior: — Mortar

[98a]   **(or mortar)** /one/ **its inferior to this; but I do not conceive there is sufficient for this purpose: — Much of the gravel is white-washed as at** ~~B B~~ **P. Praya: It struck me that the cause of this & calcareous** /matter/ **was owing to the rock being stretched at the base of the Andes:** pumice stone, conglomerate
The gravel likewise in places contains concretions varying from ~~some~~ some of fist to the head of **(not rotted of course) smally crystalline gypsum (V Specimen) it**

[99a]   **is worked for burning to white wash the walls. —**
The gravel bed must have been formed at bottom of sea, & tranquilly with some (the superior shells show this) **chemical action; these half concretionary masses of Mortar & nodules of Gypsum all show this. —**
The same lime which here forms bed; to the South perhaps forms the immense Ostrea ~~of the South~~
The only Salina I visited is the smallest length about 20 ½ miles breadth.

[100a] **There are 5 great others. S of** /Colonia/ **1 S of R Negro: In this one salt dissolved every winter: forms white hard field in summer: 4 or 5 inches thick in the centres thicker beneath it invariably they find in all the Salinas. Madre del Sal (crystalation crossways)** **already after some** /dry/ **time crystallizing on banks: the bottom is sand very black & rather fetid perhaps this occurs from partial decay of green scum which must have proceeded from** /Carrena/

[101a] ~~I noticed~~ **The sand reposes on layers of the mortar (∴ Salina above gravel bed)** **In this black mud crystals (**/Madre/ **del Sal) of Gypsum quite thick together** in groups **including sand clearly formed under present circumstances.** Great evaporating dish slow — co-occurrence of gypsum & salt curious **The whole plain with prickly bushes and near Salinas (far more salt than sea) sea plants like those at B Blanca — Flamingoes (Include this in note) traces of worms. bodies preserved** /in/ **small rodenta** even in this arid canyon: **water about 3 feet deep: filled by**

[102a] rain ~~she & salt streams from surrounding plains~~ **Salt in quantity** ~~remains constant~~: **is not perceptibly decreased by working in other lakes forms crust at bottom** under water **as it cannot be redissolved: other circumstances the same, lakes 3 or 4** or more **times larger:** ~~0 a~~ **This lake occurs in a great depression 5 or 6 miles every way, bounded on all sides by larger parts excepting where the Rio Negro crosses to which the grand depression** *[illeg]* lowers **itself. the lakes lies in its lowest part.** ~~Now if an inland Salt thicker in middle~~

[103a] Sea Lake had been contained here when red: On the other side of this ridge on few leagues distant one meets /grand/ These <u>local</u> circumstances merely show probability of lake & <u>springs</u> in that spot. — Camp no prickles: showing salt is absent. — So that it appears the ground is /impregnated/ within the above limits the drainage of which at present fills the lake: They say Salt petre occurs in the caverns in Sandstone so that perhaps all the silt does /&/ Salinas are formed, where depression favours their existence: **but then it occurs above diluvium:** **About 1/16th part of the lake is worked & about 30000 fargas annually** /raised/: **Price at Jan 4 reales a Farga** — 3rdly worked with waggons. —

[104a] **I** ~~understand~~ **have seen** ~~from people who have~~ **in the houses & beneath a rocky bed that particular** /layers/ **which are soft & continually in damp weather coarse contains much nitre (& probably other salts) & are therefore in wet weather damp & spoil the back of houses from when plastered with mud** **also** with **much crystals of the Saltpetre in side of soft** *[illeg]* **of Sandstone** /Clay bed inland/ the Cuevas[302] when candle is brought in spark **The Country is sandy & gravelly plains arid all way to Port St Antonio** /Harris/ **Respecting the Salina, it**

---

302 Spanish for caves.

[105a] is stated that the little streams which flow into it are not salt. But that this is a strong salt springs on sides of lake which flows chiefly in the winter. Hence the salt is explained as being above the gravel bed, The Gypsum must be now crystallising as being on & in black sand above the mortar & different from Gypsum in gravel & gravel:

8 9th [August 1833] day wasted, one of the prices for undertaking any

[106a] expedition. Young male Indians in a Schooner work well fine young men dress cleanliness hair person &c &c very tall. brown statues

9th 10th [August 1833] Bad day so would not start: ~~Several Gaucho in Company~~ Long tailed shrike[303] Callandra different habits, habits much wilder Traversia[304] (Saliferous Sandstone Traversia ??): sits differently on twig: alights in summit of branches, does not use its tail so much: song infinitely sweeter: Toco Toco[305] or Taupes[306] & Aperea different from Maldonado, latter smaller tamer, appears more in day feeders frequent hedges & holes: have 2 young at a time

latter quite different more

[107a] ~~more~~ distinct louder sonorous, like distant cutting of small tree more peculiar noise double & not three or 4 times repeated only twice, said to have no tail (?) & blind (?) Inhabits same sites — more injurious than Talpe

Bird runs like animal at bottom of hedges does not easily fly. not loud Singular single

Oranges ~~flora~~ fruit ~~young trees~~ Olives ? Yes

Cactus

[108a] Salina. 18 miles up river. Is not R Salado Salt?

Do Sand Cliffs occur to South of R Negro & Salinas? de Origine /~~Concharum~~/ Little bird with pointed tail inhabits traversia, hops about bushes like Parus:[307] constantly uttering harsh shrill quickly reiterated chirp

long billed BBB. inhabits do, quiet fly about & <u>running</u> hopping very quickly on ground much like common B.B.B. & picking at pieces of dung

~~Long tail shrike~~ Carrancha song very beautiful many on thorny twigs enliven Traversia

[109a] most resembles but more powerful some of the reed warblers harsh notes intermingled & some very high ones ~~Nest~~ Nest small on sand /same/ /illeg/ Salina to our left some time before right. what /distance/ in summer

---

303 The thenca or mockingbird. Specimen 1213 in *Zoology notes*, p. 153, apparently the bird listed as *Mimus orpheus* in *Birds*, p. 60. Calandria was the local name.

304 Spanish for passage, road.

305 This animal is closely related to the taupe. The toco toco or tuco-tuco described by CD 'as a rodent with the habits of a mole'. There are about fifty species of tuco-tuco. See specimen 1267 in *Zoology notes*, p. 165; listed as *Ctenomys braziliensis* in *Mammalia*, pp. 79–82. Discussed in *Journal of researches*, pp. 59–60.

306 Taupe is French for the mole. This is probably a species of *Ctenomys*.

307 Specimen not in spirits 1469 in *Zoology notes*, p. 393; listed as *Serpophaga parulus* in *Birds*, p. 49.

~~11~~ 10th [August 1833] **Started. country same rather less spiny trees: [Aboro]
Grande** a great valley **to P Bosa running [across] the country: many valleys [&
defiform] not explicable which look as if it originally made so general small
gravel & mortar: one place Sandstone like former** Pezo Primero superior small **with
Tosca bed in** ~~valley~~ [defiform] **salt lake & plants & spring small: country very level
evidently same formation**
Slept at about 11 leagues from

[110a] **the town in N ½ E (true course): Sleeping place found a Cow. fine
stillness, dreary plain, comfortable night, like Gypsies; horses not arrived:
describe general arrangement: Passed Walleechu tree. only one I saw.** first
subsequently others **3 feet diameter long diameter low much branched. Indian god
shout when about 2 miles off. surrounded by bones of horses** covered with strings
instead of leaves **& remains of Ponchos (thread pulled from) cigar smoke upwards
spirits in wholes smaller Yerba**³⁰⁸ **& Gaucho has seen all this found &c &c —
think horses will not be tired All tribe know this God Men & Women & children
Gaucho**

[111a] **start the offerings; 9 leagues from the Town: It is perhaps because a well here
landmark & striking** ~~point~~ **objects in the plain. & as being half the ring in a
dangerous passage in summer** ~~in a dangerous & dry traverse~~
**11th (Monday) [August 1833] The next day much the same country destitute of
almost every animal — There are few Guanacos deer & Ostriches** most herds of
any animals **Here are only air. Carranchas & male** ~~Buzzard~~ Vulture
no Lachuzoas Gravel & mortar **At Pozo Secundo (9 leagues from
Colorado) a remarkable flat plain at much lower elevation is seen stretching
for many miles, vegetation & appearance the same: But did not notice
Mortar**

[112a] **only gravel. — It must be same for at Pozo there was** green to blue **sandstone &
Tosca. — Here also salt lake & banks of a well with nitre** encrusting **(Slept night as
before Foxes howling around no water) R Colorado**
12th [August 1833] **about 3 leagues from sea plain very gradually lowers onto
plain with clover & Lachuza & no spiny bushes & called Pampas.** I suspect
must be Tosca plain of other side of River green short turf character of coast the
same s

~~aid to~~ **stretches to Union Bay with flat islands & mud banks sinks soon into
(sea) diluvial plain;** ~~then into~~ **parts are salt petre marshes — with saline plants:
these places & line of sand of low hillocks. It is curious how salt petre occurs
in diluvium**

---

308 The herb *Ilex paraguayensis* used to prepare mate, the popular South American infusion.

[113a] ~~These plain from d~~ **plains — I am inclined to think the <u>Pampas</u> not diluvial although little above level of sea.**

NB Opposite Patagones in <u>certain</u> diluvial plain there are small Salinas. (ask Harris[309]) & **worked & bare plains with nitre: How is it to be explained that Salt should there occur?** — I can only imagine springs; for plain fertile ∴ earth not in salt **What is cause of low sand**-stone **plain? Abrasion?** lower saliferous plain (**50 or 60 feet lower**)

Pleasant ride Gauchos in a line, robes flowing easy seat. Spurs & sword **clanking anecdotes of riding** change horses, three horses: **Indian white horse; Gaucho Laughing &**

[114a] **& talking: arrived at the river. Sauce & reeds; about size of the Severn:** 60 yards of water mares **swimming for division of the army**

¼ mile square, encampment Banditti — Examination passage from General Rosas[310] **Extraordinary man nearly 300 thousand cattle.**

Perdrix & Scolopax[311] most numerous kind in the dry plains **builds on borders of lakes (eggs white spotted with red)** about 5 or 6 in small flocks from 2 or 3 to 30 or 40 **13**[th]) **Wednesday** [August 1833] **Nothing to do, miserable day, kill time**, Frozen with *[illeg]* **swampy plain, overflowed**

[115a] **snow water** in summer: **only amusement watching Indian families**[312] *[dirty]* & **ornamented beads. long hair beautiful children all in** */party/* **riding great number** about 400

All my days are wrong — certainly I started on Sunday. [10 August 1833] **14**[th] [August 1833] **It is clear the plains around here (including Salitras) have been lately formed in an estuary of sea — land is now flooded in summer** are islands in swamps of; **then sand dunes when yet remains as** ~~islands~~ **attesting former sea. About 10 leagues in direct lines from Rosa: This river floods in December from**

[116a] **Snow: Rio Negro from snow & rains:**

General R grave intelligent enthusiastic popular **laughing bad sign; anecdote of mad man: Mendoza trade**

15th [August 1833] **4 leagues to first Posta**[313]

direction ENE. Course of river beds diluvial:

2nd **posta (11 leagues?) direct. N by E Gaucho thinks so 12 leagues from sea: in interior 17 leagues** there is a **rough plain — Salinas —** different aspect & few spiny

309 James Harris, British trader and sea captain at Rio Negro in Patagonia. He acted as pilot for the *Beagle*, and rented *La Paz* and *La Liebre* to FitzRoy for surveying shallow coastal waters.

310 Juan Manuel de Rosas, provided CD with horses and safe conducts on inland journeys.

311 See specimens 1224 and 1273 in *Zoology notes*, p. 169; listed as *Tinochorus rumicivorus* in *Birds*, pp. 117–8.

312 Allies of General Rosas.

313 Camps or posts on the Pampas at approximately one day's ride distance apart.

bushes above *[colony]* more fertile, more grasses, different plants **Plain**
about 40 30 or feet vegetation 3 or 4 3 feet sandy earth, little

[117a] **Gravel & mortar** calcareous **white tosca 2 feet with dendritic manganese red Tosca**
red aluminous whole depth of well

Comm. Icterus.[314] **Black Cuckoos** (where there is water) (read finch yellow tree)
Plover; at 1ˢᵗ Posta *[known]* by different names. Congos. Toco Toco
Aperea[315] **2 young ones**

Avecasina many in at 20 flocks above 5 or 6 eggs in each nest long tailed Shrike
North of Tosca plain

Great range of Medanos E & W coast to coast & far in interior much water;
Medanos.[316] **with flat vallies like at B. Blanca. Sierra Ventana N by E Compass at**
3ᵈ Posta

~~Medam E & W compass~~ range

[118a]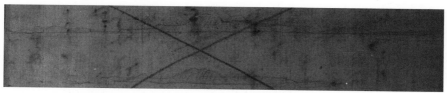

[coastlines] **WNW) Extending from very little (E by E (?)**
East of N to little beyond ~~beyond~~ East **of NNE**
(all Compass)

[119a] **From 2ᵈ posta (with many fresh water lagunas) on edge of Medanos 2 & ½. leagues**
across the Medanos with flat plains like B Blanca: till you come to escarpement (?)
much broken through by *[vallies]* *[inclined]* **(old coast) composed of very**
ferruginous aluminous Tosca reposing on a white soil (this is only upper part of hill
rest hidden **with most superior part & ~~large~~ bed 2 or three feet thick in parts, in**
others few inches of calcareous aluminous rock — often with little manganese &
minute linear cavities: hard V Specimen: —

[120a] **(16ᵗʰ) [August 1833] 3ᵈ Posta to 4ᵗʰ: 7 leagues: plain at foot ~~at~~ of escarpement same**
plain with numerous undulations looks quite flat from great escarpement **(as**
shown by Cuervas of Lievres[317] **& well at 3ᵈ Posta. red Tosca as other (S)**
side near this (4ᵗʰ) Posta the plain terminates, abruptly north edge of alumi
Limestone rock. — & covered with Medanos; now shifting which run W by
S & E by N (Compass). The edge ~~non~~ about 50 feet above, the low land (into which

---

314 *Icterus*, a genus of Oriole.
315 Aperea is the South American guinea pig or cavia.
316 Spanish for sandbanks on the seashore.
317 *Lepus europaeus*, hare, probably introduced from Europe.

salt water flows in gales) **These Medanos do not stretch very far & terminate abruptly to the East (Road round them) (white Ostriches): this plain**

[121a] **is rather higher than the one to south of great Escarpement**

gravel less frequent ~~great Escarpement not above 80 feet high~~

From this (4ᵗʰ) Posta Pueblo bears NE (½ East) Compass — & Sierra do Ventana NNE (compass) & high ~~land~~ plain stretches from ~~NNW to ENE~~ WNW to North, then lower behind the town other side of town.

It forms a large land basin with immense number Salitras, lakes, marshes & streams from Sierra de Ventana into Bahia Blanca

Both BB Birds build in holes

2 /varas/ long: Casara³¹⁸

[122a] **(Smell deer at ¼ miles distance) last night comfortable little Ranchita: Black lieutenant. Indians marched by in numbers; /dilema/ to dine with six gauchos pleasant man & certainly by far best conducted** excellent asado³¹⁹ **little ditch: —**

Long tailed Shrike only comes in Summer to R Negro!

Yet saw one 2ᵈ Posta N of Colorado —

Whole ground soft with Taupas never leave their holes ~~Casara~~

The distance from 4ᵗʰ Posta to town ~~to~~ **about 3 leagues in a direct line, but more than**

[123a] **6 in the road — whole road through mud swamps very wearisome on** ~~other~~ **North side mortar plains with fragments of quartz from Sierra de Ventana:**

5ᵗʰ Posta again mud swamps /Souzed/ in black mud. — — & Posta 5 men murdered Heard of cannon Indians close; kept close to the mud to escape, left the road believe friendly Indians: =

/Curious in/ land swamps **with little evidence of sea lately being** there, yet **millions of small Turbo in** obscure lines **Medanos in swamp**

Young ones sit on Capincha bark: weight of some animals of Maldonado. is in this book. *[4 or 5 illeg words in pencil deleted]*

[124a] **Sunday (17ᵗʰ)** [August 1833] **Don Poncho thinks Sierra runs E & W for about 30 or 40 leagues & to the W very large Salinas**

day spent in killing time no clean clothes, no books. I envied the very kittens playing on the mud floor. —

Pichey. Mataco. Paluda.³²⁰ all inhabit same plain. first wonderfully abundant, buries itself with great celerity has 2 or 3 young at one time

---

318 The 'house builder' or rufous hornero *Furnarius rufus.*

319 Spanish for burnt or roast meat, Barbeque.

320 Local names for different species of Armadillo. Pichey: listed as *Dasypus minutus* in *Mammalia*, p. 93, current name *Zaedyus pichiy*. Mataco: Specimen in spirits 403, cited in *Mammalia*, p. 93, as *Dasypus mataco*. It is the Southern Three-banded Armadillo (Apara), currently named *Tolypeutes matacus*. Paluda: (CD had no specimen), cited in *Mammalia*, p. 93, as *Dasypus villosus*. It is the Large Hairy Armadillo, *Chaetophractus villosus*.

Molita[321] does not go south of Tandil

Taupes different note ~~hear~~ here single repeated at equal times, or accelerating very noisy in evening ~~till~~ after sunset, quiet at night. —

[125a] Harris about Salinas & R. Negro Don Juan ostriches breeding

Salt from this Salinas

Base of high plain

18th [August 1833] Commandante lent me horses & soldier: soldier lot nobody leaves the fort to hunt /for/ with him two men killed, he wounded & horse balled: now are close, coast road safe. it was enough to make one watch deer running away fast. as if frightened by some other object. — ship not arrived: picked up ~~our~~ 2 fresh horses yet so very bad back & thin one left behind. after returning: eat a Pichey breakfast dinner & supper horses quite tired. not more than 25 miles & walking: Slept in Camp: Salt petre ~~whole Tosca plain~~ Taupes: Zorilla:[322] in morning returned; horses miserable

19th [August 1833] hardly able to crawl; I walked another horse left behind; killed kid

[126a] no water: all Salt: suffered so much from hot day. could not walk. dreadful inhumanity riding such horses: delicious drink of fresh water: arrived at Pueblo after a miserable ride: — The places where salt-petre are every where. water in Cart ruts. — The places where it occurs are muddy & bare & look as if sea entered at high water. — In mud walls the Camara makes its holes not being aware how thin they are actually make hole quite through:

20th [August 1833] Bought a good young horse for about 4£..10s, went out riding the plain north of the town

[127a] is an immense one & stretches away to the great apparently to the foot of the Sierra de Ventana. — It is /I should/ ~~150~~ or 200 feet above the swamp & perhaps higher than great escarpement. — The summit is covered with bed several feet thick of a Limestone or Mortar generally containing minute extraneous matter & generally porous. — V 2 specimens it seems general & forming horizontal bed. — There are no pebbles not even quartz: vegetation sterile & little grasses. resembling Tosca plain: I do not know whether the whole

[128a] bank is composed of it or whether diluvial: Mortar, or calcareous bed in Tosca. — I incline to former opinion: The low plain in which town is, is about 30 or 40 feet above swamp. covered with quartz & apparently composed of fragments of mortar including quartz pebbles: with only 3 feet sandy earth: — The high plain forms a sort of Coast line of barrancas The whole plain strewed with bits of mortar. —

The number of Bulimas[323] (especially) & other land shells on high plain very curious

---

321 Local name for a species of armadillo. Specimen not in spirits 1413 in *Zoology notes*, p. 390.
322 The hog-nosed skunk *Conepatus* sp. See the *Beagle diary*, p. 177.
323 A genus of land snail.

[129a] every square inch has 3 or 4 I never saw such numbers The whole surface for leagues was like unto to sea – beach: V. Specimen except one all dead ones

21ˢᵗ [August 1833] So tired of doing nothing started to Punta Alta; on the road had alarm of Indians. coolness of the mans method. not at once galloping. crawling on his belly: women: worked at cliffs & bones: beautiful evening: Ship not arrived: very bad night, excessive rain so much of a gaucho do not care about it: next morning also rain therefore

[130a] 22ᵈ [August 1833] therefore started back: on road observed fresh tracks of Lion. commenced unsuccessful hunt & dogs seem to know: not ambitious to see Lion. have no individual name: all cowards from the use of balls: because so few pebbles: on return found Harris arrived night before. uneasy about ship: met on road Indians from this place, supposed to have murdered the post master: the Generals message about their

[131a] heads & division of army to follow the Rastro³²⁴ — if guilty. to massacre them: ostriches. males certainly sit on eggs easily distinguished: stray eggs first laid: many females, said, I know not on what evidence to lay in one nest. about 50 eggs in the belly: analogy to African method manner of laying: —

Avestruss Petises:³²⁵ colour oveiro

My opinion of geology of Punta perreika is quite

[132a] altered: observed following facts: vegetable covering or diluvium; with much pumice & sea & land shells, same as now exist: pumice in upper gravel: all the pebbles much larger than the Southern pebbles & almost entirely quartz white or coloured: also rounded pieces of the Argillo calcareous rock so characteristic of the Tosca. — This might easily have escaped my notice before I was aware of

[133a] its importance: The Tosca contains fragments of shells, quite different from the true Tosca. — I have no doubt that this bed results from destruction of Tosca beds with bones. Describe locality of big skeleton such as M: Hermosa. there there are no fragments of shells: bones in groups calcareous concretions. &c From the height of barranca The gravel has evidently been hardened by calcareous matter such as upper Tosca bed would produce)

[134a] & the form of beds is such as produced on a beach: under smally different circumstances (such as more open sea) I can easily believe such a bed would form at present: but then the Tosca bed has such horizontal white lines in it. — the lower diluvium has big quartz

324 Spanish for trail.
325 Avestruz petise, the small rhea, or Darwin's rhea, of Patagonia, ('petiso' being a slang word used in Argentina for a very short person); listed as *Rhea Darwinii* in *Birds*, pp. 123–5, plate 47, current name *Pterocnemia pennata*. See *Zoology notes*, pp. 188–9, 101–2, and 'Ornithological notes', pp. 268–77.

pebbles arranged so /truly/ ((the different state of bones in the two sites is by this hypothesis well

[135a] explained)) That I think it necessary to have been formed beneath the sea & at some, (though perhaps shallow depth): if so a change of level is necessary which perhaps the low plain of town of P̶u̶ Bahia Blanca requires & East & West Medanos:

The gravel here is evidently quite different: different size points out different distance of origin. One Cordilleras[326] (?) or Port Desire (?) the other Sierra Ventanas. The Ventana gravel ancient. The most northern site of small pebbles

[136a] is low saliferous plain South of R. Colorado: — Formation of town of Bahia Blanca. rubbly bed the Calcareous rock i̶n̶l̶a̶n̶d̶ with /interstitous/ marly; I should think certainly formed from detritus of upper beds of main plain. with quartz pebbles: —

The non occurrence of quartz pebbles in high plain remarkable. — Some headlands lower than main plain are covered with bed. some feet thick

[137a] of a mortar looking rock: (modern bed (if mortar is diluvial) of Punta alta is shown to be modern. by not having mortar) Is this owing to whole cliff being so: to different height of elevation: (i̶m̶p̶r̶o̶b̶a̶b̶l̶e̶) or superficial covering of diluvial: (?! if the latter it is probable the covering is very modern for then the mortar in low plains contains quartz whilst high does not ∴ same of of land. !?) The more compact, crystalline purer nature of mortar or calcareous superficial bed *[illeg]* against its diluvial origin. —

[138a] My alteration in view of geological nature of P. Alta is owing to more extended knowledge of country it is principally instructive in showing that the bones necessarily were not coexistent with present shells, though old shells: they exist at M: Hermosa pebbles from the beds of which occur in the gravel: Therefore such bones if same as those at M: Hermosa must be anterior to present shells: How much so. Quien Sabe?

[139a] 23ᵈ [August 1833] Ship seen over the horizon of mud banks

24ᵗʰ Sunday [August 1833] Rode to the Boca[327] but a NW wind wind was too strong to allow ship or boat to approach nearer

Arrival of troops against murderess of the post

Heyque Leuvu. Falkners[328] name for Sauce grande.

Indian name of Fire wood?

C̶o̶l̶o̶n̶e̶l̶ ̶O̶ ̶B̶r̶i̶e̶n̶.̶ ̶N̶a̶t̶u̶r̶a̶l̶i̶s̶t̶

All the dates from time of Starting on the 10ᵗʰ are wrong: this ought to have been 11ᵗʰ & Sunday 25ᵗʰ instead of 24ᵗʰ —

---

326 Spanish for mountain range. In South America generally synonymous with the Andes.
327 Spanish for mouth of river.
328 Thomas Falkner (1707–84), Jesuit missionary in Patagonia, 1740–68. Falkner 1774. An almost identical entry occurs in the *B. Blanca notebook*, p. 1a.

[140a] Mud banks called Crangeijo: when will not carry: horses inclination of dip in sandstone plains great as shown by rapid current: of R. Negro

Agency of Volcanoes in forming Sandstone plains shown at Patagones by Pumice Conglomerate

Long-tailed little bird of Patagones — found at Bahia Blanca runs very quickly

[141a] Monday 26th. [August 1833] — The boat with M[r] Chaffers in command, having in rain tried to beat up; slept on the water & arrived this morning — I rode down to the Boca: returned, accompanied them to Commandante not thought safe for Sierra Ventana arranged plans for B. Ayres with comm: Miranda:[329] & very civil; returned on board, waited for Cow to be killed: started after dark for ship: fine moonlight: calm: ship moved: arrived on board ½ after one oclock. —

[142a] 27th. [August 1833] — Whole day consumed in relating my adventures & all anecdotes about Indians to the Captain.

28th [August 1833] Very actively employed in arranging everything for Buenos Ayres: Most delightful the feeling of excitement & activity after the indolence of many days in the last fortnight, spent in the Spanish settlements!

[143a–166a blank]

[BACK COVER]

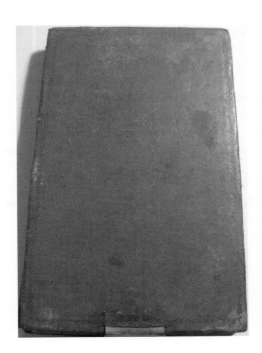

[INSIDE BACK COVER]   C. Darwin
                                        H. M. S. *Beagle*:

329 Commandante Miranda was a subordinate of General Rosas.

2/—

Thalassidromus[330]

   Dollars

Stokes 10

Harris 30

Rio: Tapes. N. of Maldonado

Rio de la Plata

May ~~14~~ 15th (rainy day) 1833:

S W dip

N N W

or do

W

Slaty limestone dip to SE (by /S/?) from great ~~Quartz~~ Slaty Quartz H. with lines white & red quartz WNW

W N W W

[1b] Mem: junction of escarpment

balls with Fagus. T del F[331]

Are not the climate in S America — later than the northern Summer ?

~~Is not it hotter in~~

Serro de las animas

[2b] /Tape/< >

/short th/< >

also much in usual great< > shorter /nyroe/: Shield reaching half way down body: & leaving a little of the neck exposed when animal[332] crawls: Brancheal orifice at side about ⅔ of whole length of shield ~~do~~ from anterior part of shield:

tail moderately pointed Shield with concentric

[3b] furrows of oval shape corresponding to form of shield: Body slightly wrinkled: ~~crawls.~~

When it crawls arches body like caterpillar in small degree

Tuesday 14th [May 1833] R. Tapes

Sierra de Tapalca 9 leagues Camino real & Buenos Ayres fragments of Quartz:

/Cabilda/[333] del /Rey/ Limestone

[4b] Spirit bottle?

Big knife

Prometheans

---

330 Specimen not in spirits 1349 in *Zoology notes*, p. 389; listed as *Thalassidroma oceanica* in *Birds*, p. 141.

331 Possibly the fungus found growing on the Antarctic beech, then called *Fagus antarcticus*. See specimen not in spirits 1155 and specimen in spirits 528 and 532 in *Zoology notes*, pp. 386, 337. See *Beagle plants* pp. 178–9.

332 A land slug, specimen 614 in *Zoology notes*, p. 151.

333 Spanish for town-hall.

Medicine Calomel

Pistol, balls, powder:

Letter to Commandante

Map: Note Book: soft paper. Spare pencil; small hammer: Compass: Stockings: gloves: handkerchief: wine flask? Comb: 2 handkerchiefs nightcap: Passport. Poncho. Salt petre

Allmanack for /Ellsgood/ for 1834

[5b–6b excised]

[7b]   San José 1877

Patagon 79

Poblado

Inhabitants of San José murder in Church

There were many Indians. Cattle & Horses: hero at first arriving

River called Negro from the Cacique:[334] only 20 & 30 first *[illeg]*: ~~from~~ 2 or 3000 Indians came & stole every thing often obliged to escape & hide in rushes: after /sack/ of the town but by other tribes: & in the Cuevas: saw old door with /loop-holes/: burnt all the houses

All the Indians in the neighbourhood died of Small Pox & a sort of plague in throat: Indians — balls Lasso & Chascas[335] a very few knives & dress & paint the same

[8b]   100 league. Sierra del Diamente: this branch river (at head /well/) about 18 feet: went about 3 or 4 ~~feet~~ leagues

The ~~river~~ Negro ends in a lake. — 3 or 4 leagues in circumference & the little river flows into it.

Laguna (/Launchen/)

He says about 10 leagues from coast of Baldiva This Sierra Imperial is 5 leagues Ships can see it 70 leagues distant: nobody but God could climb it — & if upon it shake hands with

upon being asked if on it Ave Maria Santissma Jesu Christe

150 /married/ 150 185

[9b]   Don Pedro= who keeps the Keys of Heavens:

Pines trees of the Diamente No volcanoes: stream emptys itself into N shore — very distant from Imperial:

before Diamente may Sauces

Then /Cypresses/ →

Both sides hilly: Much snow on Imperial:

8 months about (1783)

[10b]   Plenty of /hawzana/ at of foot of Andes: lower Sierra so called ~~English Tuvega~~; 34 men in all

---

334 Indian chief.

335 'Chascas (messengers or ambassadors) of the Indians', *Beagle diary*, p. 180.

In 1810 murdered the men at S Joseph & remained 2 years with them
Indians like so many the sand of Caque (sailed to R Negro) protected them:
The river is in a Box, very rapid: forced
[11b] to dig ditches: inland the vessel: same vessel: vessel about 14. tuns: & now 20: So
shallow, Laguna no vessel could enter from big stones did not enter (other account
with boat:)
often been there since a prisoner: nothing but this river: close to the Sierra
Imperial:
That the Indians go to Baldiva to trade: in breaks: buy wheat & Indian corn often
wished him to go:
[12b] lake too cold to go in from snow water —
Indians treated him very well: sold at R. Negro
About 40 leagues to San Luis for junction of Diamente:
First man who gave account was English: — & all his account wonderfully true
In the interior bad country where /passed/ the Islands of Churchills [Churichoel]: 70
leagues:
Indians line in small vallies & plant little Wheat & Indian Corn very bad country
[13b] rocky country of hard rock
Owe Harris 58 Dollars
[14b blank]

Textual notes to the *Falkland notebook*

[FC]   c vol of humanity *[2 words illeg]* 49] *pencil. Not in CD's handwriting, not
transcribed.*
[IFC]   1.14.] *Down House number, not transcribed.*
88202334] *English Heritage number, not transcribed.*
3] *added by Nora Barlow, pencil, not transcribed.*
[5a]   *a small mark in ink appears to be a nib test.*
[8a]   (Is ... migration)] *added pencil.*
[14a]   22^d ... observ] *added pencil in left margin.*
[18a]   *sketch drawn perpendicular to the spine.*
[19a]   *a small ink mark after* 'lava' *appears to be the testing of a nib.*
[29a]   *[illeg]] written over* 'somewhere near'.
[39a]   the] *overwritten by* '1st'.
[40a]   Mon] *ink.*
Monomenos] *ink.*
[76a]   *a few ink marks after* 'formations.' *appear to be nib tests.*
[84a] | Gran Bestia |] *ink.*
Capt ... me] *ink.*
[85a]   *page in ink.*

Bones] *circled in pencil.*

No] *added pencil.*

Mineral ... Nitre ?] *pencil.*

[86a] What Fruits ... 2^d] *ink.*

Standain?] *pencil.*

differ species] *pencil.*

Light ... house] *ink over pencil.*

[87a] one days ... birds] *pencil.*

[88a] *page in ink.*

[89a] *written and sketched perpendicular to the spine. 'A' in diagram, ink.*

white with hard plates of sandstone] *light brown ink over pencil.*

sca Gypsum ... cavities] *light brown ink over pencil.*

[90a] *sketch in pencil, page in light brown ink, written and sketched perpendicular to the spine.*

[93a] cliff ... mixed with] *ink over pencil.*

Now ... middle] *pencil.*

[103a] These local ... spot. —] *ink.*

but then ... waggons. —] *ink over pencil.*

[114a] Frozen ... *[illeg]*] *very faint pencil.*

[118a] *sketch drawn perpendicular to the spine.*

[121a] great Escarpement not above 80 feet high] *pencil, overwritten by other entries.*

NNW to ENE] *pencil.*

[122a] Yet ... Colorado —] *pencil.*

[124a] Pichey ... at night] *in ink.*

[125a] 18th ... kid] *ink.*

[126a–138a] *pages in ink.*

[139a] Arrival ... post] *ink.*

All the ... 24th —] *ink.*

[141a–142a] *pages in ink.*

[IBC] C. Darwin] *ink.*

[1b] Serro ... animas] *not in CD's handwriting. bottom third of page excised.*

[1b–2b] *lower third of leaf excised.*

THE *B. BLANCA NOTEBOOK*

∽

The *B. Blanca notebook* takes its name from the port city of Bahía Blanca (White Bay) on the coast of Argentina. The name of the bay derives from the typical colour of the salt covering the soil along the shore. The notebook is bound in red leather with the border blind embossed: the brass clasp is missing. The back cover of the notebook has a label of cream-coloured paper (55 × 16 mm) with 'B. Blanca to Buenos Ayres' written in ink and a few barely legible words in pencil, not in Darwin's hand. Perhaps Darwin added the label to the back cover, against his usual practice, because, with the clasp missing, the covers appear almost indistinguishable. The pencil holder has been repaired with two small stitches near its opening. The notebook contains 52 leaves or 104 pages. The notebook was written in two sequences, pp. 1a–88a and pp. 1b–16b. Almost all pages are cancelled, in pencil.

It was the fifth notebook used by Darwin during the *Beagle* voyage. In one respect it is radically different from any of the other field notebooks in that three-quarters of its pencil entries were overwritten by Darwin in ink. This was done for almost all of the three-week inland expedition in September 1833, pp. 2a–68a. Perhaps his pencil was too faint, although the pencil writing when occasionally not overwritten does not seem fainter than usual. It would have cost considerable time to write over so many pages a second time. The entries for 5–7 September and 13 September where written only in ink. The pen (and possibly also the ink) used for those four days seems to differ from the one used for inking over; it is distinctive and somewhat finer.

The *B. Blanca notebook* covers two distinct time periods, firstly 29 August to 21 September 1833, pp. 1a–68a, pp. 1b–9b; then, secondly, after a seven month break for the southern summer when the *Beagle* went south, rather patchily from 14 April to August 1834, pp. 69a–87a. It has thus a rather complex relationship to no less than six other field notebooks, as follows: the earliest entries seem to overlap with the last in the *Falkland notebook*; then the *B. Blanca notebook* was supplanted from 21 September 1833 to 13 April 1834 by four other notebooks, *St. Fe notebook* up to 13 November, then the *Banda Oriental notebook* to early December 1833, then the *Buenos Ayres notebook* briefly around Christmas, then the *Port Desire notebook* to

13 April 1834. The *B. Blanca notebook* overlaps between April and June 1834 with the *Banda Oriental notebook*.

Darwin was greatly influenced by some of his discoveries during the southern summer of December–January 1833–4. Within the space of a few weeks he collected not only the remains of a 'new' kind of rhea which was different from the rhea with which he was already well acquainted (see below and introduction to the *Rio notebook*) but also unearthed a tolerably complete skeleton of a strange new mammal. At first Darwin was very uncertain what sort of creature it might be, but once Richard Owen examined it in England in early 1837 Darwin started to see it as an extinct giant llama (see below and introduction to the *Port Desire notebook*). These two discoveries assumed immense importance to Darwin in providing him with natural links between species in space and time. He juxtaposed these two finds in one of the most important series of entries in the *Red notebook*, pp. 127–9, penned in the Spring of 1837, which scholars now see as pivotal in understanding his path to a workable theory of evolution.

In April–June 1834 Darwin also read the third and final volume of Lyell's *Principles of geology*, received before but read after the Santa Cruz river expedition.[336] This volume consolidated Darwin's commitment to Lyell's gradualistic geology and clarified a number of previously confusing geological issues, such as whether 'Primary' formations were always older than rocks of Secondary and Transition age. Lyell showed that they were not always oldest, and should be renamed 'hypogene'. He renamed 'unstratified primary' rocks (e.g. granite) as 'plutonic', and 'stratified primary' (e.g. gneiss) as 'metamorphic'.

In his third volume Lyell also paid particular attention to countering the theory of 'paroxysmal elevation of mountain ranges' of the French geologist Jean Baptiste Élie de Beaumont (1798–1874), who cited the Andes as the most recent example. As Herbert 2005 has shown, Darwin's reading of Lyell's third volume, just after seeing what to him were proofs of the gradual elevation of Patagonia, and just before experiencing at first hand the potential elevating power of many successive earthquakes in the Andes, provided just the right focus for his theoretical energies. Darwin had heard Sedgwick praise Élie de Beaumont's theory before the voyage, so he must have realised that his rapidly approaching opportunity to see the Andes for himself might be crucial in deciding whether to follow Lyell or Sedgwick.

In terms of places described, the *B. Blanca notebook* mainly covers Darwin's long treks across the Pampas which started at the Rio Negro and eventually finished in Buenos Ayres. The first, from El Carmen crossing the Rio Colorado to Bahia Blanca (*c.* 250 km, 155 miles) in August was covered by the *Falkland notebook*. The second, from Bahia Blanca to Buenos Ayres (*c.* 750 km, 466 miles) in September is covered

---

336 See CCD1: 371 note 5, and Herbert 2005, p. 69.

here in the *B. Blanca notebook*. The third is covered by the *St. Fe notebook*, the fourth and fifth by the *Banda Oriental notebook*.

Other localities found in the *B. Blanca notebook* are as follows: brief descriptions of the mouth of the Santa Cruz, pp. 69a–75a, Port Famine [Puerto Hambre, near Punta Arenas], and the Magdalen Channel, off the Straits of Magellan. Port Famine acquired its name following the abandonment in *c.* 1586 of the Spanish fort caused by the failure of food supplies and consequent starvation of the inhabitants.[337]

The last few pages of the *B. Blanca notebook*, which may contain Darwin's first words on the geology of the west coast of South America, seem to relate to the island of Chiloe [Isla Grande de Chiloé], off the coast of Chile. These pages must date from the second half of 1834, or very early 1835.

Bahia Blanca to Buenos Ayres, August–September 1833

It is difficult to date the nine pages of notes at the back of the notebook, but the entry 'Cusca places eggs in other birds nest' on the inside back cover seems linked to the note about the chusco on p. 53a which is dated 13 September 1833. The entry is suggestive of the *Molothrus*, or cowbird, with its cuckoo-like egg-laying habits. Darwin later discussed 'brood parasitism' in the *Origin of species*, p. 216. He published a paper on *Molothrus* a few months before he died.[338] Thus the *Beagle* voyage provided him with materials used for the rest of his life.

The other entries at the back of the notebook seem to date from 1833. The last line of the inside back cover and p. 1b is in FitzRoy's handwriting. The entry includes a telegraphic memo about '(Emulation) M. Barral — ? Naut. Almanac 1834 — Letter — under cover M^r G'. This refers to the French survey ship *L'Emulation* and FitzRoy's apparent requests to Darwin that he procure a map of Buenos Ayres, perhaps from Monsieur Barral, a surveyor in the French navy, to speak to Philip Yorke Gore, the Chargé d'Affaires in Buenos Ayres, and to obtain the *Nautical Almanac* for 1834. On p. 2b is the note 'Don Many great bones (Lorenzo)' which seems to refer to FitzRoy's vague request in October 1833 for Darwin to speak to Gore and to 'Señor — Don — or Colonel Something, or Somebody', who could be Don Lorenzo.[339]

There follow more memos, and a charming sketch on p. 4b of the paw of a 'Paluda', or Large Hairy Armadillo. There are some torn pages and some jotted memos of ideas to be considered further, judging from the fact that they are in ink. A series of blank pages completes the 'b' sequence. Further study of these entries might reveal a more precise dating.

---

337 See Winslow 1975, p. 349, and the watercolour reproduced in Stanbury 1977, facing p. 184.
338 Darwin 1881; *Shorter publications*, pp. 451–2.
339 CCD1: 336. See Keynes 1979, pp. 77, 79, 160, 162, 232.

On the inside front cover almost the first entry is a mention of Samuel Richardson's last novel *Sir Charles Grandison* which Darwin apparently bought in Buenos Ayres in September 1833. This book is also mentioned in the *St. Fe notebook*. Before the voyage Darwin thought Richardson's *Clarissa* 'the most glorious novel ever written'.[340] The next jotting on the inside front cover is the Spanish name 'Gillermo' and its English equivalent 'William', in a hand which does not seem to be Darwin's.

The entries over-written in ink start on p. 2a. On p. 3a Darwin recorded the date Thursday 29 August 1833 and the note 'Very successful with the bones, passed the night pleasantly' at Punta Alta, and subsequent pages are an exceptionally readable description of his geologizing there. At this point it is worth summarising Darwin's total collection of fossil mammals as it can be very confusing working out which fossil came from which locality.

Firstly, the finds can be grouped according to the names given to them by Richard Owen, and their present day classifications. Most are edentates; these are the armadillo *Dasypus*, and the extinct *Megatherium*, *Hoplophorus*, *Mylodon* and *Scelidotherium*. There was also the 'giant rodent' later classed as the notoungulate *Toxodon*. Next there was Darwin's 'giant llama', subsequently classed as the litoptern *Macrauchenia*. Finally there is the horse, *Equus*, which is an odd-toed ungulate, and some rodents, called *Ctenomys*. The fossils mentioned at various places as *Mastodon* and *Megalonyx* seem to have been either misidentifications or very poor specimens.

Darwin found his fossils in three main regions, but in various locations within each of those regions, some of which he visited several times. He also purchased some key specimens, rather than excavating them himself. He described all these localities in the geological introduction to *Zoology* and at various points in *South America*, p. 106, where he also provided a summary of these and other fossil localities known to him.

The three main regions were, in the order described in *Zoology*, proceeding southwards, firstly: a vast area around the Rio de la Plata, including Bajada de St Fé/Banda Oriental/Entre Rios/Parana and the Rio Salado; secondly: Bahia Blanca/Mt Hermoso and Punta Alta; thirdly: Port St Julian. Simplifying considerably, the fossil genera he found from these regions were as follows: region one: *Mylodon*, *Megatherium*, *Toxodon*, *Glossotherium*, *Hoplophorus*, *Equus*; region two: *Megatherium*, *Mylodon*, *Scelidotherium*, *Hoplophorus*, *Equus*, *Ctenomys*; region three: *Macrauchenia*.

The importance of the fossils to Darwin is made clear in the second edition of *Journal of researches* (1845), written after he had formulated most of his theory of evolution by natural selection, but before he published it. He declared on p. 173 that this evidence was crucial for understanding the origin of species:

---

340 CCD1: 96.

This wonderful relationship in the same continent between the dead and the living, will, I do not doubt, hereafter throw more light on the appearance of organic beings on our earth, and their disappearance from it, than any other class of facts ...

There are some lyrical passages on these opening pages of the *B. Blanca notebook*: **'quiet little retired spot, weather beautiful & nights; the very quietness almost sublime'**, pp. 4a–5a. On p. 7a, on 1 September, he was **'wandering about with my gun & enjoying sunny day'**.

There is a very clear geological section, showing the gravel at sea level, then **'Real Tosca'**, then **'impure Tosca'**, then a **'whitish line'** and a capping of 'Diluvium', this word not inked over perhaps indicates that by the time Darwin was reworking these notes he had abandoned that term. Herbert has shown that during the voyage Darwin gradually dropped the term 'diluvium' (used by Buckland and Lyell), with its Noachian connotations, for deposits such as the one at Punta Alta which he was sure were of marine origin.[341] The section clearly shows that the 'Diluvium' is covered by **'vegetable mould Sea shells & land'**, p. 6a. Comparing these notes to those he made at Punta Alta in September 1832, it is clear that Darwin's new section is virtually the same as that published in *South America* (reproduced below).

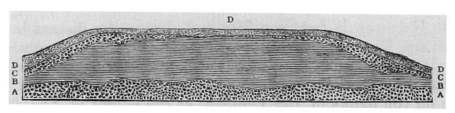

'Section of beds with recent shells and extinct mammifers, at Punta Alta in Bahia Blanca.'
from *South America*, fig. 15, p. 82.

On p. 10a he remarked **'gran bestia all nonsense: The animal of which whole skeleton was lying in pieces of stone, tolerably in proper position & imbedded in sand'**. Then on p. 12a, 4 September, **'Cruel ennui found books exquisite delight'** which relates to his reading of a Spanish book on the trial of Queen Caroline;[342] **'nobody knows pleasure of reading till a few days of such indolence.'**

---

341 Herbert 2005, p. 157 note 60.
342 See Barlow 1945, p. 196. The trial was the unsuccessful attempt by King George IV in 1820 to pass a bill through the House of Lords annulling his marriage to Caroline of Brunswick (1768–1821) on grounds of her behaviour. Caroline, who enjoyed great public support, was brilliantly defended by Henry (later Lord) Brougham, but she died only one month after being excluded from George's coronation, an exclusion which resulted in riots in London at her funeral. The *Beagle* played a role in the coronation by being the first fully rigged man-of-war to pass under the (old) London Bridge. At that stage she was a brig (two-masted). She gained a mizzen when refitted for surveying in 1825. See Thomson 1995.

On p. 13a Darwin seems to have found some whale bones, with barnacles (*Balanus*) attached, proving that the bones had lain on the sea bed long enough for the barnacles to live and grow on the bones before both were enclosed in the surrounding matrix. In such cases '**the matrix is not possibly Tosca & animal diluvial**', p. 14a. The fossil mammals, therefore, had not been killed by any 'diluvial' flood; rather they had floated out as carcasses into a giant 'proto-Plata' estuary, as happens to this day. Here were mundane natural causes still active at the present day. In *Fossil mammalia*, p. 6, Darwin described the vast numbers of skeletons which had resulted from this process: 'As their exposure has invariably been due to the intersection of the plain by the banks of some stream, it is not making an extravagant assertion, to say, that any line whatever drawn across the Pampas would probably cross the skeleton of some extinct animal.'

There are many pages of considerable interest concerning General Rosas's war against the Indians, massacres being justified 'because they breed so', p. 16a. Darwin, understandably, had mixed feelings about his trek: on 6 September he was 'Drunk from mattee & smoking from indolence & anxiety about starting', p. 26a. He saw a flint arrow head and was told that in Pre-Columbian days the Indians did not have horses, p. 27a. In the *Beagle diary* there is a retrospective entry for this date which shows he was confused about this, having found a horse's tooth with *Megatherium* remains at Santa Fé. There was much of zoological interest and on p. 28a he recorded what he was told about how the 'Avestruz Petise' (later named *Rhea darwinii* by Gould) differed from the northern rhea. A few months later, at Port Desire, this 'good information', p. 29a, came back to him just in time for him to save the remains of a specimen of this highly significant 'new' species of bird.

The next day he approached south of the Sierra de la Ventana, on his way to Tandeel [Tandil], and then to Tapalguen [Tapalqué]. The notes are mainly geological, but not entirely: '**hunted & killed beautiful fox**', p. 35a. There are some interesting sketches in the notebook of what Darwin called the 'mammillated plain', pp. 41a, 50a, and on p. 54a he wrote a brief summary under the heading 'Geology' of his arguments for the marine origin of the Bahia Blanca deposits. Darwin developed this style of argumentation far more fully in the *Santiago notebook* in entries which appear to date from early 1835.

There follows page after page of detailed description as Darwin travelled from Posta to Posta: '**Night at Sierra very cold, first wet with dew then frozen stiff**', p. 37a. The going was tough, through country that he sometimes likened to the Cambridgeshire fens: '**road rather better. like Cottenham Fen**', p. 57a; '**I have now for some days eat nothing but meat & drunk mattee: long gallops in dark. eat Lions meat was very like a calf**' p. 60a; '**wife of old Cacique not more than 11**', p. 61a; '**many quinces & peaches**', p. 65a. On the shores of a lake near

Guardia del Monte [San Miguel del Monte] he found part of a *Hoplophorus* carapace, p. 66a. Finally he reached Buenos Ayres on 20 September where he stayed with Edward Lumb.[343]

Santa Cruz, April–May 1834

The notebook entries jump to 14 April 1834, p. 69a, but the rest of the second half of April, and the first week of May, was the Santa Cruz expedition covered in the *Banda Oriental notebook*. This remaining section of the *B. Blanca notebook* is mostly in pencil, p. 82a being the exception. The notes are much less discursive than they were for the September 1833 Pampas expedition, and consist largely of lists of birds, a somewhat cryptic sketch of what appear to be concretions on p. 71a, and barometer readings, for example, for 8 May on p. 73a.

There is a curious note on p. 74a to the effect that uplift of a continent, if made of hard rocks, might be hard to detect. This and a few other entries date to mid May. On p. 78a there is a sketch which is the prototype for one of the diagrams which eventually appeared in Darwin's 1842 article on erratic boulders (reproduced below).[344]

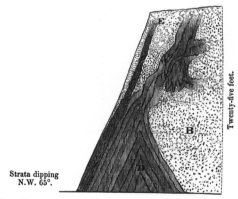

B. Yellow sandy mud.
D. Blackish mud or sandy clay.

E. A stratum which gradually disappeared in the surrounding mass; its included layers were much convoluted.

Gregory Bay in the Eastern Part of the Strait of Magellan.

'Fig. 2, which is traced from an outline [of a cliff at Gregory Bay] made upon the spot' from Darwin 1842a, p. 422.

---

343 See p. 5a and the introduction to the *Buenos Ayres notebook*.

344 Darwin 1842a; *Shorter publications*, pp. 147–62. There is a draft of this and the following notebook diagram in the geological diary DAR 34.158A.

Since the *Beagle* was at Gregory Bay only very briefly at the end of May 1834 this allows the diagrams to be dated.

The next dated entry, p. 77a, is 2 June, in Tierra del Fuego: 'splendid day. — Sarmiento appeared. — theory of views — brought savages. — Skirmish Bravery. — slings & arrows'. Barlow 1945, p. 223, linked the phrase 'theory of views' to a section of the *Beagle diary* for the first week of June in which Darwin gave an explanation for the tendency to underestimate the heights of mountains in that part of the world. On p. 81a Darwin waxed lyrical about man's insignificance in such a spectacular landscape, and mentioned the 'Niagara of ice' which seems to be the glacier depicted in Martens' watercolour engraved by Thomas Landseer as an illustration in FitzRoy's *Narrative* 1, facing p. 359 (below).[345]

Mount Sarmiento (from Warp Bay) by Conrad Martens. Published in *Narrative* 2.

Chile, late 1834 or early 1835

Finally there are six pages of jottings from the west coast, starting on p. 82a, which are difficult to date. Here is a whole new series of geological issues to consider, such as volcanoes and earthquakes, mines and fossil plants, caves and mineral springs. At this point in the voyage Darwin was a seasoned field geologist, ready to tackle some of the real burning scientific issues of the day. Top of the list was the origin of the mighty Andean Cordillera, and the link between volcanoes and coral reefs.

345 See Keynes 1979, p. 113, for the original watercolour, dated 9 June 1834.

[FRONT COVER]

[INSIDE FRONT COVER]

Charles Darwin
H. M. Ship *Beagle*

Field Sports
Prairie Ch. Grandison[346]
46
180
226
|Glass Bottles|

Sierra Ventana	3344
Above Horizontal	1837
Below	1450

W Wesos hu
Wilsoz
Wilson
Gillermo William

---

346 Richardson 1781. Volumes 3–7 of CD's copy at the CUL are inscribed 'Chas. Darwin Buenos Ayres Sept. 1833'.

[1a]   Heyque Leuvu, Falkners[347] name & now for Sauce Grande
        Is there a list of of Postas: Ventana & Spring
        Thursday 29th [August 1833] (31 days)
        Sun rises 6° 29' A M.
                sets 5° 31'
        Thursday (3)
        In a ~~fortnight~~ week 6° 18'
        Sun rises /0 1/
                sets 5.42'
        In other week (12th)
        Sun rises 6.13
                sets 5.47

[2a]   **Falkner says that the vallies which cross the Sierra Tandil open to the N or NW**
        Last point Westward of Tandeil range is called Cayru.
        West of plain between Ventana Hurtado. Casuhati: Vuta Calel & Tandeel.
        Huecufu Maqui or Sand desart.
        narrow pass west of the Ventana: also T

[3a]   **see road in old charts**
        What Spanish name of prickly fire Wood Sanquel?
        Horses **Thursday 29th** [August 1833]
        after dinner in the Yawl we started on ~~an~~ point expedition
        I staid at night at Punta Alta in order for 24 hours of bone searching
        /Mem/ the Captain Compass

        [sketch of houses?] **Very successful with the bones, passed the night pleasantly**
[4a]   **Friday 30th** [August 1833] **at noon proceeded in the yawl amidst the intricacies of the**
        mud banks arrived at Pilots House, proceeded /old/ horses & got to Guardia at 9 oclock
        Saturday 31st [August 1833]
        My Vacciano not having arrived, started to Punta Alta to superintend bone searching
        quiet little retired

---

347 Falkner 1774. An almost identical entry occurs in the *Falkland notebook*, p 139a.

[5a] **spot, weather beautiful & nights; the very quietness almost sublime even amongst mud banks & gulls sand hills & solitary Vultures: Saw here beautiful little Parus tufted partridge: common plover** ~~oriole~~ **Casicus, fieldfare rose starling very abundant.** 3. B. Blanca birds (long billed all common. NB. Avecassina was last year near Salitras

[6a]
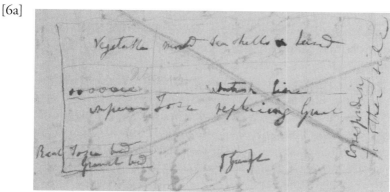

[section of Bahia Blanca] **Vegetable mould Sea shells & land** Diluvium **whitish line impure Tosca replacing Gravel Real Tosca bed Gravel bed Gravel corresponding to other side sea**

[7a] **The Tosca replaced replacing gravel in channels & abruptly**
September (1ˢᵗ) [1833]
After wandering about with my gun & enjoying sunny day: at noon started back for the Town: Saltpetre forms a thin crust over those extensive Lakes which were a few days ago with water appears like Snow & edges like drift. Is the blackish mud cause

[8a] or effect in Salina at Patagones sand fetid. ground I believe always low, first observed it on Pampas South of R Colorado: Saltpetre appears to effloresce on the surface: near town great body of horses, though[t] dust was great fire, very difficult to drive, if lion comes or even fox amongst them at night: || Toriano[348] Chilian: Indian with white horse

[9a] defeated near Churchil Toldes being there they having travelled from B Blanca — || Son of Cacique negro is now at Tandeel chanel or chanily
Rat nearly as large as common grey English rat hinder feet demi-palmated lives on edge of brooks, **Commandante has often seen & heard of large Paluda scales in cliffs Large gull often goes far inland, 40 or 50 miles**

348 Toriano, the famous Araucanian Indian chief (*Narrative* 2: 106).

[10a]   miles attends the slaughtering house, the cry the same as common English gull, when approaching rookery: gran bestia all nonsense: The animal of which whole skeleton was lying in pieces of stone, tolerably in proper position & imbedded in sand ? is this piece of Tosca washed out of ~~whole~~ old cliff or modern

[11a]   reformation, examine matrix
Monday 2ᵈ [September 1833]
Spent in doing nothing The water on plain where town stands has ~~not~~ very little saltpetre formation then is more but in valleys same The salitra in this plain has above 3 feet of diluvial sandy earth. 3 sorts of Cassicus[349]
one wanting the sparrows around the solitary Estancias as little supply the cheerful place of domesticus as the Cuorvos in /gorged/ flocks do rooks. —

[12a]   Tuesday 3ᵈ [September 1833] Harris & Rowlett started for the Boca to be picked up by the Yawl which has been surveying the head of bay: will pick up bones at Punta Alta return on board & next day the ship sails — I have since heard ship did not sail
Wednesday (4ᵗʰ) [September 1833] Cruel ennui found books exquisite delight time gallops: Spanish Edition Barcelona of the Queen's Trial[350] & ~~Shenstone Fire~~ & Spanish story book: nobody knows pleasure of reading till a few days of such indolence.

[13a]   Fieldfare. chace & catches in the air large Coleoptera. 113 40 25
NB. Muchissino Dung of Horse & cattle very deep (from earth) with holes beneath like Geotropes under horse dung. I have one specimen another from cow: omni-stercivorous.[351]
When lying on the plain the Carranchas come & ~~settle~~ soaring over you settle at about 50 yards distance and watches you with an evil eye
The black fragment of bone at Punta Alta like M. Hermoso
Balanus from Whale & pieces of Whale bone: If the bone

[14a]   of skeleton are encrusted with marine animals in those parts now covered the matrix is not possibly Tosca & animal diluvial: —
The Zorilla so conscious of its power to injure wanders about the open plain in day time does not attempt to escape dogs will not attack it except when much encouraged: froth violently & running from nose & efforts at

[15a]   sickness: often smelt on fine evening: —
/Benon/ change horn every year
Lievre have 2 youngs ones in hole; made by themselves
Thursday (5ᵗʰ). [September 1833]

---

349 A genus of birds in the Icteridae family (Orioles).
350 Desquiron de Saint-Agnan ed. 1821, a Spanish edition of the trial of Queen Caroline, consort of George IV.
351 Barlow 1945, p. 196, noted 'Perhaps Darwin's only attempt at coining a new scientific term on a classical basis;=eating all sorts of dung.'

Don Juan Leon ill. — Cannon fired to celebrate a victory obtained over Indians
in very rough mountains /bastarte arriba/ between Colorado & R. Negro 113 Indians
in all: ~~nearly~~ all taken, ~~46~~ 48 men killed, 2 Cacique; one escaped with good
horse

[16a]   one prisoner is not sure, if useful traitor that he will be killed: all women above 25 or
26 murdered: excellent authority: man would not allow it was bad but necessary
because they breed so: Orders have come from the general to send party to small salinas
as a party of 30 or ~~40~~ 50 Indians are there, only few leagues from the road. — It was not
Bernantios Indians

[17a]   who killed the Posta, but some strangers whose track leads to the Pampas. —
~~Ptera Ptera~~ <u>Long legged</u> Plover[352] have very pointed oval eggs. olive brown. with dark
brown patches at obtuse end
That the distance between Colorado & R. Negro only 12 or 14 leagues. horse travels
easily on a trot: with two fresh water lakes; Mountain very rough half as high as Ventana
only one Christian slightly wounded Indian when taken almost bit ~~fin~~ the thumb clean
off suffering his eye to be

[18a]   nearly pushed out: Sham dead with knife under skin:
It was prisoner Cacique who told of the three other Caciques being at the Little Salinas.
with only few men each. —
They recovered many Bahia Blanca horses; Don Pablo amongst others when they took
the Indians. — Partly discovered from dust of Horses. — Peons[353] want to take
horses & ∴ do not attack ~~Christians~~ Indians

[19a]   Large Maldonado Partridge is at B. Blanca: I do not know for the other —
The Indians at the small salinas which are to be attacked have on average 3 or
4 horses each some more some less: all horses, private property: divide soon after
robbery: —
Cacique with white horse battle fought near Churichoel
some of his people gave information respecting /layer/ in mountains NW of
Chundril & North of

[20a]   Colorado. amongst mountains ½ as high as Ventana could see Cordilleras, like Ventana
from Medanos. — Upon attacking these & (other Indians) they disperse in every
direction & stop neither for women or children because know it is death: Christians
200: when dispersed five Carbines once then & always trust more sword or spear: if
cannot come up use the balls which stops horse: my informer chased one man. who
cried out Compañero no matar me

[21a]   at same time could see him disengaging balls to dash his brains out sabred him & cut his
throat with knife: There were two pretty white captive girls from Salta. could not speak

---

352 Possibly specimen not in spirits 1420 in *Zoology notes*, p. 390.
353 'workmen or Peons' *Beagle diary*, p. 170.

but Indians. — other captives with Gen: Pacheka: — NB. Indians of Salta small. no
horses slings & bows & arrows

Four men run away separately, one killed, 3 others taken, turned out

[22a]   to be Indian Chascas, they were on the point of their council, great mares feast.
/Fandarja/ & next morning all ready to start: the 2 first were asked (~~all~~ the three being
put into a line) to give information about errand &c — refused. shot one after the other;
the third being asked, said as others no sé: + adding fire I know how to die: noble
patriots. not so prisoner Cacique

[23a]   His information will be important as relating to grand reunion. Chascas were coming
on to these Indians at small Salinas (Cacique gave information of these being there,
hence present expedition). The chasca were young men under 30. 6 feet high & white
very fine men had come from Cordilleras good way to the North. immense
communication from thence to Salinas at B. Blanca —

[24a]   Grand point of reunion supposes 5 6 or 700 then nor will be in spring about 1500:
present system kill wanderer, drive to centre for great attack: — one of killed Cacique.
Chilenian Toriano sons: all storyes about Toriano true. taken by commandante alone:
Indians treat Christians; just same as treated, all with beards killed: great consternation
will be at Cordilleras

[25a]   Chascas killed: (only one christian wounded). — Sandstone beyond Chundril few. —
These Sierras N of Colorado Travel towards Sierra hard primitive rocks, (granite?)
grey coloured: Plains between Colorado & Negro. similar with Spines &c. —
Fine camp about Chundril on borders of river: Women taken at 20 years old never
content:

[26a]   Tehuelches very tall, informer talked of them in strongest term looking up to ceiling.
generally one foot more than me: Friday 6th [September 1833] — Drunk from
mattee[354] & smoking from indolence & anxiety about starting: constant reports about
Indians made important by firing guns: air hazy. thought gale was coming one they say
fire far in the camp; now so

[27a]   dry: —
Saturday (7th) [September 1833] —
Saw piece of opake-cream-coloured flint; remains of arrow head. had barbs ∴ not
Chusa.[355] — twice as large as Te de Fuego; often found at Churichol Indian
antiquarian; before horses & balls changed manners: — No horse Indians have arrows:
Do not fight with all only Chilenians & Auracarians. When Cacique with white horse
~~was killed~~ escaped between 20 & 30 ~~with him~~ killed:

---

354 Mate: an infusion prepared by steeping dried leaves of yerba mate, containing stimulants including
     caffeine.

355 'The only weapon of an Indian is a very long bamboo or Chusa ornamented with Ostrich feathers
     and pointed by a sharp spear head', *Beagle diary*, p. 170.

[28a]   Saw in Pulperia[356] a boy Empeño. sent by Bernantio (resident friendly dealing tribe) as pledge for some spirits for a dance: Yet these friendly Indians it is not considered proper for one or two to go by themselves. — Has son Christian Teniente, educated by General Rosas Avitruce Petisses[357] frequent sea side. South of Colorado overo. feathers same structure body & neck & head similar legs rather shorter, covered

[29a]   with feathers to claws. has sort of fleshy 4th toe without claw. — eggs a trifle smaller: Head with scattered hairs. cannot fly. — good information. — my prospects are now better, gracias a dios start in the morning — hazy air
Sunday 8th. [September 1833] — **Started 1st Posta about 4 points west of Sierra, 4 leagues**
Layer of friendly Indians
B 350,5: high camp with diluvial sandy earth covering sloping valleys: new sort grass. no spring /plants/ bushes

[30a]   **First saw ridge behind ridge of Sierra. Only one point of view. Near Posta** good salt **Salina in depression small: & Salitras in depression but far above level of sea. — I have formerly guessed at height of plain: was told first could see Sierra Ventana, 2 leagues N of Colorado. — with Indian put hand to head & grunted in same manner as near Wallechu ∴ what for?**

[31a]   **/Misar/ Sierra: — 2d Posta 6 leagues supposed direction of town. B 237. Ventana B 325 Boca B 165? Salitra at last Posta above diluvium with bits of mortar. —**
Plain I should think 200 or 300 feet above level of river; river deep rapid
12 leagues to Boca & 6 8 to Sierra: not good information
R. Sauce about 20 feet across deep banks. with much Sauce impassable except with many f w. shells

[32a]   **in two places & mouth good: against Indians Met on the road great troops of cattle & horses, many lost: 15 soldiers a short time ago officer came, with 500 horses all lost except 6: afterwards other troops soon perceived by bare** hos **heads & long streaming hair they were Indians: going to Salinas for salt Eat salt like sugar curious difference with Gauchos, habits so** different **similar**

[33a]   **Near the Sauce Posta Turnips doubtless like** *[illeg]* **different I think from European sort longer more stringy & acrid; variegated thistle: Plenty of Lievres, I was here told by B. Ayres man that they only use Biscatchas holes, is it not that these parts where Biscatchas are plentiful, they then do not make holes; hence different account:**

---

356 A 'drinking shop, which also sells a few other things', *Beagle diary*, p. 155.
357 Avestruz petisse.

[34a] Was told, that in Sierra Tandeel in 3 months 100 Lions were killed, from being destructive to calves, they kill colts or small calves by turning back & breaking necks, good information: When I arrived at S. Posta procured horses & started for Sierra, had difficulty in finding water because streams bury themselves & not even wood enough

[35a] for Asado, ~~har~~ & therefore half spoiled, hard to find any place so barren; on road hunted & killed beautiful fox & armadilloes: immense numbers of deer; few Guanaco: Plain level with longish brown withered grass; vallies or quebradas rich, valley of Sauce two miles broad with Turnips & fertile: Sierra rising from plain, without any trees or gentle rising (or very little)

[36a] has strange solitary appearance & deserves the name of Hurtado:[358] The plain abuts against & amongst the hills & it is remarkable that above a mass of ~~small~~ little rounded fragments, there was a bed with coloured lines as near the B B. of mortar with large pebbles: the plain as far as I can judge has about 12 to 20 feet of ~~gritty~~ calcareo-argillaceous ~~mortar~~ siliceous rock without pebbles In one place an inter mediate stony

[37a] bed. These from one place earthy Tosca plain very same as near B Blanca. I can hardly tell whether mortar above pebbles is same as main plain if so very modern which agrees with remarkable absence of quartz pebbles on plain near Sierra. Night at Sierra very cold, first wet with dew then frozen stiff, water in Caldera;[359] about quartz rocks; see a gaunt against parts of vicinity of Sierra

[38a] **Monday (9ᵗʰ)** [September 1833]
In morning started early & proved MacKintosh by carrying water to summit,[360] reached summit of highest part of back difficulty of ascent from ~~great~~ immense numbers of steep abrupt vallies, general character of Sierra; sides steep: when on top was obliged to descend near camp to horse pass; saw horses thought they were Indians: having thus lost much time started again & with immense labours & cramp from new muscles

[39a] being brought into action arrived at a point ~~on~~ nearly as high as summit he about one hours walk from it; not good ~~not~~ to proceed, ~~few~~ view hazy. dry clear, could not see sea; ascent unfortunate might have come to foot on ~~foot~~ horseback: The universal rock is quartz generally with laminae sometimes coloured, also little micaceous slate where mica is not visible occasionally cleavage tortuous: stratification like cleavage runs (true) NW by W & SW by E: hills about NW & SW ∠ generally 45 to SW, or more ∠

358 Separated.
359 Spanish for a large crater.
360 'The ice which in many places coated the rocks was very refreshing & rendered superfluous the water, which I actually carried to the summit in the corner of a cape of the Indian-rubber cloth.' *Beagle diary*, p. 184.

[40a] **on the first few hundred feet above plain the rock is covered with hard breccia as sometimes is seen on coast base ferruginous former sea? Cleavage connection with T. del F: Vallies on Sierra remarkable for size & numbers effects of sea, & then more wet than now dry climate: now look like land islands: to S. few low hills in same plain: N mammillated**[361] **plain:** (Biscatches good to eat. meat white:) contradiction extreme ruggedness absence of detritus

[41a]

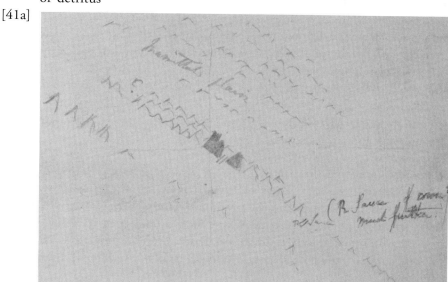

mamillated plain R Sauce (**R. Sauce of course much further:**)

[42a] **It is hard to understand how a bed of solid ½ semi-crystalline mortar, should be formed above gravel & not be contemporaneous with great plain formation yet in** the **former fragments of quartz: I incline to think it is different because in other gravel bed, the with land shells, there was** rounded **fragments of mortar, yet bed somewhat resembled true mortar: Many new plants (some same as M Video) &**

[43a] **birds & much Ice on rocks in middle of day on S side: Returned tired & rather disappointed having traversed so much, nothing to see except pure quartz: yet scene novel & little fear like Salt made fire fearlessly: Hurricane night, much mattee & smoking good for cramp: very much fighting in Sierra Chinas run on top & fought with big stones: To the west**

361 Covered in numerous small hillocks.

[44a] Sierras with one road & water at top, Indians drove horses up & Christians although more than a thousand, could not touch them. On Sierra white Shrike & Woodpecker:
(Young Toco Toco all say so: Gato Pajero,[362] (of straw))
Tuesday 10th [September 1833] —
Riding slowly & looking for rocks, returned to Posta at Sauce: scudded before wind: R Sauce travels to the north of Ventana into

[45a] the Interior: has floods in middle of summer & floods now from Rain & ~~rain~~
Falkner true: commendante son never was prisoner.
I was to have waited for an officer: letter from General Rosas but good opportunity with Post-Master. so started
Wednesday 11th [September 1833] —
Pass of Sauce
Bearing to 3d Posta: 46 (good)
to Sauce Posta 75 ?
to Ventana 310.7
Distance from pass to S Posta 2 ½ L[eagues]

[46a] distance of Sauce Posta to 3[d] Posta 15 leagues but said to be 13 in direct line: before & at Pass the river crosses the silex formation of Ventana & the hills gradually lowering are seen stretching away for few leagues to B100; Quartz here pure: at pass Barranca of pure reddish soft Tosca: *[illeg]* Amongst the mamillated hills to our left (for they do not extend to East) pieces of square land showed remaining

[47a] of great camp — formation: whole camp with stony mortar & same vegetation: Rather stupid ride, crossed a rivulet half way R del Indio Rico /Fuentes/ Sauce Gaucho talent of observation on one side of steep hill showed me his house distant about 30 miles: in 1 & ½ mile from bank with immense horizon again showed me by Kater[363] only differed by 3° in the direction in which he said it might be little more:

[48a] arrived at 3[d] Posta. —
Thursday 12th [September 1833]
I waited here for the officer: & rode to the Sierra south W of the Posta: — This Sierra is composed chiefly of Feldspathic rocks V. Specimen with specks of mica, which in upper parts passes into slate generally pale, or reddish purple, not fine or much laminated: The Sierra extends but

---

362 Specimen not in spirits 1443 in *Zoology notes*, p. 392: 'lives amongst the thick straw at Bahia Blanca, also found in Banda Oriental.' See *Felis pajeros* in *Mammalia*, pp. 18–9, plate 9.

363 A Kater compass, named after Henry Kater (1777–1835), physicist and instrument maker.

[49a] **little to the East of Posta but stretches far to NW. — it is here narrow but gradually widens by succession of short parallel chains. — The cleavage & hills run (true) NW & SE or NW by W & SE by E & dip Southward.** 3^d Posta 2^d ... or vertical

[diagram of hills]

[50a]

[Pass of Rio Sauce]

3^d Posta Camino Pass of Sauce

The rock is very similar in different places: & hills strikingly resemble those of Maldonado:

[51a] **The stream at foot is called Guitru-queque & entro Guegen. — (Saw beautiful oriole.) The Sierra as seen from north shows four points gradually lowering, /mine/ the third is but very little less in height**

Description of **Post**a: miserable shed open at both ends, of straw: about 50 horses & sometimes a cow, game: thistles for road, 15 leagues from one & now (after murder) 30 from other: only amusement smoking & mattee: barren <u>looking</u> long withered grass camp: Partridge

[52a] **in evening like frogs: few Vultures — watching for them to be killed: too much appearance danger (like salted meat) dog bark jump up: Pteru-Pteru** when playing at cards by firelight **cry heads inclined horrible looking men:**

Game at balls distance 35 yards but about one in 4 or 5 times: <u>could</u> throw them at between 50 & 60:

Carranches do not run like Cuervos[364] eggs in cliffs. colour oveiro **cry like Spanish G & N: at Sauce saw other sort, legs & bill blue, feathers light brown. except crown of head & eyes darker.**

[53a] B Friday 13th [September 1833]

Bird called chusco. lays in sparrows nest: ostrich 4 or 5 in one nest proved by /bet/: run against wind as well as deer: Ostriches lie hid amongst straw: noise of ostriches

Belly of Peecheys with various Coleoptera & Larva: Foxes in immense numbers: Lions never roar even when taken, catch deer by day; saw young Viscatcha half eaten, live amongst straw & in holes: immense number of Aphodia[365] (V. Specimen but not about horse dung, although plentiful: Went out hunting. no sport. pleasant gallop: 2 sorts of partridges: Barranca at Sauce Pass diluvial: at 3[d] Posta do.

[54a]   Geology

That the mortar S. of B Blanca similar to great camp formation & even that of great Escarpment. — That the mortar owes its lime to same cause as beds on Sandstone: that it is cotemporaneous with Porphyry pebbles: that it overlies talus of S. Ventana: nothing has happened since from absence of superior pebbles but /in/ the south time for few similar shells that of short duration from no organic remains: that Tosca is beneath it from great Escarpment: M. Hermoso (& Sauce Posta ?) impossible to ascertain: that where it does not abut against Sierra horizontally is owing to modern diluvium: Formed of course beneath sea:

Saturday 14th [September 1833] —

Leaving long chain of Low hills stretching out of sight to the

[55a] **NW; & mortar formation at foot great swamps: then flat plain like sea; could not guess formation slept in camp, at sunset having started at ½ after 12 oclock.**

(Mataco & Paludo at St Jago) Supposed Bearing to 3[d] Posta 215 15

Domingo 15th [September 1833]

First saw great cranes: Carranchas & Pteru & Pteru eggs: latter sham death: (late mem) like peewit. — **whole road with many swamps & good long grass: arrived at middle day at 4th Posta: to having passed 4th where men were killed:**

[56a] **From this 5th to 3[d] Posta: said to be 30 leagues perhaps 20: — From 5th Posta mean bearing of many fire made in our track between this & sleeping place was 208: Even in this road dry islands in swamps hard mortar: Cranes carrying bundles of Rushes. (Man who was killed at Posta 18 wounds) Fish in shallow water**

Long legged plover cry like little dog mounting bank numerous not inelegant eggs like Pteru Pteru **many snakes with black patches in damp swamps 2 yellow lines & tail red: also specimen**

---

364 Spanish name for the black vulture, also known as the Gallinazo; listed as *Cathartes atratus* in *Birds*, p. 7.
365 Specimen 1492 in *Darwin's insects*, p. 76.

[57a] **21 men in this Post: hunting 7 deer 3 ostriches 40 eggs Partridges & ~~armadillo~~**
Peecheys: mortar **slept in open air**
Monday 16th [September 1833] —
Road to 6th Posta: black peaty plain swamp with long grass heavy riding near Posta
mortar **Island. Posta near lake enlivened** by **many black=necked swans & beautiful**
ducks & cranes:
To 5th Posta 205 } B supposed
To Pass of Sierra Tapalken
47
Distance 5 or 6 leagues — to 6th Posta 5 leagues, road rather better. like
Cottenham Fen: large **graceful soaring flight flocks of glossy**

[58a] **Ibis:** Mortar. **at South foot of Sierra Tapalken: Last night remarkable hail storm**
(20 deer hides ~~ostriches~~ **already found dead & about 15** ~~deer~~ **ostriches saw hides &**
flesh of ostriches: man who told me had head tapped up from blow: Carranches
also killed Ducks: eat Partridges, with black mark on back [hailstones] **as big as**
apples broke Corrall: Molita & few Hares: ostriches now *[illeg]*

[59a] **Direction of Pueblo Tapulken from Sierra B 7°**
6th to 7th Posta 10 leagues (short ones) camp very fine black mould. much mortar.
north & south of Sierra Tapalken. close to foot: Sierra table pieces, granular
quartz general white nearly horizontal strata dipping to NW ??? general range E &
W: only one road up to this little table: was assured rock was as hard in Corrall
~~near Tandeel~~ **at Vuulean only reddish**

[60a] **Gauchos seek white pieces for striking fire. I have now for some days eat nothing**
but meat & drunk mattee: long gallops in dark. eat Lions meat was very like a calf:
thought with horror they were eating young calf not born: — very curious if great
Table land of Falkener Vuulean is quartz.
informer says Barranca 30 or 40 feet high:
Tandeel Bears 119 from this Posta (supposed)
Tuesday 17th [September 1833] **fine fertile camp**
7th to 8th 8 leagues by

[61a] **the side of Rio Tapalken pass it two or three times considerable stream at Town of**
Tapalken: bought biscuit. curious plain covered with horses & Toldos:[366] **always 2**
Chinas[367] **on horseback: wife of old Cacique not more than 11: Pulperias: Black-**
headed gulls, breed in Fens: Ostriches lay eggs in middle of day: black & white
Fly-catcher & long tail bird at bank of River Pure mortar (like Hermoso
concretions, with manganese & called Tosca) causes many small rapids, it lies in
horizontal beds above. —

366 Dwelling made by Patagonian Indians from animal (usually guanaco) skins and wood.
367 Indian women.

[62a]   palish Tosca which latter contains nodules of a darker Tosca V Specimen — All the
mortar I have seen doubtless is the same as this: & therefore intimately connected
with M: Hermoso, these nodules, there a bed:
8th to 9th Posta all to East of River; country & Fen
There are not here Hares or Peecheys more owing to country than Latitude for
Peecheys occur to NW: (Some of Rondarios monks are now officers; Indian with
sleep, families of

[63a]   men at Colorado: families beautiful
9th Posta 6 leagues East of R. Tapalken
Wednesday 18th [September 1833]
9th to 10th Posta 8 leagues Great turn to East swamp & Camp: 3 rivers enter into
Laguna Des aguadero, with Barrancas & Peecheys:
10th to 11th Posta. — 8 leagues galloped in 2° . . 50' : flock of golden Cassicus:
Little Mortar
11th Posta first Estancia some ~~white~~ white Salitras. — Camp moderately
good. —
passed Indians & Chinas going to trade with Yergas[368] to Monte. like English goods
wool

[64a]   *[illeg]* ~~11 rings in tail then scales~~: Paludas Noctural: No Peecheys:
11th to 12   3 leagues
12 to 13   6 leagues
Much water: crossed the Salado
~~General Rosas Estancia~~
~~Thursday 19th~~
about 40 yards wide: very deep crossed in canoes: The banks Mortar & whitish
clay: arrived at dark at 13th Posta. General Rosas Estancia. very large long building
fortified: 900 faragos maize.

[65a]   immense herds 74 square leagues: 200 Peons, formerly safe from Indians:
furniture almost outside Estancias yet best of all Thursday 19th [September 1833]
13th to 14th Posta 4 leagues started early to camp. level with Clover & Thistles
in great beds. Viscatches hole. Camp like B. Ayres : & Fennel near the Guardia :
Arrived at Guardia at 9 oclock nice scattered small town with many quinces &
peaches

[66a]   ~~in Blossom~~: passed by great lake & other near town: with cliff about 4 feet reddish
Tosca — with ventral stalactites of Mortar: vesicular & paler Tosca in
concretions. — I found plates on beach evidently near proper place as Tosca shows
also many of the bone fragments of rotted bones not worth bringing

368 Cloths woven of wool by Indian women.

[67a] **14th — to 15th 6 leagues: first saw the Acanthus very few tufts**
15th to 16 5 leagues
16th to 17 6 leagues
Much Rain: many Estancias on horizon & cattle marked by Ombu tree[369] ~~Ombu~~
Slept at house of half Indian Said no at first from Robbers repeated everything I said:
Friday 20th [September 1833]
15 to 16th 5 leagues a Much Acanthus
16th to 17 do fine camp
Approach to B Ayres: very fertile Olive Agave hedges. Willows in leaf *[illeg]* &
Pantanas : —

[68a] **Saturday 21st** [September 1833]
Various Business
Paludas & Matacoes at Cordova
at 15th Posta 40 yards deep red Tosca — nearer the city harder pale Tosca exactly
resembles that of Guardia del Monte
60
 6
———
360
of above 400

1408

[69a] April 14th: — 1834
Falkland Lark here (double band Kelp bird here) I do not think true B Blanca bird
comes further south than R. Negro:
Black & Brown bird St. Julian bird finch no further ~~than~~ north than St. Julians:
rises & utters a peculiar noise when doing this flight peculiar soaring

[70a] Condor secondaries
East range of Kelp
Kelp bird not present where Kelp absent
Great numbers of Bathengas & Apercas & Pumas
Short billed snipe
Little Hawk female
T. del Finch male
St. Julian Finch female:

369 Ombú: herbaceous tree of the Pampas *Phytolacca dioica*. See specimen 839 in *Beagle plants*,
   p. 164.

[71a] Long (R. Negro) tailed bird — male
Wren female

[diagram which may show the concretions referred to in *Banda Oriental notebook*, p 47 'at ship'] 3 or 4 feet largest 2 or 3 yards

[72a] Little mouse (male)
~~5 drams 59 gr~~
5 dram 29.81 (apoth)
Gerboise (male)
3 oz — (24 apothecary
<u>Avecasina</u> female
Young T. del Finch less brilliant. head less blue back less green belly more dirty orange
437.5
　　3
‾‾‾‾‾‾
1312.5
　24
‾‾‾‾‾
2336

[73a] Red-throat & red tail creeper both males?

60　　　534
　5　　　621
‾‾‾‾‾‾‾‾‾‾‾
300　　　913
　29
‾‾‾‾
329

May 8th　—　　　　　　　　　　　*[401]*
AM
9.50　　30.556　　　　A 43[370]　D 35

---

370 This and similar tables in the *Beagle* notebooks record barometer readings. 'A' = 'attached', i.e. the thermometer attached to the barometer. 'D'= 'detached', a separate specialized thermometer.

10	— .155	do	D 34.5
	N & S edge of plain		
12	30.024	A 48.3	D 39
12.30	29.763	A 49	D do?
1.7	29.621	A 48.5	D 41
5	30.512	A 44	D 39

556
512
44
556
22
534

[74a] Gr. part of coast degrading

proof shape of continent where hard rocks in present Hence small elevations would leave no signs behind

35 / 25.000
     135
    1150
    1080
     700
18 / 36 36
36
18
288
36.
6.48
160
135
25
1350
5 / 36
  .57

Patella[371] Voluta[372] in Plain

[75a] on Board

80
 6
(480)

---

371 The limpet, a genus of gastropod.

372 A genus of neogastropod snail, common in tropical seas. It was a volute shell said to be from a gravel pit near Shrewsbury which in 1831 Adam Sedgwick so 'utterly astonished' CD by declaring without hesitation must have been 'thrown away by someone into the pit'. *Autobiography* p. 69.

9. AM.	30	214
12.10	—	203
PM 4.40	—	176
	—	.038

May 13th <u>noon</u> Lat Long?
Lat 51° 47' Long 3°. 25' E S. Cruz
135 mile from shore
May 14th 52° 8' Long 3° 55' E
— 15th Lat 52° 28 Long 1° 36' E
(1.18 ~~inch~~ yard in a mile)
   2
63.60 70
23.60

[76a] /in/ 18. 3.4 /ft/
oval — flat /topped/. —
with /corner/ like Cordilleras Rock (another twice as long at furthest point ∠ n 65 w
(comp) 55

[77a] June 2^d [1834] Adventure arrived. — splendid day. — Sarmiento appeared. — theory
of views — brought savages. — Skirmish Bravery. — slings & arrows Guanaco
(Anecdotes about Hawks)
~~Chimango~~[373] ~~at P. Famine~~
Adventure bad times at the Falklands

[78a]

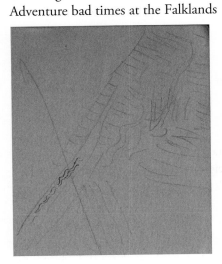

---

373 One of the carrion-hawks, smaller than the carrancha; listed as *Milvago chimango* in *Birds*, pp. 14–5.
    CD notes in his *Zoology notes*, p. 161: 'if a person lies down in the plain, one of these birds will soon
    appear & patiently watch you with an evil eye.'

[Gregory Bay in the Eastern Part of the Strait of Magellan][374]

65° ~~W N W~~ NW by W

[79a]   1.18

         3
      ─────
      35.40

         3
      ─────
      106
      418
      ─────
      524

[strata at Gregory Bay]

[80a]   June 2^d. [1834] —

The most NE point of Clarence Island does not look like Slate. —

hills about M. Tarn N̲W̲ ??

8th [June 1834] Bad day for all but Sailor; curious scenery constant dirty cloud driving clouds peeps of rugged snowy crags: blue glaciers: rainbows squalls — outline against the lurid sky: not ~~fit for residence~~ of has no claims no authority here

[81a]   man. — How insignificant does wigwam look — /The/ Fuegian man does not look like ~~man~~ the lord of all he surveys —

~~Sarmiento man~~ The inaccessible mountains wider power of nature despise for control seem to say here we ~~reign~~ are the sovereign. —

9th [June 1834] Solemn stillness of peaks: gradually unveiled: saw whole height, 7/8 snow & Niagara of ice: —

Failed in getting anchorage

[82a]   Caves. F. W. Fish & Huevos de los Gigantes. ~~T~~ Coal vegetable impressions. — organic remains. — Earthquakes, times, nature of undulation. affects on building: wave cracks, Springs: Mineral Springs: effects on neighbouring Volcanoes

374 See the fair copy of this and the following sketch in the geological diary DAR 34.158A. See note 344.

Road to Valdivia; Englishmen there? . — Concepcion country? .
rise in ground at same time in Valparaiso? —
Postman where lives &c ?? —
Mine? Slate? Granites? Limestone? —

[83a] ~~Estacillos~~ Estacillos   Volcanic Dust
Cheucau[375]   Earthquakes not felt at Valparaiso Lima
Granite   Birds &c
Volcanos
Castro
Coal   Nature of Breccia
Apple trees   Geotropes.
Cave 10 ft. —
Lice: —

[84a]   Valparaiso
SSE to S½ (& one to West of South.)
Angular patch of Syenite Black fine gneiss —
Masses of Feldspar & Quartz
Green veins
Hornblende long crystals

[85a] Old buildings
Fort which could not anciently be seen
Water bursting out through the ground
9° 30'   30.466   A D 44 A 53
10° 30'   29.742   A D 49 A 55
11   29.356   50 ?
Flag staff
12   29.008   D 53

[86a]
1.45   29.05 Trochus[376] Patella &c
2° 15'   28.72   A 56
D50 Top of highest hill
2 40'   28.684 very highest
4 PM   30.450

[87a] Sandy gneiss SSE dip ~~N. W.~~ —
Veins SE & NW

[88a blank]

---

375 See specimen 2127 in *Zoology notes*, p. 235; listed as *Pteroptochos rubecula* in *Birds*, p. 73. See also
Journal of researches 2d ed., pp. 288–9.
376 Top snails.

[BACK COVER]

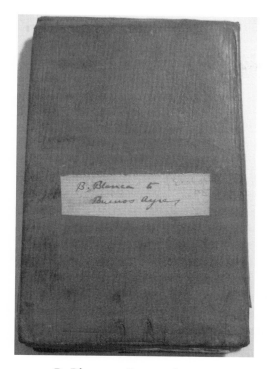

B. Blanca to Buenos Ayres

[INSIDE BACK COVER]

C. Darwin

From Port to /Posta/ (100) B 215 to 3ᵈ P

Guitru— gueýu.

Cusca places eggs in other birds nest

qʳ Shoal off C. Corrientes. ESE. 5 or 6 miles. —

[1b] qʳ Mouths of rivers & Indian names between Monte Hermoso and Cape Corrientes. — distances from either or Each place — do they open into the sea — or are mouths closed generally by sand banks — how wide — how deep — ever Entered by boats or vessels — marks — approach — ?

Bˢ. Ayres

Mʳ. G.[377] journal — Rio Negro — Villarino[378] — Chart of Bahia Blanca — how used? — French Survey of river — (Emulation)[379] M. Barral — ? Naut. Almanac 1834[380] — Letter — under cover Mʳ G. —

377 Philip Yorke Gore.
378 Basilio Villarino (1741–85), Spanish naval explorer. See *Narrative* 2: 314.
379 *L'Emulation*, a French survey ship.
380 Anon 1833.

[2b]   Salitra. South of R. Negro?
       Ostriches
       St Josephs expedition
       Meeting of Caciques
       Up the river
       Rincon Gordona Estancia
       Don (Lorenzo) Many great bones.
       Washing
       Poncho big Bag
       Bread (12?) Sugar Yerba
       2 packets of Cigars
       Salt
       20 Pesos
       7|350
          50
          58
       5|108
          21

[3b]   Accounts
       Harris[381] 59   Place of rest
       Captain 80   (339)
       Stokes 10   180
       Thistles Turnips Lievres   159
       much cattle loss
       Indians on road
       Big Bottles: Note Books

[4b]   Paluda[382]

[paluda paw] 4 5 1 2 3 inside

1.2. middle ~~toes~~ nails equal grand flat first with large ball — 3 long narrow:

---

381 Possibly James Harris sealer of Del Carmen on Rio Negro who acted as pilot to one of the *Beagle*'s hired attendant vessels, *La Paz*.

382 A species of armadillo, listed as *Dasypus villosus* in *Mammalia*, p. 93. It is now known as the Large Hairy Armadillo, *Chaetophractus villosus*. See *Zoology notes*, p. 180.

4th like 1 & 2 : 5 shorter hind ~~legs~~ leg similar. all nails shorter: belly with rows of stiff hair: back with 8 bands (& 9 soft): tail half length of body: long hairs on back: 9 teeth in upper jaw: 10 in lower: 3 lines <u>as big</u> as Peechey

    18. 20

[5b–6b blank]

[7b] Head man C. Cangallo 98

    M Gore

    Margrave / [tear]

[8b blank]

[9b] Mem: [Capt] King Cat:

    Mem Wickhams³⁸³ shells:

    Lowe. Kelp. on S. Islands: Western N. limit: —

    Water communication between Valdivia. Concept: Valparaiso

    Southern birds

[10b–16b blank]

Textual notes to the *B. Blanca notebook*

  [IFC] 4] *added by Nora Barlow, pencil, not transcribed.*

           W … William] *not in CD's handwriting.*

    [3a] Horses] *pencil.*

    [6a] *page written perpendicular to the spine.*

    [9a] lives on edge of brooks] *ink.*

  [11a] The salitra in this plain has] *ink.*

  [12a] — I have since heard ship did not sail] *ink.*

           ~~Shenstone Fire~~] *cancelled in ink and pencil.*

  [13a] very deep] *ink.*

  [15a] Thursday … horse] *ink.*

[16a–28a] *pages in ink.*

  [29a] with feathers … morning —] *blackish grey ink.*

           Sunday … bushes] *blackish grey ink over pencil.*

[30a–52a] *pages in blackish grey ink over pencil.*

  [37a] bed] *blackish grey ink.*

           about quartz rocks] *blackish grey ink.*

  [39a] horseback] *blackish grey ink.*

  [40a] (Biscatches … detritus] *blackish grey ink.*

  [41a] *page written perpendicular to the spine.*

  [42a] *two ink marks appear to be nib tests.*

  [48a] *three ink marks after '3^d Posta. — ' appear to be nib tests.*

  [49a] 3^d … ~~vertical~~] *pencil.*

---

383 John Wickham (1798–1864), First Lieutenant of the *Beagle*.

[50a]   The ... Maldonado:] *blackish grey ink over pencil.*

[52a]   when playing at cards by firelight] *added ink.*

       eggs in cliffs. colour oveiro] *added blackish grey ink.*

[53a]   *page in blackish grey ink.*

       noise of ostriches] *added ink.*

       Barranca at ... do.] *added pencil.*

[54a]   Geology] *pencil.*

       That ... sight to the] *blackish grey ink over pencil.*

[55a–68a]   *pages in blackish grey ink over pencil.*

       (late ... peewit. —] *blackish grey ink.*

[63a]   families] *added ink.*

       like English goods] *added ink.*

[67a]   Slept ... said:] *added ink.*

[68a]   *small ink marks after 'Cordova' appear to be nib tests.*

       60 ... 400] *blackish grey ink.*

[72a]   437.5 ... 2336] *in ink.*

[73a]   60 ... 329] *ink.*

[74a]   Gr. ... behind] *written perpendicular to the spine.*

[76a]   (another ... long] *added ink.*

[82a]   *page in ink.*

[83a]   Geotropes.] *added ink.*

[BC]   8 Vol of [DD]] *not in CD's handwriting. not transcribed.*

[IBC]   1.11.] *Down House number, not transcribed.*

       88202331] *English Heritage number, not transcribed.*

       C. Darwin] *ink.*

       q^r Shoa ... 6 miles. —] *in FitzRoy's handwriting.*

[1b]   *page in FitzRoy's handwriting.*

[5b–6b]   *lower half of leaf excised.*

[7b–8b]   *lower half of leaf excised.*

[9b]   Lowe ... Valparaiso] *ink.*

THE *ST. FE NOTEBOOK*

∽

The *St. Fe notebook* takes its name from the town of Santa Fé in north eastern Argentina, near the junction of the Paraná and Salado rivers, opposite the city of Paraná. The notebook is bound in brown leather with an embossed border: the brass clasp is intact. The front cover has a label of cream-coloured paper (68 × 25 mm) with 'Buenos Ayres. St. Fe and Parana — Cordillera of Chili' written in ink. A piece of woven cotton string is tied around the hinge of the clasp; it has two knots at its outer end and one knot in the middle. This might have been used to secure a pencil. A piece of fine string, c. 20 mm long, is stuck through a hole in p. 1a. The string presumably secured the pages through the facing pin holes to p. 60a. The notebook has 119 leaves or 238 pages. It was written in two sequences, pp. 1a–238a and pp. 1b–5b. The entries relate to 27 September–21 October 1833, 2–6 November 1833 and 18 March–20 April 1835.

St. Fe is one of the most interesting of all the *Beagle* field notebooks. Indeed, in some ways it can claim to be the most precious of them all, as it spans what was perhaps Darwin's geologically most prolific period of the voyage, and the one during which he committed to a Lyellian, or essentially gradualistic, view of the geological history of South America. By the time he had crossed the Andes he had seen proof of their complex vertical oscillations, to be measured in thousands of metres over vast periods of time. The notebook is also significant in that it contains, in the 1835 section, some remarkably lengthy entries, in fact those written around 1 April are possibly the longest daily field entries Darwin ever wrote. Some of these entries were obviously written when Darwin had leisure to reflect on the day's observations and it is here that one starts to see the emergence of Darwin the habitual theorist.

After the voyage Darwin was recognized as an authority on the geology of the Andes and he was confident in 1846 in asking rhetorically 'how opposed is this complicated history of changes slowly effected, to the views of those geologists who believe that this great mountain-chain was formed in late times by a single blow'.[384] Introducing his account, Darwin wrote with characteristic modesty: 'Considering how little is known of the structure of this gigantic range, to which I particularly

384 *South America*, p. 248.

attended, most travellers having collected only specimens of the rocks, I think my sketch-sections, though necessarily imperfect, possess some interest'.[385]

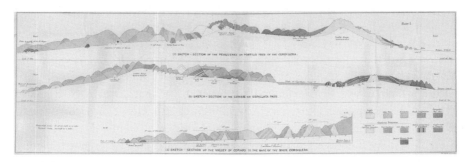

Plate 1 from *South America* showing geological sections through the Andes. The first two show Darwin's southern and northern traverses between Santiago on the left and Mendoza on the right (*c.* 200 km, 120 miles), the third showing the Copiapò [Copiapó] valley in northern Chile (*c.* 100 km, 60 miles).

Thus Darwin justified publication of his classic geological sections (above) through the Andes in *South America*, which have been shown by subsequent research to be remarkably accurate.[386] The original field sketches, drawn on the spot, come from the *St. Fe notebook*. The notebook sections only needed to be 'stitched' end to end with minor adjustment to form the published sections.

Two of the most important field sketches from the *St. Fe notebook* are on pp. 125a and 146a.[387] These two are instantly recognizable as the left and right halves of Darwin's published section no. 1. The published sections are coloured. Several pages of the *St. Fe notebook*, around p. 195a, are smeared with watercolour paint of exactly the colours which appear in the published sections. So around 1845, when Darwin was preparing his fair copy sections (such as the one in DAR 44.33 reproduced in Herbert 2005, plate 7) he apparently tested the watercolours by dabbing his brush on the *St. Fe notebook*.

The first part of the *St. Fe notebook* covers expedition no. 3,[388] Darwin's 1833 trek from Buenos Ayres to Santa Fé and Paraná and his return by boat. The second part of the *St. Fe notebook* covers expedition no. 7, Darwin's great 1835 circuit across the Portillo and Uspallata Passes of the Andes, which he labelled prosaically 'Cordillera of Chile'. In fact most of the highest mountains are in Argentina.

The *St. Fe notebook* presents the most dramatic contrasts between the many aspects of geology and natural history which Darwin encountered on both the

---

385 *South America*, p. 176.
386 See Morton 1995. The section is reproduced as a fold-out colour facsimile in van Wyhe 2008a, p. 39.
387 Reproduced in Barlow 1945, plate II.
388 We refer to the expedition numbers used by Barlow 1933 and her map reproduced on p. iv.

Atlantic and Pacific margins of South America. It is significant also as a record of his health, as it covers the episode, from 2 October 1833, when he had an illness sufficiently protracted to force him to alter his travel arrangements. His plan was to ride from Santa Fé to Monte Video across Entre Rios and Banda Oriental, but by 10 October he abandoned this idea in favour of taking a balandra (a type of barge) down the Río Paraná to Buenos Ayres. The *St. Fe notebook* also records how on 26 March 1835, when in the high Andes, he was bitten by the Benchuca [Vinchuca] bug from which it was first suggested by Adler 1959 that Darwin might have contracted Chagas' disease, although Keynes 2003, p. 284, cites convincing reasons for doubting this. Paradoxically, to judge from Darwin's exertions in the Andes, at the moment he was bitten he was perhaps fitter than at any other time in his life.

Perhaps the most extraordinary aspect of the *St. Fe notebook* is the break of about 16 months between the first period of use, around 21 September to 13 November 1833, and the second, 12 March to 20 April 1835. This is by far the longest break in use of any of the notebooks, and means that in terms of content *St. Fe* is practically two separate notebooks. During the break in use there is a complex sequence of use of six other notebooks and some loose notes.[389]

At *c.* 22,000 words, *St. Fe* is by far the longest of the notebooks. It is twice as long as the next longest notebook (*Coquimbo*) and seven times longer than the shortest notebook (*Sydney*). Not only does it have many more pages than the other notebooks — but the number of words per page is also much higher than the others. Furthermore, there are far more diagrams in the *St. Fe notebook* than in any other notebook except the *Port Desire notebook* which has slightly more. All the diagrams in the *St. Fe notebook* are in the 1835 part.

Buenos Ayres to St. Fe and Parana and return, September–November 1833

The earliest dated note in the *St. Fe notebook* was made in Buenos Ayres on 27 September 1833, p. 9a, and follows on directly from where the *B. Blanca notebook* left off. This note, which marks the start of expedition 3, is preceded in the front sequence by six pages of names, addresses, memoranda, shopping lists and sundry bits and pieces of information, and there are more such notes which seem to date from the same period on pp. 1b–5b. It is clear from the *Beagle diary* that at this time Darwin was staying with Edward Lumb at Calle de la Paz.[390]

389 The notebooks which fill the gap in use of the *St. Fe* are as follows: firstly *Banda Oriental* which followed on from the 1833 part of *St. Fe* directly with expedition 4; then *Buenos Ayres* very briefly, then *Port Desire, B. Blanca* briefly, *Banda Oriental, B. Blanca, Valparaiso, Santiago, Port Desire, Galapagos*, then some loose sheets of paper which are now in DAR 35.328, then the *Santiago notebook* before the *St. Fe notebook* was used again in 1835, albeit with another brief interruption from *Galapagos*. This sequence is based only on *datable* passages and may not be the complete sequence. After the *St. Fe notebook* was finally put aside, the *Coquimbo notebook* took over for field notes, with the *Santiago notebook* used for theory.

390 See Winslow 1975.

There are references in these early pages to numerous contacts, such as the bookseller Steadman on p. 5a. There is a reference to Darwin's assistant Syms Covington (1816?–61) on p. 4a.[391] Darwin was assisted by Covington perhaps as early as September 1832 but this became his full-time position from June 1833. There are various reminders, e.g. 'Tailors bill', p. 4a, and to buy items that would be needed on the planned expedition into the interior, e.g. 'Mice & Rat Traps', p. 5a. Darwin also intended perhaps to revisit the fossil collections in the Buenos Ayres Museum which was then as now one of the finest in South America.[392] Darwin noted a 'Megatherium found at R del Animal', p. 7a and on p. 8a there is the first known mention of the Galapagos by Darwin.

As soon as Darwin got under way on his 500 km (310 miles) trek to St Fé, '3 leagues [i.e. about 18 km, 11 miles] from Luxan [Lujan]' he met a gaucho who had travelled with Captain Francis Head (1793–1875). Head was a colonial governor who travelled in South America as manager of the Rio Plata Mining Association in the mid 1820s and Darwin referred to Head's description of Mendoza when he got there in 1835.[393]

Immediately Darwin started to note anything of interest, such as 'Biscatchas run badly', p. 9a. Very soon he was 'rather unwell', p. 10a, though still commented on the vast numbers of horse and ox bones lying around. By p. 13a he was at Rozario [Rosario], future birthplace of the mid twentieth-century revolutionary Ché Guevara (1928–67). The turbulent political history of the then emerging post-colonial states of South America visited by Darwin is described in the excellent but somewhat neglected Hopkins 1969.

There is a strange story to be told about the entry on p. 12a: 'horizontal variations in colour & hardness & some Tosca rock'. In *South America*, p. 87, Darwin described the geology of the banks of the Rio Paraná at San Nicolas de los Arroyos: 'when on the river I could clearly distinguish in this fine line of cliffs, 'horizontal lines of variation both in tint and compactness' and in a footnote to that page he added: 'I quote these words from my note-book, as written down on the spot, on account of the general absence of stratification in the Pampean formation having been insisted upon by M. d'Orbigny as a proof of the diluvial origin of this great deposit.'

It seems likely that by substituting the more poetic words 'tint and compactness' for his field words 'colour and hardness' Darwin was, perhaps subconsciously, creating a more striking description of the layering of the sediment, thus bolstering his view of gradual deposition against d'Orbigny's catastrophic interpretation. As discussed in the introduction to the *Buenos Ayres notebook*, this is not Darwin's only mis-quotation of his field notebooks.

---

391 See Young 1995.
392 See Parodiz 1981.
393 See Head 1826.

It is as well to bear in mind that, quite apart from natural hazards, Darwin was at considerable risk on this expedition from 'very bad people', p. 16a. It was no joke when he 'found pistol stolen' at Rozario, p. 13a.

On 1 October at the 'Aroyo Saladillo', p. 15a, on the Rio Tercero [Carcarána] he discovered in a soft sandstone 'a large rotten tooth & in the layer large cutting tooth', p. 16a. Owen, in *Fossil mammalia*, pp. 17–18, identified one of these as from a *Toxodon*.[394] It fitted precisely into the appropriate socket of an almost perfect skull Darwin purchased about six weeks later 300 km (186 miles) away in Uruguay. Soon Darwin encountered 'immense bones of Mastodon', indicating the vast numbers of fossils which existed by the Rio Paraná.

By p. 19a on 2 October Darwin noted that he was 'Unwell in the night, to day feverish, & very weak from great heat'. He was also in the midst of the Indian eradication programme: there was a 'dead one', p. 20a, and 'Lopez other day killed 48', p. 21a.

Darwin eventually arrived exhausted at Santa Fé, p. 22a, and was obviously quite unwell, though 'much better' by 6 October, p. 25a, having crossed the Paraná to Santa Fé Bajada [Paraná]. He found a limestone packed with fossil molluscs, brachiopods, fish bones and so on. On 7 October he carefully recorded, p. 29a, the behaviour of some spiders, one of which provided the following balletic spectacle: 'Shoot several times very long lines from tail, there by slight air not perceptible & rising current were carried up-wards & out wards (glittering in the sun) till at last spider loosed its hold, sailed out of sight the long webs lines curling in the air'.

By now Darwin had decided to sail back down to Buenos Ayres but was delayed by contrary winds and 'very timorous navigators', p. 32a. On 9 October the temperature was 79 °F (26 °C) at 8 p.m. Darwin was struggling to keep up with all the fossil mammals he kept finding: 'clearly two sorts of Megatherium … cotemperaneous with Mastodon: case of latter two or three inches thick', p. 30a. Obviously the 'case', i.e. carapace was from an armoured edentate, such as *Hoplophorus* and certainly not a *Mastodon*, and it is interesting that by p. 32a Darwin was going to see a very large 'Paludas [armadillo] case'.

He could not excavate the case from the bed which was 'unquestionably … above the Limestone' but in compensation he 'found tooth of horse', p. 34a. This was a puzzling find. It was well known that there were no horses in the Americas when Europeans arrived in the sixteenth century. Darwin wondered if the tooth was 'washed down'? Sadly 'the Barranca being inclined precluded the final certainty of the question'. This little tooth had great significance for Darwin, especially as he found it soon after pondering on evidence from arrowheads which clearly showed that the Pre-Columbian Indians did not have horses.[395]

394 On display in the lobby of the Exhibition Road entrance to London's Natural History Museum is a cast of a complete skeleton of *Toxodon platensis*, the original of which is in the La Plata Museum which is almost certainly the museum visited by CD. The skeleton was not one of the 'petrifactions' he would have been able to see in 1833, as it was excavated later in the century.

395 See the introduction to the *Falkland notebook*.

In *Journal of researches*, p. 149, Darwin was at pains to show that the taphonomic evidence — the state of preservation — 'compelled' him to believe that the horse was contemporaneous with the extinct *Mastodon*. Owen, in *Fossil mammalia*, confirmed that Darwin had indeed found remains of *Mastodon*, but also that the horse tooth, and one he found in Darwin's collection from Punta Alta, was a Pre-Columbian *Equus curvidens*, proving that horses had existed but gone extinct in the Americas before re-introduction from the Old World. This discovery, Owen wrote in *Fossil mammalia*, p. 109, was 'not one of the least interesting fruits of Mr. Darwin's palaeontological discoveries'.

But why had horses disappeared from a continent which, to judge from the way they had prospered in the short time since re-introduction, was an ideal environment for them? Had horses and all the other mammals been wiped out by some catastrophe? Or maybe 'species senescence' was the answer: 'as with the individual, so with the species, the hour of life has run its course, and is spent'.[396] Darwin later used the tooth as the centrepiece for his mature discussion of extinction in the *Origin of species*, p. 318.

> No one I think can have marvelled more at the extinction of species, than I have done. When I found in La Plata the tooth of a horse embedded with the remains of Mastodon, Megatherium, Toxodon, and other extinct monsters, which all co-existed with still living shells at a very late geological period, I was filled with astonishment; for seeing that the horse, since its introduction by the Spaniards into South America, has run wild over the whole country and has increased in numbers at an unparalleled rate, I asked myself what could so recently have exterminated the former horse under conditions of life apparently so favourable. But how utterly groundless was my astonishment!

Darwin explained in the *Origin of species*, p. 319, that according to his theory of evolution, extinction is exactly what would be predicted. It did not, as 'some authors' supposed (presumably he meant the Italian palaeontologist Giovanni Battista Brocchi (1772–1826)) require species to have *internally limited* longevities in the the manner of individual organisms. This did not mean that Darwin denied that different groups of organisms (e.g. molluscs and mammals) endured for different lengths of time, as it was common knowledge that species in some groups on average came and went more rapidly in the fossil record than did others. In fact the fossil groups, such as ammonites, with the shortest species longevities, were the ones most useful for dating rocks. Neither, in Darwin's view, did one have to postulate some catastrophe as the only explanation for extinction, even when it appeared to happen to large numbers of species simultaneously.

Thus extinction for Darwin in the *Origin of species* was merely what happens when the number of organisms in a species dwindles to an unsustainable level due to unfavourable conditions of life (he did not complicate the discussion by mentioning 'pseudo-extinction', when one species has evolved into another species and

---

396 *Journal of researches*, p. 212.

therefore ceases to exist). Furthermore, Darwin argued, it is usually impossible to be sure exactly what the unfavourable conditions were, and this argument must apply in the case of the horse in Pre-Columbian America. In other words, Darwin had come to accept Lyell's gradualistic view of extinction, in which the case of the horse in America was not unexpected. By 1859 where Darwin differed from Lyell was in his view of how new species originated in the first place.

In the second edition of *Journal of researches* (1845), p. 176, Darwin had already extended his discussion of extinction from the one he gave in the first edition but dropped any mention of species having fixed life spans. He concluded one of the most obviously evolutionary sections of his book, which included the enigma of the extinction of the Pre-Columbian horse, with a remarkable analogy:

> To admit that species generally become rare before they become extinct—to feel no surprise at the comparative rarity of one species with another, and yet to call in some extraordinary agent and to marvel greatly when a species ceases to exist, appears to me much the same as to admit that sickness in the individual is the prelude to death—to feel no surprise at sickness—but when the sick man dies, to wonder, and to believe that he died through violence.

Ironically, perhaps, palaeontologists today believe that the more or less simultaneous extinction of about 80% of the larger South American mammals some 11,000 years ago *was* due to some extraordinary event, and not just a series of co-incidental individual extinctions.[397]

Darwin's considerations of extinction are further discussed in the introduction to the *Port Desire notebook*, where one of the most intensively studied documents of the voyage is analysed.[398] This short essay, dated February 1835, in a file labelled 'scraps to end of Pampas chapter' in DAR 42.97–9, contains Darwin's earliest known reference to the origin of species: 'This correlation to my mind renders the gradual birth & death of species more probable.'[399]

The *St. Fe notebook* continues with many pages of observations, such as 'fresh & indubitable sign of tigre' (i.e. the jaguar, *Panthera onca*) on p. 38a. Curiously Darwin spent 13 October 'in bed because cannot sit up'. Fear of tigers it seems had destroyed 'all pleasure in wandering about', p. 39a, and tigers had replaced Indians as the main topic of conversation with his fellow passengers. At one point he took a boat and rowed it himself. This might mean that he did not even have his servant, or 'peon', with him, p. 41a. He watched the scissor bill bird *Rhynchops nigra* ploughing the surface of the water for fish. Darwin's description of the bird's 'wild rapid' flight is sharply observed but is not as beautiful as the published account.

---

397 See Benton 1990.
398 See for example Hodge 1985.
399 See van Wyhe 2007b.

Being at anchor in a small vessel, in one of the deep creeks between the islands
in the Parana, as the evening drew to a close, one of these scissor-beaks
suddenly appeared. The water was quite still, and many little fish were rising.
The bird continued for a long time to skim the surface; flying in its wild and
irregular manner up and down the little canal, now dark with the growing
night and the shadows of the overhanging trees.[400]

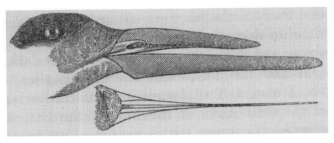

The scissor-bill bird as depicted in *Journal of researches*, 2d ed., p. 137.

On pp. 43a–45a Darwin described various fish which can be recognized as the
new species of *Tetragonopterus* described by Leonard Jenyns (1800–93) and exqui-
sitely drawn by Benjamin Waterhouse Hawkins (1807–89) in *Fish*, plate 23 (below).

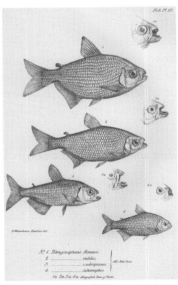

Various species of *Tetragonopterus*
from eastern South America.
Plate 23 from *Fish*.

On pp. 45a–46a Darwin recorded what he
was told concerning the 'gran Seco' or great
drought of 1827–30, when it became difficult to
use the river because of the stench: 'hundred of
thousands carcases dead on banks (fall down barrancas) float in water: could not pass many of the
streams for smell — it would be said some great
flood had killed all, especially as after it all rivers
were very much flooded corresponding deposit'.
In *Journal of researches* he made the point that this
drought, followed by floods, had within a year
washed down and buried thousands of skeletons:
'What would be the opinion of a geologist, view-
ing such an enormous collection of bones, of all
kinds of animals and of all ages, thus embedded in
one thick earthy mass? Would he not attribute it to
a flood having swept over the surface of the land,
rather than to the common order of things?'[401]

400 *Birds*, p. 144.
401 *Journal of researches*, p. 157.

By p. 47a he seemed to have heard of the uprising by Rosas' supporters against the Governor of Buenos Ayres, General Juan Balcarce (1773–1836), and when Darwin arrived at the outskirts of the City on 20 October he had great difficulty in entering.[402] He had to get ashore at Rio Las Conchas [Reconquista] and make a detour to Quilmas. 'I am in bad predicament', p. 52a, as even though Rosas' passport would eventually get him back into the City as if by magic, his collections were on the boat and might have been lost.

After a great deal of inconvenience, however, Darwin took the packet boat to Monte Video with 'Many passengers: women & children all sick', p. 55a. Finally on 4 November, p. 55a, he got back to the *Beagle* but because it was not sailing for another month he took up residence on shore. By good fortune, Darwin's collections and other belongings somehow found their way to him in Monte Video.

On 6 November Darwin went on a short excursion, crossing the Rio St Lucia on horseback, to geologize, but he did not find any bones. By the following day he was back, and he then switched to using the *Banda Oriental notebook*, not using the *St. Fe notebook* again for almost a year and a half.

Valparaiso to Mendoza, March 1835

The bulk of the *St. Fe notebook*, amounting to more than 180 pages, dates from about 12 March to 20 April 1835. These pages record geological observations and occasional notes on natural history and ethnography, of Darwin's great traverse of the Andean Cordillera from Santiago[403] [Santiago de Chile] to Mendoza via the 'Peuquenes or Portillo Pass', and the return via the more northerly 'Cumbre or Uspallata Pass'. These traverses are described in *South America*, pp. 175–87 and 187–207 respectively.

The Andes stretch some 9,000 km (5,600 miles) down the western side of South America. They are the result of a continental plate moving westwards against an oceanic plate which is moving eastwards. Detailed structure varies considerably from north to south, but generally the Andes are at their widest and highest in the north of

---

402 See Parodiz 1981.

403 CD seems to have been uncertain what to call the Chilean capital, since he used 'St Jago' in his field notes and various *Beagle* diaries until at least May 1836 (*Animal notes*, p. 12; see Sulloway 1983, table 2). CD mingled 'St Jago' with 'Santiago' in the ink lists of heights and distances in the *St. Fe notebook*, pp. 60a–65a, in one case on the same page, p. 63a. These notes seem to have been written in Valparaiso and the question 'Was the great wave which destroyed Concepcion quite sudden' on p. 65a obviously post-dates 20 February 1835. The reminder 'to draw 261:2 at St Jago' narrows this to before he arrived there on 14 March 1835. On the same page CD wrote 'Pencil note - Book some in town', suggesting that he intended to buy a new notebook in Santiago. Curiously he used 'St Jago' on the label of the *Valparaiso notebook*. It is unknown when CD labelled the 'Santiago Book.', presumably it was after the book was put aside after the *Beagle* voyage as it was labelled on both sides like his post-voyage notebooks. Unlike the latter notebooks, however, the *Santiago notebook* is named after a place described in it. Even more curiously he used the spelling 'St Iago' in his correspondence during the voyage. He used 'Santiago' in *Journal of researches*, but reverted to 'St Jago' in *South America*.

the continent, with an average elevation of 3.5–4 km (2.2–2.5 miles). Tectonic development of the Andes began in Palaeozoic times with accretion but accelerated in Mesozoic times with the opening of the Atlantic. A section or slice across the Andes which would include the sections examined by Darwin in the *St. Fe notebook* (and to a lesser extent in the *Copiapò* and *Valparaiso notebooks*) is broadly comparable to the rest of the chain, but differs in many details from sections to the north and south. From the Pacific Ocean eastwards there is the Peru–Chile Trench, then the Coastal Range, built largely of granite batholiths, beneath which the Nazca Plate is melting. These are the areas where most major earthquakes would be expected, a point noted by Darwin.[404] There is then a well-defined Central Valley where Santiago is situated. About one third the way into the section are the Cordillera Principal (Darwin's Peuquenes Range, which he sometimes called the 'central range').[405]

The Chile–Argentine border runs down the middle of this range which has some of the great peaks such as Aconcagua, at *c.* 6,960 meters (22,840 feet) the highest mountain in the world outside the Himalayas. Darwin described the porphyries, granites, slates, sandstones and Mesozoic fossils of this range. FitzRoy determined the height of Aconcagua and was able to prove that it was higher than Chimborazo in Ecuador. Aconcagua was one of a chain of volcanoes that Darwin reported to have suddenly and simultaneously erupted on 20 January 1835.[406] The middle of the section are the Cordillera Frontal (Darwin's Portillo or Eastern Range), which can perhaps be considered as a southern extension of the Altiplano plateau consisting mainly of granites, andesites and other igneous rocks partly thrust eastwards over the South American Shield. Darwin reported conglomerates in this range derived from the Cordillera Principal, implying that the latter was uplifted first. Next there is the fold and thrust belt of the Precordillera (Darwin's Cumbre and Uspallata Ranges) with a granite core and many complex structures. There are sections of spectacularly coloured Mesozoic and Tertiary sedimentary and volcanic rocks, such as the Puente del Inca section and the complex stories of crustal mobility exemplified by the petrified Agua de La Zorra trees. Darwin returned westward at this point but if he had continued east through the section he might have encountered some of the lower Pampean Ranges, a series of basement uplifts.[407]

The first dated page in the *St. Fe notebook* from 1835 is actually to be found on one of the pages which span the transition from the first sequence and the second sequence. Inevitably, since there are no blank pages, such a distinction in a notebook in which the pages were not numbered by Darwin, is somewhat arbitrary. The page is dated 12 March, was written in ink in Valparaiso, and among several other

404 *South America*, p. 185 note.
405 *South America*, p. 180.
406 Darwin 1838; *Shorter publications*, pp. 40–5.
407 For a summary of the geology and an excellent general account of the expedition see Keynes 2003. For a more detailed modern geological account of some of CD's localities see Morton 1995.

interesting notes seems to record Darwin's first interest in tropical corals: 'Corall in sea', p. 233a. Sulloway 1983 discussed this issue in detail in connection with the 'coral passage' in the *Santiago notebook*.[408]

The first of the front pages from p. 60a are written in ink and appear to be notes in preparation for the trip to Mendoza. There are lists of places with what seem to be their heights in feet and distances in leagues. It is clear from the *Beagle diary* that by 14 March Darwin had come from Valparaiso to Santiago by coach, and was staying with Alexander Caldcleugh (d. 1858), who was probably the source of much of the information (e.g. p. 64a).[409] Darwin was already familiar with Santiago, having spent a week there in August–September 1834.[410] Caldcleugh was a Fellow of the Royal Society and was a private secretary to the British minister at Rio de Janeiro, as well as a promoter of the Anglo-Chilean Mining Company. He was the author of what Darwin told his sister Susan (1803–66) were 'some bad travels in South America'.[411] Caldcleugh's book contained a beautiful but very basic map of the geology across the precise part of South America covered by the *St. Fe notebook*.

In these early pages there are references to places on the Pacific coast such as Coquimbo and Concepcion, where the earthquake of just a few weeks before on 20 February, which almost certainly had its epicentre under the ocean, was felt severely. Pencil came back into use on p. 65a and Darwin asked about the tsunami: 'Was the great wave which destroyed Concepcion quite sudden?'. Darwin also made some notes at this time in the *Galapagos notebook*, including observations from his trip to Santiago from Valparaiso.[412]

The expedition seems to have begun by p. 66a which is dated 18 March. Here begins possibly the longest and most detailed geological note-taking sequence of the entire *Beagle* voyage. Barlow 1945, pp. 232–3, said she could 'give no impression of the pages of geological argument'. She did, however, make a brave stab at summarizing the geology, and created a vivid picture of Darwin in the Andes: 'clambering over the rocks hammer in hand and with shortening breath from the great altitudes, riding in the icy winds and sleeping on the bare earth'. In fact Darwin wrote to his sister Susan on 23 April that he had carried a bed with him.[413]

The first of many small diagrams is on p. 71a but the more serious section diagrams start on pp. 75a–78a and date to 20 March 1835. The characteristic alphabetical listing of specimens begins to appear, as for example on p. 79a, with

408 See the introduction to the *Santiago notebook*. Presumably the Sulloway 1983, p. 376 note 17, date of May 1835 for 'Corall' on this page should be March as no other evidence has been found for a May date for this page.

409 See Herbert 1995.

410 See the introduction to the *Valparaiso notebook*.

411 Caldcleugh 1825; see CCD1: 446 and Sulloway 1983, note 10.

412 See the introduction to the *Santiago notebook*.

413 CCD1: 445.

specimens J, G and H. Later in the notebook these become extensive lists which we have compared with the four specimen notebooks at the Sedgwick Museum, where Darwin's rock samples are kept. Very often it is easy to translate the alphabetical lists into the numbered lists in the Museum, which were compiled when Darwin had leisure after his expeditions, and it is a great privilege now to be able to go from these lists directly to the actual specimens Darwin collected and recorded in the *St. Fe notebook*.[414] It is also a straightforward matter to compare some of the notebook lists to those published in *South America*, as for example the list in the notebook on pp. 196a–197a with *South America*, pp. 190–2.

There is a fascinating entry written on 20 March which begins on p. 83a 'Valley very curious higher up'. This is followed by a diagram and a description of a huge mass of 'Alluvium enormous angular fragments' separating two valleys on the approach to the Valle del Yeso. Darwin was 'greatly perplexed' by this mass which he could not believe was laid down by river action: 'I hardly dare affirm these hills are alluvium', p. 84a. He described this 250 meters (800 feet) thick mass in his geological diary, but it was not until after the voyage that he realised that it was a glacial moraine.[415] In the surviving part of his 'big book' chapter on geographical distribution, written in 1856, *Natural selection*, p. 545, Darwin referred to this moraine which was 'thousands of feet below the line where a glacier could now descend'. He quoted from his geological diary (DAR 36.460) in a footnote, and cited the moraine as an illustration of the global Ice Age, which by the 1850s was an accepted fact, and must have had a dramatic affect on biogeography in geologically recent times. Darwin made a rare direct reference to his *Beagle* geology field work in a very clear short description of the moraine in the *Origin of species*, p. 373.

Zoological notes occur from time to time in the notebook, such as what Darwin was told about condors, but high in the mountains there was 'very little vegetation, no birds and insects', p. 88a. He noted the 'Red snow' which showed up in the mules' hoof prints and was relieved to be told that as the winter was beginning to set in unless one heard thunder it was unlikely to snow, p. 94a. The 'red snow' was described in more detail on p. 126a and in the *Zoology notes* Darwin recorded that he placed some of the spores 'between the leaves of my Note-Book'. In *Journal of researches*, p. 395, Darwin changed the phrase to 'pocket-book'. The exact page where Darwin inserted the spores was probably the excised lower left corner of p. 226a. Unfortunately the facing pages have been excised so that no trace of the spores remain in the notebook.

Darwin was amused when he overheard the peons' explanation why the breakfast potatoes were still uncooked after a night's boiling – the cooking pot chose not to co-operate (rather than the thin atmosphere at high altitude) p. 95a. On pp. 103a–104a Darwin made the momentous discovery in some black calcareous shales at Peuquenes

---

414 See Herbert 2005.
415 *Journal of researches*, p. 389, states that it had dammed a lake but did not call it a moraine.

of fossil *Gryphaea* oysters, a univalve (snail), *Terebratula* brachiopods and 'a piece of an ammonite as thick as my arm'. These were later dated by Alcide d'Orbigny as of Neocomian (early Cretaceous) age, now dated as about 120 million years old.[416] Around p. 115a Darwin started to use numbers to label his specimens.

The peons' 'strange ideas' about 'puna', or the affects of altitude, are noted on p. 129a. Darwin compared the 'tightness of head & chest' to 'running on frosty morning after warm — room'. 'Fossil shells forget' is explained in the *Beagle diary* as the way the excitement of finding the fossils completely overrode his awareness of difficult breathing. He noted the 'magnificent wild forms' and the 'resplendently clear' air. He said the view from the '1st ridge' was 'something inexpressibly grand', and on p. 131a, in one of the most moving lines in all the notebooks, Darwin thought he 'never shall forget the grandeur of the view from first pass'. The view may be the one he drew on p. 128. Alexander von Humboldt, in a letter to Darwin dated 18 September 1839, singled out the passage in *Journal of researches*, p. 394, derived from these notes, as one of several 'belles pages'.[417]

By 22 March on p. 131a Darwin began the ascent of the eastern (Portillo) range, where, at an altitude of about 4,000 meters (13,000 feet), he had a 'fine view of crater of Tupungata'. He was prevented by a snowstorm from collecting some of the rocks.[418] Within a few pages he was back down among 'flowers like Patagonia' and birds and 'very many mice', p. 133a. On 24 March he killed with his hammer a viviparous lizard with babies which 'soon died', and on p. 136a he caught a young snake. The lizard was mentioned by Darwin in his letter to Henslow of 18 April 1835.[419]

By p. 144a Darwin had a 'view of Pampas' from the west. Perhaps the 'line of glittering water lost in immense distance' was the Rio Plata, some 1,000 km (620 miles) away. On p. 145a there seems to be a reference to Indians being used as trackers, and Darwin noted his bivouac for the night of the 24th, after passing the only Estancia in the area.

Darwin later realized that the differences between the botany and zoology on the opposite sides of the Cordillera were highly significant. In *Journal of researches*, pp. 399–400, he drew attention to these differences in a remarkable passage. He wrote of the animals on the east side 'We here have the agouti, bizcacha, three species of armadillo, the ostrich, certain kinds of partridges, and other birds, none of which are ever seen in Chile'. He stated that the differences between the species made no sense as the environments on both sides of the Andes seemed essentially identical. A note on p. 400 (which remained unchanged in later editions) clearly stated that the belief in the immutability of species was merely an assumption: 'This is merely an

---

416 See the appendix to *South America*.
417 CCD2:219.
418 *South America*, p. 183.
419 See CCD1: 445 note 7.

illustration of the admirable laws first laid down by Mr. Lyell of the geographical distribution of animals as influenced by geological changes. The whole reasoning, of course, is founded on the assumption of the immutability of species. Otherwise the changes might be considered as superinduced by different circumstances in the two regions during a length of time'.[420]

Darwin now turned north towards Mendoza and a vast cloud of locusts flying in the same direction near Luxan [Lujan] is described on pp. 150a–152a. That night he had the 'horribly disgusting' experience of being bitten by the 'Chinches' (Benchuca bugs). On p. 153a there are descriptions of the 'sad drunken raggermuffins' of Luxan and the next day Darwin was in Mendoza which was not worth describing because 'nothing can be added to [Francis] Heads description' (Head 1826). On 29 March Darwin mentioned the 'very fine grapes' which are today the basis of Argentina's wine industry. On p. 156a he slept at Villa Vicencio and on the next page started to record the geological section there.

Mendoza to Valparaiso, April 1835

Darwin was unimpressed by the 'very tame' scenery at Villa Vicencio which 'Mr Miers' had made notorious, p. 163a. John Miers (1789–1879) was a mining engineer who, like Caldcleugh, was scientifically minded. He was interested in botany and travelled in South America in the 1820s and 1830s. He was author of Miers 1826 which was in the *Beagle* library and referred to several times by Darwin.[421]

Darwin noted the date of 1 April on p. 164a, forgetting that March has 31 days. The section continued with a rapidly growing list of specimens as he visited some of the mines and wrote pages of extraordinarily detailed descriptions. The entries around pp. 172a–174a are remarkable for demonstrating that Darwin was starting to write as if for an audience who might need persuading, in this case of his belief that the rocks had been changed by heat. Entries such as 'We will recur to the subject', p. 172a, 'as I shall show', 'It seems a bold conjecture I firmly believe a true one', p. 173a are clearly the beginnings of Darwin the 'prospective author', to borrow Sandra Herbert's phrase. It was at this time that Darwin began to use the *Santiago notebook* for more developed theoretical writing. Darwin was at the easternmost extremity of his second published section.[422]

On the actual 1 April, p. 178a, Darwin discovered the petrified forest at Agua del Zoro [Agua de La Zorra]: 'looking for silicified wood found in broken escarpment of green sandstone 11 silicified trees' standing about 20° to the vertical. Darwin was

---

420 This is an additional instance of CD's frank references to his belief in transmutation before publication of his theory in 1858/9. See van Wyhe 2007a.

421 See Herbert 1995.

422 Herbert 2007, p. 317, reproduces a page from CD's geological diary which is derived from these pages (DAR 36.508).

immensely proud of this discovery which he reported in detail to Henslow in a letter dated 18 April 1835.[423] This is one of Darwin's discoveries reported in the letters which Henslow decided to abstract and read to the Cambridge Philosophical Society on 16 November 1835.[424] Henslow's action took Darwin completely by surprise when he found out about it while in the Atlantic in 1836.

In *Journal of researches*, p. 406, Darwin stated that Robert Brown (1773–1858) identified the fossil trees as coniferous, and explained at great length how the trees prove the complex history of uplift, submergence and re-emergence of the Andes. The age of the trees is now known to be late Triassic (about 245 million years old) and Darwin's discovery at this classic locality was commemorated in 1959 by the erection of a monument at the site.[425]

There follow many pages describing the traverse over the Uspallata range until on p. 188a Darwin entered the 'grand valley of Cordilleras' with 'Ostriches, Toco Toco. Apereas Pichy Paluda'. The landscape was 'extremely sterile'. On p. 189a he was told to beware the dangerous passes and to 'carry thick Worsted stockings'. On 3 April he was at the 'Paramillo', p. 190a, which is the Paramillo de Las Cuevas, one of the late eighteenth-century refuges on the Argentine side of the border, where Darwin slept one night.[426] By p. 196a Darwin was near the Puentes del Inca [Puenta del Inca], the famous, but in Darwin's view overrated, 'Inca's Bridge', a natural mineral archway over the river. Darwin recorded the Puenta del Inca section in great detail on 6 [actually 5] April. On p. 201a he sketched the bridge itself, showing the 'crust of stratified shingle, cemented together by the deposits of the neighbouring hot springs'.[427] This diagram was later copied into the margin of the *Beagle diary*, p. 319. The bridge is today a significant tourist attraction south of the highway from Mendoza to Valparaiso, just east of the Argentine–Chile border.

Darwin's magnificent section through the mountains at this point, which forms the left central part of his published section 2, is illustrated by Morton 1995 who calls it 'the most classic section of the High Andes.' For comparison with Darwin's section Morton provided more recent sections with a wonderful photographic panorama of the mighty Aconcagua fold and thrust belt.[428]

On p. 199a Darwin mentioned the 'hot springs (and much gaz)' which he later described in *South America*, pp. 189–90 note. At this point in the notebook Darwin also made some zoological notes such as the Lion and the humming bird and other birds Darwin saw at the bridge, and the high altitude 'Indian huts', p. 201a, which he speculated in the *Beagle diary*, pp. 320–2, implied a deterioration in the climate

423 CCD1: 442.
424 Darwin 1835; *Shorter publications*, pp. 2–31.
425 Morton 1995.
426 See Morton 1995, p. 193 and fig. 7.
427 *Journal of researches*, p. 409.
428 Morton 1995, p. 192, figs. 4 and 5.

since the time of the Incas. On p. 209a he noted the 'remarkable electricity' and the 'transparency of air'.

Darwin made an intriguing but rather cryptic comment on p. 210a: 'immense degradation & consequent talus striking feature: *arguments of length of world from deluge applicable in each part since sea retired*' [italics added]. The remark may mean that older arguments about the length of the world since the time of the Biblical flood, by Darwin's time no longer accepted by geologists, were sometimes based on the thickness of diluvial deposits. Darwin, by working out how the mountains had risen from beneath the sea and been eroded into valleys filled with immense amounts of talus, could see that each small event in his long chain of causes was equal to the old biblical timescale.

As Darwin made the gradual descent back to Santiago he noticed the cacti, p. 211a. He spent all of 7 April, p. 214a, searching for a lost mule, and on the next page he noted a 'Big Rat' in a 'high tree'.

On p. 218a Darwin made a precise literary reference in ink, showing that he was back at Santiago, probably at Alexander Caldcleugh's house, and there is an out-pouring of place names, dates and references to follow up. Then what must surely be a retrospective entry: 'all this time to Valparaiso not quite well saw nothing enjoyed nothing', p. 218a. On 15 April he 'Started for Valparaiso — dead mens heads on the poles', pp. 218a–219a. On p. 222a are numerous references to insects he collected such as those near the warm spring of the Villa Vicencia.

Finally, around 20 April, Darwin was back at the house of his old Shrewsbury School friend Richard Corfield (1811–87) in Valparaiso. Corfield had looked after Darwin there when he was very ill, perhaps with typhoid, in 1834.[429] The last notes on p. 232a seem to be payments to Mariano Gonzales, Darwin's guide in Chile, '6 dollars' and to Covington 'on account 8 dollars'. The entries in ink on p. 233a, already discussed, seem to be dated 12 March 1835, so appear to predate the previous page.

Thus ended the longest continuous sequence of notes in any of the field note-books, in which Darwin recorded the first complete traverse by a geologist across the longest mountain range in the world. These notes were eventually published in greatly expanded form in his *South America* and remain a classic account in the history of geology. They also provided Darwin with the raw material for his great theoretical paper linking earthquakes, volcanoes and the vertical movements of the earth's crust, and, perhaps most importantly, convinced him of the almost unima-ginable immensity of geological time.[430] Darwin stated emphatically at the end of his account of the Portillo Range: 'What a history of changes of level, and of wear and tear, all since the age of the later Secondary formations of Europe, does the structure of this one great mountain-chain reveal!'[431]

429 See the introduction to the *Santiago notebook*.
430 Darwin 1838; *Shorter publications*, pp. 40–5.
431 *South America*, p. 187.

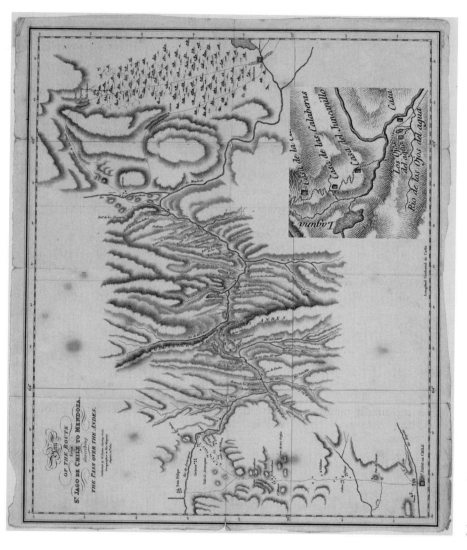

'Plan of the route from St Jago de Chile to Mendoza, describing the pass over the Andes'. Map from Darwin's collection and possibly carried with him in his 1835 traverse of the Andes. St Jago (i.e. Santiago) is at bottom left, Mendoza at far right. Published by W. Faden, London, 1823. A detail panel has been added on the lower right. (DAR 44.12).

[FRONT COVER]

Buenos Ayres. St. Fe
and Parana —
Cordillera of Chili

[INSIDE FRONT COVER]
[1a blank]
[2a]
[3a blank]
[4a]

Mr Lumb	338	Peon for B Blanca
2800	406 or 58	silver Harris
	100	Peon to St. Fe.
C. Fitz Roy & Stokes	630	
	25	(for horses first stage)
		Tailors bill
		Passage of Covington
Dixon /Price/ &		Jars & Box
/do/ Valparaiso		Money
Recon		/agent/ M Video
Endorsements		John Gouland
Katers Compass		Compass
/Terallos/ letter		Fleming
		watch
117 /Big Hat/		/Turtle/

Peon (86) 7 / 6 before starts for Books? Silver Rials

[5a] Tailor, Shooting Jacket mended. & Trowsers: Big bag: Lindsay Calle Piedad.
Big bottles: Large Clasp knife. 130D *[Protractor]* Bees wax Rosin. Corks for Jars
Hargrave 55. Piedad Steadman 30 Cathedral Field sports *[Prairie]* Ch. Grandison[432]
2 dollars Mr Loudan Mr Lumb 28 *[Caypur]*
denture mended[433]   Stirrups

58
  7
406
  5
90
  7
630
80
  5
400
  7
2800

Jar for Molitas Market fish:
Write Bills   Pen knife
Repair the book   Mr Tweedee *[seeds]*
Paper for plants: Tow: Bladders Cotton: Woollen Stockings
Bottle small with large mouth: Gold Leaf.
Mice & Rat Traps Blank silk handkerchief
Pills Druggist. Bishop *[Reconquesta]* opposite church
museum see petrifactions
M^r G chart of B Blanca French Survey of River
[new chart] 1832
Calle Piedad Watson Wood Snuff Box
Mackintosh *[Water]* + Washing bill:
Seeds of grass. M^r *[Flint]* *[Tylor]*

[6a] Wood chopped by axe
J Britain direction   Sending map
Snake Mr Hind   Snake Lindsay
Indigo French Bookseller
Gunpowder & shot: Note Books
Chili Letters

432 Richardson 1781.
433 On CD's teeth see Colp 1977, p. 135.

Letters from Mr Hooker
Grand Seco^434 ???   Snuff
to M Video   Gun
Formation Sugar. Bread. Cigars
at Corrientes Letters of Recommend
If possible paper. (brown) soft:
obtain them money clothes. passport
30 leagues up Shells in cliffs
Great Seco Load Pistol
(/From/ in 1828 for 3 years supposed one million died in the Province)

[7a]   Water not fresh ab**ove 50** miles Island Marineras.— Laguna de los Patos About
       50 miles from the mouth
       Said in the country where no Capincha fear of Jaguars: not occur S of Plata :
       Biscatchas drag things to holes
       At Punta Gorda. Matanza
       /Cat/? Barquin: Limestone:
       Hard rock at Mr Hooker Estancia
       Megatherium found at R del Animal
       Paper for Plants. M: Video

[8a]   Captain Robertson
       Part of Co**piapò** — Wasko
       Cliff beach of Shells
       Hill like /bubble/ Quiloata;
       2 men deserted from Conway: Abingdon, Gallapagos.— Sail-maker
       Frezier. 1715 - translated by Dr Halley.^435 South Sea

[9a]   Friday ~~28~~ 27th [September 1833] —
       passed Capella Moron & slept on ~~the~~ about 3 leagues from Luxan great bed 6 leagues
       of young thistles; same Gaucho with Head: Biscatchas run badly like rats. tame:
       infinitely more numerous than to South:
       no geology except on surface apparently pale hard Tosca:
       Started not before 1 oclock
       Saturday 28th [September 1833] —
       passed Luxan on river by bridge nice church & Cabilda. from thence to Areco smaller
       town. country generally flat

[10a] yet there are great inequalities; for from some places very large horizons are in view —
       nearly whole country thistles & clover: miniature forest clumps — rise green out of the

---

434 Spanish name for the great drought of 1827–30 which caused the deaths of vast numbers of cattle on
     the Pampas and an explosion in the numbers of mice. Hearing of it made a great impression on CD,
     who referred to it in *Natural selection*, p. 181.
435 Frézier 1717.

very dust. — few birds. whole country Estancia: distant one from other: — Passed R Arrecife — on barrells. & slept rather unwell yet paid for 31 leagues

[11a] In several places, saw pale & red Tosca, with a very little mortar: Luxan & Areco Rivers
       Sunday 29th [September 1833] —
       About 7 leagues North of R Arrecife: well in reddish Tosca, with some Mortar country bare, as road, thistles not having sprung up: Biscatchas ~~drag~~ drag bones & thistle stalks to their holes.
       In evening first saw Parana at distance. Woody Islands: at St. Nicholas large straggling town
       Many of the brooks in whole journey paved with bones of horses & oxes

[12a] San Nicholas on river: large schooner & vessels. many islands with bush & Entre Rios tambien: barranca 30 or 40 feet high. perpendicular Many sorts of Indian figs. chief change in vegetation: cliff pale Tosca, horizontal variations in colour & hardness & some Tosca rock no mortar; For future Pistol in hand: not leave guide: —
       Monday 30th [September 1833] —
       Both Bahia Blanca birds; Sparrow, small pidgeon, & scissor bill stay whole year build in marshes all this near

[13a] St Nicholas; the road whole day, very near the river, camp very level, with precipitous valleys, cliffs red, lofty, 50 or 60 feet: Rozario on a point close to river striking appearance very many bushy Islands, bushy & Entre Rios. do: striking view: camp with few thistles, good grass: many cattle: new trees mimosa, more flowers: Rozario nice town, level plain hospitable man, found pistol stolen: Peon very weak: today only 20 leagues: yesterday 24: Colegio St. Nicholas, striking

[14a] looking from large Church, in very fertile & flat plain: passed Arrozo del Medio & entered St Fe: much reddish Tosca with little Tosca rock: Aroyo del Pabon very much Tosca rock: precisely resembling that of R ~~Salado~~ Tapalken, impossible not to recognise same formation: some with very numerous serpentine cavities lined with black, others much harder & semicrystalline V Specimen: this rock

[15a] forms pretty cascade of about 20 feet plenty of water: Aroyo Seco same geology
       Aroyo Saladillo, I saw the phenomenon of rapidly running brook with brackish water bank with Acanthus thistles: I do not much like the Inhabitants; civil d—d rogues: The views of river; 3 or 4 miles across very different from anything I have seen, from number linear shape if islands not like lake
       Barrancas most picturesque:

[16a] Tuesday 1st October: [1833] Our sleeping place I think had very bad people: started by moon-light & arrived on coast of Carcaraña by sun-rise: cliffs composed of pale, ~~da~~ Tosca with very many Tosca rock stalactites: above it layer with small Tosca concretions marked with manganese, & above it earthy very soft sandstone: in the former there was a large rotten tooth & in the layer large cutting tooth: Procured fresh horses & started for the Parana views quite pretty lake scenery: found two large straggling deposits

[17a] of immense bones of Mastodon.[436] V Specimens. very rotten in perpendicular cliffs, heard of many other bones: when it is considered that these are only sections of an immense plain, how very numerous these animals must have been: at Parana barranca: bottom pale yellowish clay with curious numerous ferruginous cylinders above great bed (50 feet in /whole/) of earthy reddish, stalactite & nodules of Tosca rock

[18a] numerous, especially in lower part: in this part were the bones sticking out of cliff — returned by noon to the house & started again: Cormorants at Parana & many beautiful new birds. I think sea birds enter the camp more readily from its openness: scissor bill eat fish, no mud banks, sit on grass camp, as on mud banks: BB Bird at Cordova: Biscatchas place dung bone &c on flat above

[19a] entrance of holes: are abundant where no thistles: All four Armadilloes at Cordova Peechy Port St /Elina/ Passed R. Monge, brackish as the Carcarana; but little: Camp north not so dead level, yet regularly uneven: never /saw/ camp, but with different length of horizon arrived at good Estancia at Sunset:
Wednesday 2ᵈ [October 1833]
Unwell in the night, ~~next~~ to day feverish, & very weak from great heat: every thing shows the great change in

[20a] small change of Latitude dress & complexion of men: Oranges & ~~gr~~ immense Ombus beautiful birds & flowers. reminded me of Brazil: About 3 leagues south of Coronda entered the /Montes/ Home Guardia Monte, flat diluvial plain with Mimosas, lawn scenery, ~~took Escort~~ — Indians, saw a dead one in road & Estancia desolated

[21a] This forest extends some leagues South of Coronda took Escort (in this part Indians). — Lopez other day killed 48 Indians in an Island — (Tyrant) Time of old Spaniard (no) Coronda prettiest village I have seen; from many Ombus & oranges. Saw simple Top fashion of spinning. —
Crossed in a canoe

[22a] a Riacho[437] & arrived very much exhausted at Santa Fe: obtained an empty room & bed & made ourselves tolerably comfortably. —
Thursday 3ᵈ — [October 1833]
Very unwell in bed: — Santa Fe kept in very good order, large, straggly, every house, with Garden

[23a] Town looks green & clean
Friday 4ᵗʰ [October 1833] —
Unwell in bed. —
Saturday 5ᵗʰ [October 1833]
Crossed over to the Bajad passage of about 4 hours, winding about the various Riachos generally as broads as the Severn & much deeper & more rapid
gave me a great idea of size of river: at last crossed blowing fresh. the main stream: saw

436 A type of extinct elephant.
437 Spanish for a narrow arm of a river.

[24a]   several large vessels one which ~~drew~~ had drawn 17 feet water had entered: Barrancas /high/ 6̶0̶ 70 or 7̶0̶ 80 feet at least. muddy water Continually falling covered with luxuriant vegetation, picturesque: with humming Birds: Town very straggling, but rapidly increasing from fine position /of Province/: half a league from the Port, on account of Paraguay Indians formerly:

[25a]   Sunday 6ᵗʰ [October 1833]
Much better. rode to the bottom of cliff; lower great bed pale yellowish earthy clay; this rest on horizontal or variously curved strata of butiminous clay or sand, deposited by streams, in /some/ /case/ the vegetable /films/ are visible. V Specimen in others not: the clay bed contains ~~few~~ shells in irregular layers chiefly of ~~two~~ 3 sorts. V Specimen. it contains also plenty of Gypsum: Above this is a great bed of limestone more or less pure when

[26a]   pure Crystalline, when less so containing v̲a̲s̲t̲ number of large ostrea, few Pectens[438] impression of Terebratula & other shells; also very few pebbles of coloured silex: certainly fragments of big bones & fish bones: above this ~~occurs~~ the lime becomes very impure with many ostrea, Above this great bed (½ altitude of all) of reddish earthy Tosca

[27a]   in lower part in has a bed of aluminous /calcareous/ concretious masses with manganese V Specimen like at M Hermoso to which Tosca resembles, also few smaller concretions in upper parts: in some case beds of yellow indurated sand in this Tosca:
As far as I recollect Tosca rock a[t] Pabon resembles that at R. Tapalken

[28a]   Monday 7ᵗʰ [October 1833] —
Walked to the Barrancas, found myself much tired. Quantity of Limestone varies, exceedingly in some parts nearly absent — In one place fine white sand above Limestone: all beds subject to change: —
Found black Epeira, amongst bushes in society of some hundreds (all same size ∴ age) main threads very strong common to many vertical webs. each web one or two feet from the

[29a]   other: spider black with ruby marks on side of back:[439]
Saw a largish (running spider) Shoot several times very long lines from tail, there by slight air not perceptible & rising current were carried up-wards & out wards (glittering in the sun) till at last spider loosed its hold, sailed out of sight the long ~~webs~~ lines curling in the air:

[30a]   clearly two sorts of Megatherium ~~to~~ found at Carcarana & Arroyo del Animal cotemperaneous with Mastodon: case of latter two or three inches thick:
The Limestone at Punta Gorda (I am told) de- is near waters edge: Shells are found at R. Hernandrias, & by Brooks a valley is known to run across the country: country west of Santa Fe all low: In Entre

438 The scallop *Pecten* sp., a bivalve.
439 See *Journal of researches*, p. 42.

[31a]   Rios Biscatcha — Molita Paluda
Tuesday 8th [October 1833] Thirty thousand Inhabitants in Entre Rios
6 in the Bajada. 1825:
Beneath limestone bed there is other one of greenish fine greasy (Magnes) Clay. where
dry exfoliates & shrinks & falls. then limestone falls & is worked: bed about 12 feet in
some places resting on the yellowish sandy clay:

[32a]   Wednesday 9th [October 1833]
Delayed by bad winds, very timorous navigators: weather most oppressingly hot at
8 oclock at night 79 outside house, with many fire flies: Pleasant idle time extreme
hospitality: one of few men I would trust :
Thursday 10th [October 1833] —
Blowing a gale of wind from the South most unwillingly delayed another day:
At noon went to see some of a Paludas case

[33a]   in a Barranca of red Tosca, one league East of town on R. Tapas, which enters
R. Conchas. —
Tosca with calcareous & Tosca rock concretions: The shell formed a case well between
4 & 5 feet across entire, but soft no bones except a lump: the bones were said to be less
than full grown cow. — very many bones in various parts of Barranca, chiefly small

[34a]   Excepting a large piece of shoulder blade: —
Unquestionably this bed is above the Limestone: In one place found tooth of horse
in red compact Tosca & well buried, it being a Horse, only cause of doubt of real
position, after long examination, I came to conclusion that the Tosca might have been
washed down

[35a]   & rehardened; but not very probable: the Barranca being inclined precluded the final
certainty of the question: How wonderful number of bones — Great bone was not well
covered yet I think belongs to Tosca: Smooth surface of [case] a [internal]: (Saltpetre in
Entre Rios) with shells & gypsum Marble at [Cordona] & Modern Formation

[36a]   The House rat of Maldonado is very common here, in out-houses: crawls much in
hedge — 6 mamma on each side, the third is placed as far distant from the 4th — as first
from third:
Friday 11th [October 1833] Indolence of the master did not start: great misfortune to
me: in evening went out shooting
procured specimens — Oven bird called Casara or house-maker how

[37a]   well is BBB called Casarita:[440] Captain F. Certhia here: yellow breasted Sylvia do:
Callandria:
Saturday 12th [October 1833] Started Gracias a dios: Tosca above Limestone divided
into two beds, lower one pale, other upper bright red. ½ gale of wind in our teeth: beat
amongst the low islands, at last fastened the bark to the

---

440 A bird, the 'little house builder', see *Zoology notes*, p. 158.

[38a]   trees of one. — I started for a scramble, two sorts of trees, commonest willow covered
with creepers & other plants, swamps covered in floods, muddy sand: fresh &
indupitable sign of tigre:
Sunday 13th [October 1833]
Meat bad, fine fish 4 sorts, rain & gale whole day, in bed because cannot sit up: at
Bajada red sturnus & common oriole, black & white fly-catcher

[39a]   then man informs me that a South wind here always clears weather as SW
of B Ayres. — ∴. Current changes hence wind does not result from impetus but
suction:
Monday 14th.[October 1833] —
Ten thousand curses, wind yet SE & dirty weather: Many especially Maldonado one
Kingfishers: & Cormorants: Rather better weather so that I could fish & in boat pull
about the creeks: all pleasure in wandering about the Island

[40a]   is destroyed by fear of tigres. — in this journey main conversation rastro del tigre as
before rastro de los Indios: met my Peon running at great rate, attack people in vessels:
Tuesday 15th — [October 1833] after some delays, so very cautious, we started passed
Punta Gorda with its Indian colony chiefly red Tosca — with bed of Limestone at
base. — Sailed quickly down stream amongst the intricate

[41a]   Islands: we came to anchor, (from foolish fear of bad weather in a narrow Riacho, I took
the boat & proceeded up it for about a mile — narrow with willow & creepers &
winding, deep, slow stream the cry of birds & fire flies: saw to my delight scissor bill just
at night (perhaps at Laguna at Maldonado at day, because extraordinary number of fish)
flying in that wild rapid manner as at Bahia Blanca & ploughing the water amongst
jumping fry:

[42a]   Maldonado kingfisher builds nest in trees: Slept on deck on account of muskitoes &
heat :
Wednesday 16th [October 1833] Started, with wind a beam arrived at Barrancas on
West coast above Rosario: Barranca very even about 40 feet high, generally with
irregular band of yellowish clay beneath as described at Gorodina:
at Rosario with Tosca-rock: Passed the mouths of the rivers which we crossed generally
near the Barrancas

[43a]   straight-backed fish[441] — silver bands in back irridescent greenish brown — dorsal fin
pale dirty orange — tail fin central part black — above & below /this/ bright red &
orange: — hump back fins pale orange — tail with central black bluish black spot
behind branchiae silver band: — back colour do
salmon.[442] — blueish above gradually shading down on sides fin tipped with fine red
especially tail but with black central band

---

441   Specimen 748 in *Zoology notes*, p. 178; listed as *Tetragonopterus rutilus* in *Fish*, p. 125, plate 23.
442   Specimen 747 in *Zoology notes*, p. 178; listed as *Tetragonopterus abramis* in *Fish*, p. 123, plate 23.

[44a]   Salmon[443] grows to 2 or 3 feet long: sharp belly about twice along.

The wind not being <u>quite</u> fair came to anchor: thousands of muskitoes, difficult
to sleep —

Thursday 17th [October 1833] —

Gale from SW & Rain. remained at anchor about 5 leagues above
St Nicholas. —

Fish[444] with low eyes, upper part of body & those fins with faint tint of yellow but
stronger on head with dorsal clouds of black, tip of tail do:

[45a]   beneath snow white; sometimes little bigger; usual size pupil black, iris white
not very common.

Little Indian boy our passenger value one ounce:

Friday 18th [October 1833] —

Night very cold, started early. Barrancas same height much broken down // Curious
equality & lowness of society: sons of officers commandante of B.B. officers count as
Representatives: general dishonesty. G. Rosas weighs everything one cannot
understand it at first: // In great Seco many people lost 20,000. others less: brought
live cattle to eat

[46a]   to San Pedro. Barandero — hundred of thousands carcases dead on banks (fall down
barrancas) float in water: could not pass many of the streams for smell — it would be
said some great flood had killed all, especially as after it all rivers were very much flooded
corresponding deposit // Carranchas muy picaros,[445] steal eggs //

Almost Becalmed sailed on till 2 AM. Passed San Pedro at early night

[47a]   Saturday 19th [October 1833]

Heard of revolutions: very little wind: Ennui

// yellow-breasted bird at Maldonado sings well //.

I should think the difference between Tosca rock & mortar is owing to the latter being
nearer point of origin of Lime — Cotemporaneous with Gravel of South ∴ not
subsequent as not broken: ? The alluvium /not/ extended cotem. with /part/ alluv
/rocks/ of T. del F.

I remember that the mortar on North side of great Escarpement North of Colorado —
is <u>Mortar</u> not argillo-calcareous rock. //

[48a]   a little South of Arroyo Cruz great valley. greatest interruption I have seen in Barranca a
mile or two wide doubtless connected with those which I noticed between Luxan &
Areco. — This is the only vallies which are not immediately explicable &
proportionable by present brooks: height of Barranca remains the same. — the river
here leaves the Barranca on the Southern point <u>of great valley</u>

---

443 Specimen 749 in *Zoology notes*, p. 178.
444 Specimen 746 in *Zoology notes*, p. 178.
445 Spanish for crafty, sly.

[49a]   Came to an anchor middle of the night, near the mouth.

Sunday 20th [October 1833]

Changed vessels, as the one which anchored near us drew less water, in this proceeded to the bar then changed again & proceeded a very narrow Riacho. — with many wild peaches & oranges on each side & some large black gallinaceous birds; tide being against

[50a]   us — hailed a canoe & proceeded to La Punta de St Fernando. — There first heard of great revolution, could not take boat: went all evening from one great man to other great man & at last got permission in the morning to go to General Rolor's camp.

Monday 21.st [October 1833] —

Arrived very early at the Camp; horrible looking set of men. — Rolor traitor:

[51a]   got order to go to General chief. difficult to procure horses. — often obliged to show Licence: water over horses back. Arrived at rebel Camp. General Rosas brother at last got permission to go with a party with white flag to bridge. — from thence on foot leaving recon &c behind to proceed if I could on foot to the city, passed centinel by pulling out

[52a]   old passport & making circuits at last reached the city. — They are down right scurrilous set & the town is much alarmed about being ransacked. I am in bad predicament. my servant peon is out in the country with my goods. the vessel coming down the river has my collections one million cattle died in grand seco Salinas have inhabitants from Flamingoes.

30 leagues above Bajada shells in Cliff

[53a]   M Video. big bottles paper. cork, Iron rust 8 by 8 inches Spirits Bladders:

M[r] Maclean

Capt. Fitz Roy write to Admiral Otway[446]

Specimen [s]

[dentist]

Wilson. Rector. Islington

Bishop of Calcutta

Price 2 & Dixon 1 Lumb. Valparaiso Chili

[54a]   Stevenson South America.[447]

Dolores names of Schooner in Bahia Blanca

5 dollars [Stuart]

Measure big bottles

Cigars Spanish bookseller opposite M[r] Waldegrave

446 Rear Admiral Sir Robert Waller Otway (1770–1846), Commander-in-chief of the South American Station.

447 Stevenson 1825.

[55a] November 2$^{d.}$ [1833] With difficulty got on board packet, heavy musketry || General utter profligacy character absolute government, History of revolution: at night foul winds
Many passengers: women & children all sick
/3^{d}/ [November 1833] Anchored in sight of Colonia
3^{d} fowl winds, intelligent German
4th [November 1833] arrived on board, *Beagle* does not sail
5th [November 1833] Took up residence on shore

[56a] 6th [November 1833] Started early with party to St Lucia, swam the horses ¼ of mile passed on about 5 or 6 leagues. — came to Barranca 50 40 or 60 50 feet high: composed of coarse particles of white quartz from dust to goose shot in size loosely cemented, with narrow bed of clay in one part, no organic remains. Covered with reddish Tosca which contains particles of quartz

[57a] & stalactites /&/ plates, /forming/ hexagons of calcareo-argillaceous rock
rock evidently deposited where Tosca has contracted but the substance contemporaneous, because there are parts forming ½ the mass & beneath that without scarcely any. — It appears to cover the sandstone filling its irregularities: Is this Tosca same nature with B. Ayres formation. — the

[58a] concretionary Tosca rock looks like it, but presence of quartz pebbles marked difference. It is curious never finding a fragment of bone — extreme modern from covering sand bed just like new forming beach of river. —
Recrossed river St Lucia at sun set. & slept at a Ranch returned rested early to city

[59a] Found gneiss cleavage near city running W by S & E by N M. Video
Saw great Iguana kill large green Lizard[448] —
Molita at M. Video /regular/ rocky soil: Tapalka
Nest of bird shot by Maldonado lake; sort of B B Bird

[60a] Dry beds of lakes of Coquimbo. —
Great earthquake of 1751 destroyed Concepcion?
The land rose. —
Shells found 500 ft elevation on Concepcion
English D^{r} at Mendoza has the head of a Megatherium.
Road to Coquimbo. Barranca with shells
Las Vacas gold-mine. Meirs Conchilee.
Uspallata plain 5970 ft. Meirs.[449]
The Canota pass from Mendoza. to Uspallata far most interesting:
R. Quinto bed. horizontal strata

---

448 Mentioned in *Zoology notes*, p. 381: 'I sent home a skin of a large lizard Iguana.— I know not its number. I saw it one day catch & kill a green lizard 7 or 8 inches long, & shake it like a dog'
449 Miers 1826, 1: 277. The following line on the Canota pass is also a note on this page.

(Cliffs at river of St. /Luis/ & north of it
(Los Gigantes. — Traversia between San. Lucia & Mendoza —
Mendoza 2600 ft.
I should think Luxan village good section S. of Mendoza
Pass of Canota in the Paramillo. horizontal beds over the strata. —
El como de los *[illeg]*

[61a] Castle Hill. — Calcareous & gypseous Tufa Incas bridge. —
R. de los Horcones much gypsum between Breccia.
Paramillo de las Cuevas. some say. Volcano
Red & green snow
M. Entre

‖ Post house Pindo	1773
Santiago	1691
Post house Chacuturo	2139
Cuasto of do.	2896
Villa Nueva.	2614
Primera Quebrada	3215
Guardia	5148
Ojos de Aqua.	6874
Casucha[450] of Juncalillo	7730
— — Calavera	9450
Cunha	11920
Las Cuevas	10044
near Estero de S. Maria	7928

W. foot of Paramillo de Juan polices on river bed 7380.

[62a]

Cuesta. Paramillo —	7888
Tambillitos	6250
Uspallata post house	5970
Villa Vicencio	5382
Mendoza	2602. —

D. Felipe Banza

Santiago	2620
Casa de las Calaveros	10603
— Cunha —	12709
— Cineras —	
— Puquiros —	
Mendoza —	4474

450 Small round towers with raised floors formerly used as store-houses by the Indians.

		Leagues	
Mendoza to Luxan.		5	
	Caracal	5	
	Estacado		12
	Arbolera		3
	Cenoza de Alvarez		4
	Capilla		3
[63a]	Chacaro		4
	Portillo		3 (ridge)
	Los Puquenos		3
	Casa de Piedra		6
	S. Gabriel		8
	Melacolon		5
	S. Josè		3
	Guardia		1 ½
	St. Jago —		9. ½
Total:			80. Spanish leagues
Santiago to Colina		6	
Villa de S. Rosa		19	
Guardia de Resquando		13	
Ojo de Aqua		5	
Laguna del Inca		5	
Calaneras		1 ½	
Carrula		1 ½	
Los Cuevos		1 ½	
Paramillo		1	
[64a] Puquiros		5	
Punta de las Vacas		3	
Uspallata		13	
Villa. Vicencio		15	
Mendoza.		15	
Total		104 ½ leagues	

Chlorite slate. Colina !. Caldcleugh

Salto de Agua

Gold mine in Iron Pyrites on road to Valparaiso. — Limits of Traversia Tosca rock
Cordonese geology:

ask Mr Caldcleugh about metallic veins. — Caldcleugh Mastodon direction of Lamina
of slate at St. Luis. —

Proceeding in straight direction to Rincon on bank of river good section of St Jago
plain — Badge —

/Perfect/ — Limit of snow Height of passes French: Naturalist Muscles near Mendoza

[65a] Fox Biscatcha

Was the great wave which destroyed Concepcion quite sudden? — Elevation of land at present day

I have to draw 261:2 at St Jago

2 dollars worth of /Iron/

Mr Smith — lake running E, W narrow but some leagues long. ~~ran~~ wave about a yard high: gutters with same direction stopped running: —

If the land really does oscillate 20–30 ft effect very trifling —

Pencil note - Book some in town *[illeg]*

[66a] 18.th [March 1835] ~~Rode over the burnt plain~~. Noticed clearly at the Almeira & Cuesta in front of St Jago, that the strata dipped internally, judging from large fragments, stratified greenstone worth examination. — Travelled till we came near the Maypo, struck up a not broad valley with high mountains on each hand. — Method of travelling &c &c luxurious water melons on road, provisions. Custom house civil on account of strong passport: pretty valley trees loaded with peaches bending & breaking with the weight — grapes nectarines & large apples. — Well within the valley, Cliffs of pebbles

[67a] 2-300 ft high appear to abut into valley as part of plain lowering higher up: river flows over bed of same. —

On first entering the valley pap of fine-grained syenite; after this there came an extensive line of low hills composed of very fine white granite with black mica — These run about N & South & were capped by great mass of stratified red rock, dip East: certainly porphyritic Breccia we afterward as some distance crossed this rock, much of which was of a greenish tinge & materials much blended,

[68a] in such a variety, I saw a band of coarse breccia about a foot wide. — Before however arriving at this, there was a rough track composed of semi-crystall. greenish Trappean rocks; part of which was slaty & what I have called "altered" generally however more crystalline parts slate coloured porphyritic with delicate crystals: I imagine this forms a subordinate ~~mass~~ strata between Porph-Brec. & granite — anyhow it is certain that the latter reposes & appears filled by the former: there was also

[69a] a mass of high crystall-porphy-greenstone, the origin of which probably is distinct from the greenstones & altered slates. This is the very first range

19th [March 1835] In front of sleeping place a hill of soft, whitish brecciated rock, (A) & nature not certain, as we passed along there appeared a transition by the varieties (B) (C) (D) into a compact greens porphyry — hence I presume it is merely a variety in Porph Breccia, as far as I could see from stratification it overlies the more crystalline varieties; part of the latter was partially columnar. —

[70a] our road turned for long time south & then about SE: on each side bounded by great mountains with very abrupt escarpements divided into many strata all I believe

Porph=Breccia — the coarsest fragments 6–12 inches long, separate sooner than break, chiefly perhaps in highest parst [parts] — Strata shown by a more /stony/ crumbly variety interposed between compact sorts of which (E) is specimen: Such wherever I could see, is the constitution, the strata are almost always inclined but not at great ∠ & apparently various direction, generally however the

[71a]   main valley, has  basset[451] edge on each hand, in some, same dip on each side, but angle various; one ravine synclinal — I also saw two very clear mantle shaped spots. — In few places were there dykes: I saw however this structure

[section showing dyke]

where (at great distance, a rock appeared to have been injected: The Porph-Breccia immense thickness, the summit of mountain of S. Pedro Nolasko is composed of it — I judge from great fragments remarkable color & stratification. — Probably 6-7000

[72a]   ft: at least. — I saw one spot dip at about 50° — whole scene great scene of violence & subsequent excavation in cracks — perhaps N only most prevalent. — As we road up the valley. — fragments announced: approach of Granite: a range of white hills is composed of fine white syenite with very little quartz & some ~~feldspar~~ mica (F) — also angular black patches, which I forgot to mention yesterday — also the stellated Black mineral of Yaquil — These mountains

[73a]   appear to the ~~South~~ E (running about SE & NW to be capped by grand mass of P. Bre. dip from, & the group is surrounded on all sides by much higher hills of smally inclined P.B. =
Mem. Gay — Cauquenes — Granite the metamorphosing agent. —
Roar of Maypo from stones every where. /its/ branches plain 2-500 ft high. — where ravine does enter. rays of ridges —
(Mem: Condor of Concepcion). — Hence cultivation extends so far. —

[74a]   Rode till we came to almost the last house; ~~plain~~ fringe more thinly populated: Scenery grand /almost/ faces, stratified, color purple, no wild forms, cloudless sky a remarkable scene if not very beautiful. — Time of year late, bringing in the animals. — Muddy Maypo, like sea — mountain I suppose 3-5000 ft high on each side very massive. — The granite /formed/ not very far from the active Volcano of Maypo. —

---

451 Edge of stratum cropping out (OED).

[75a] 20th [March 1835]

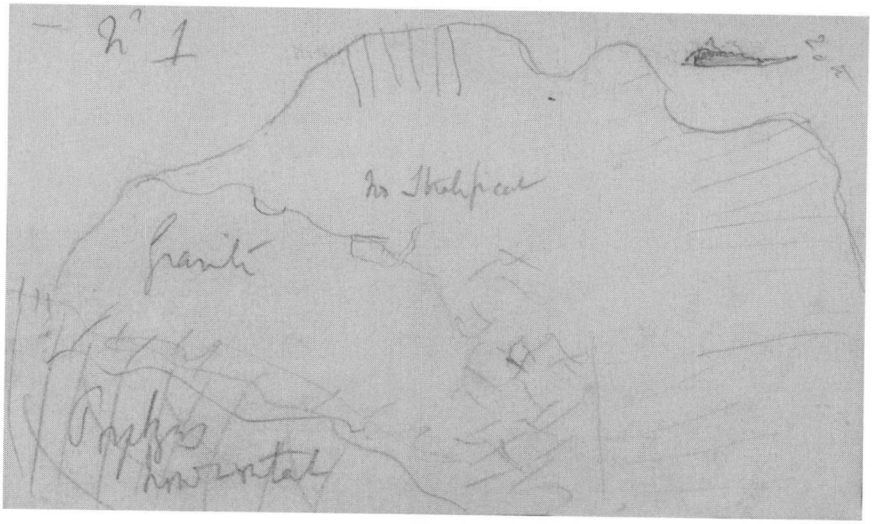

n^{r} 1 Granite no stratificat Porphyry horizontal[452]

[76a]

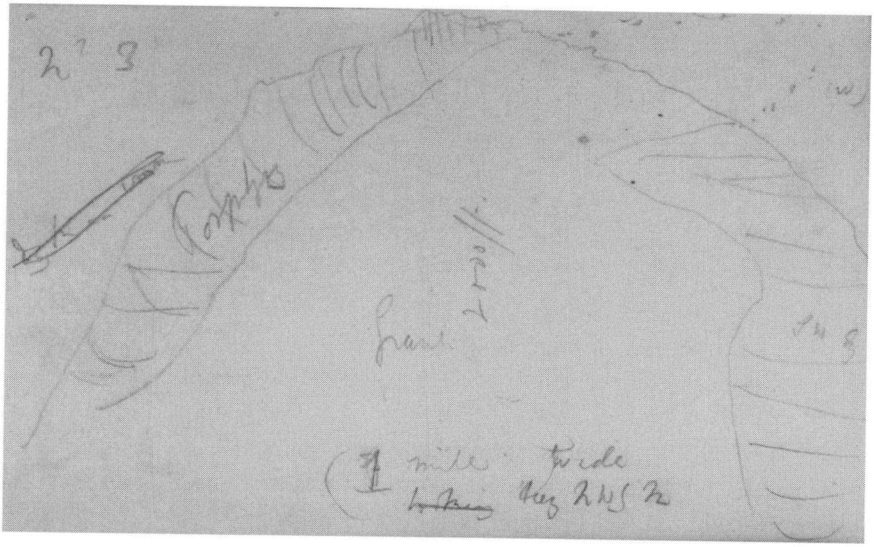

n^{r} 3 ~~dyke imm~~ Porphyry Granite 2000 ft (W) (1 mile wide ~~looking~~ NW by N SW */by]*

452 See the fair copy of this and the following sketch in the geological diary DAR 36.473A.

[77a]

n^r 2 70 Granite 80°[453]

[78a]

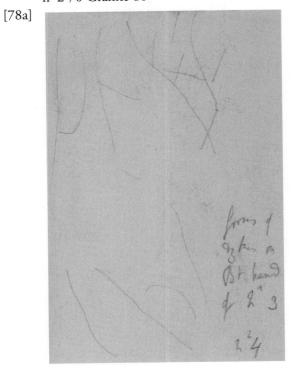

n^r 4 /forms/ of dykes on Rt hand on n^r 3

453 See a possible fair copy of this sketch in the geological diary DAR 36.474.

[79a]   (20th—). [March 1833] A little above our sleeping place is the junction of the M R. del
        Valle del Yeso & the R: of Volcan: with the Syenite some true Granite occurs & there
        are nests of the green mineral of Quillota & black of Yaquil. — So The granite passes
        into (J) variety — judging from fragments is directly covered by black hornblendic
        rock (G) which in parts is very obscurely brecciated (H) there is some true, hence
        perhaps origin — The granite affects pap forms hills, & from the

[80a]   dip is covered by Brecc: Porph: Beyond this, (just) the valley is partly formed & partly
        traverses the strata. There dip is about 40° & bends up to vertical as at Cauquenes: I see
        in the Porph. Brecc — most beautiful & numerous alterations of finer sediments &
        some included large white beds: we approached a group of lofty very sharp points:
        /H/ near /it/ side of hill traversed by very many dykes. — though diff formation:
        [illeg] near

[81a]   (& for a hours ride shown by fragments), saw appearance as in n^r 1; where immense
        mass of granite joins above in a very undulating (more iron pyrites) line, (& veins)
        inclined with superior Porph. it is fronted at some little distance by porphyry nearly
        horizontal — The very peak is inclined at about 70° & curved, the side is only a little
        inclined: the lower parts appears without stratification. I judge from enormous hills of
        fragments.— The above section is seen from the west — another main valley

[82a]   beyond Cuesta del Indio, shows, the granite (n^r 2) ( ) a mountain within another capped
        by almost vertical Porph: (hence new forms) & curved lateral section. — The line of
        granite appears to run N by W & S by E. — On opposite side of valley perhaps seen:
        (n^r 3) west side <u>whole mountain</u> dip of about 50° on East side inclined at about 80° to
        same direction, but tops appear turned over: Traversed by extraordinary net work
        of green dykes — These are granite dykes traversing the highly inclined strata

[83a]   from the mass — (n^r 4 large dykes). — magnificent examples of elevation. — explains
        all rough tops, above snow level. — Fragments of saccharine limestone.
        Valley very curious higher up:⁴⁵⁴

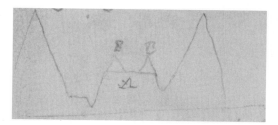

[valleys in the Valle del Yeso] B B A

A central plain of Alluvium enormous angular fragments separate a ~~valley~~ ravine
nearly 1000 ft deep from other 400–500 ft — which run nearly parallel: the mass is
generally plain, but for some extent forms valley between two rows of hills of Alluvium
as shown

454 See the short description of this moraine in *Origin of species*, p. 373.

[84a]   ascending, again we have a plain, which terminates in a mass of lofty hills perhaps
        1500 ft above river, as if pitched by water spout, & hollow round cavities these are
        same as the lateral hills (BB). the ~~left~~ Rt hand & smaller river arises between these hills,
        the ravine loosing its depth — The left, larger & Southern stream Valle del Yeso, flows
        on round their foot — I hardly dare affirm these hills are alluvium — yet fragments
        chiefly granite & some others, no true rock —
        To the Eastward of the plain just like bed of lake through

[85a]   which the latter river flows

[section showing plain and river] grand plain plain (B) alluvium river /river/
The structure of intermediate plain sometimes a valley requires either diluvial action or
quiet sea — No proof of quiet sea — either in levelness of hills on head or rounded
stones — Properly these hills ~~in~~ (B) ought to exist in all parts. Alluvium of same height
seen on all sides of grand valley. —
In bed of lake. river has taken small effect passes out by narrow entrance. — Most
striking geology Porphyry. white granite green dykes — Gypsum —

[86a]   where dip is mentioned as about 40° — we have following appearance

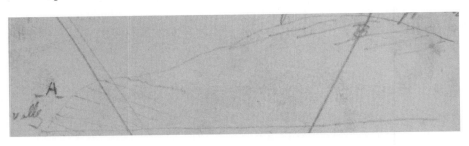

[sketch section]⁴⁵⁵ valley A B
in one hill on opposite side of valley strata continue bending up to vertical — at B at
very little inclined dipping directly towards latter. — I believe (B) are the hornblendic
rock tilted by granite already mentioned The tilt of A is quite separate although in same
hill & dips SW direct NW & SE. — ⸮ parallel to granite ? —
Some of the breccia exceedingly

---

455 See the fair copy of this sketch in the geological diary DAR 36.470.

[87a] pretty patches of green & various coloured fragments: — Where I first noticed granite (n^r 1) on the plain fragment as big as house of granite <u>one</u> side large fragments of porphyry united by granite. — Saw ~~some~~ much of the curious rock & feldspar with the porph: Breccia —

Lost mules: very hot: party from Mendoza — hills very lofty patched with snow. — rode over some natural ice-house purple color — fine peaks from vertical stratification — enormous piles of rubbish — Mountains few separate, not many lines

[88a] enormous masses — still resplendent clear — dark blue cloudless sky (some little Puna[456]) — very little vegetation, no birds or insects. — Old Indian house. — after crossing great pile of alluvium Cuesta del /Indio/. striking bed of lake: cattle: Slept with Vaccaros. Valle del Yeso. — Cold — Summit purple rock. —

Condor. 2 large white eggs no nest lay in November or December — a whole year before can fly. are called condors

[89a] with black ruff. white Huitre:

Have seen nothing yet which looks like Lava

[90a]

N^r 5 W Pebbles &c Porphyry G gypsum G Sandstone B /river/ Breccia fine specimen 1000ft–2000ft Gypsum Porph N.[457]

---

456 'The short breathing from the rarefied air is called by the Chilenos, Puna', *Beagle diary*, p. 308.

457 See a possible fair copy of this sketch in the geological diary DAR 36.475A.

[91a]

n^r 6 E P P W Gypsum S & without Strata

[92a]

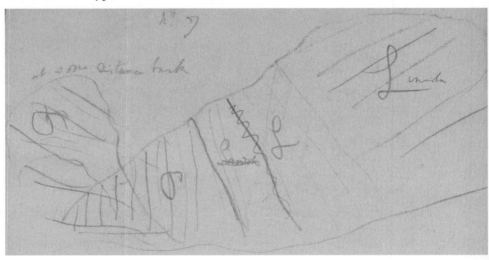

n^r 7 at some distance back G L Sandstone Limestone

[93a]   21st [March 1833]–[458]
         L. Breccia (Red Sandstone) section n^r 5
         (M) Impure gypsum <u>do</u>
         (N) Slaty sonorous Limestone <u>Pass of Puquenos</u>

458 This date appears to be repeated on p. 127a.

Val del Yeso

(O) Gypsum with lines of black crystals —

(P) Impure gypsum? ~~more abundant~~ not worked ~~with~~

(Q) Large <u>pap</u> of <u>statuary</u> marble

(R) Concretions in gypsum generally not so dark coloured marble

[94a] (S) <u>Green</u> . sandstone

(T) Ferruginous sandstone

(W) . do with green Epidote[459] (abundant

(V) Porphyry (more crystalline than the rest from protruding pap

(W) Carbonaceous? shale with do *[3 or 4 words illeg]*

Tupunjeta: scenery: condor. cold wind

Puna — scenery: colours /form size/ profound /despise/ Valle del Yeso. Section n^r 5. 3 vallies — Puquenos — slaty sandstone Stratification beyond —

Red snow — mules. Peaks of snow glaciers — cloudy night mercy of Elements no thunder no danger: picturesque party of travellers from Mendoza

[95a] beautiful sun rise, showing peaks already bright with sun. — Sulphuretted Hydrogen Potatos not boiled ~~water~~ /pot/ did not close The point above the granite is one of the highest in the country. the dip although about 70° is not directly from the granite, but seems influenced by some common cause which has tilted all the neighbouring peaks from 80°–90° to the West. —

The gypsum formation is first seen on South side of valley (of lake) if only seen there great doubt would remain concerning its position: we see a very extensive side of hill great mountain entirely composed of masses which pass into each other suddenly without regular strata. a large patch of one many hundred feet every

[96a] way will contain other varieties or bounded by them. naming the rocks in abundance perhaps an indurated Ferruginous Sandstone (T) which passes into a Quartz rock, & most generally much green crystals (W). We have perhaps in equal quantities an indurated bright green Sandstone (L). & the gypsum formation This consists of fine white saccharine stone, with lines of black crystals (O). & water lines This occurs in enormous masses: it contains large concretions of ~~fine~~ coarsely crystallised marble either white or slate coloured (R). The surface is /rough/

[97a] & marked sometimes with water lines is intimately blended or joined with gypsum: some of the concretions are many feet in each way & of an irregular figure. — There are also large masses of beautiful white masses paps of statuary marble (Q). — ~~Surface~~ curiously traversed by fissures, as if baked, like loaf of bread. — Judging from other places. <u>green</u> Sandstone most regular companion of Gypsum: There is much impure gypsum which is not used for the wine (/Lastos for agua diente/): this occurs in a baked or concretionary

459 Silicate mineral often found in metamorphic rocks.

[98a]   forms — I saw interstratified in thin & contorted layers with green sandstone. — I only
        saw one fragment of gypsum crystall. in transparent plates. —
        From general state of all the rocks there can be no doubt action of heat. —
        This whole great formation seemed entirely to underline the B. *[illeg]* as judging from
        stratification of neighbouring mountains. also such soft beds might be seen traversing in
        a vertical direction the grand peak above the mines: a little to East of Granite Peaks. —

[99a]   yet at the very mine a little ridge or point of vertical Porphyry (V), crystals scarcely
        perceptible & a black slaty rock (W) to the height of 200–300 ft projects right in the
        gypsum: which latter rests uncomformably on the top of strata. being separated by
        broken ~~mass~~ fragments of greensandstone such as occurs contemp with gypsum, I feel
        no doubt now that it has bodily been forced upwards.
        Before following sections I was quite at a loss: the gypsum here is not clearly stratified
        but clearly at right angle. —

[100a] A little higher up in the valley on the north side we have the section (n^r 5), where we see
        an enormous mass of gypsum. I should think at least 1000 ft resting on & covered
        conformably by the red P. Breccia. — This probably represents the whole mass of
        sandstone & gypsum described as a mile or two to the SW. — The gypsum is not very
        pure, & contains a good deal of Carb of Lime. it is much marked by water lines (M): it
        rests on a Breccia fine grained

[101a] Very little altered dark red as usual (L). The ~~par.~~ section is not quite a straight one, as
        where brook enters there is a angle. the lower Breccia bends up, & becomes highly
        inclined about 70° degrees dips Easterly. conformable to it there is stratum of fine green
        sandstone, & then again the gypsum. as seen by the water lines is ~~parallel~~
        conformable & ~~side by side~~ but angle increasing almost to vertical — These three beds
        are beautifully seen in contact: The gypsum may be 2-400 ft thick. — Again to the left
        or west we find a very

[102a] Porph-Brecc dipping at *[about]* 45, in opp direction *[ver]* to west — the intermediate
        space is covered. —
        on the South side anticlinal line of this valley & directly in S. line at a great elevation
        (section 6) in the mountain side the gypsum is seen between nearly vertical strata of
        Porph. Breccia to the west the Porphyry is without stratification & probably
        corresponds to the opposite dip mentioned in last ~~dip~~ section —
        I should imagine the gypsum formation is inferior to the grand P. B. *[in]* the outer

[103a] Cordilleras: — In this neighbourhood I noticed some slaty beds, probably a sandstone
        with the P. B. — we soon began the long ascent of the talus which forms the pass of
        Puquenes: there many enormous fragments of a pale brown limestone with immense
        quantities of Gryphaea: also a piece of an ammonite[460] as thick as my arm which formed
        part of a spire — Higher up we came to a black hard compact Aluminous Limestone

---

460 Type of fossil mollusc, usually with a coiled shell, important for dating Secondary age rocks.

which alternated with innumerable layers of a harder slaty sonorous kind (N). This abounded with very imperfect

[104a]   impressions of Terebratulata & many bivalves[461] one only univalve[462] except some few ammonites. These Limestones all dipped to the East at ∠ *[for]* 30° to 45° & some of the points considerably higher up to 70° or 80° — It rested on the red P. Breccia; contained a bed of gypsum, with ~~Limestone~~ Marble & green Sandstone perhaps 2-400 ft thick, which was capped by this Limestone with shells. The very pass is thus formed & neighbouring peaks —

[105a]   I shall presently show that probably this Limestone is capped again by Porphyry — Unfortunately I cannot tell the relation of this Limestone with the grand Gypsum formation: Probably the great thickness of the ~~former~~ latter is a local formation: The Limestone must I imagine be at least 3000–5000 ft thick: so difficult to judge: — I imagine from color in some of the mountains of yesterday in the higher parts I saw the Limestone but not to so great thickness. following section will show what I think of

[106a]   the superposition. —

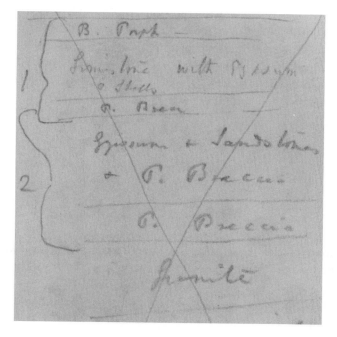

[stratigraphical section] 1 B. Porph. — Limestone with gypsum & shells P. Brecc.
2 Gypsum & Sandstone & P. Breccia P. Breccia Granite
The relative position of two groups 1 & 2 is uncertain, but the beds in each known —
If they do not replace each other, such

461 Molluscs with two shells (e.g. mussels, oysters).
462 CD used the word univalve for snails.

[107a] must be their superposition:

(W) Limestone A (E Lime Gypsum Limestone Porph B

(A) In a W & E section (A) is the pass of Puquenes: beneath which beds will be *[seen]*. on the west side lower down are anticlinal beds of Limestone will be seen forming a little N & S ravine. —

To the north of the pass on west side. *[many]* most lofty peaks with the wildest forms will be seen with the strata

[108a] dipping from 70°–80°. These strata generally bend up in a small curve from the ridge where inclination is rather less. — These peaks & ridge formed either of Limestone or P. B. form, as (on W. side) 3 or four anticlinal ~~ridges~~ lines, N & S ravines occurring both in synclinal & anticlinal parts. —

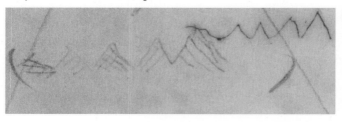

[section of hills]

I have omitted to state, all these lines are about NNE & SSE

[109a]

W Gypsum Limestone BP BP E

Having crossed the pass — we rode over Limestone dipping, as before regularly Easterly: /where/ to the north of the main valley (of degradation in which road lies E & W) the Limestone may be seen gradually bending till it becomes nearly Vertical: across a little ravine we see very red Porph. Brecc with same dip, which ~~wh~~ is /yet/ a little Easterly — again there is a broader valley & the Porphyry dips at about 45° to west —

[110a] (Section n^r 7) is on the opposite side of the valley of degradation /a/ to the South. The gently inclined Limestone meets by a fault that of 70° degrees. the inclination of which increases till it is nearly vertical when it lies along side some of the bright red Porphyry: These sections prove, I think that the Limestone is properly covered by the B. Porph. — I may mention that the Limestone on this slope also contained fossil remains —

[111a]

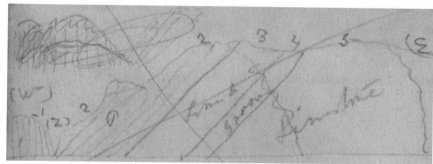

(W) 1 (2) 2 2 3 4 5 P Limestone Gypsum Limestone (E
Section on North side of valley proceeding Eastwards. — Beyond the nearly vertical: we see mountain of very red P. Breccia: beneath which there appeared to be ~~(2)~~ Limestone, which contained a bed of same thickness as last section of gypsum (4). beyond this we have again the Limestone (∠ about 45° exactly same nature: all this is identical, with the section of the last pass & is clearly owing another parallel

[112a] line of elevation (Z) being Zynclinal line. —
Limestone (5) had /gr/ ∠ perhaps 60°

W Limestone Congom C (E

Again to the East of last hill of Limestone: we have for the space of some miles a succession of hills of Conglomerate bed from 1500 to 2000 ft. thick, color reddish sufficiently hard for small pebbles to break before extraction, large one will come out

[113a]   it is very hard — The dip was almost everywhere about 45°, as shown by line of pebbles: Excepting that close to the base of the Limestone the dip was a little less 20° or 30°. from every point of view it appeared most unquestionably to dip directly under the Limestone: Seeing this, & colour & Brecc. Conglom. structure, I came ~to~ first to conclusion that this was the case: when I examined it I found there was not Porph: structure in matrix, & on examining the pebbles few were crystalline but the black Limestone, with <u>its shells</u> & the peculiar

[114a]   Sandstone which accompany the Gypsum. — This made the case clear. — The Conglomerate appears to have undergone action of heat: rises into lofty mountains, /fronting/ the Escarpements, (at less however elevation) of the Limestone: It will be seen perhaps owing to upheaval of granite of Portillo chain. —

Marine deposit. —

(X) This new conglomerate

(3) Hill of Porphyry in do

(2) Slaty micaceous Sandstone

[115a]   (1) do converted into Quartz rock

(2) Protogene[463] W of Portillo

(3) & ~2)~ Reddish /Eurite/ in granite

(4) Serpentine (?)

(5) Gneiss on East Slope

This new conglomerates fill up the space between Limestone at base of Puquenes range & the Portillo range. /Only/ fringe round one hill of Porphyry (nearer to the latter range) which Porphyry (4) is one of the rare instance /where/ for what I can see to the contrary may

---

463   The 'earliest' rocks. At the time it often meant gneiss and other crystalline rocks. Closely related to Primitive and primary. See notes 207 and 241.

[116a] may have flowed. —

W 1 conglomerate 2 ? 3 4 Granite E (1)

The conglomerate rests conformably on some soft (2) & variously coloured beds the nature & age of which I am ignorant of they appear soft & like wacke in the manner in which they crumble: — These rest on inclined beds of a very evenly & much laminated (z) micaceous hard Sandstone; this, have been traversed by some soft

[117a] porphyritic dykes: & containing some hard. black clay-slate: This rests on the grand granite range — Where near the granite it is converted into stratified granular quartz rock (1). — It is traversed (near here) by grand system of granite dykes, well seen by red colour of Granite. —

Is is this Sandstone same age as Conglomerate, with which it is conformable, or much older. — The granite extends a long distance to the

[118a] west of Portillo. it is almost all Protogene (2). large red crystals of feldspar & do of quartz in lines (some Epidote), little Chlorite. Hornblende or Mica — Porphyritic with large crystals of Feldspar There is some true Granite. Forms pointed conical hills in enormous groups —

At the the Portillo it is chiefly a grey syenite or granite fine grained. is traversed by great dyke shaped masses of hard flesh-coloured Eurite,[464] hard, strike *[illeg]* (3) & ~~som~~ also serpentine? (4) Also a soft

[119a] decomposed. substance like wacke. — Such forms the very Portillo. Also on one side saw near this spot a stratified mass reposing on granite. of some soft substances of same color & structure as noticed on other side as overlying: the slaty Sandstone: But the granite as I have said extends right across the ridge. On the eastern slope judging from fragments there is both ferruginous quartz & gneiss — at the Manantiales where first vegetation begins, there was much fine white granite containing often times prisms

464 At the time this meant a variety of syenite which showed evidence of having flowed freely.

[120a] of Hornblende. This chiefly was on the N. side on the South Protogene: which lower down the valley was universal of a fine red color, forms peaked cones: peaks owing vertical planes of fissures: in some places this protogene passed very much into the nature of the compact. flesh-colored dyke of the Portillo. Having ~~often~~ seen in many places the true white granite traversed by red dykes near the Portillo. & the general distinctness of the hills of the two sorts. I very much suspect the Protogene is of posterior injected origin: descending

[121a] by the Mal Paso The protogene is capped in very many places by grand horizontal mass of a gneiss, with <u>numerous</u> /wavy/ lines of quartz exactly as in Chiloe. — The Protogene projects in little points through it: has pieces sticking on the sides; is entirely naked or covered in horizontal manner: In an enormous mass which had fallen down, I examined the junction; the cleavage of gneiss is at right angles to it. — the Protogene sends off small veins into it & includes fragments. —
Lower down the valley the gneiss seems to pass into a grey

[122a] harsh ~~stratified~~ laminated mass of granular slate (6). — The (Los Arenales) ravine in which all this is seen runs almost directly from little below. Portillo to Pampas in East line. — whereas the road in the other side is zigzag. crossing apparent from one *[illeg]* valley to other. by the water courses. — ? ~~I can well im~~ ʿ Is the slaty Sandstone, which dips conformably with the Conglomerate of the same age or much anterior: ~~it contained~~ no fossil remains I could see: ? — The question is an interesting one. — I feel no doubt that the whole Portillo

[123a] range has been elevated subsequently to the Limestone one having been partially elevated into dry land. —
In the <u>last section</u>, there was a fact showing the complication of the stratification:

(A) Granite Sandstone ? Conglomerate Conglom
A hill (A) a little to the South apparently composed of B. Porph. dipped at about ∠ 20° right towards the granite range. —
In the valley between Portillo & Puquenes ridge there was some very cellular Tufa. — Probably connected with neighbour Tupungata — no Lava —

[124a] The granite extends at least mile to the west of Portillo:
/Ribbon/ jasper with first Gypsum formation in fragments: —
<u>The last section</u> shows that probably there has been more than one line of elevation in space between two ranges

At foot of ~~granite~~ [of] Protogene, where covered by slaty ~~qua~~ sandstone, there was a small arch of do reminded me of main range in Falklands. —

W sandstone granite

Early in journey, noticed in many places in P. Breccia. the [em] spherical - - <u>columnar</u> <u>structure</u>

[125a]

(W) number of lines anterior P G P P G P gypsum 10 Lime at least 3 lines of elevation of B.P & Limestone (E) 1s Pass Limest P P Limest Gypsum = R2 Limestone Conglomerate Porphyry [Hill] [true]? Slaty Sandstone Granite Portillo X is level of mouth of valley of Maypu

[126a] Red Snow[465] — seen on both of the highest ridges. — (little spores, ~~rather more than~~ twice their diameter apart

above limit of perpetual snow. — appeared like bits of ~~dirt~~ brown dirt scattered over snow — partly optical deception seen through the globule of ice: appeared of all sizes. to at at 1/8th of inch — When picked up appear to disappear: Examined in lens are

---

465 The red snow is discussed in an entry dated 20 March 1835 in *Zoology notes* pp. 286–7 and *Beagle diary*, p. 309, *Journal of researches*, pp. 394–5, and *Beagle plants*, pp. 207–9. Keynes identifies the 'spores' as the alga *Chlamydomonas nivalis, Zoology notes*, p. 288.

groups of 20–40 little circular balls — through both lens. appear like eggs of small
molluscous animal: — Crushed stain fingers, & paper noticed by hoofs of mules &
where thawed: thought it dust of Brecc Porph: although remembering Miers:[466] color
where mules had trod, beautiful rose with slight tint of brick

[127a] red: examine paper
At great height — saw condor — common sparrow & grey bird very high of flock of
plover — also black Furnarius — P. Julian Finch. — many plants & bushes the same
21st [March 1835] We left our friends the workers of gypsum: & continued up the flat
valley. — curious deception seeing a flat valley with inclination towards one, appeared
inclined opposite direction. — Flock of Condors [gogged]: ascent of first range of Puquenes.
zigzag very steep: few yards. mules breath: admirable mules: wild party long string crys.
appear so diminutive no vegetation to compare with: passed over much perpetual

[128a] view looking westerly up valley. from below Mal Paso. showing Protogene & gneiss. —
shaded parts latter. —

Granite Z (no.9) (m) ⊥ planes of fissure (a) [Fore] Ground[467]

[129a] snow — red snow — curious peaks: dead horse beyond Portillo: upside down. — Puna:
strange ideas about all the water here has Puna. — tightness of head & chest: like
running on frosty morning after warm — room: — running fifty yards deep & difficult
no other sensation: imagination much to do with it — fossil shells forget — people die
of it? graves — Resplendently clear — piles of Talus, bright coloured rock. —
magnificent wild forms. — (view from the 1st ridge) something inexpressibly grand:
would not speak: despise taste of those: like Thunder-storm: splendid contrast of
colours — snow. —

466 Miers 1826, 1: 322–3.
467 See the fair copy of this sketch in the geological diary DAR 36.490v.

[130a] profound valleys — no insects few birds — Condor — little pretty plants — many torrents: excessively cold wind — Slept at night at the foot of the first pass: where little vegetation appears, headache — bad fire: pot all night boiling Peons conversing, "The cursed pot does not choose to boil Potatoes" amusing conclusion — Cloudy night: after excessively cold evening: mercy of elements: no thunder no danger on awakening the Arriero — Saw Tupungata immense beds of perpetual

[131a] snow. — a whole district goes by that name. — Small glaciers I think: Arriero has seen smoke. — Never shall forget the grandeur of the view from first pass. —
22^d [March 1835] Hilly intermediate country just within limits of vegetation: in two months of year send Cattle: dare not do so now — nearly all the Guanaco have gone —
[illeg]
Cloud disappeared showed us the peaks already bright in the sun. appeared through gaps in mist of stupendous height: smell of Sulp.- Hydrogen said only to be perceived in the morning.

[132a] Began ascent of Portillo: wild red granite peaks: fine view of the Crater of Tupungata & the Escarpements of last ridge: perpetual snow — frozen spiculae fell thick all mist. — Portillo narrow — descended to vegetation:
good protection beneath big stones — clouds disappeared severe frost — moon & stars — excessively bright — several parties anxious enquiries about snow — ~~mis~~
23^d [March 1835] Descended ravine, much steeper & shorter than ascent: grand plain like sea of

[133a] white clouds. beneath brilliant where we were: hid the equally level Pampas. — Entered the /band/ clouds the region of bushes & stopped at one oclock. — In mist whole day. — (Los Arenales)
24th [March 1835] — Vegetation spring bushes. many flowers like Patagonia: Blue & Orange finch long-tailed tit: tufted do. red-tail Furnarius. Guanaco dung in heap: just the same in appearance (Ulloa):[468] very many mice:
Biscatcha on a peak: very different aspect: more bushy tail. tinge of red in breast: <u>Viviparous</u> (<u>Autumn</u>) Lizard: centre of back. scales. black edged narrowly with dirty yellow this band broadest in centre: on each side of this ash-colored space:— sides scales blackish brown: rather more broadly edged with yellow — Belly pale ash color; legs & head

[134a] do. /furnished/ with few black spots: killed by blow of hammer: young are protruded: soon died: amongst fragments of mica slate, high barren mountain first limit of bushes. —
Mr Gay Valdivia elevation 6-7000 ft
(6). Gneiss
(7) Black Clay Slate with impressions of crystals of Iron Pyrites
(8). Grand system of Porph. Eurit. dykes in do.

468 Ulloa 1806, 1: 440–1.

(9) one dyke in do.

(10) dyke between protogene & slates

(11) Lowest bed of Greystone[469] Lava with Olivine

(12). do higher up

(13). a variety of Basaltic Lava

(14). Intermediate bed of Scoria

In water course <u>nothing</u> but gneiss & granite & Protogine latter most abundant

Pass of Puquenes, divortium aquarum: & Mendoza thick

[135a]

n^r 10 1 2 3 4 5 6 7 8 dykes (z)

[136a] Snake, sandy plain. Chaquaio — color primrose yellow with broad jet-black bands, which contain bright scarlet red square marks: belly black, except beyond tail where rings of black & scarlet are continued all round — Scarlet brightest near head — young one —

I have already said that the gneiss regularly reposes on the Protogine & that its dip of cleavage is Easterly: n^r. 9 represents a view where shaded parts are gneiss. — cannot fancy a more curious one: contrast of colors

highest peaks just covered, some base & lower down from other point of view exact correspondence

[137a] on opposite sides of profound valley in height of horizontal line of Protogine, no limit of excavation of valleys. — n^r. 10 is continuation of /Panorama/ to the South: (z) in each case being same lump of granite

on a slope of gneiss perhaps. 1200 ft is a cliff of Lava from 3-400 ft thick: this Lava declines a ~~little~~ to the East. but not so great angle on the gneiss. —

---

469 A general term for lavas intermediate between basalts and trachytes.

The surface is not level, is seen resting on, backed to the West & to the South by gneiss in this latter quarter, great bare hills of Protogine project up through the gneiss, close on to the ~~gran~~ Lava. (7) are these hills, which cannot be seen, from this point of view: I traced the gneiss almost up to the Lava but not junction: the gneiss Spec (6) contained in one place black clay slate (7). —

[138a]   The formation is remarkable by 8 great dykes of porphyritic Eurite (Spec 8). the lowest is greenish with fine large crystals of Feldspar, the others are all exactly similar, rock of pale color, where the base is now yellowish, apparently originally of a pale greenish with small crystals: — These dykes are parallel to the layers of gneiss — I only know they are dykes by examining the junctions in two places, which is waving & violent, each rock perfect character in contact: some of the upper ones ~~are~~ quite break their parallelism: most of the dykes are 20–30 ft thick & in the middle of hill are nearly as abundant as the gneiss. — I have said, how completely the gneiss forms a scale over

[139a]   the Protogine (where I examined junction behind Lava, Protogine contained some white variety): who will say that these dyke do not come from the Protogine? There was one black irregular dyke (9): & another ferruginous sort, traversed the junction of Protogine & gneiss: It is manifest even from the Pampas: that the whole of the Protogine has once been covered up & much clearly remains to the South — from color so much bare Protogine is an unusual phenomenon. Returning to section (10) 2 is pap of granite as in (7) rising in gneiss (1) in lowest bed of Greystone Lava with olivine: laminated these laminae often curved & nearly vertical (Spec 11) (2). bed 25 ft thick of broken pieces of red compact scoriae, cemented together apparent by an aqueous cement

[140a]   whitish (Spec of scoriae 14) are pebble of basalt: bed very uneven: undulating (3 & 4) two beds (as they appear) of greystone (Spec 12) (5). Covering of cemented red scoriae as before — There is some basalt, for I saw much large fragments (Spec 13) — (6) Covering of alluvium, thick some rounded pebbles, cemented in places together by mortar: — Descending valley to the Guardina, we there meet form: of Tufa: light stone compact, scales of mica numerous fragments of Pumice & some few bits of hard rock (used as dripping stone in Mendoza), is continued for some extent in little water worn hills & on sides of valleys

[141a]   The Lava has clearly been deposited before the present valley has been excavated: it must have come from north side, for we have /traced/ old rocks on W & S side & it declines to East side. — Could not see any crater-formed hill. — After the valley had partially been excavated, the Tufa must have deposited, which has subsequently been cut through the underlying gneiss — The tufa forms a line of cliffs beneath the Lava — At the base of the Protogine ~~hills~~ mountains: the foot of Cordilleras is composed only, a few little water worn mole hills — Beneath the Tufa saw a pap of white granite coated by gneiss. — The Lava continues to descend. (columnar in one spot) is at last stopped (but must have

[142a] been before present transverse opening) by a line of low hills which dip NW by
        W & SE by E. —
        The Lava regularly abuts up against these hills being perhaps 200 ft thick. — The
        extreme ranges of low hills, mole hills as I have said are oblique to main ranges. — there
        are several of them, but none /had/ much more than 500–600 ft high above /br/ sloping
        plain: The first I examined is entirely composed of bright purple & red Porphyry (with
        grains of quartz) only in one place. I saw Breccia structure; they run NW by W & SE
        by E; are traversed by river. — The next was a Porphyry color like protogine, almost
        composed of brick-coloured feldspar —

[143a] perhaps ejected. I omitted the first had a very obscure dip of about 60 70° to SW. —
        Beyond this there was one other range, which I believe the to be Porph. Breccia: After
        what we have seen cannot doubt that there are paps of granite not far beneath: perhaps
        probably they appear as in some places:
        The recent upheaval of Protogine explains great deficiency of Porph. Brecc. at this part
        of Eastern slope: proved by Patagonian pebble bed not general the case. —
        white-tailed Callandra: white-tailed humming bird — little parrot:
        These low hills are separated by broad flat valleys: where Porphyry: pebbles are white
        washed exactly as in Patagonia these valleys perfectly unite

[144a] with the slope which for a distance of nearly 10 miles forms a considerable angle with the
        Pampas — vegetation exactly like R. Negro. — Where Porphyry low ranges were nothing
        but pebbles from them. — Further in the plain /large/ rounded blocks of Granite:
        NB. Beside (ejected?) red Porphyry, there were other hills of whitish do —
        did not start till midday, ascended volcanic beds — view of Pampas: dark blue, from
        rising Eastern sun: striking line of glittering water lost in immense distance: very level to
        the South. considerable undulations to the North: passed Guardia. heard of much rain!
        day before, the clouds which formed beneath our

[145a] feet — true Pampas Indian: kept to hunt: always one: some few years. came across
        tracks of one who had passed on one side. hunted him whole day; over dry stony
        mountain & found him concealed. — on Horseback?
        Entered broad flat valleys: view of Pampas again — I think more striking, certainly very
        like the sea — termination of view: passed Estancia of Chaquaio. —
        entered Sloping plain, like R. Negro: appeared no distance, but very long. bivouaced for
        night. —
        Little oblique water-worn ranges, send little spurs into the sloping platform on plain. S.
        [illeg] Patagonia. no water
        25th. [March 1835] — Sunrise — intersected by dead straight line — in a paralellogram
        figure on rising —
        Spider:[470] habitation in centre web (refine. segments of regular Vertical net work being
        attached? to them from which strong web lines go in all directions /very/ numerous: Los
        Arenales

---

470 Specimen in spirits 1234 in *Zoology notes*, p. 359.

[146a]

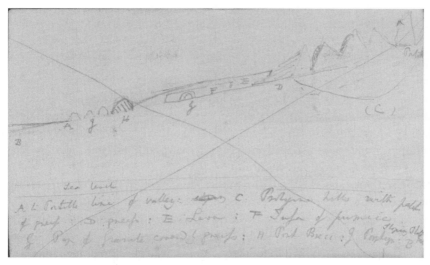

B A J H G F E D (C) Portillo Sea Level

A. to Portillo line of valley: ~~sloping~~ C. Protogine hills, with patch of gneiss: D gneiss:
E Lava: F Tufa of pumice.

G. Pap of Granite covered by gneiss: H Porph Brecc: J Porphyry: B sloping platform

[147a] struck right east. till we left the sloping platform: & came to level marshy ground which
again was succeeded by low plain: cooled with much saltpetre & plants same as
B. Blanca: before this passed two or three Estancias: To the East I see low escarpments.
is this low ground, owing to the want of upheaval or removing action of water: country
formed of horizontal layers of sandy earth apparently <u>overlying</u> gravel of Porphyry. —
In a cliff I noticed that a thick efflorescence extended as far as soil would be damp with
floods. — no vegetation on this talus: ℈ what does Humboldt say about the effect of
draining? Day very hot — stupid ride: stopped at

[148a] Estancado one Rancho hardly saw a person: Gallinazo[471] bend of river & Black &
White Muscicapa: Can see Pampas from Portillo.
26th. [March 1835] — Dew last night in Traversia in Cordilleras scarcely any: this days
ride is called 17 leagues through a Traversia — with only one house where there is
water — quite level very hot — less so on the mule. — On our right hand we had for
many miles a regular low escarpment, perhaps 80 ft high — perhaps 15 miles from
Cordilleras: with valleys entering on the lower plain: upper plain composed entirely
(15) 2646

[149a] of Porphyry pebbles, cemented by light friable calcareous matter. not to be distinguished
from R Negro — pebbles many as big as mans head — above this other escarpment —
clearly an island! — Lower plain — some Tosca — a little Tosca rock — lines of small
Porphy pebbles — fine white sand & do /~~dun~~ / aluminous powder: —

---

471 The black vulture, also known as the Cuervo; listed as *Cathartes atratus* in *Birds*, p. 7.

I fancy much Protogine all along Cordilleras. — 3 Tupungalos — extensive
secondary[472] ranges near Luxan — Tufted Partridge ostrich Biscatcha Peechey —

Horse have not passed. —

[150a] long wearisome ride, scarcely met a person. — Near Luxan: noticed as I thought heavy
smoke — turned out to be locust — cloud quite impervious ragged — reddish brown:
all flying north: many scattered outlyers resting on ground — in these advanced guards
sky like Mezzotint engraving — main body about 20ft above ground perhaps 10 miles
an hour — with light breeze from South perhaps 2000 3000 ft high. — Noise that of
strong breeze through rigging of ship

[151a] Where ma a cloud had /alighted/ far more than leaves in the trees — fields tinged with
their color. — people sticks & shouts: had been coming for many days past — but
curiously had never crossed the River till this day — Poplars stripped of their leaves —
from Traversia: affect not explicable; avoided being struck: after the clouds had once
alighted they appeared then to fly E & W or any way. — distant red cloud so like heavy
smoke that we disputed for some time

[152a] common pest: the greater number appeared resting than eating. = crossed river entered
Luxan (NB. the clouds of Locusts gradually thickened & thinned) refreshing rows of
Poplars & Willows & artificial brooks — very small village. — At night good to
experience every thing on[c]e. — Chinches[473] the giant bugs of the Pampas: horribly
disgusting, to feel numerous creatures nearly an inch long & black crawling soft in all
parts of your person — gorged with your blood. —
27th March 1835] — [Luxan — to Mendoza 5 leagues quite level — like Chili —

[153a] beautifully cultivated square mud walls — house with roofs of do — immense orchards
of figs peaches vines olives — celebrated with fruit. —
Inhabitants sad drunken raggermuffins — Pampas continues ⅔ Indians & reckless
manners, but not the elegance of further East. — All horsemen: Village nearly all the
way —
At Luxan immense water melon for ½ penny a piece: half a wheel Barrow full of peaches
for 2 ½ —
(28th) [March 1835] Mendoza — nothing can be added to Heads[474] description;
people themselves it is a good place to live in, but not to prosper

[154a] Alameda — pretty, very straight. Cordilleras tame: scenery no comparison with
St Jago: — very quiet.

---

472 Rocks of an age intermediate between Transition and Tertiary. More or less equivalent to the current
     Mesozoic but extending down to include rocks now called Upper Palaeozoic (e.g. Carboniferous).
473 Local name for bed bugs (*Triatoma infestans*) or 'Vinchuca' bug, often spelled 'Benchuca' by CD. See
     Beagle diary, p. 315, and *Darwin's insects*, p. 89.
474 Head 1826.

Big bones. R. — Quarto & Los Gigantes Governor polite old man — everyone polite in these countries the commonest peons looking at a old black woman with large goitre touched his hat respectfully evidently as an apology for looking at her: — (29th). [March 1835] no Blue Sparrow = Aparea = Queriquincha[475] very fine grapes: extreme heat & dust of plain. — Goitres

[155a] Plain from Mendoza to V. Vicencio. Greywacke pebbles: hillocks of alluvium (white washed) horizontal stratified: distant escarpements obscure: near the entrance valley small outlying ridges appear rather wedge shaped than ridges. —
On entering greywacke, some little rather coarse, greenish /dusky/ color, generally fine almost passing into clay-slate but little laminated excepting in parts — strata (or cleavage?) nearly vertical: perhaps most general about NE & SW strike: some contorted & some great faults: all first range composed of this; alternation of more & less laminated rock. —

[156a] Rode all day, Traversia, no water or horses (NB. Greywacke must contain a grey crystalline limestone, for there is a quarry in a hill on road: this greywacke from fragments, forms whole face of Cordilleras in this part) — level. particularly desert. Guanacos — Entered valley slept at V. Vicencio — Head & Miers.[476] — brooks larger as we went up as S. Ventana explains dryness of sloping talus. — Mendoza all town & /Cheucsas/: Locust never go north of Mendoza.[477]

[157a] 30th — Villa Vicencio [March 1835]
(16) — Greywacke — moderately coarse
17 — Clay — slate fine laminated /soft/
18 — Breccia conglomerate above purple slate
19 — Pebble from do of porphyry. —
20   Sandstone with yellow water lines
21 — Purplish do —
22   White coarse /substance/ with foreign   particles
23 — Do finer grained harder tinge of green

Baths mile from house. in Greywacke & Slate — appears to be anticlinal band at spot: four walls diff. Temp. most hot only tepid (Elmis Colymbetes Tadpole) slight taste of Sulph of Soda? & other brooks higher up — purgative: Greywacke (16) alternates with fine blue laminated clay slate (17), which certainly is more abundant more

[158a] in the interior: dip of nearly all the slate is high 45° to 80° — I really cannot tell whether cleavage or strata. — the main band runs N & S — a great formation, the lower ban ridge chiefly runs NE & SW: the dip at in these high hills dips all

475 Nine-banded armadillo. See specimen in spirits 375 ('the Taturia Pichiz') in *Zoology notes*, pp. 179, 332.
476 Head 1826; Miers 1826.
477 '3152. Locust v. private ground P. Mendoza.' *Darwin's insects*, p. 90.

westerly or towards the hills; ascending (Blue clay slate in very regular strata
traversed at rt ∠'s by many quartz veins) found the Clay Slate covered by purple
variety in all parts, & this again by a coarse softish white Breccia Conglomerate
(18) particles of quartz, the pebbles chiefly

[159a] Porphyry (19); of such varieties as are found on other side; we then have (neither of
these beds are very thick) & appear conformable in one section): an immense mass of
a sandstone cemented (20) with white particles & curvilinear ferruginous water
lines: also a purple sandstone which (21) passes into a fine varieties which cannot
be distinguished from those beneath the conglomerate: also <u>much</u> softish white
aluminous rock with foreign particles (22) & a compact greenish tinge fine
grained of same nature (23) besides those there are varieties of greenish & <u>Blueish</u>
rocks

[160a] which are closely connected with nature of clay-slate:
I saw above in one locality the conglom a little common blue slate: All these varieties
seem blended in /one/ hills without much order, in another section they alternated in
<u>very numerous</u> strata: these formations extended far to the west, dipping to west with
tame outline: I must remark that certainly the dip of these curious white soft beds is less
than the laminated blue slate: in one case they appeared quite conformable, in another
they appeared not

[161a] so. I almost traced the very junction: the blue thinly laminated blue slate;
became more rotten, cleavage slightly curvilinear; & crossed by other system at
right angles so that it became a mass of splinters; & vertical cleavage hardly
perceptible changed color becoming greenish & brownish, this was close to the
Purple, /the/ the stratum of which was inclined at less angle now in other case, the
purple & blue were conformable — I strongly suspect that the mechanical beds
above are influenced by that law noticed in the Falkland Isd. — However this may
be I feel sure the soft beds are superior parts of <u>same formation</u>: the purple rock
being the connecting

[162a] link — appearing above & below, & indeed other varieties of the Slates: ~~From the~~
Further to the West much reddish rock is seen to succeed. — The conglomerate is in
some places so loose that pebbles of hard Porphyry can be <u>extracted</u> of all ~~sizes~~ degrees of
rotundity. there are no slate pebbles. — The fact is important as connection with the
rocks & conglomerate overlying clay-slate (conformably) West of Portillo. It tends to
prove that clay-slates ~~is here~~ are more recent than the Limestone & Porphyry: I strongly
suspect the white, purple & brown

[163a] soft strata between Sandstone & Conglom. — W of Portillo are of this nature: (Clay
slate have no Organ. remains, splintery fracture) (In all the
valleys much Tufa). Villa Vicencio notorious from Mr Miers Scenery very tame) —
(There is no such great difference of the Greywacke /of/ the Clay Slate & the Pale
Sandstone with the Aluminous beds. —

[section 'lofty hill']

hill very lofty upper part drawn too large A B at least 2000 ft above V. Vicentio
(A) The <u>conformable parts</u> — (B) where they <u>appear</u> /not/ be so. Slates altering as
described dotted line /division/ /when/ between two classes: —
This distinctness of ∠ of dip is also seen in general scenery of country

[164a]   April 1ˢᵗ [1835][478]

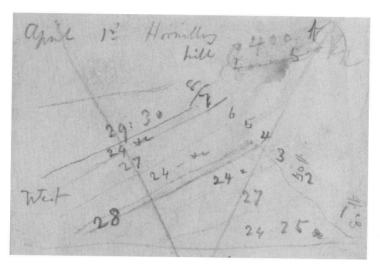

[section with heights][479]
Hornillos West 29:30 24&c 27 24-&c 28 24& 27 24.25 8 7 6 5 4 3 2 1 50ft 80ft
(hill 400 ft 4 or 5)
24.25 — Aluminous Sandstone
26 (do) in close contact with 27
(27) Feldspathic white greenstone Lava
(28) Pitchstone[480]

---

478 The actual date was 31 March 1835.
479 See the fair copy of this sketch in the geological diary DAR 36.508 and Herbert 2007,
        p. 317.
480 Glassy acid lava.

(29:30) Augitic Lava, mixed with some [Toadstone];[481] parts slightly concentric structure & little Pitchstone, imperfect.

Bed (1) is superior to the Various Porphyries is in places fine grained & compact is covered conformably by fine crystalline

[165a] sonorous rock, angular structure of a white Feldspathic Lava (27) the inferior junction is not well defined the sediment appearing to contain balls of the Lava & being harder & altered in contact (26): This Lava is again capped by fine sediment which contains a bed of Pitchstone generally with brecciated structure (28) scarcely 2ft thick. Above this the sediment (ferruginous water lines in all places) sediment containing some layers of large rounded crystaline pebbles. one or two larger than my fist & well rounded. they are of the same white Porphyry as in Conglomerate (18) of yesterday: Again we have varieties of (27) Lava. — Sediment & then the Augitic

[166a] Lava (Bed 8) (Spec 29: 30) This section is /right/ West of the houses. — I omitted to state the junction of Augitic & its inferior sediment is defined the rocks 2 miles apart possessing their true character

The Mines

(31) Hardened white aluminous rock at base of Granite: metallic fumes.

(32) Granite

(33) Fine white Granitic dyke percolating the Sandstones which are converted into Quartz, as may be seen, above Granite

(34) Veta in ; Granite

(35) pieces of the best auriferous[482] /Gia/ in do

The Boque Mine

(36: 37) the curious white rocks in which veins occur

(38) one of the best ores, said to be Phosph of Lime

[167a] (39: 40) The regular gold ores

(41). (like clay) a veins just opened; contains some Pyrites deep down thought to be good sign.

At the houses

(Blue /Baize/): grand curious Porphyry beds of Lava: (42) a loose fragment but I do not doubt from same locality (43) — The very limit where crystalline rock joins to a greenish rock like following: (44: 45) occur with the white aluminous sandstones: organic impressions comes from bed (1) in last section. —

(46) Metallic vein said to contain silver in above bed: (47: 48) parts of do bed stained by metallic fumes (49: 50) Sandstone &c beneath the curious Porphyry. —

(51) do very abundant —

(52) white rock blasted by fumes, near a vein

481 A term for various igneous rocks. Derived from Tödstein (rock with no ore).
482 Containing gold.

[168a] (53) Lowest Lava bed

(54) White dyke in the Clay Slate

Passing from V. Vicencia to the Hornillos, we pass Clay-Slate & Grewwacke: where we last see fine laminated Clay Slate: the cleavage is variable, chiefly about 80° to the S & SW ∴ running nearly E & W: on the west extremity of the plates: a red greywacke, so compact, as almost to resemble Porphyry & dips to the West & certainly appears unconformable (I omitted to state the Clay Slate is traversed by nearly vertical of dyke of soft white rock (54): the undulating edge shows it is not bed I believe if dyke it may be fissure filled up) although such it appears the cleavage of slate becomes tortuous & less inclined & seems to contain some of the very red rock: a short distance further on we

[169a] see purple & blue clay slate apparently conformable right beneath the white rocks dipping about 30° westerly. — I am still more strongly inclined to suspect that Falkland case here occurs: following the valley we have the white rocks (such as top of yesterdays mountain, & this is just in the N & S line seen running up the country but rather lower; there is not here so much of the purple rocks: these beds are many hundred feet thick; we then come to a bed of crystalline rock, with a slight curved columnar structure (53): where it rests on the sediment the latter is harder & has assumed a slightly spherical structure & of a purplish color — there are many large pebbles of hard, sonorous purple conch: fract Porphyry with few crystals but grains of quartz one is inclined to attribute the purple color to degradation of Porphyry rocks.

[170a] The crystalline bed is perhaps about 100–200 ft thick; it is covered conformably by a repetition of the white sediments; all dipping about 25°–30° westerly. These sediments consist of sandstone with ferruginous lines, more or less coarse & more or less cemented by aluminous white matter.

Spec (49. 50 51). These beds perhaps 400ft thick: Close beneath the superincumbent crystalline rocks are very curious. I do not think 100 specimens would show all varieties; spec with /base/ will show some; the bed is perhaps 250 ft thick (all these measurements very inaccurate) is abundantly amygdaloid: & every where highly crystalline, varieties pass into each other without law): the inferior point of junction is distinct

[171a] on a large side, the very point the two substances seem rather blended: (43) in the lowest bit of crystalline rock — This most extraordinary bed of rocks is covered with the white sedimentary rocks of which (24. 25) (44: 45) are specimens and many other varieties. the latter contain organic remains: The rocks above this are seen in a hill a little way up the valley: the section of which is given: it must be remarked that a fault must run right across the hill, for to East the dip is as seen shown at the back of it or to the west the dip is at small angle to the East: This fault is seen a few hundred yards to the north as thus: —

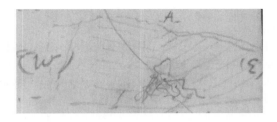

[cliff with talus?] (W) A (E)

The point A separates all the westerly dip already described. —
The Easterly dips, will <u>presently</u> be described. —
The whole of these sedimentary beds & the crystalline are traversed by a remarkably
system exceedingly numerous of metallic veins; the base would seem to be iron; they
seem

[172a] generally nearly vertical, & forms intricate net works. — They contain gold & silver
in disseminated particles (46) in specimens of such veins said to contain a little silver;
the effect on the strata to a considerable distance is remarkably (indeed whole masses
of strata) being blackened hardened, & heavier; the effect is most palpable in the
white sedimentary rocks (47: 48: 52) will give some idea of this: The appearance of is
of a shattered rock all the fissures burnt & blackened by metallic fumes. —
We will recur to the subject: With respect to the crystalline rocks, interstratified; the
appearance of the junction, although not quite even, being rather blended is certainly
not that of injection: none of the crystalline rocks show any signs of aqueous
deposition. — I firmly believe they are a succession of Lavas subaqueous
time of clay Slate <u>most ancient</u> !

[173a] (Mem: Toadstones of Derby); /infiltered/ with aqueous deposits; altered by fumes of
metals; & as I shall show probably have undergone, during time of tilting great heat.
(also greater alteration of sediments near the crystalline rocks). It seems a bold
conjecture I firmly believe a true one. == To return to the section at the beginning of
this day; above the highest augitic bed, I believe there are to the west several more
alterations. I had not time to examine. — From the point of fault (or synclinal line), for
a mile or two to the west, we have similar beds all dipping at small ∠ 20°–30 to East —
till we come to the highest Eastern ridge of the mass of the Uspallata range. — a height
I should suppose at least 7000 ft; — we have beyond this a tract of undulating country,
of no great irregularities, with broken dips — excepting two or three great mountains of
Granite; the one which I examined consisted of white feldspar, with

[174a] crystals of Fel black mica & Hornblende (32). very little quartz; similar to grey granite
of Portillo. — another great granite mountain /base/ NNW of this (The granite
contained veins of Epidote).
Rock granitic form & appearance). — These mountains are coated around their base
by the <u>white</u> rocks, tracts of which are singularly blasted by the metallic fumes (31):
The very highest gravel mountain is capped by a stratified mass of strangely altered

quartz, /flaky/ & Jaspery rocks — These are traversed by thin dykes. of a fine grained granite? of which & the altered sides (33): in only a few places in /the/ cap could I trace the white aluminous sandstone so as to recognise my old friends, indeed in the plain, there were black jas siliceous stones, which may be owing to altered

[175a] Clay-Slate: — To the West of the granite mountains in the distance a fine westerly dip is visible, we have seen an Easterly one on the East side: grand anticlinal bands — It is in the white rocks & underlying granite where all the best mines are worked: the main veins appear all to run within a point of NW & SE. — They are <u>excessively numerous</u> chiefly auriferous. — The first I examined were high up & in the granite: The ~~veins~~ /vitus/ were about a yard wide & composed of several such rocks several mines (34); in these are included the ferrug. = aurife. gias[483] of which (35) are specimens of size of best quality it is worked in a crucero. —

(The Boque mine)

Some hundred yards lower down, amongst the curious white rocks (& entirely over-lying the granite) (36: 37): there are auriferous veins (V 38–41): Also one of copper. Further the westward there are silver mines & northward Copper ones, said to occur in same

[176a] kinds of rocks. — It is manifest <u>all</u> the metallic veins come from underlying granite. — There can be no doubt the granite, breaking through the clay-slate, has tilted & uplifted these beds, as at the Portillo; the age has already been presumed the same. — the metallic veins are probably connected with heat of cooling granite:

Granite (W) E A F G B G C (E)

E white rocks dipping East (A) sunclinal point (F) white rocks dipping W; Lying upon the clay slate (G). (B) /Portillo/ anticlinal line in clay slate & grewwacke (C) outlying hills apparently dipping (E) perhaps from color containing some "<u>white</u> rocks". —

[177a] April 2ᵈ [1835] road to Uspallata. —

(55). What I called Pitchstone in patches I believe is Lignite[484]

(56) A common sort of trap-Lava as in W /slope/

---

483 Veins.
484 Fossil wood turning to coal.

(57) Vein in Lava & Sediment. —

(58) Green Sandstone of trees

(59) (60) 2 common varieties above

(61) uncommon do. — do. —

(62) Piece of tree so much altered not to be recognised

(63) Metallic dyke near Trees

(64) Augitic lava above do beds

(65) Blank Laminated flinty slate minute traces of organic remains

(66) Brick red bed with pebbles small

(67) Bed of Lava

(68) One of the white beds

(69) Grand Porphyritic dyke. SW & NE

Rode from the Hornillos to Uspallata: fine view of mountains: curious colors = Avecasina: several parties —

[178a]   Summit of Uspallata range <u>is</u>, as I have said, capped with Lavas & Sedimentary: on the west side, the dip is constantly westerly within a point: the inclination of any line of strike, not being constant for any great distance: $\angle$ from 20 to 30. = I saw one spot about 45°. = The Lava on this side seems in greater proportion: has more the appearance of Lava. — spherical structure: chiefly black sorts: the junctions are quite distinct, ie the superior ones (56) is a common variety. I found both it & sediment, traversed by vein (57) (what I call the Sandstones, are never true ones: cemented by the white substance generally variously sized particles;) I found a considerable quantity of fine layers of less glistening Pitchstone: & a good deal in patches in layers (55). I believe Lignite in layers & bituminous shales; also single isolated angular patch in Sandstone in other locality: Looking for silicified wood found in broken escarpment of green sandstone (58) 11 silicified trees & 50

[179a]   or 60 columns, (Lots wife) of Sulph: of Barytes:[485] /drusy/ cavities: form completely /straight/ either entire silex or Barytes: nearly all same diameter; little more or less 18 inches: in silicified centre of tree evident & all the rings: impression of bark in Sandstone: in Barytes only analogy makes me know what they are: the 11 are within ~~60~~ 50 yards of each other: the most remote not above /120 ft./ <u>no where else</u>, did I see a trace: The strata incline 20°–30° WSW: <u>All</u> the trees incline about 70° to ENE: I expect 2 silicified pieces as thick as my arm & smooth which are imbedded: horizontal: ~~trees~~ some trees only a yard apart. many two or three: appear vertical: Barytes one traced seven feet: silicified 4 ½ ft: sandstone consists of many layers in color & texture. which embrace trees: In sandstone vein of Sulph. of Barytes. ~~37~~ & in other place. nodules of agate: few fragments of Volcanic rocks ~~La~~ Green Sandstone, covered superiorly by <u>brownish ones</u> (59: 60) & compact purplish ones, & contain some odd

---

485 Barium sulphate (a heavy white mineral).

[180a] varieties (61) (NB one tree converted into column of stone without structure (62)) —
These varieties rather pass into each other & are not very constant in their position:
The Sandstones may perhaps be 4-500 ft thick they repose on Lava, are capped by
immense conformable bed — 1-2000 ft thick of black compact Lava (64). Looking to
the west. I could see at least 5 grand alternations several 100 ft thick of ~~these~~ such /two/
formations — close to the trees there is a broad metallic dyke, which runs NW (63)
affecting the strata — I saw other close beneath the Lava: in a brownish sandstone an
infinity of parallel threads curved which united & parted in groups. — Another in
the green Sandstone has been worked, from situation gold or silver which runs nearer
E & W. — situation before Agua del Zoro: & before the ruin /by mine/ S. side of
road: — Still following down the valley we have jet black fine <u>Laminated</u> flinty. slate
(65) by wetting one side traces of organic remains will be visible: between the layers,
[181a] concretionary lumps of same substance occur. The layers are curved & seem to follow
over uneven surface of Lava: I may mention in all the following less altered beds a
ringed or concretionary structure is evident: In a coarse brown sandstone, fissure
brown rings dependent on them on large scale as at Chiloe: I mention a few of the
more remarkable varieties in the alternations: quite white bed, soft with particles of
quartz, like in Chiloe &c: those compact kinds, as with organic impressions of
yesterday: a great quantity of extraordinary red: crumbling substance (66) which lay
upon a brown substance & alternated with Lavas: I here first saw an /interlac/ /nest/ of
Volcanic dykes, traversing Sediment. Near a ~~mountain~~ hill chiefly composed of
Volcanic rocks & higher than stratification render probable perhaps eruptive: ~~200~~
100 specimens will not show all the Lavas): a red conglomerate of crystalline pebbles
as large as nuts: (67) bed of Lava, by the Agua del Guanaco. —
Again beyond this we have quite white sedimentary beds (68): As far as Agua del Guanaco
[182a] the dips, with few & little irregularities had been Westerly; here a line of low hills
dipped due East a valley & beyond some higher ones west; these had a $\angle$ considerable
from 45° to 70°: a N & S line of Elevation at western foot, as at Eastern: (NB top of
granite hill at least 8000 ft): the last low hills, from very limit of plain of Uspallata: a
little to the South there is great irregularity ~~to~~ the dips apparently owing to oblique
lines of elevation: one of these (in the road) runs about SW & NE & seems caused
by grand dyke or chain of hillocks of a Porphyry (69) which seems injected amongst
white stone (68). These dykes probably anterior to the Granite: To the South, hills all
highly inclined strata some nearly vertical, great confusion Hills much water worn —
conical view, white, black volcanic <u>red</u> crumbling: purple green Sandstone Lilac
injected Porphyries: every shade of
[183a] brown — Valleys <u>broad</u>[486] & flat: a mere furrow present water course: enter grand
inland plain of Uspallata: to North nothing beyond horizon. — April 3ᵈ. [1835] —

---

486 According to Sulloway 1983, p. 364, this is the first instance in which CD spelled 'broad' instead of
'broard' during the voyage.

Uspallata ½ the day (70) altered clay slate: (71) compact purple Porphyry — injected in do (72) White do mixed with latter: these north of house Uspallata

(73) altered, almost crystalline slate:

(74) (75) injected Porphyries in do. —

(76) Sheet of Lava over the Slate

(77) Sedimentary bed ~~above~~ in same section

(78) Silver? ore from last slate, said to come

(79) — Argentiferous[487] lead

(80) Gold ore also from mines of Uspallata: ~~NB Porphyry in the road~~

Before arriving at the house: the plain is bordered by a band of clay slate, with veins of quartz; cleavage vertical, or lightly inclined dipping to the East (NB. Salitra at the Agua Guanaco: Porphyries on the road S of the section): The Slate, exceedingly, harsh sonorous (70) brittle, laminated, but the laminae not fissile; some partly crystallised

[184a] (like 73). crossing a N & S ridge the Slate became green & purple, /splintering/ & was succeeded by band of soft-purplish Porph (71): generally much softer than specimen, this is mingled in veins & large masses of a soft white sort (72) passing without any rule. — the junction is nearly straight. I did not see point of contact; & was much puzzled; there seems a dip of whole mass to the East; following plain to the South: the slate is seen on the East side; & the formation of Porphyry is shut in to the South: at mouth of pass of Caviota, a broad flat valley by degradation exposes extraordinary scene, a cluster of Porphyry hills of all colors distinct & blended from dark Porphyry, lilac to quite white & yellow, of the latter (74: 75 Spec[n]) these have burst through the slate (V map) or through a mass of slate fragments, which seem to have overlied the slate; they are firmly cemented together: domes of Porphyry underlie

[185a] these rocks; others are capped by them. I found one mass of white, specimen, Porph inclined at about ~~80~~ 70° on a mass of fragments; it sent a small vein into it. & the sides were marked by the indentation of fragments some of which could be extracted, other broken, remained in — The rest seems to have hardened the fragmentary mass. — These hillocks from basal parts of (& evidently near point of eruption) a line of strike of grand mass of beds which dip to NE 50°–60° — These beds consist of almost same Porphyries, but are evidently Lavas separated by layers of Sediments & lying on the slate & subsequently upheaved instead of bursting through it: The slate forms a triangular mass, is all composed of a greenish almost crystalline kind (73): quartz veins: cleavage NE: SE: (upheavals): bands of Grewwacke also altered: In the Slate said to be silver mines. Spec (78): To the South, there is a corresponding immense escarpment of Porphyry Lavas, dipping to SSW, instead of SW $\angle$ 70° — thus the two lines of strike

---

487 Containing silver.

[186a] are not parallel, but converge: Beyond again Slate mountain — Plan will show best. — This escarpment looked exactly like the <u>painted</u> Geolog. sections of mountains:

I omitted in the ~~very~~ intervening slate mountain: 2 points of purple & red, show injection. — To return to first escarpement, describing from Slate, a pale purple rests on red breccia. full of large green fragments of Slate, on a pure Breccia of Slate (as in the injected hillocks) or on the Slate itself: above this a very heavy compact Lava (76) sphaerical structure a great mass of purple & these curious white masses of Porphyry, divided into some divisions: a thin greenish (77) white sedimentary much stratified mass: grand quantity of cream slightly columnar — colored Porphyry, divided into three divisions — (dyke crossing obliquely the sedimentary bed): a grand mass of the Lilac & white Porphyries: In the injected Porphyries veins of Cryst. Carb of Lime.

[187a] do not differ from the Lava: in some Brecciated structure: never conglomerate, different aspect from. Brecc. Porph: appear more of one sort: It is clear, Porphyry has on a N & S line, burst through Slate has overlaid it & alternated with sediments, has been upheaved in a point (Granite hill) by the Slate which is altered: Although some outburst has taken place here: main situation on E true slope of Cordilleras: The Porphyries extend far East, amongst the Slate hills. — Argument that Clay-Slate, distinct form (& Slate hill here): — // Uspallata mines. between (79) the 2 road; in clay slate: gold ore (80) like Hornillos at least equal probability belong to Tertiary period we have seen the injected Porphyry is subsequent — in some cases to the Tertiary form:) The Grand S — Escarpement does not cross the river:)

Plain, islands. — water worn hillocks: alternations of white & bright red sands; slightly agglutinated, with lines & masses of pebbles; of the green altered Clay-Slates Porphyries & Granite: (surface scattered over with pebbles, much protogene) Beds dip

[188a] ~~all westerly 45° or more, with some irregularities~~ capped by great beds of horizontal gravel, which near river is universal in enormous barranca: subsequent residence of sea — which side did the altered green Clay Slate come from? — Pebbles, size, nuts, eggs 2 fists —

(B Alluvium in plain) W B A A Lava & less inclined sediment Gra E
In morning rode to mountains; old kill Christ not very civil, after noon crossed plain entered grand valley of Cordilleras; side of river of Luxan. much larger here, than there: sudden fall after sunset: extraordinary colors after sunset;

In this plain Ostriches, Toco Toco. Apereas Pichy Paluda — Extremely sterile. —

[Grand valley]

The lens shaped red covering from grand valley of red Porphyry: over true horizontal gravel of Barranca of River

[189a] Extraordinary confusion, from injected Porph. Lavas, Sediments & Alluvium, all most brilliant colors. — April 4ᵗʰ [3 April 1835] (81) Protogene like Porphyry: (82) altered slate with pebbles. (83). much altered slate (84) Fragment picked up near junction of granite & slate with crystals: (omitted faults in sand alluvium. Slate exceedingly contorted in the intermediate mountain: also at the Portillo, white Porph dyke in micaceous sandstone). With respect to last little section: must have been some general cause separate from the excavations of each valley: could a debacle, round so many pebbles. leave horizontal bands of the finest white sand, white-wash pebbles: cement the with Carb of Lime. This last fact noticed in high valley of Maypo: — started early; gale of wind, fine barren valley: only one resinous bush. not so lofty or so ~~barren~~ wild as the Portillo road. — Bad passes; could walk backwards: was told to carry thick Worsted stockings: if mule stumble probable death no chance[488]

[190a] had to unload at P. las Vacas; 5 £ expensive repair pass: not so on the Laderas: found Estero de las Vacas too much water, alojamiento,[489] there; no pasture for poor mules: before arriving here crossed a Pass of Jaular, the worst I saw considerable hill of Alluvium, is the Paramillo, (dead animals exaggerated) tuft of grass in mouths Juan /Pobre/ Bushes to the top: ~~not many~~ parties: — ~~Valley~~ Cajon of Luxan, enters Cordilleras almost N & S bends gradually SSW — SW, till we came to the Estero de las Vacas; thus in the section too much of each form: is seen. The mountains on each side of mouth, seen at distance, form an enormous platform 2-3000 ft thick, cut by great square valleys: consist of a strange mass of various colored in patches of Porphyry which appear rarely divided in strata. perhaps by sediments such strata incline at very small ∠ to E: all colored Porphyries seem to have been injected into

---

488 'I have been quite surprised at the degree of exaggeration concerning the danger & difficulty.' *Beagle diary,* p. 318.
489 Lodging, or steerage in a ship.

[191a] each other; interlaced by grand, red dykes: a dark hill of rock, was backed, as if split & filled by which rock, which sent off dykes, branching ∠ 45°, thinning downwards, rare case. —

[dykes?]

Where valley bends a little more westward, we have on north side a grand range of peaked mountain, sometimes reaching over to the South. rock is Protogene in a basis of Porphyry (81): is intimately connected with the pale porphyries, but almost  always inferior: appear sometimes however to be injected with white sorts & even stratified; one place, where closely joined on to white Porphyry, the latter almost consisted of its own fragments joined by fire: on the south side appear chiefly to consist of stratified Porphyries, which apparently have flowed: I cannot doubt these sections are the basis of band of volcanoes, from which

[192a] the stupendous streams of Porph. Lavas have flowed: scale of chain of Volcanoes. there would be cones of crystall: rocks which have not flowed: — In a pass beyond the Jaula; first clearly noticed. Porphy. Breccia — certainly different, although alternating with white beds; these blended into each other: many parts could not be told from the injected, but the large round, distinct pebbles; not only round, but if those various, rounded many sided fragments which we see on beach: pebbles *[illeg] [4]* sizes of being arranged in bands — Claystone or Jaspery parts with no or few crystals; masses of green Porphyry with narrow bands of purple, brecciated sort, a stratified structure & general aspect <u>Porph: Greenstone dykes</u>: I am certain these are injected ejected & baked Porphyries in these

[193a] Cordilleras to be distinguished: These Porphyries appear tilted on each side regularly by a SW & NE line, hence Valley. — They are traversed by grand dykes & masses of Porphyry, so only remnants of the ribbons of escarpment remain: I saw large mass, which had thus burst through the strata, of a white porphyry. it self containing dome & dykes of a compact reddish black Porph: in the form of fragments cemented by fires: & this again traversed by Porph: <u>greenstone dyke</u>: Consider the confusion. — At the Pass of P. de las Vacas there is a great hill of Protogene Porphyry: thus also. upbursting: — From this Point of the R. de las Vacas, we have clay slate, underlying the Porphyries (Valley rather more W than SW): at first, the clay-slate is somewhat similar to that of Uspallata range with vertical contorted cleavage

[194a] we then have it more crystalline (83) from which it passes into true greenstone with grains of quartz. Cleavage even yet perceptible sometimes, about NW & SE: judging from fragments there is very much, conglomerate, whited by same cement, as pebble (82). most undeniably true pebble: specimens show state of Slate, a degree more altered than at Uspallata: Specimen does not show Conglom: struct: quartz veins:

Slate traversed by some dykes & masses of soft white rock; after riding for a mile or two
at the base of such hills, we find true granite. white, but rusty by air (traversed by green
dyke) underlying & protruding into the compact altered Slate: capped by the
B. Porphyry joined by net work of dykes: at base of extensive junction, grand fragments
of granite, ~~containing~~ almost composed

[195a] of fragments of green crystall: rock, joined by granite, as in other pass. Porph Breccia =
Saw amongst fragments at base, mineral (84):

[valley of Luxan] NW P.B Slate granite Valley of Luxan Luxan P.B X Porphyries
Lavas & sediments A alluvium ~~R de las Vacas~~
This section is imagined to be east & west across the mountains, between the Punta &
R. las Vacas: the dips are uncertain & not to S & W, but perhaps NW & SE: the part of
section (A) is of course further to the North — (x) ~~contorted~~ injected hills of Protogine
Porphyry. —
I omitted to state. I conjectured I saw Lavas over the P. Breccia where first
met with. — it is clear that Granite is the axis of the slate, & that the
Protogine Porph is only accidental & concerned with Volcanic emission similar
colors in injected, ejected the source baked sediments, alluvium, Porph
formations:

[196a] Section of mountain S of Puentes del Inca — 2000–3000 ft high. —
April 5th [4 April 1835]
+ 2 Lowest bed I visited is a coarse white conglomerate: many particles of quartz, almost
blended together matrix not Porphyritic: There is to the W a lower bed. doubtless
conglomerate:
3. White Sandstone, very quarzose, almost /quartz/
forms conglomerate or sandstone
4) Red Bed & white bed not visited
5 owing to talus & want of time
6/85 yellowish fine thinly stratified Limestone nodules sometimes /domes/ of Carb
of Lime: figures like shells sometimes
7/86 20–30 ft Feldspathic Lava
8 Same as (6) more compact, part stained purple. —
9/87 dirty purple, partially crystalline rock amygdaloid with Carb of Lime

10/88 /ridges/ red sandstones covered by grand bed 300 ft of coarse red purple Conglom pebbles — nut to mans head: all red excepting white quartz: the coarsest I have ever seen in Chili

[197a] 10 contains a bed of white Limestone? in centre
11/89 Grand wall of Feldspath Lava (highest part of section seen from the Puerta): upper parts cellular lined with yellow: This perhaps is about half way up the mountain
12/90 a crystalline dull red sandstone which passes by
13 compact red Sandstone into a very coarse white conglomerate base chiefly particles of quartz
14 Alternations of conglomerate, compact red sandstone & such as (12) & much (9)
15/91 Greenish crystalline Sandstone
16/92 Many strata of compact or more laminated red sandstone
17 Gypsum 18 Red Sandstone (as 92) & layers of white Lime (93) 19 Gypsum 20 same as (18). — 94: 95: 96 Gypsum bed.
These strata all dip about 30° to W by S in some places 45°. They are traversed by many faults. The lowest bed of Conglomerate & white lowest Sandstone is most altered, or rather the same = Porph. fine grained Sandstone: all are compact

[198a] & have undergone action of some heat. but I have seen none so little: on the north side of valley, the lowest red beds are seen on the altered Slates: There is no passage to the two beds of Feldspathic (with mica) beds: I can have no doubt they are Lavas: the passage of the Sandstones & those same Porph Kind is certain & proves heat: At the foot of mountain much white Limestone with impressions of shells. did not happen to hit on site: The gypsum is precisely like in Valle del Yeso: concretions of white marble & others lined with black. — coarse crystals — lines of the black crystalised: Gypsum apparently in lines with Lime & Selenite & pure soft light Gypsum: The intervening & capping rock is curious specimen is not fair, because laminae are generally more even: like the contorted layers with green sandstone

[199a] in the Valle del Yeso: (This rock forms some of the very highest points): The line of Strike of these (N by W) strata is very important & forming some of the highest hills: in this very line hot springs (& much gaz) of this Puente issue: — To return to the R. de las Vacas. from it to the unison of rivers we have much altered Conglom & Slates penetrated by whitish Porphyry: & this extends some way up where the road bends to the N of W: near the Casa de Pujios: S of river granite is seen, in paps upheaving the altered slate, & superincumbent strata, (such as section, NB. section illustrates in /color/ place where the P. Bre. was seen in this pass) & tilting them to SW, from this they curve to the above grand dip. — Now /from/ the mountain to the S. & N of this basin of /illeg/ the colored strata are seen horizontal forming, enormous snowy mountains, & here we have simple conglom: these Porph. = Conglom. I

[200a] must remark, that size & much rotundity of pebbles in two states quite agree: — To the
        W. then at last section we have this. —

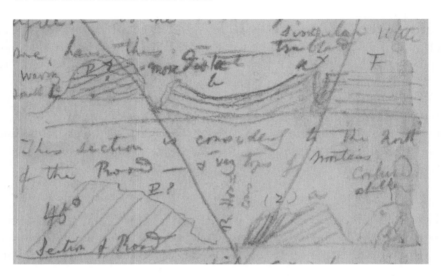

[curved strata and section of road] singular little troubled waving small ∠ D ? more
distinct b a F

This section is considerably to the north of the Road — & very tops of mountains
45° section of Road D? R. Horcones (2) a confused /altered/ (aa) part of same strike (a x) of
course /most most/ lofty (2) strata inclined 70° instead of 30° as rest of mountain. (a) is the
same as section seen on N of Road: The horizontal strata is very lofty ridge very unusual. —
The part a F, can also be seen. (b) removed by valley of Hercones. Some way south of
road: — Have forgotten — I think each Gypsum bed & intervening bed 300 ft each. —
Rode for half day, to the Puenta saw, humming bird, & Lion, long billed Thrushes

[201a] Red breast creeper — Bushes, at least 1000 ft above Bridge: ascended mountain
        without very much Puna 2-3000 ft: tame ride large valley & much Tufa, over
        alluvium. — Yesterday, several groups of Indian huts, look as if great party had
        migrated & been obliged to stop for some time in most barren spots: doors very low. 3 ft
        huddled together, square, — passed 2 Casuchas. — Mules, pass rivers /but/ with riders
        argument against "Rational". —

        (April 6ᵗʰ) [5 April 1835] I do not believe Strata above Granite above 6000 ft: the
        upper Gypsum strata removed on outer parts of Andes. — Incas bridge irregular hilly
        plain of valley filled up with pebbles & detritus

                                        [Puente del Inca] (B) x (S)

a fan of ferruginous cellular Tufa covering a part: the river having cut as far as (x). continued to scoop out to the Southward; rubbish (B) fell down from plain & supported

[202a] (m) whilst river, continued forming arch. — the oblique junction is very evident (horizontal & confined) plain generally horizontal gravel this not so; hence rubbish: — My hypothesis of Tufa is that it was deposited after valleys excavated & just before sea retired; matter before that generally deposited — hence Tufa from these

Springs extends above then leve in the slope, above their level: Springs hot — violent emission of gaz:

Concretions, where water drips, in heaps Guanaco could not well descend there & Springs above: — Travelled to day to below the Casucha of Jununcillo /looks/ curious, miserable dungeons: very cold wind, no bushes by C of Cuevas or Calaveras: road crosses no snow — I should think the Portillo rather higher

[203a] view not comparable: —

Amongst enormous heaps of debris, especially Gypsum, large & curious funnels: Strange forms of Heaps of blackened fragments of Porph Brecc; have given name of Crater of Paramillo de las Cuevas In last great section, beneath Bed (2) we have thick bed of a true, reddish dull purple (common color) Porphyry, coarse & not well crystallised, yet a Porphyry is very obscurely brecciated (97) has in its upper parts /lines/ [illeg] contains small pebbles, matrix Porph: above this we about 80 ft of a compact grey Limestone, almost white; lowest bed a conglomerate of Limestone; contains bands in all parts of cemented breccia: lower parts all fractures weather red (99): upper parts yellow: (NB. in Porphyry vein like lines of Cryst. Carb of Lime). most obscure marks of Shells; on plain fragment of same sort

[204a] of S. Stone, with more perfect generally a trace of gryphite[490] (100) the most perfect shell an oyster (101). There was a white Limestone with line of little quartz pebbles — white or red, like shot: — above the Limestone same thickness of same color harder heavier (than those above) Conglomerate, matrix with slight trace of crystal (98). above this I believe Limestone, which /on/ the first contains a hardened blue marl (102) with nodules of Limestone: we then have (Bed 2): I doubt whether this other is quite constant: also, whether /1ˢᵗ/ Porphyry immediately covers the altered Slate: — Above the highest bed described I see there is again Gypsum, again white & red in lines, containing large bed of green sandstone; all the neighbouring highest summits are thus formed. the quantity of white lines

[205a] in the highest cap being variable: To the West I noticed important fact that all the various red beds above main Lava (N B. this bed variable in thickness) which in section contain no gypsum, receive wedges of this substance, so that at last they are entirely lost & replaced by mass of gypsum: so that one end all red beds at other all white: whereas some of the upper gypsum beds seem converted into a yellow sandy Limestone

490 Fossil oyster of the genus *Gryphaea*.

(103) & some black, traversed by thin ferruginous veins: I think whole section above granite at least 4000 ft — & gypsum formation above the main Lava) above about 2000 ft: — Excepting the Cumbre & altered peaks of P. Breccia all seem composed as thus shown — The proportion of Gypsum & Red stone never remains constant but always superior

[206a]   no 1) South of Valley

[section to South of valley and Section to North of Cumbre]
E A B P 2 Pas de los Cuevas D E Cumbre F G H J W II North of Cumbre E) A B (Z) (P) P 2 3 4 E G H J W SSW

Valley of Cumbre nearly E & W good, section on each side: rather different: in each beginning with the horizontal strata & neglecting the oblique patch of anticlinal confusion owing to granite: First the South side: B /has/ the grand described section

[207a]   the dip being from 25–30° & beds great thickness & some faults, (∠ 25. thickness 4000 ft how long would the horizontal section be) — We have the upper strata forming the hill (P.) of Gypsum &c. — Here there is a bed of pale calcareous Sandstone? (103): In lower part will be seen a patch of contortion. this will be mentioned in section to North of valley. At (C) we have either injected rock, or I believe vertical strata: Beyond this hill (D) dips at about 45°. to W by S, is composed of slightly Porph. Breccia & the Limestone: ∴. lower beds. more altered than the section: again (E) consists of red thin layers of Sandstone & Limestone (104) which do not appear to belong to hill (D) are vertical & contorted each way, chiefly however dip to W. traversed by extraordinary system of inclined & vertical dykes. green yellow & red. — the Cumbre passes obliquely over north margin of this (F): upon this rest dark purplish Porph. B. again a whitish sort: to the west of this, to the Casucha of Calavera, there is a grand mass of altered rocks, forming where road

[208a] runs, by excavation, on double row. apparently dipping to the West: Section to the North: Between B & P, parts of same of dip, is the valley of R. Horcones. where a patch of 70° W dip has been noticed connected probably with an Easterly dip seen to the north in distance: — The 2 P's: similar. Beyond this, we have mass of strata dipping from 60°–80° ~~running~~ to SW up the country to some distance; the S. point of this line has caused, the contortion S. of valley.

Behind & to the west of this there is a E & W range of lofty (as Cumbre) rugged peaks formed of Porph & Porph. B — with little stratification (hence valley full again of Crystall fragments) (hill 4) a /tight/ W curved dip is clear, corresponds to (D) but much more altered. these dark Porphyries are threaded in extraordinary manner. by grand dykes & very white Feldspathic crystalline

[209a] Porph: — (might well come from the white syenites) This trouble & the Porph. nature of conglom: again well agree. — The valley of Luxan corresponds to (E). The Cumbre similar excepting there is not an intermediate ~~valley~~ ravine of excavation: These hills very much altered — dykes high dip where seen — Probably (E) as (C) marks distinct line of upheaval — But I do not understand these hills. — at C of Juncalillo we come (from depth of ravine, no other cause) to the white syenite, with very little quartz — which has played devil with a line of N & S hills, which appear capped by remnant of Gypsum form, dip small ∠° W a — (Probably Gypsum f. has been removed from ridges of the Cumbre) — Ridge clearly N & S. — // Remarkable electricity in high regions: shirts, sheets, leather straps: dryness — lips, heads, of hammer = absence of Vegetation & dark color of sky, transparency of air

[210a] three great features cloudless owing height; colors of mountains to Andes: airy /in/ everlasting look. The ranges certainly run to the W of north a little: In the W Cumbre range perhaps some injected Porph; there is much true dark greens & purples: extreme regularity of weather: small Cumuli such as seen from S. Cruz: — Valleys owing to Stratification, but immense removals: — immense degradation & consequent talus striking feature: arguments of length of world from deluge applicable in each part since sea retired: Alluvium at Puenta del Inca, filling up base of valley has lost regularity of outline: rivers only constantly remove small pebbles & grind to dust large ones: Excepting high up it is rare to find the rivers deepening the solid rock: Silicified trees in nearly same level. I do not think above 20–30 ft diff: — Perhaps beneath escarpment. roots: no /lines/ to the foot:

[211a] April 7th [1835]: Half day — lost mule, to the old Guardia — Between this & the ojos de Agua — Chili begins with dotted bushes & cylindrical cacti especially at latter place: rather pretty — the range beyond Cumbre at the Juncalillo is just capped with Stratified Strata: We then have at the Ojos, a range of altered rocks supporting great mass dipping 25°–30° to W of various white purple, red, harder softer more or less porph. more or less coarse conglomerates: — Part of this range appears to dip NW — (hence perhaps valleys — This is at the Guardia

[section of Juncalillo] /W/ Guardia + C de Juncalillo [Jacuncillo] Cumbre

[212a]   The petrified trees (taking inclination of Strata into consideration very much on a level
not more than might well happen growing on a plain. — 20–30 ft above road — so that
I suppose they grew in same sandstone, but confess I am not sure, did not sufficiently
examine — of course soil would be much altered carbonaceous matter removed, silex
deposited. — perhaps — 100–150 yards apart most distant ones — Lava examined to
the North — The great bed perhaps 800 ft thick seen further to the South: not lake —
great changes to make unmake remake

[213a]   again — similarity with marine deposits — Form of land before upheaval not admitting
one & the Lava having flowed from the W — being then thicker. & Volcanic dykes. —
Perhaps section of a large wood — No trees in the reddish Sandstone above at a short
height Believe the strata below to be same varying Sandstone /as/ described above. —
Beneath volcanic rocks & Lava all conformable (plain of Uspallata two Entrances) but
do not know the very variety. —
Not lacustine: because we see plain at near V. Vicencia marine nearly same elevation as
Uspallata —

[214a]   8 7th [April 1835] lost whole day — hunting for mule
9th 8 8th [April 1835] To the W of the West dip at the old Guardia: another mountain
much altered capped succeed on West side by same stratified mass dipping to West:
very many dykes: some of them white I believe granitic —
Beyond this mountains with little stratification, based on white Porphyry, which passes
into true Syenite fine grained with quartz very abundant like Cauquenas /Portillo/ —
some little hillocks in the mouth of valley — the lowest rocks near mouth of valley to
/day/ were chiefly greenstone Porphyritic, owing to Slate? Saw no where high dip or
Easterly

[215a]   one — very tame scenery — Valley of S. Rosa — certainly looks very flourishing after
the stony mountain sides —
I suspect whole side of mountain one dip, with some faults & much altered: — White
dykes of Mr Gay — not quartz: but then Porphyry so similar in constitution to
Syenite. — as Protogine

Porphyry to Protogine (105) in specimen of this Porphyry with angular blacker patches of included rock: flat valleys — all forms of dykes peculiar to Andes: — (Big Rat[491] of Aconcagua in high tree: —

If the sea has covered the base of the tilted strata in Uspallata. it surely must have done so before upheaval: —

[216a] Rather if we prove the sea has /bee cone/ covered to within 1000 ft of these upheaved Strata: surely /it/ must have covered them before doing their tilt. —

Cleaveland Geology of United States: in upper parts especially of Alluvium with shells & sharks teeth beds of gravel. p. 638[492]

Primitive formation generally incline to SE at gr $\angle$ than 45°

Pyrenees 1763 Toises — St Gothard 1431 — M. Canis 1807. —

Height of Patillo — Caldcleugh p. 310[493]

Calculated from Humbold[t] 12.800

[217a] Prussian voyage round the world 1830–32: Some time in Chili

Mine: Saw Pedro de Nolasko runs E & W: argentiferous lead copper — Silver &c.

V. Specimen

Mr Broderip[494] — has shells from from Coquimbo given by Mr Caldcleugh

Shells at Illapel

[218a] All the days are wrong from their being 31 days in March days Reached St Jago

10th (owing to 31)

11 9th [April 1835] From the Villa Nueva to to near Colina. — noticed that /certainly/ the Porph. Greenstone & greenish Porphyries are more abundant here than in the Cordilleras owing to Lowest bed of Slate: (NB Mem: Limestone of M. Gay distinct from Porpicio)

(All this time to Valparaiso not quite well saw nothing enjoyed nothing)

10th [April 1835] From Colina to St Jago pleasant city — Mr Caldcleugh nothing but the Porphyries on the road—

(15th) [April 1835] Started for Valparaiso — dead mens heads on

[219a] the poles. —

Mr Caldcleugh says veins run NNW & SSE. — but in thickest part of mountain = better near the surface — upper parts Carb of Copper — /thick/ in Crucero = richness of = Anthracite found in the mountains

The very first ranges of low hillocks to the W of St Jago, consist of the plate-like Porphyry mingled with an amygdaloid rock almost composed of Epidote (106) —

---

491 Specimen not in spirits 2207 in *Zoology notes*, p. 406; listed as *Abrocoma bennettii* in *Mammalia*, pp. 85–6.

492 Cleaveland 1816, pp. 638–9.

493 Caldcleugh 1825, 1: 310 gives 12,800 feet.

494 William John Broderip (1789–1859), barrister and conchologist who assisted P. P. King with descriptions of molluscs and cirripedes from the first voyage of the *Beagle* in *Narrative* 1: 545–56.

These hills a little more to the Westward contain & ~~N &S~~ N by W & S by E veins of Copper which has been worked for few months: At the base of the C of /Prad/ we find

[220a]   a ~~soft decom~~ strange mixture of rocks of harsh nature & generally in very decomposed state (107) & these rocks contain an infinity of veins like masses of decomposing Trappean rocks some of them Porphyritic (such as Z d Z d).

At the foot of the Zapata we have the decomposing gneiss (— mica generally replaced by greenish mineral) with large concretions — This occurs in all parts & makes me suspect that the (107) rocks supply its place — This rock is traversed by many veins — like dykes of Trappean

[221a]   rocks — which differ from most injected ones I have seen in splitting & uniting in many thin ones — & in one case sending off a long vein which thinned to nothing downwards. — The gneiss /clearly/ /included/ is altered: yet I am quite in doubt concerning the nature of these dykes. —

Higher up we have an infinity of veins which like quartz veins appear contemporaneous of which (Z) will show specimens

I remain in doubt is there a gradual alteration from

[222a]   the gneiss of the low countries to the greenstone & Trappean rocks — important point overlooked

Elmis & Colymbela warm spring of V Vicencia

Cicindela — Saline banks of river of Estacado[495]

Buprestis[496] — Curculio — Traversia of Mendoza — : Cryptocephalus Chili
 600 ft

Bembidium[497] Mendoza —

[plant leaves and doodles?]

Coprophagous Traversia Mendoza abundant

Heterom do & /High/ East Valley of Andes —

[223a]   Was told at Mendoza large bone found at Los Gigantes & R. Quinto —

Mastodon — Paraguay

Sediment of Uspallata. regaining crystals of Feldspar curious probably owing their origin to degrading of Feldspathic rocks

Thick bed of shells at Quinquin mentioned by Frezier[498] as about 200 ft elevation

Frezier remarks that Llama & Guanaco dung in heaps, useful for fuel for Indians. —

495 Specimen 2841 in *Darwin's insects*, p. 89.
496 See specimen 3227 in *Darwin's insects*, p. 90.
497 See specimen 1025 in *Darwin's insects*, p. 71.
498 Frézier 1717.

[224a] Biscatcha, bones & round holes foot of Pampas = Close to the mountain Biscatcha = Ostriches Peecheys in Uspallata

/Huming/ Bird plentiful Valparaiso

April 20th. [1835] — Saw first on road for St Jago — 2d days journey — Perhaps earlier

N B there are step-like cliffs in some of the N. American lakes

Mem Mendoza

[225a] I hear of all four kinds of armadillos near Mendoza

[226a] /Potato/ 30d

/owner/ 17:2

Mariano[499] 2

Corn 2

Pistols

[illeg] 1 oz

[227a–236a excised]

[237a] Mendoza — Mariano 6 dollars

Covington 8 dollars

wild Potato seeds: Shells at Corfields Strong Boots [illeg] as well as Box: Trousers Shooting Jacket — Seeds

S. Covington on account 8 dollars March 12th — 1835 Valparaiso

[238a]   Valparaiso —

Position of Shells with respect to hard Breccia. over gneiss. — Reexamine existence of such shells. — glass — evaporating dishes

Quicksilver mine at Campario important relation with Porphyries of Hungary & Mexico

Reexamine dyke. direction & that of Alison:[500] Veins.

Corall in sea. — Mr. Green. —

Does not Corallina[501] emit gass in sun's rays? — Important. —

Greenstone of Cascade & Laguna.

The Porphyry in vein road through /Flora/ left of Flag staff. ? —

Posterior orifice of Planaria.

Are Balanida or Serpulae[502]

499 Mariano Gonzales, CD's hired guide and companion on his expeditions in Chile.

500 Robert Edward Alison, English author and resident of Valparaiso and later managing director of a Chilean mining company who wrote on South American affairs.

501 See *Zoology notes*, p. xxvii note 49.

502 A genus of marine worm which makes and lives in a calcareous tube often attached to shells and pebbles.

[BACK COVER]

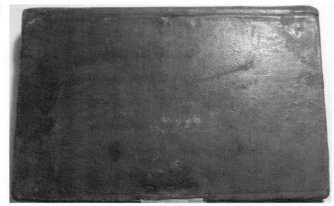

[INSIDE BACK COVER]

Charles Darwin
H. M. S. *Beagle*
September 1833
Buenos Ayres

A B / A C D
-1
-2
-3
-4
-5
-10
-15
-20
H9
Send
Dean Funes Historia[503]
owe 280 for coal to Lumb Harris?
Rowlett

[1b] In the Rincon of Gordona near St Miguel many great bones also at Lorenzo

---

503 Gregorio Funes (1749–1829), Dean of the Cathedral of La Paz. Funes 1816–7.

Direction of Bookseller for Books

[sketch of dykes?]

Baja
Shells about Between Arroyo[504]
St. Salvador & Agracindo. loose shells burnt for lime
Juan *[Tate]*. Boot maker. Mercedes
Calle de St Domingo. a Lady
Don Alexandre Martinez
Maestro Brownero —
Fossil bone *[Pedazo del Mano] [illeg]*

[2b] R. Palmar.   *[Bugglen]*
       Yerma   Katers Compass Indigo
       [Yuquia]   Banc. de Rozario

Brown paper: Old Box
Pix axe
Bread. Yerba Sugar:
Cigars
Jar with Spirits
[License]
Fish hooks (Cart & Canoe

$\dfrac{25\ |500}{20}$

[3b] The Investigation of Guano on coast of Peru (Avica) interesting (Dung): —
Indigo — Carmine. Sap Green good for *[Mierloz]*:[505] infusion of black pepper &
cinnamon
// Boué Journal de Geologie t.II P.205 talks of Tertiary formations of B. *[P.]*
St Salvador.[506]
Epsom Salts at Melincue, Azara

---

504 Stream or gully, also aroyo.
505 Possibly the same plant mentioned in *Narrative* 2: 166: 'A curious plant is found in Patagonia (and at
     the Falkland Islands), somewhat like a very large and very solid cauliflower. It is greenish, or
     yellowish-green, tough, and very abundant. It grows upon and close to the ground, forming a
     lump like a large ant-hill overgrown with moss and grass. From the succulent stalks of this plant a
     balsamic juice or sap exudes, which is particularly good for healing wounds.'
506 Boué 1830.

Salt between St. Fe & Cordova. Encyclopaedia Britt[507]

Helms[508] (German Mineralogist travelled from B. Ayres to Lima: —

[4b]   in 1793 Lightning struck Buenos Ayres in 37 places (killing 19 people in one storm.

Encyclopaedia Britanica

Molinas History of Chili[509]

To forward bones from Don Antonio /Rios/

Set of Port Saws: 2 Tenon Saws — Joiner

Hot Baths ~~near~~ 3 leagues Mendoza

Letters to Chili

[5b]   Molin/a/ <>

/Move/ <>

Bird <>

/Mol/ <>

/can/ <>

I /m/ <>

Mention<>

Fish<>

[6b]   shoes & <>

one <>

Mao <>

Textual notes for the St. Fe notebook

[1a–60a]   *a hole was punched through these pages and bound with string. The knotted end remains through the hole in 1a only.*

[2a]   *page not in CD's handwriting.*

ă/I] *not transcribed.*

H $] *not transcribed.*

1.13] *Down House number, not transcribed.*

5] *added by Nora Barlow, pencil, not transcribed.*

88202333] *English Heritage number, not transcribed.*

[3a]   *there is a red stain, possibly watercolour paint.*

[8a]   Captain Robertson] ink.

[13a]   level plain] ink.

[28a]   :] *ink over pencil in both instances.*

[47a]   ? The alluvium…del F.] *added in ink.*

[60a–64a]   *pages in ink.*

[61a]   M. Entre] *in pencil.*

507 *Encyclopedia Britannica.*

508 Anton Zacharias Helms (1751–1803) German mining director in Peru 1788–92. Helms 1807.

509 Molina 1794–5. CD acquired a copy when he arrived in Valparaiso.

[64a]	Cordonese…Tosca rock] *added pencil.*
	Caldcleugh Mastodon] *added pencil.*
	[Perfect]…passes] *added pencil.*
[67a–215a]	*pages are stained grey, apparently from being wet and wiped dry.*
[69a–70a]	*leaf has a tear in it, which has been repaired.*
[78a]	*sketch and caption upside down from other entries.*
[81a]	*there is a red mark after '*fragments.*' which is possibly watercolour.*
[90a–91a]	*sketch continues across following page, large watercolour stain mirrored on both pages.*
[125a]	*sketch and caption written perpendicular to the spine.*
	= R2] *ink.*
[128a]	*sketch and caption written perpendicular to the spine.*
[135a]	*sketch and caption written perpendicular to the spine.*
[146a]	*sketch and captions written perpendicular to the spine.*
[176a]	*sketch and* 'E white…rocks". –' *written perpendicular to the spine.*
[179a]	37] *crossed ink.*
[194a]	*there are stains of two shades of blue watercolour.*
[195a]	*there are stains of rust, three shades of brown and purple watercolour.*
[206a]	*sketch and caption written perpendicular to the spine.*
[216a]	Cleaveland…45°] *ink.*
[222a]	600 ft] *written perpendicular to the spine over* 'saline'.
[223a]	Sediment…rocks] *ink.*
[225a]	*lower right corner of page excised.*
[226a]	*lower left corner of page excised.*
[238a]	*page in ink.*
[IBC]	*on the left edge of the page there is a 1/8 inch scale, written parallel to the spine, in ink.*
	owe 280…Harris?] *upside down from other entries on page.*
	owe…Rowlett] *written upside down from other entries on the page.*
[5b–6b]	*leaf excised, two fragments remain.*

THE *BANDA ORIENTAL NOTEBOOK*

〜

The *Banda Oriental notebook* takes its name from the country through which Darwin travelled called Banda Oriental or 'East Bank' of the Rio de la Plata. Following independence in the late 1820s it gradually became known as the Republica Oriental de Uruguay. Today it is called Uruguay. The notebook is bound in brown leather. The brass clasp is intact. The front cover has a label of cream-coloured paper (60 × 20 mm) with 'Banda Oriental S. Cruz.' written in ink. The end papers and edges of the pages are marbled. The notebook has 121 leaves or 242 pages, all written in one sequence from the front cover. A piece of cream-coloured paper pasted on the inside back cover (25 × 115 mm) secures a brown leather pencil holder presumably added by Darwin.

Banda Oriental covers two distinct periods of the voyage. The first period, pp. 1–37, November 1833, covers the Banda Oriental expedition (expedition no. 4). The second period, pp. 38–111, 240–1, was April–May 1834 and covers the Santa Cruz expedition (expedition 5), and some weeks which followed in Tierra del Fuego. As usual, the actual sequence of Darwin's note-taking was quite complex. The November section dates from the 14th, p. 5, to the 28th, p. 37, at about which point Darwin stopped making dated notebook entries, until 27 December in the *Buenos Ayres notebook*.

Banda Oriental, November 1833

The first few pages of the notebook are blank, except for 'Charles Darwin H. M. S. *Beagle*' on p. 3. On p. 5 he recorded heading west from Monte Video with the intention of seeing the Rio Negro and Rio Uruguay, to take advantage of the extra month available to him once FitzRoy decided to take more time to work up the charts, before rounding Cape Horn.[510] Darwin's plan was to visit the estate of an Englishman, about whom we know very little except that his name was Mr. Keen, near Mercedes.[511]

A few days before Darwin started his expedition, on 12 November 1833, he wrote to Henslow and dispatched a large consignment of specimens.[512] Darwin was anxious

---

510 See Keynes 1979, p. 170.
511 Spelled 'Keane' in *Journal of researches*, p. 181.
512 CCD1: 351.

to know what Henslow thought of his endeavours. Darwin did not receive the feedback he desired until the following March, just before the Santa Cruz expedition.

According to the *Beagle diary*, on 14 November Darwin slept 'in the house of my Vaqueano', i.e. cowboy, in Canelones, about 40 km (25 miles) northeast of Monte Video. The next morning they 'started early', p. 5, but were hindered by flooded rivers. The following day Darwin recorded 'stomach disordered' so he had to spend a second night at a place called Cufré, but this did not stop him confirming that the rocks were 'granite and gneiss' and noting the minerals present in the granite. He also noted how this area had suffered in the 1820s during the war between the Argentines and the Brazilians.

Darwin saw an 'Owl killing snake' mentioned in *Birds*, p. 31, and he noted 'woodpecker nest in hole'. This was the campo flicker, *Colaptes campestris*, which was famously cited in the *Origin of species*, p. 184:

> on the plains of La Plata, where not a tree grows, there is a woodpecker, which in every essential part of its organisation, even in its colouring, in the harsh tone of its voice, and undulatory flight, told me plainly of its close blood-relationship to our common species; yet it is a woodpecker which never climbs a tree!

By the third edition of the *Origin of species* in 1861 Darwin added the support of Azara 1809 to his assertion that the woodpecker never climbs trees. In the copy of the fifth edition of the *Origin of species* (1869) on *Darwin Online*, a sharp-pencilled previous owner wrote in the margin against Darwin's claim on p. 220 that the campo flicker never climbs trees, 'How could it, poor thing, when it is in a place where "not a tree grows"'.

In 1870 the ornithologist and popular writer William Henry Hudson (1841–1922) claimed that the campo flicker did climb trees and suggested that Darwin had blundered.[513] Darwin responded with uncharacteristic severity.

> Finally, I trust that Mr. Hudson is mistaken when he says that any one acquainted with the habits of this bird might be induced to believe that I "had purposely wrested the truth in order to prove" my theory. He exonerates me from this charge; but I should be loath to think that there are many naturalists who, without any evidence, would accuse a fellow worker of telling a deliberate falsehood to prove his theory.[514]

In the sixth edition of the *Origin of species* Darwin modified the sentence to read: 'in certain large districts it does not climb trees.'[515]

---

513 Hudson 1870.
514 Darwin 1870, p. 706; *Shorter publications*, pp. 365–7.
515 *Origin of species* 6th ed, p. 142. See Winkler *et al.* 1995, p. 326, and Steinheimer 2004, p. 310.

On 16 November 1833, in Banda Oriental, Darwin noted with amusement that the postman arrived a day late but was only carrying two letters. Darwin seems to have delivered these for him when he reached the town of Colonia [Colonia del Sacramento] the next day, 17 November: 'delivered my letters', p. 6. The old part of Colonia was established by the Portuguese in 1680 and is today a World Heritage Site. Darwin was also amused by the pride of local people in their political representatives who 'could all sign their [own] names', p. 7.

Darwin was intrigued by a mass of 'muscles' (i.e. mussels), p. 10, near the harbour and could not decide how they were deposited '15 feet above high water'. He mentioned them in *South America*, p. 2, as evidence of elevation. On pp. 11–4 of the notebook he described the Tertiary Pampean Formation and the 'Primary' rocks which it overlays at Colonia, before making some fascinating notes about the cattle on the estancia where he was staying, owned by the Chief of Police.

On 18 November, on p. 13, Darwin recorded how he was able to recognise the same 'tosca' formation that he saw far to the south at Bahia Blanca: 'eyes shut think I was in Patagonia'. On the 19th Darwin saw a white limestone quarried at the 'Calera de los Huerfanos' (Limekiln of the Orphans) which is marked on a modern map as a tourist attraction. From there he passed the Arroyo Las Vacas, a 'straggling thatched town' on the Riacho, p. 17, and then on to the Arroyo Las Vivoras to spend the night at the home of an American who worked at the Carmacho limekiln. By the 20th Darwin reached the Rio Uruguay at Punta Gorda and there occurs in the notebook a cryptic quotation, p. 17, '"Sylvester Lellow complete collection of Banda Oriental." ?? !!'. Presumably Lellow was the American.

Darwin stayed the night of the 20th at the large estancia where a 90-year old woman lived who 'positively states that very early in her life no trees?! no trees except one orange tree', pp. 18–9. Darwin was told of quicklime bursting into flames in the quarry, causing great consternation amongst some superstitious locals. He searched for jaguars: 'Jaguar (went out hunting) cut trees on each side with claws sharpening'. As he later wrote in *Journal of researches*, p. 160:

> One day, when hunting on the banks of the Uruguay, I was shown certain trees, to which these animals are said constantly to recur, for the purpose of sharpening their claws. I saw three well-known trees; in front the bark was worn smooth, and on each side there were deep scratches, or rather grooves, extending in an oblique line, nearly a yard in length. The scars were of different ages. A common method of ascertaining whether a jaguar is in the neighbourhood, is to examine these trees.

Darwin recorded on p. 21 how the men at the estancia could perform extraordinary feats of killing and skinning mares (which they would not ride). In the evening Darwin went on towards Mercedes, and stopped at another estancia, which was in the charge of the landowner's nephew. Together they visited an army captain.[516] There ensued one of the most delightful conversations recorded during the voyage. The army captain expressed 'great surprise at being able to go by land to N. America'. He then asked Darwin to '"answer me one question truly" are not the ladies of B[uenos] Ay[res] more beautiful than any others"', p. 22. Darwin replied 'Charmingly so'. The interview continued, pp. 22–3: 'One other question do ladies in any part of world wear bigger combs. — I assured them not: they were transported & exclaimed — "Look there, a man who has seen half the world says it is so. — we only thought it to be the case.' The following day Darwin hired horses and 'Passed through <u>immense</u> (no Biscatchas) beds of thistles … often as high as mans head' with 'Very uncomfortable riding', pp. 23–4. The geology was out of the ordinary: 'soon after leaving white mortar rock — arrive at bed of white jaspery rock marked with manganese — containing nodules of milk agate', p. 25.[517]

Darwin spent the night of 21 November at a 'small Ranches' then 'arrived very early at the Estancia of Mr Keen on the R Berquelo near Mercedes', p. 26. Almost immediately Darwin made (but at some point deleted) what was perhaps one of his earliest references to the relationships between animals on continents and neighbouring islands: '(are there black rabbits on West Falkland)'. Perhaps this was the subject of a discussion with Keen, who although out during the day returned that evening.

The following day, the 23rd, Darwin geologized. The next day he rode with Keen to a place called Perica Flaca, where there was a 15 m cliff with extraordinarily coloured sediments containing some bones: 'The question is whether Bone occur in Tosca contemporaneous with Punta Gorda bed, or with Bajada [i.e. St Fé Bajada]', p. 31. Falconer 1937 explained that this section was of great significance as it led to a dispute between Darwin and Alcide d'Orbigny.[518] D'Orbigny doubted Darwin's observations, first published in *Journal of researches*, p. 171, that the white limestone (a marine deposit), first seen at Calera de los Huerfanos, overlay the Pampean Formation.

In *South America*, pp. 87–95, Darwin was at pains to substantiate his observations. Falconer reported that various geologists had revisited the sites in question in the early twentieth century in an attempt to determine the case. The majority view was apparently that Darwin saw a formation under the limestone which only looked like the Pampean Formation. Falconer praised Darwin's pioneering geological description of the area, however, and he stated that Prof. Karl Walther, of Monte Video, had erected a granite obelisk at what he took to be the key section which he named Rincón Darwin. Winslow 1975 reported that a nearby village was then called Villa Darwin, but Green 1999, who provides a photograph of the obelisk, states that the local people call it Sacachispas.

516 See Parodiz 1981, p. 57.
517 See *South America*, p. 93.
518 d'Orbigny 1842.

Darwin noted that the view of the Rio Negro from the cliff was 'decidedly most picturesque for the last four months' and that the river was '2 severn' i.e. twice the width of the Severn, p. 32. On 25 November he 'Rode to dig out bones of giant' which were found in place but had since washed under water. Owen, in *Fossil mammalia*, p. 57, described a rather battered skull as that of *Glossotherium*, a new genus of ground sloth, and there were pieces of what seemed to be a carapace. Darwin found the circumstances 'Interesting as connection between Casca & big bones', p. 33. Casca most likely meant armadillo-like case, as it is Spanish for 'shell' or 'helmet'. In *Journal of researches* Darwin reported that he found such a case near the *Glossotherium*.

The next day Darwin went to a house 'to see large head & bones washed out of Barranca & found after a flood. pieces here also of Casca', pp. 33–4. Keen played a key role in securing this skull for Darwin for the price of eighteen pence and in arranging for safe shipment of the fossils to Lumb in Buenos Ayres, who sent them to Henslow in England.[519] The skull was later described by Owen in great detail as that of the new notoungulate genus *Toxodon*.[520]

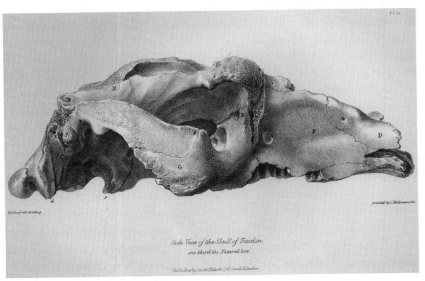

Side view of the skull of *Toxodon* from the Rio Sarandis.
Plate 2 from *Fossil mammalia*.

These pages are of interest with respect to the term 'diluvial', which Darwin gradually dropped during the voyage. On p. 34 he wrote 'most probably diluvial hence animal of Tosca & diluvial age', but then following a 'very bad night — wet through; extraordinary thunder', pp. 34–5, Darwin recorded 'Granite in immense blocks' and

519 See Winslow 1975.
520 *Fossil mammalia*, p. 16.

seems to have changed his mind about the age of the tosca 'The white Tosca bed certainly different from the usual grand covering. probably of a different date', p. 35.

On 27 November Darwin noted 'country whole distance Primitive gneiss' and on the 28th he 'Arrived in middle of day by same road to Monte Video', p. 35. The remaining two pages are difficult to date as the next date in the notebook is 18 April 1834. Page 36 appears, however, to be concerned with some fossil bones in the possession of a 'Padre' (clergyman) at Las Pietras (Las Piedras) with a tail which Darwin drew and thought was an 'extraordinary weapon'. This must be the 'dasypoid quadruped' which Darwin mentioned as seeing near Monte Video in *South America*, p. 107.

Page 37 is highly significant as it is apparently an ink essay on the perplexing sections Darwin had just seen on the banks of the rivers Uruguay and Negro. This page seems to indicate that Darwin realized that the estuarine conditions which created the Pampean Formation had occurred at least once previously, and that the last two such estuarine periods were interrupted by a marine phase, thus supporting a view of the geology of the Pampas as one of repeated elevation and subsidence. A few months after his Banda Oriental expedition Darwin took this idea much further in his first of several geological essays written during the voyage. This first essay, which he headed 'Reflection on reading my Geological notes', is now numbered as DAR 42.93–6 and was published with analysis in Herbert 1995. Herbert, p. 158, explains how the 'Reflection' essay, which she dates to around March 1834, shows Darwin beginning to speculate on how the elevation might be caused by 'swelling of the Globe', and which might result in land only recently risen from the sea having therefore only a limited stock of animal and plant species (i.e. biodiversity) due to 'no Creation having taken place'.

Darwin's second geological essay is headed 'Elevation of Patagonia' (DAR 34.40–60) and, since it contains references to Santa Cruz, was almost certainly written a few months after 'Reflection', in mid 1834. It picks up on several of the themes in 'Reflection' and is discussed below.

Eight days after arriving back in Monte Video, Darwin set sail on the *Beagle* on her final departure from the River of Silver, bound for Port Desire and Port St. Julian in southern Patagonia in company with the *Adventure*. The next notebook entries are for Port Desire on 27 December (*Buenos Ayres notebook*, pp. 87–8).

Santa Cruz River, April 1834

By the time the *Beagle* left the Falklands for the last time, at the beginning of April 1834, Darwin, aged twenty-five, was beginning to see himself as a geologist. As he wrote to his sister Catherine on the 6th:

> There is nothing like geology; the pleasure of the first days partridge shooting or first days hunting cannot be compared to finding a fine group of fossil bones, which tell their story of former times with almost a living tongue …. I long to be at work in the Cordilleras, the geology of this side, which I understand pretty well is so intimately connected with periods of

violence in that great chain of mountains. The future is indeed to me a brilliant prospect.[521]

About ten days after he wrote this letter, Darwin started on expedition no. 5, the gruelling seventeen day struggle against the fierce flow of the Rio Santa Cruz, in a brave attempt, led by FitzRoy, to find its source 300 km (185 miles) away in the Andes.

The expedition began on 18 April, with the first entries in the *Banda Oriental notebook* on p. 38, and ended on 8 May on p. 103. The *Beagle* was by then ready to explore the Pacific coast. FitzRoy chose the Santa Cruz, with its large tidal range, as an ideal place to lay the *Beagle* ashore and repair some minor damage to the keel and damaged copper sheets, before re-entering tropical waters where, as FitzRoy later wrote 'worms would soon eat through places on a vessel's bottom from which sheets of copper had been torn away.'[522]

'*Beagle* laid ashore, River Santa Cruz.' Engraving after Conrad Martens from *Narrative* 2.

The expedition did not reach the source of the river because the party was forced by dwindling rations to turn back downstream, having come tantalizingly close to Lago Argentino which was discovered thirty-nine years later.[523] However, the scientific observations made by FitzRoy and Darwin during the course of the expedition amply repaid their exertions and represent a high point of their collaboration. It took just three days for the expedition to return to the *Beagle*, the three whaleboats which were so laboriously pulled upstream shooting sometimes dangerously fast back downstream.

521 CCD1: 379.
522 *Narrative* 2: 283.
523 Barlow 1945, p. 220.

FitzRoy took a particular interest in the Santa Cruz and later published a paper on the expedition in the *Journal of the Royal Geographical Society*.[524] In *Narrative* 2 he provided a chart of the mouth of the river and a map of the river itself, showing how it fizzles out in 'Mystery Plain'. FitzRoy seemed to refer specifically to the geology seen and no doubt discussed at length with Darwin on the expedition, in his notorious chapter 28 of *Narrative* 2 'A very few remarks with reference to the Deluge', pp. 657–82.

FitzRoy's map of the river Santa Cruz from *Narrative* 2.

There are several superb drawings and watercolours of the expedition by Conrad Martens, such as 'Basalt Glen'.[525] Some of these are very helpful in understanding the geology of the valley, which Darwin discussed in *South America*, pp. 9–14, 112–7. Martens left the *Beagle* in September 1834 and arrived in Australia in April 1835. When Darwin visited Australia in January 1836 he paid Martens for a watercolour of the Santa Cruz expedition.

'Basalt Glen — River Santa Cruz.' Engraving after Conrad Martens from *Narrative* 2.

524 FitzRoy 1837a.
525 Reproduced in Keynes 1979, p. 205. See the engravings from *Narrative* 2 on *Darwin Online*.

The Santa Cruz expedition was of great importance to Darwin's understanding not only of the geology of Patagonia but of the geology of the world. By the end of the expedition he had come to see the Santa Cruz river valley as an uplifted channel which was once under the sea, like the *Beagle* Channel today. This fitted his emerging understanding of the whole South American continent as one which is gradually emerging from beneath the waves due to vertical forces. He was deeply impressed by the vast extent of the several slightly sloping plains which he could trace for hundreds of miles, and this suggested to him a series of successive step-wise elevations on a continental scale, linked to the rise of the Andes. This was not exactly, as Lyell might have preferred, gradual elevation at a uniform rate. As Darwin wrote in *South America*, p. 10: 'I think we must admit, that within the recent period, the course of the Santa Cruz formed a sea-strait intersecting the continent. At this period, the southern part of South America consisted of an archipelago of islands 360 miles in a N. and S. line.'

It was at the end of the Santa Cruz expedition that Darwin read the third volume of Lyell's *Principles of geology*. Shortly thereafter Darwin wrote his second geological essay, entitled 'Elevation of Patagonia' (DAR 34.40–60).[526] Darwin cited 'Lyell Vol III P. 64' on p. 109 of the notebook a few pages after the date 25 May; a rare citation in the field notebooks. The same page reference occurs in an insertion in his earlier 'Reflection' essay.[527] Just before this reference to Lyell, on p. 108, Darwin made a note to 'Reread Pampas Notes & copy out' which is perhaps the prompt to write the 'Elevation' essay. Darwin wrote to Henslow in July 1834 about the expedition, adding 'you may guess how much pleasure [reading Lyell's third volume] gave me'.[528]

The account of the expedition in the *Banda Oriental notebook* opens on p. 38 on 18 April: 'caught mouse: pleasant party cheerful running water'. The next day Darwin made a sketch of the river terraces which is similar to the one published in *South America*, p. 10, fig. 6 (below), although the latter represents a higher section through the valley. Darwin may not have been able to write conveniently while on the move, as he reported in what must have been a note made in the evening, 'day has been splendidly fine: but country terribly uninteresting: no living beings. insects fish &c &c', p. 40.

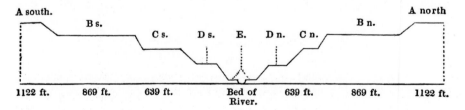

'North and south section across the terraces bounding the valley of the R. S. Cruz, high up its course.' Fig. 6 from *South America*, p. 10.

526 See Herbert 2005, pp. 160–6 for discussion.
527 See Herbert 1995, p. 33 note 39.
528 CCD1: 399.

There are several references in FitzRoy's *Narrative* to Darwin going ahead of the party to scout and to make the best use of his marksmanship, and he often helped FitzRoy and John Lort Stokes (1812–85) with mapping the valley. By p. 41 Darwin reminded himself to investigate the 'Effect of Earth quake in Chili on river courses'. After finding a boat-hook lost on the previous expedition the party went beyond what the crew of the *Beagle* had managed on the first voyage: 'Beyond this Terra Incognita', p. 41.

On 21 April they found a huge 'Quartzoic or Feldspathic' boulder, and Darwin eventually cited this and others in his paper 'On the distribution of the erratic boulders and on the contemporaneous unstratified deposits of South America', which included a sketch section through the bank of the river.[529] This section was republished in *South America*, p. 114, fig. 18 (below). On the following notebook page Darwin recorded catching a fish which was later identified as the new species *Mesites maculatus*.[530]

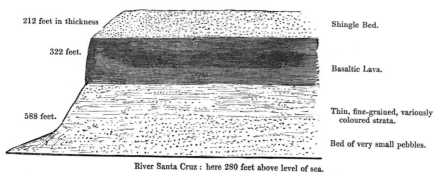

'Section of the plains of Patagonia, on the banks of the S. Cruz. Surface of plain with erratic boulders; 1,416 ft. above the sea.' Fig. 18 from *South America*, p. 114.

There was evidence of Indians keeping track of the expedition, but they were never seen. A dead guanaco was found floating in the river and eaten by most of the party, although FitzRoy recorded that some of them could not overcome their aversion to carrion.

By 22 April there were five sets of plains. Darwin caught a 'red-nosed mouse', p. 45, and noted seeing a 'Callandra', which was probably the Patagonian Mockingbird *Mimus patagonicus*.[531] The next day there was a little leisure for collecting beetles.[532]

Darwin noted the 'immense quantity of gravel!'. Cliffs made their progress difficult so they crossed to the other bank. On p. 47 Darwin recorded 'many

529 Darwin 1842a; *Shorter publications*, pp. 147–62.
530 *Fish*, p. 119. This fish is now known as the Inanga, see Pauly 2004, p. 92.
531 See *Red notebook*, p. 117, note 159.
532 See *Darwin's insects*, p. 80, which quotes the note made against the Santa Cruz entries in CD's insect list: 'where no white man probably ever before arrived'.

Ostreae & great, red <u>concretions, like at ship</u>'. On p. 49 occur the first of many barometric and angular readings which allowed Darwin to estimate the heights of the plains, by leaving his shipmates while he went climbing. On p. 51 there is a distinct fingerprint on the page, presumably Darwin's. Darwin was by now fascinated by the regular series of vast plains at increasing altitude: 'My great puzzle how a river could form so perfect a plain as 2^d & cemented even in its highest parts — draining of sea ???', p. 51. On p. 52 Darwin recorded seeing 'an ostrich about ⅔ size of common & much darker coloured exceedingly active & <u>wild</u>', which was obviously the lesser rhea. On the same page Darwin noted for the first time the lava flows which were such a feature of the higher reaches of the valley.

On the 26th, on p. 54, Darwin made a striking comparison with the volcanic scenery he saw more than two years before at Port Praya in the *Cape de Verds*. On p. 55 he was convinced that the landscape 'must be effect of sea', but two pages later 'now I can hardly think this valley was <u>formed</u> & these beds deposited in it at bottom of sea'. There is an excellent sketch map of the plains on p. 60.

On p. 61 Darwin remarked 'Boat injured bad days work: Most interesting geology, distant hills'. There follow several pages of geology and 'great hexagon column 12 feet each side', p. 67. He was obviously delighted to have 'Shot Condor! length $3^f 8^{in}$ tip to tip 8 ft iris scarlet red. pairs with young ones, female: magnificent bird: good days tracking', p. 66. These notes eventually contributed to the account he wrote in his 'Ornithological notes', p. 45.

However, geology predominated, with Darwin convinced by p. 69 that the gravel above the lava 'must have been formed beneath the sea'. On the 28th he 'found Indian tripod, first signs of man since the ferry: small grave'. He remarked that the gravel all around made hunting with horses impossible. On p. 73 he drew a diagram showing 'A curious appearance of the Lava — where perhaps currents met or were stopped. throwing up several waves. about 2. feet high; & broard as represented columnar from a centre'. These he described in *South America*, p. 116.

The next day Darwin ascended 'some still higher lava cliffs' where the rock seemed to be different from lower down, p. 76, but still in his view submarine lava. He described the irregular surface of the lava and he noted what he thought to be evidence of 'much diluvial action' in the overlying gravel. On p. 78 he 'saw distant snowy mountains' and in the *Beagle diary* he recorded how this news was 'hailed with joy'.

On 30 April Darwin recorded more giant boulders: 'one was 5 yards square & about 5 feet deep!', p. 80, and he speculated that 'Perhaps the excessive alluvial action, consequent on retiring waters from hills formed the inland cliffs?', p. 81. He was puzzled about the beds he saw containing oysters, and he confessed 'I do not understand the system of plains in this valley'. By now the Cordilleras were 'in full view' and he was very impressed by the vast spreads of porphyry pebbles which he guessed were from the mountains: 'How immense the period during which this bed of pebbles were formed', p. 87.

In *Journal of researches*, p. 218, Darwin criticised former geologists who, in trying to explain the erosion of the lava and other rocks of the valley:

> would have brought into play, the violent action of some overwhelming debacle; but in [the case of the Santa Cruz] such a supposition would have been quite inadmissible; because the same step-like terraces, that front the Patagonian coast, sweep up on each side of the valley. No possible action of any flood could have thus modelled the land in these two situations; and by the formation of such terraces the valley itself has been hollowed out .... we must confess it makes the head almost giddy to reflect on the number of years, century after century, which the tides unaided by a heavy surf, must have required to have corroded so vast an area and thickness of solid rock.

By 1 May the going was very tough, and Darwin was intrigued to find petrified wood which he thought came from the same bed as the oysters: 'if Palms from Lat 50° very interesting', p. 90. Pebbles were of sizes 'from walnut & apples & some as big as 2 fists', p. 91. He felt ready to imagine a scenario for the formation of the valley: 'Perhaps plains & opening at head of river, might be explained by a strait, at very first elevation water in mountain cut it through in the channels, & so on till elevation stopped the passage & river commenced', p. 93.

On 2 May the Andes were 'in view all day' but the river was 'tortuous', p. 94. Darwin recorded the peculiar way that guanacos revisit the places where they have left dung; he published this observation in *Journal of researches* and referred to it in *Natural selection*, p. 522. There were great blocks of slate and ancient conglomerate and Darwin could 'see a gap in the mountains', p. 97. By now he was sure that the plains had a marine origin. On the 3rd there were 'signs of Indians' such as a pointed stick and some ostrich feathers. By the 4th FitzRoy decided to take a party of fifteen armed men a few miles further but supplies were low and it was 'very cold' so they could go no further. They began their descent down the river on the 5th and on that one day covered the ground it had taken them '5 & ½ days tracking' to ascend, p. 101.

Darwin wrote that 6 May was a 'pleasant day' with 'many guanacos' and 'many ostriches', and since the latter had not often been seen before this was 'proof of extreme wildness', p. 103. On the 7th Darwin drew a sketch section of the valley, and on the next day they 'arrived on board' the *Beagle*, which as Darwin recorded was 'repaired. False keel masts up' and described in the *Beagle diary* as 'fresh painted, & as gay as a frigate'.[533]

---

533 See the complete transcription by Rookmaaker on *Darwin Online*.

A gap of four days follows in the notebook which is partly covered by some entries in the *B. Blanca notebook*, while Darwin and various crew members 'killed a lion & curious wild cat & 2 foxes & condors', p. 107. The 'curious wild cat' was probably the Gato payaro. A later specimen was described thus in *Zoology notes*: 'A Cat; in a bushy valley: did not run away: but hissed'[534] The Santa Cruz specimen was mentioned by Darwin in his 'Reflection' essay.[535]

Felis pajeros from eastern South America. Plate 9 from *Mammalia*.

Tierra del Fuego, May 1834

On 12 May the *Beagle* put to sea 'hunting for L'Aigle rock' and the weather was severe. Darwin was 'sea sick as usual & miserable'. On the 16th they anchored 'close outside C. Virgins', p. 107. Keynes 2003, p. 229, explains how Darwin here found a new species of bryozoan (*Caberea mimima*) which he called a Coralline and which showed some extraordinarily complex anatomy and behaviour. The next entry is dated 25 May in Tierra del Fuego and Darwin reminded himself to 'Reread Pampas notes & copy out Gen observation Color T. del F map', p. 108. Perhaps the map referred to is the one now in DAR 44.13.[536] On p. 109 Darwin noted Lyell's

534 *Zoology notes*, p. 401.
535 See Herbert 1995, note 38.
536 Reproduced in colour by Herbert 2007, p. 315, and as a fold-out facsimile in van Wyhe 2008a, p. 19.

discussions of gypsum ('Vol III P. 64') and of Etna ('P. 77'), both references trans-ferring directly to the 'Reflection' essay.[537] Darwin also quoted Lyell's new word 'hypogenes' for the first time.[538]

There follows a long gap of blank pages in the notebook until p. 240. There seem to be some field notes from the Santa Cruz expedition, as Darwin clearly discussed the 'old diluvium' and 'new diluvium' which were the Pampas beds below and above the lava respectively, p. 241. There is a reference to the same boulder mentioned on p. 88, suggesting that p. 241 was written on 1 May.

As Darwin reported to Henslow in a letter of July 1834, two months after the Santa Cruz expedition and just into the second half of the *Beagle* voyage, his scientific notes (i.e. the notes he wrote up on ship, not his field notebooks) amounted to some 600 foolscap pages. Up to this point in the voyage the geology and zoology notes were of almost equal length, indicating that Darwin acted as much as a zoologist as a geologist. Clearly he did not see himself as a botanist as, apart from specimen lists, Darwin did not keep separate botanical notes during the voyage.[539] The Santa Cruz expedition contributed significantly to Darwin's emerging percep-tion of himself as a geologist, as it was in the months after the expedition, during the southern winter of 1834, that he started to generate far more geology notes than zoology notes. By the end of the voyage there were about four times more geology notes than zoology notes: 1,383 geology pages, mainly in DAR 32–8, compared to 368 zoology pages in DAR 29–31.[540] Having walked from the Atlantic to within sight of the Andes, and having unravelled at least in his own mind a very plausible geological history for the southeast part of South America, Darwin was beginning to formulate a grand theory linking elevation to mountain building.

As Pearson 1996 and others have pointed out, a key part of Darwin's success was his ability to see how a series of observable phenomena, if reiterated for sufficient time, could produce profound changes. Darwin achieved such a vision during the voyage in his understanding of the bewildering variety of igneous rocks, and in his brilliant explanation for the gradual series of coral reef formations. This was due not only to the influence of Lyell, but also to the natural phenomena before him which provided striking instances of multiple former causes. The investigation of the Santa Cruz valley revealed how the southern portion of South America was rising

---

537 See Herbert 1995, p. 33, note 39.
538 See Pearson 1996 and the introduction to the *B. Blanca notebook*.
539 According to Duncan Porter, in *Beagle plants*, approximately 20% of CD's notes from the voyage are on botanical subjects. It is interesting to note that CD's scientific life may be seen as a gradual transition from geologist (1830s–40s), to zoologist (1840s–50s), to botanist (1850s–70s). Although he claimed he never considered himself a botanist, in the latter decades of his life CD published a series of books on plants and made many fundamental contributions to plant science (Allan 1977).
540 See Gruber and Gruber 1962.

from the sea due to massive subterranean forces, so that what began as a chain of islands was becoming successively a range of mountains and ultimately a continent. This sequence due to elevation should, according to Lyell's view, be compensated for by a reverse sequence, as some other hypothetical region of the world gradually subsided beneath the waves. It is evident in the *Santiago notebook* that before Darwin even left South America he guessed how corals would form a series from fringing reefs to atolls which would allow one to reconstruct exactly such a sequence.

[FRONT COVER]

Banda Oriental
S. Cruz.

[INSIDE FRONT COVER]
[1 blank]
[2]

Charles Darwin

H. M. S. *Beagle*

[3–4 blank]
[5]

14th [November 1833] Started in the afternoon & arrived at Canelones
15th [November 1833] Started early — very much delay at the 2 rivers. passed the town of St Lucia. — grand <u>green</u> undulations, country does not appear flat to me = country between St Lucia & St Jose no rock — St Jsé granite with large crystals, afterwards some gneiss — José — Lucia — Canelones nice Towns — every-person armed atrocious murder. — crossed Pabon & slept near R. Cufré — saw in Plata fine green undulating Pampas view —
Owl killing snake.[541]
General season of day-feeding
[6]   woodpecker nest in hole:
16th [November 1833] — Stayed all day in fine old hospitable Portuguese; stomach disordered & river of Rosario much swolen: nice Estancia; granite & gneiss former in large crystals with silvery mica & garnets.
17th [November 1833] unfrequented camp.

541 See specimen 1293 in *Zoology notes*, p. 161; listed as *Athene cunicularia* in *Birds*, pp. 31–2.

crossed Rosario deep & rapid. no boat (yesterday Postman with 2 letters principal towns) breakfasted village of Colla said to be Lima — much granitic rocks. — Country more /uneven/ arrived at Colonia. at ½ after one delivered my letters; Head of Police received me in his house son agreed to go his Estancia on following day, so I staid Town pretty looking from irregularity few vessels ruined church

[7] much injured by Brazilian war: general injury to country from same cause. making so many officers — good sign general & extreme interest in representatives. — heard person say that representative of Colonia — were not men of business. but could all sign their names: Town on a point. harbor made with islands (anecdote about French scurvy tell Captain). — like M. Video Church very curious ruin 8 years ago 11 killed: not much powder, very massive walls — so completely shattered
rock of Colonia (V Specimen) very fine grained. <u>dark</u> black gneiss. /siliceous/ much resembling that

[8] that class of rock at Maldanado — the cleavage run E & W vertical or to N. crossed by parallel & approximate lines of fracture. —
In one of the streets; fragments of the great oyster, <u>dendritic</u> with manganese. — Was told a French-man was making a mine & that Government stopped him; said to be entire & juntas forming a bed. !! Are they a fragment of the real St Jose bed, or washed in a heap by some <u>former</u> rush of water — decidedly only occur in this one spot about 15 feet above high river. — not washed

[9] certainly by present river or they would be more universally abundant. — /S̶u̶n̶o̶r̶/ It is certain the Gneiss must be a few feet below. —
(18th) [November 1833] Whole country surrounding the town red earthy Tosca (as at M V.) with very numerous concretions & honey comb plates. — One side of harbor has barranca. prescisely resembling those St Gregorio — base a coarse sand — or rather: minute fragments of quartz — above a marked line of separation. — white sand, pale Tosca & common reddish with

[10] numerous white Calcareous concretions. — In one place in line of separation some few small muscles shells. such as at M. Video even retaining partly their colors. — Is not this important as connecting the very modern shells beneath gravel <u>with Calcarious Matter</u>, with their showing gravel & great Tosca formation co-temporaneous: Height of line above s̶e̶a̶ high river about <u>15 feet</u> & always fresh water: —
From Colonia to R. St Juan country is undulating little rock like that of Colonia — In the Sierra de St

[11] Juan some slaty gneiss cleaving N & S. other irregularly NE & SW. Some beds of quartz & some gneiss more granitic. —
In the pass of St Juan there is some earthy pale Tosca rock. —
But what is remarkable along whole side of Sierra & not much inferior in height are beds of pure white mortar & of semicrystalline often p̶u̶d̶d̶i̶n̶g̶s̶t̶o̶n̶e̶ Breccia with fragments of Granitic rock (many large) (no specimen) some opaline rock all mingled together evidently facing the side of granitic rock & almost surrounded

[12] by hills of granite, chiefly on low hill (these phenomena in several places)
The hill is clearly formed by vallies being formed at base of granite hills — I should
think had formed a plain interfolding amongst hills of granite & not posteriorly lifted
cannot say. — evidently lies on granite & beds not thick Formation <u>precisely resembling</u>
those at foot of Sierra Ventana & Tapaken only rather more extraneous pebbles. — How
is it ~~this~~ formation is better seen near the
[13] hills? Has been burnt for lime seems doubtfully to answer. — *[vey]* very

curious finding this formation here. —
shaded ones Lime. —
Degrado same colour outside as Patagonia in <u>west</u> precisely same formation. eyes
shut think I was in Patagonia: —
On surface fragments of Granite much alluvial action:
Rock about Estancia Gneiss. — Near the ~~Sea~~ River, 6 or 10 feet above level of swamps
are sand banks of same form many Medanos
[14] as those at Bahia Blanca, with very numerous Bivalve shells, said to be still more
numerous at mouth St *[Francis]* if they are fresh water small *[ex]* charge is necessary
if salines. requires greater change in configuration of Sand. but does not relate to
muscles under Tosca. —
In evening had *[pleasant]* ride about Estancia. Exception Mercedes 2 & ½ leagues square
excellent rincon ∴ water, much wood for exportation — Lime *[illeg]* horses very good
Corall & Garden 3000 Cattle — 600 Sheep 800 Mares 180 broken horses Harbor. —
When Limestone is pure *[genus]* shells. — the Aluminous matter injurious to shells.
[15] has been offered 2000£ wants 2500 (or perhaps less) How very cheap. —
Cattle driven twice a week to centre spot & then by the union of Tropillas counted.
Tropillas recognised by curiously marked animals. —
(19th) [November 1833] WNW of Sierra de St Juan Mortar — Hills appear to run N &
S between the head of two streams. —
<u>Calera</u> de los Huerfanos 4 Leagues before arriving at Las Vacas Mortar formation.
(V Specimen)
Generally more white & pure: a few more hundred yards further on — a <u>highly</u>
ferruginous sandstone with specks of quartz
[16] mingled irregularly with a paler & less sandy sort — This occurs irregularly till their are
some low hills near Las Vacas –
Here the same rock occurs very abundantly. — & in section of "A tres Bocas" is seen
lying over pale Tosca abounding with calcareous matter & concretion — The summit

of the hill granite — in this manner there were repeated alternations of gneiss granite & this Sandstone. — The latter here contained large fragments of Quartz. — ~~Wood~~ Much wood like Corinda. — Las Vacas

[17] straggling thatched town ~~wit~~ on Riacho with many small vessels: much delay & trouble. — No Biscatcha. Cuervos soar in flock: — The very red sandstone evidently is of same formation of the mortar — extends very far (is said) to the sources of Las Vivoras: Slept at the Calera de Carmacho
(20th) [November 1833] — Narrow bed of Limestone extending from near Mouth of Las Vivoras to Punta Gorda & crosses to R /Nankry/ in Entre Rios —
"Sylvester Lellow complete collection of Banda Oriental." ?? !! —
Much of the Limestone

[18] is grains of quartz cemented together. — generally pretty pure with casts of many shells & sometimes large oysters. — beds separated by beautiful white sand. Considering the gradual change of Huerfanos. — the great change of red Sandstone, I think this same formation with mortar; it looks as if this bed with respect to red Tosca does not remain constant: the Lime requires same time /as/ &c with Bajada —
House 108 years old. —
old woman of 90 years old positively states that very early in her life no trees ?!

[19] no trees except one orange tree
Lime Kiln fairly covered up by diluvial bank so as not even to be guessed at. — trees grown over mud Time of revolution Lime left /in/ in some kilns for 18 years quick in middle yet, burst out in flames constant occurrence superstitious fears — vegetation above: —
Curious subterranean arch at Colonia. —
3500 small green parrots killed in one field of corn near Colonia. — Jaguar (went out hunting) cut trees on each side with claws sharpening

[20] plagued by Foxes barking never return to dead body: Gato pajaro[542] inhabits Banda Oriental. —
Coast of R Vivoras, R. Uruguay regular red Tosca with numerous balls & oblong pieces of hard white Tosca rock. At Punta Gorda this occurs beneath a pale clay with ~~the~~ large Oyster shells & above the regular white sand & Limestone. the next Point further up is pure Tosca —
Here we ~~ele~~ clearly

[21] have same bed beneath Limestone as is above at the Bajada; only in the latter case the Tosca does not contain so many nodules: One man lassoed 22 mares on <u>foot</u> in Corrall, tied their front legs, killed, skinned & staked them.

542 See specimen not in spirits 1443 in *Zoology notes*, p. 392.

Other man will but stand at mouth of Corrall & throw every animal by lassoing legs as he leaves: Other man will skin 50 mares in one day good work to skin & stake 16 — After our pleasant ride to Punta Gorda in the evening

[22] started for Capella Nueva Mercedes: only rode a few leagues, through Coronda like wood, & arrived at <u>very large</u> Estancia. immense land owner — The nephew of owner & a Captain was resident there. — After asking me if I really knew that a hole under foot would come out side. ball all you catch take thin people in by land, where there is 6 month of night — & expressing great surprise at being able to go by land to N. America: "answer me one question truly" are not the ladies of B Ay more beautiful than any others." "Charmingly so" — One other question do ladies in

[23] any part of world wear bigger combs. — I assured them not: they were transported & exclaimed — "Look there, a man who has seen half the world says it is so. — we only thought it to be the case. —

Water of Uruguay <u>very</u> black: rapid current: In riding disturb male ostrich from nest: sometimes very savage chase men on horseback — caught by men lying down covered with a Poncho. — (Paludas) Nest of ~~green~~ small Parrot. —

(21ˢᵗ) [November 1833] Started very early & rode quietly whole day. with hired horses. — Passed through <u>immense</u> (no Biscatchas)

[24] beds of thistles: cattle lost. roads closed up: generally as high as horses back; often as high as mans head. — as the geology here resembles that of Buenos ∴ we have the variegated thistle. both sorts ~~in~~ almost invariably conjugate variegated the worst from height — Very uncomfortable riding obliged to make great turnings. — Before arriving at the town of St Salvador — much mortar, white rock & Tosca rock — also the ferruginous dark red sandstone —

All this forms bits of horizontal beds occasionally *[illeg]* Limestone up concretion Uruguay

[25] to be ~~seen~~ passed one after /the/ other (red rock first seen a league or two before crossing R. 3 Bocas)

Near source of St Salvador soon after leaving white mortar rock — arrive at bed of white jaspery rock marked with manganese — containing nodules of milk agate — externally decomposing to great thickness into soft white rock — Something of same nature as one specimen found in Mortar of Sierra de St Juan — ~~In one part of road~~ These small pieces of granite rock. — Then white Mortar. Then red rock

We slept at small Rancho near to where I was going

[26] it being late in the evening. — Next morning

(22ᵈ) [November 1833] arrived very early at the Estancia of Mʳ Keen on the R Berquelo near Mercedes —

Many highish hills all flattish tops & low Barrancas of this red rock — which lower becomes whitish with red streaks & concretions

(V. Specimen) still lower it is white sand cemented loosely together: These flat topped hills run NE & SW interrupted by wide transverse vallies. — Pietras. Maldonado (~~are there black rabbits on West Falkland~~)[543] Lower in the vallies there is a hard sandstone

[27] or rock consisting of minute fragments of quartz of various degrees of purity & very hard. — This in parts contains beds of Flints, with veins & conchoidal fracture.
V Specimens — in sinking wells here & Mercedes, this flint is the lowest rock. —
In one place sandstone contained large pebbles of quartz & granite. — it here contained inferior layers of mortar. — & some whitish Tosca. — I have no doubt such beds separate the sandstone & subordinate flints from the ferruginous red sandstone (23^d) [November 1833] On road to Mercedes

[28] The grit sandstone occurs of various /puentes/ with more or less Mortar passing into more or less pure Limestone. — At Limekiln this side of Mercedes, Limestone occurs beneath a siliceous bed with much botryoidal[544] quartz (V Specimen). —
At Calera Daca. — the rock is fine hard white mortar with siliceous /veins/ lying on a gritty sort (or grit sandstone cemented by lime (? is not all Sandstone thus cemented:) No organic remains —
It is said that near there the red ferruginous

[29] sandstone occurs above the Limestone — which is as I should suppose. for certainly the grit sandstone & Limestone are all of a bed — occurring above or in mortar & Tosca beds & here occurring above. — at /Tres/ Vivoras above clay. — But I think from general local high position the red sandstone is a superior formation — R
R Negro fine river with fine blue water, & well wooded; pretty valley. — poor straggling town. —
(Curious pebble/s/ from coast of Uruguay. — Rincon de las Gallin/aco/)

[30] (24th) [November 1833] Started for Pedro Flaca, went wrong road travelled through very long grass. (uninhabited) above horses back like oats. — on road saw with mortar, some fine red Tosca — & afterwards the very red sandstone above Mortar:
at Perica flaca cliff about 50 feet high. — upper bed prettily coloured flint (V. Specimen) or agate. mixed with the very pure tallow — looking Limestone — actually mingled together in masses so that bothryoidal silex on the Lime. — grit sandstone & finer sandstone largely conglomerate, rounded pebbles coloured pink (by Mangenese). — grains

[31] of sand cemented by Lime. — & ¾ of lower cliff various degrees of purity of mortar with some bed & masses of reddish Tosca hard. — calcareous lowest bed pretty pure mortar. — This is contrary order to what is general — for flint generally appears to be lowest. —
In very many parts before arriving at Perrica Much fine red Tosca. I must think this beneath red sandstone

---

543 This may be one of CD's earliest references to island endemism.
544 Bunched like grapes, a habit of some minerals such as haematite.

The question is whether Bones occur in Tosca contemporaneous with Punta Gorda bed, or with Bajada Monte Video

// Punta Gorda very satisfactory seeing Tosca beneath solid rock //

[32] view of Rio Negro from cliff very pretty — river 2 Severn — ~~fa~~ current very rapid. banks well bushed winding rocky cliffs —

decidedly most picturesque view for the last four months: Horses tired left Peon behind. — excessive heat Pantanas. — Late at night:

(25th). [November 1833] Rode to dig out bones of giant; found in place ~~evide~~ washed down & covered with sand & clay, when first discovered were in dry bank. — now under

[33] water. — bones scattered broken lying close together about 20 yards distant bed of Tosca. — with small line concretions like bed at the Bajada; Interesting as connection between Casca[545] & big bones:

Wheat Pubrilho[546]

In evening saw a Domidor[547] Mount two colts tame describe process.[548] — excessive fright. — horse died yesterday. —

(26th) [November 1833] Started went round by a house to see large head & bones, washed out of Barranca & found after a flood. — pieces here

[34] also of Casca —

Barranca. Whitish Tosca — Sand — which occurs for a considerable distance, *[is]* most probably diluvial hence animal of Tosca & diluvial age — 4 or 5 leagues to South of Mercedes plenty of Quartz — then little grit sandstone — plenty of primitive rock — then large bed of flint — finally all great blocks of primitive rock. — Granite in immense blocks near sleeping place — Sleeping place 3 leagues beyond R Perdido — very bad

[35] night — wet through; extraordinary thunder —

// Bones from the Sarandis[549] will not have paper in the box: //

~~(27th)~~ The white Tosca bed certainly different from the usual grand covering. probably of a different date.

(27th) [November 1833] country whole distance Primitive gneiss running W by N & E by S & ~~NW by~~ in other locality NW by W & SE by E: Bad Indian Gallego:[550] — Slept at one stage beyond St. José. —

---

545 Spanish for 'shell' or 'helmet', most likely meaning armadillo-like case.

546 Specimen not in spirits 1593 in *Zoology notes*, p. 394, and *Beagle plants*, pp. 174–5. CD's notes have not been found and were apparently given to Henslow who published an extract in Henslow 1844; *Shorter publications*, pp. 176–7.

547 Spanish for a horse-breaker.

548 Described in *Beagle diary*, pp. 204–5.

549 A small river entering the Rio Negro.

550 'One of the Post-houses was kept by a man, apparently of pure Indian blood; he was half intoxicated. — My peon declares that he in my presence said I was a Gallego; an expression synonimous with saying he is worth murdering. — His companions laughed oddly: — & I believe what my Peon

(28th) [November 1833] Arrived in middle of day by same road to Monte Video —

[36] At Las Pietras red sandstone in horizontal beds near to which /in/ fine granite specimen is finer than the generality —

Padre with a tail very heavy & solid, fragment 17 inches long, circumference (longest)

11' ½ at end before blunt 8' ½   Vertebra within attached to the case — extraordinary weapon. ~~Solis~~ /Arroyo/ Seco — Tooth. Solis grande. — & Tala about 10 leagues N Branch of St Lucia,

Arroyo Tala

I believe name Don Damasio de Laranhaia. Griffith[551]

Vol 133 /~~Palvith~~/ p 133

[37] I think, the limestone & clays with oysters, the sandstone, flints all are variations of Tosca bed with concretions (∴ red sandstone above such varieties & above Tosca & Limestone above Tosca) That the red sandstone is above these: That possibly the bones are found in an earthy Tosca above this sandstone (∴ more mineralogically like beds at M: Video) But that fossils to the W of Parana & B. Blanca belong to bed below limestone or if not such perfect coincidence in formation shows circumstances very similar. —

I have little doubt respecting sim: of sandstone here & at Maldonado & Gypsum to the N: V Mawe.[552] —

[38] 18th [April 1834][553] Started pleasant sail to /nearly/ above tidal influence <u>Armadillo</u> — noticed several successive plains; some quite low 20 feet high: yet with regular gravel & apparently earth —

∠[r] Blocks

sleeping place oysters arca & 2 Univalves gravel white-washed (⸮ last character characteristic of original gravel?) — plains to the N. fresh higher series: caught mouse:[554] pleasant party cheerful running water:

⅓ & mile

said was true; when I remonstrated with him on the absurdity, he only said, "you do not [know] the people of this country". — The motive must have been to sound my Peon, who perhaps luckily for me was a trust worthy man. — Your entire safety in this country depends upon your companion.' *Beagle diary*, p. 204.

551 Cuvier 1830 (edited by Edward Griffith), p. 133, discussed Don Domasio's announcement of additional bones of the head assigned to the *Megatherium*.

552 Mawe 1825. See CCD1: 562. This page is one of the earliest ink essays in the field notebooks.

553 The second sequence of entries begins here, in pencil.

554 Possibly one of the specimens not in spirits 2066–7 listed on p. 402 of *Zoology notes*.

(19th) [April 1834] 3 parties tracking many islands like Parana: bad walking rapid very clear river over pebbles shallow hence troublesome. few Guanaco or ducks. no fish or animals or fertile ground
Horses head: Callandra

[39] singing: flocks of Sturnus Ruber: cliffs with Turitella: Slate pebble little E of 2[d] sleeping place: larger pebbles, more truly porphyritic: 2 elevations (first one most weathered & an apparent third one. old valley of river chiefly gravel & like a bank.

[valley section with terraces][555] (2) z (2 old valley) severe frost
20th [April 1834] same fact of series of Plains: at sleeping place nearly ½ the pebbles dark compact /littly/ fissile slate amygdaloid & a fragment of gneiss: curious small travels if pebbles into sea from so pebbly a river:

[40] small salitrales: country rather more open: river rapid banks pebbles tracking on N. shore: good general arrangements: drift tree: large herds of Guanaco: Ostrich swim. (river generally 17 feet deepest & ⅓ of mile wide — here narrow: about 5 miles an hour — Alarm of indians: horses bones great smoke, parties kept close: but quietly found good sleeping place: day has been splendidly fine: but country terribly uninteresting: no living beings. insects fish &c &c Same plants, same bushes growing on same formed land

[41] (Effect of Earth quake in Chili on river courses) Beyond this Terra incognita boat-hook: 21st [April 1834] First boulder 7 feet circumference depth 18 inches perhaps as much below: Quartzoic or Feldspathic? (V Specimen) Fish light greenish brown above with small blank transverse irregular bars — belly snow white: numerous same size in inlet. Fringilla common & sparrow good success No condors but Carrancha
Indians tracks with Chusa thought they had reconnoitred us: then mark of mens feet & dogs as if crossing the river (here 200 y[d.] & 30 & feet deep current 6 knots)

[42] marks on other side: Children: granitic pebbles: drowned balled Guanaco — eaten: 22[d] [April 1834] (Tuesday) many slate & some granitic pebbles I think former, perhaps solely connected with the lowermost sets of plains: I could trace 5 sets of plains — lowermost perhaps only 20 feet above river: as the highest is 500? feet: the lower uneven stone is a white sandy marly with angular cleavage like 2[d] bed

[43] at P. Desire & St Julian (specimen 2). contains some Turritella and Ostrea &c: at one place cliff 120 upper half this the gravel — & within ¼ of mile cliff of marl only 20

---

555 A very similar diagram, reproduced in the Introduction, was published in *South America*, p. 10.

covered by about 60 of gravel belonging to 2 sets of plains: gravel often covered by earth bed:

country more broken

[44]   river more rapid, banks bad. two spells. smoke to the South: weather beautiful country of Patagonia extraordinary similar:

burial places of Guanaco. 10 skulls together

pebbles in lower plains white-washed: curious (sleeping place) infertility by a little stream. little life analyse marl — salitrales. which are present perhaps cause:

[45]   // red-nosed mouse // long tailed grey creeper: // of the 5 plains perhaps 2 highest best marked: Callandra:

23^d [April 1834] (Wednesday) Rested till noon to mend clothese, clean arms &c. — caught Notonecta.[556] Colymbetes Staphylinus & Bembidium:

The bushes of R Negro owing to gravel: the present river is only a gut[557] without valley. the three lowest plains, are not very level & perhaps are owing to former less rapid river: some of the plains differ in

[46]   height solely from thickness in gravel. a cliff of 80 feet only had 10 or 15 feet of marle & this not horizontal, but worn: gravel cut by earth ∴ immense quantity of gravel! slate pebbles. certainly more numerous: many Callandra: many rodentia:

proceeded little further slowly owing to cliffs

[47]   obliging us to cross river &c

24th [April 1834] — Finely cellular irony scoriae pebble: stream from quantity of slate like Tierra del F: Cliffs with white (calcareous?) lines: other cliffs with many Ostreae & great, red concretions, like at ship: more Slate pebbles in stream than in lowest plains, some of these latter plains are perhaps owing to present river, but I think only few, from little floods I do not understand how & plain 200 feet high could

[48]   be formed: also in some places it clearly is a gut between marl: — perhaps the highest plains are separated from the 2^d highest by old valley 10 or 20 miles wide for here action of sea is less probable than at coast: killed Guanaco, goods days tracking; country more level: I do not see that the plains are at all higher

[49]   than at coast: illusions about lakes & mountains: Guanaco when wounded always comes to water ∴ when ill: — pebbles being whitewashed in low plains look not like rivers? —

(25th) [April 1834] *[illeg]*   29.995. A ther 56
                29.550. A — 60

---

556 A genus of aquatic insects in the order Hemiptera, commonly called backswimmers.
557 'a narrow passage, a channel or run of water, a branch of a stream; a sound, a strait'. OED

Noon (Latitude) 2^d highest plain — one higher perhaps 150 or 200 higher: ~~gravel~~ slate pebbles: cliffs composed of yellow or greenish earthy clay easily disintegrated with white bands: large red same concretions:

[50] no shells (I saw): Before this I ascended table land with gravel pebbles & as here noticed 2 great plains the upper perhaps & lower certainly follows the valley

the three lower plains are very obscure: there /was/ one (at noon) of some elevation, perhaps not explicable on principle of river forming low plains & then cutting a deep ~~valley~~ gut & commencing other plain Noon valley between the 400? cliff & other side (which perhaps are the highest series?) 5 miles wide. The period between 1 & 2 old plains much shorter

[51] than between 2^d & present (other elevation intervening) Gravel in same places almost cemented by (marl?): white matter. —

My great puzzle how a river could form so perfect a plain as 2^d & cemented even in its highest parts — draining of sea ??? —

How was valley at Port Desire formed before elevation sea ?? : At this (noon) cliff many of the stone was coarse sand slightly agglutinated by ferruginous aluminous matter: one pebble of dark reddish brown amydgaloid (with agate on brown cavities)

[52] has for the last three or four days been very abundant it is quite impossible to overlook it at coast:

Saw an ostrich about ⅔ size of common & much darker coloured exceedingly active & <u>wild</u>

very goods days tracking

[The figures in this table refer to the explanation on p. 53.]

26th		A	29.883	A	59	D 59.5°
		B	596	A	63	~~D do~~
		C	454	A	67	D do
waterside		D	29.904	A	71	/Katers/
		E	− 516	A	71	1° or 2° /dip/
		F	− 629	A	69	− (62°)

Spring at foot of Lava 49.5° —

water edge 29.902 A 64 D 57

[53] a third lower plain alluded to yesterday is about ⅔ height of Lava plain (B) it abounds with great masses of Lava therefore valley of river cut through these by ordinary means: between noon of yesterday & todays Longitude first noticed great blocks of this Lava. (or amygdaloid which from its less specific gravity float further)

[The following list corresponds with the table on p. 52.]

(A) at Waters edge

(B) — Low Lava cliff (there was one higher)

(C) High plain at back of (B) & probably of same height with that of noon of 25th

(D) waters side the two first observations were taken nearer to (A)

(E) Main lava plain

(F) Spring at foot of Lava

— (G) Waters edge (near F observation)

[54] The lava cliff commences half way between noon of yesterday & to days Longitude only on north side, on South side cliffs of pure marl to waters edge like those at noon: The Lava is black augitic sometimes with olivine, the upper part very vesicular (Like Port Praya. valleys precisely same sterile scenery): it is in upper parts regularly hexagonal: there are more than one plains which Barom will show: there are lines 2 or 3

[55] lines of separation, where the Lava is more vesicular & looks like separate flows: hence distinct plains: Columns of Lava filled & separated by mortar (V specimen): The plains to the north of yesterdays noon & of (C) are so much intersected & broken into escarpements that I almost think it must be effect of sea, which will explain gravel cemented by Calcareous matter if the the great range of cliffs South of river was as it looks same elevation of noons yesterday & (C)

[56] Then these highest cliffs could not form side of valley; & the third plain (& formerly 4th) might be explained by altered height of river: The Lava plains are all lower than (C). The thickness of Lava is at least that between E & F: if & probably the total for the spring probably breaks out where lava overlies marl: But this is not quite certain: in two

[57] places in one narrow valley, there was in an inclined surface yellowish sand slightly agglutinated & without any pebbles, as at noon: in another pale bed I was surprised to see few small stalacti on concretions with manganese: now I can hardly think this valley was formed & these beds deposited in it at bottom of sea:

(some porphyry pebbles on Lava plain) From horizontal stratification

[58] & considerable elevation of site of these beds (good way below Lava plain (B)) perhaps at equal height with spring there (earthy yellowish Tosca as it would be called in the North) must be coincident with basal beds of yesterday's Noon:

But why a Hence I imagine this lava is intermediate, but high, bed in marl formation: It cannot be superior because surely then

[59] the Lava plains, would be the highest: That these lower lava plains should be present is likely because superior beds might be denuded they remaining: perhaps it actually is in middle in plains like (C), indeed it almost must run under them, for they were not more than 1 & ½ at its back:

[60]

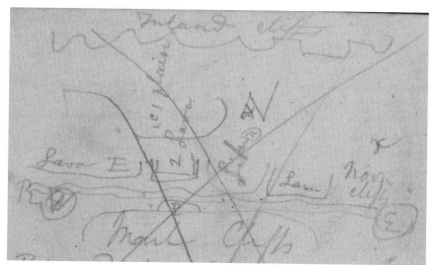

[sketch map with marl cliffs][558] Inland Cliffs Lava E (C) plain 2 Lava Like x x Z Lava Noon cliffs x W R P E Marl Cliffs (P) 3rd lowest plain

My puzzle why Lava escarpement is not seen at (Z): — Perhaps the Lava terminates in points: The course of river has been determined by South limit of Lava:

Condor is present solely where mural escarpements legs pale. Lump on

[61] head: Dark kelp bird:

If the marl in valley is subsequent, it must be anterior to gravel, but surely it would contain volcanic particles:

Boat <u>injured</u> bad days work: Most interesting geology, distant hills:

The high<u>est</u> table land must be over the Lava: certainly appearance looks as if Lava capped some soft talus-forming mass. (Character of sand is such that with wet cakes & cracks): impossible not to be struck with resemblance of mortar in Lava &

[62] gravel at R Negro:

The Tosca rock concretions are curious as showing recurrence of same mineralogical substances at long distances, under somewhat similar circumstances, not as any proof of similar formations:

(A) Cellular superior Lava

(B) Columnar Lava (common sort beneath cellular

(C) Calcareous matter between columns of (B)

(X) Fine grained compact marl lower bed some days since

---

558 See a fair copy of this sketch in the geological diary DAR 34.134v.

[63] Sunday 27th [April 1834] Globular Lava passes into columns

Waters edge:	29.834	A 41	D 41
Lava Cliff	29.332	A 46	D 47
Little Hill	—.264	do	do
Waterside	29.672	A 58	D 55

This fall ought not to [be] allowed or only about ½ of it:
measurement of Lava plain (E) is best, but difference of Little Hill must be correct: If as I suppose this little hill is part of plain (C) & there were other appearances further on of same fact) Then difference & height of ~~the~~ either of two observations of Lava plain ought to equal (C): This little hill in vege[t]ation & gravel

[64] exactly resembles the plain, there were no Lava pebbles: Having seen this gravel cemented by Calcareous matter explains the veins between columns of Lava: I believe Lava flowed ~~thro~~ between the formations, was partially or entirely denuded & ~~cover~~ when inland & highest cliffs was formed: the gravel of little hillocks being deposited (or even perhaps some marl?) its present position accounted for by resisting

[65] denudation: I think it must be thus, & that Lava extends beneath highest cliffs: Lava highly sonorous, where cellular very tough: one clear separation of line of greater cells & large cavities & resting on a surface of balls — (this appearance is seen on the plains occassionally) above this line, which is less than ⅓ of whole Lava bed Lava columnar. other lines but less distinct: this main line shows two streams

[66] cracks where water entered might be traced by cellularity: That the Lava rests on marl is proved by great slips & lines of springs:
Valley now simple, straight 5 or 6 miles wide, South cliffs same height with what? — Shot Condor! length 3[f] 8[in] tip to tip 8 ft iris scarlet red. pairs with young ones, female: magnificent bird: good days tracking.

[67] (D) Highly sonorous. common lava
Projection of coast from C. Blanco to Port Desire owing to Porphyry St Elena do? — High land between /Herr/ Bay & St Joseph. —
28th [April 1834] Surface of Lava generally rather undulatory: great hexagon column 12 feet each side: earthquakes great slips inclined towards cliff nearly ∠ 45°: —
E. The common white or yellow slightly agglutinated sand; beneath Lava
F. Fragments of a Boulder mentioned before
G. On plain Pitchstone or Obsidian

[68] ~~12~~ Noon

Waters edge	29.466.	A. 64	D 59
Plain	Z 28.944.	A 59	D 55
Crater:	28.734	A 54	D 52 (?)
Waters ed 6 oclock	29.520	A 52.	D 48

Perhaps one third of this rise, ought to be allowed for the Plain (Z): seated close to the river & scarcely anything between the Plain & Crater. —

This plain is situated like (Z) in the former plan on each side there are Lava cliffs & behind it likewise: I imagine it to be same height as it is in constitution as the Noon cliffs: perhaps it may

[69] be considered to belong to the Southern cliffs: it is <u>remarkably</u> level & gravel <u>much</u> white-washed so I that I think it must have been formed beneath sea: I am doubtful whether the South cliffs are higher or not than this: The Lava cliffs are <u>here</u> exactly same height proving a thickening in the bed to the Westward: If these plains have been formed by denudation of higher ones, of course the Lava would put a stop to this action. hence levelness of these soft plains & those of Lava: This plain rises <u>gradually</u>

[70] To the hill which I have called Crater forming a sort of plain without hillocks of Lava <u>smoothed</u> with gravel: the lava plain gradually rises to this. Is this Crater or lava flowing over other field: if it is not continuous to the West, this is form more probably (from inclination of bottom would cause a greater East, than W flow) This hill is about half as high as some neighbouring <u>pieces</u> of table land, (called the highest inland cliffs in other places). I continue very uncertain concerning their origin: observations

[71] to the <u>East</u> before Lava will now be the more necessary: & especially height of South Cliffs: —
Before this I satisfactor/il/y saw the yellowish agglutinated sand (E) & white bands no org. remains underlying lava, but not in contact: the stream of Lava must have had a fringed edge. such plains as both (2^d) occurring in indentations, valley of rivers line of fringe: At the sleeping place (of yesterday??), & further West valley opens: & 2 or three lower plains may again be seen. — The Southern line of cliffs most regular

[72] I may notice in lower part of <u>marl</u> cliffs some harder much laminated sandstone: Saw a Skunk: found Indian tripod, first signs of man since the ferry: small grave: drift timber: immense herd 1000 ! of Guanaco horses cannot hunt for gravel. — Guanaco sleep tail in centre, in same places in different nights then dung: dust in saucer shaped cavities

[73] (V other page)

[lava 'waves']559 (double)

---

559 See a fair copy of this sketch in the geological diary DAR 34.139v.

A curious appearance of the Lava — where perhaps currents met or were stopped. throwing up several waves. about 2. feet high; & broard as represented columnar from a centre. therefore easily seen seen in a valley.

[74] At the sleeping place there was a most deceptive appearance, for ¾ of mile hillocks of marl dipping at 45° to cliffs & vertical & even from them, covered with blocks of Lava — yet only a great slip cliffs ½ a mile inland — layers of calcareous sandstone together with odd square masses of indurated laminated do. — & some very calcareous strata. V specimen

[75] (29th) [April 1834] The plain (Z) of yesterdays seems to have divided the stream of Lava — in its back parts perhaps containing some: at the foot of it there is one of the lower plains ⅔ height, level, white-washed gravel, lower parts marl: Having passed plain (Z) The lava is seen in several plains, one above the other, the backs one most clearly beneath the inland cliffs, they are not level but rise to the West
They cross the river forming

[76] plains on that side, which seem of no great extent, but perhaps some distance beneath great S. Cliff:
After dinner ascended some still higher lava cliffs, probably much higher than Crater hill, the lava is different, contains much crystals, is greyer more compact, less cellular probably formed under less pressure, yet under sea from porphyry pebbles & many blocks

[77] of slate! Lava laminated. (Before this I found Lava with same crystals) The was lava was much thicker; as seen by springs but rested on same as before: Surface very irregular many hills & central depression with lake — much diluvial action — lava in one place like wave V drawing — In the NW irregular hills probably Lava — This (plain?!) perhaps as high as inland cliffs, of which traces were yet present:
How far connected together ?

[78] Saw a regular eyrie of 20 or 30 Condors — 2 young one with the old: saw distant snowy mountains
(H.) Calcareous rock in marl sleeping place (28th)
(J K) Grey Lava with many crystals from highest land on the 29th —
(L). common still greyer more laminated variety
(M). common black vesicular Lava with do crystals
(N B shells in coast cliffs must have been pelagic. —
Curious no mixture of Porphyry pebbles with marl. yet requiring such ages to manufacture them:

[79] 30th. [April 1834] — I think the lava overlies the greater part of the gr. Oyster bed: but from appearance of yesterday & before is covered by the upper parts. perhaps marks the changes so often noticed on the coast. it is certainly covered by the Porphyry pebbles. which must be altogether independent of it: —
The lava yet rises, irregularly towards the West & again occurs to the S always overlying the white horizontal striped beds But I do not think these + gravel rise much above

[80]   in height than last (Z) plain: (Noon). Valley to the West opens much & lowers & has been (like St of Magellan?) subject to excessive alluvial action. hence on the yet. remaining cones are great blocks of lava & of Feldspathic rocks (similar to 1ˢᵗ boulder) one was 5 yards square & about 5 feet deep! another 2 yards square, pebbles of serpentine

[81]   These rocks & slate & pale porphyries outnumber the yellow & red sorts. — Perhaps the excessive alluvial action, consequent on retiring waters from hills formed the inland cliffs?

Waters edge	29.156	A 52	D 48
X	28.546	A 48	D 44

X Line of springs & <u>apparent</u> highest limit of soft beds: site at sleeping place: above this <u>black</u>, compact Lava specimen (n) (Q other variety much laminated)

(May 1ˢᵗ) [1834]

Waters edge	29.046	A 42	D. 40.5
Alluvial Hill	27.873	A 42	D 40.
Lava limit	28.070.	do	do
Waters edge	29.042	A 47	D. 46

[82]   1   Calcare[ous]
       2   x
       3   Calc. <u>white</u> specimen (O)
       4   x yellowish iron clay
       5   green (P)
       6   x
       7   3 Calc & brown
       8   green   horizontal strata no org. remains
       9   Brown calc
      10   x
      11   ~~Calc~~ Soft Sandstone
      12   2 Calc & green
      13   large calc
      14   green

? Soft sandstone or rather very fine conglomerate, with slate & white & green conglomerates

[83]   This is section of above measurement, (3) is a <u>broard</u>, white strongly marked bed of light Calc or Agillaceous matter & resembles those of Port St Julian: perhaps occurrence here may be attributed to the source of such beds being to the North; whilst beds at the coast more exposed to currents from the Southern & less volcanic mountains: (5) is a pale green earthy clay; specimen (P): of same nature but more sandy ///// is the yellowish brown: the lowest bed is of the most new character

[84] as containing slate & Feldspathic (most minute) pebbles, it 1 & 2 lines long). it is as might be expected from being nearer mountains:
These beds do not from ½ a mile remain constant only in general points, in some places great concretionary masses of Sandstone as at Ship. — These cliffs certainly are very different from the coast — but then the difference has been gradual, first white lines appeared in clayey sand, & which contained Ostrae. /Thes/ now like true calc matter

[85] which will not support life: but ye[t] the beds called <u>marl</u> unite those at ship with these here ∴ formation similar: it might be expected that more variation would be met with in beds (& gravels) in this section than on the coast, perhaps this observation generally holds good. — (We know there is little Volcanic action to the South)
I am convinced (though often torn) that both N & S of river the Lava underlies upper part of Oyster bed: I confess I do not understand the system of plains in this valley.

[86] (Though little) I suppose the Oyster bed does thicken elevate itself, if as I it is, as I imagine: flow from the Andes the above Lava will be higher than Crater hill. —
Lions scratch ground: killed 2 Guanacoes, Condors eat one — Cordilleras in full view. Clouds & Climate — River, blue <u>narrower</u>. — Lava streamlets:
The general absence of pebbles in all beds. except this one lowest, shows how entirely the Porphyry pebbles depend on

[87] some violence, & different nature different point of origin. — How immense the period during which this bed of pebbles were formed. (anterior to Oyster form:?)
I see no more of the highest inland cliffs, all is now irregular Lava Hills. —
(May 1ˢᵗ) [1834] V. supra for water side <u>Old</u> Alluvial plain 28.730 A 51° D do 50°?
The measurement of the Lava is I daresay good: it is varieties (N & Q). it is covered by great mass of Alluvium, where the blocks are immense of Lava.

[88] & the Feldspathic rock & the one with quartz veins & slate from 1 to 2 to 3 yards square & porphyry pebbles: These doubtless are the main rocks of the Andes: one Feldspathic boulder (same sort as before) was 20 yards in circumference, & 6 feet above ground how much beneath? must have been formed beneath sea; from Porphyry pebbles, & cellular Lava. — Surface very irregular & weathered I think it very likely that the inland

[89] cliffs or table land, mentioned since meeting with the Lava is of this nature. if so Lava superior to all great The first inland cliffs can have no relation with these, else 2ᵈ plain would have some blocks from the immense pile of boulders. —
At the foot of these cliffs 2 Voluta's yet partially retaining their color, & a Patella; now these must at least have come from highe upper parts, proving modernness — for on plains I have seen none certainly near coast one would expect in parts last elevated:

[90] May 1ˢᵗ [1834]
How will Hypothesis that Lava flowed in old strata arrive
I ought to have mentioned that the lowest bed in yesterdays section, was some hundred feet above sea. — Capt Fitz found many <u>great</u> Oysters at foot of cliff. — Also in many

places pieces of of petrified wood are lying on lowest plains. which I think must have come from same bed with Oysters — if Palms from Lat 50° very interesting. — I imagined the vast mass of

[91] Alluvium is contemporaneous with Tierra del alluvium. — The valley opens; Lava seems to cease on S. side, & the whole [a] plain; one plain which was in line of junction of the two Lava plains seems to represent what generally is the case.

                      + 15 feet
Waters edge   29.042   A 47.   D. 46
This plain   28.730   A **51**.   **D**. 50

it is composed of pale yellow sandy earth containing pebbles from walnut & apples & some (as big as 2 fists), slate & green Feldspathic such as in lowest bed in yesterdays section

[92] & the great Boulders & white porphyries there were some great oysters. some rolled: & beds of gravel: & some with none: This is about 6 miles to W of great section. but it certainly is lowest bed of great cliff range. which indeed lies to north of it with cap of Lava. —

plain not very level, slightly inclined to river. —

~~I at~~ There is not one Lava pebble, therefore no Red Porphyry. I at first thought this was an old valley, but if so sides, ~~with Lava must~~ before Lava must have been uplifted ∴ Lava not under water

[93] which we know to be the case & lastly Lava certainly would have filled up old valley: this curious old alluvial bed was formed before the great oyster bed. —

Perhaps plains & opening at head of river, might be explained by a strait, at very first elevation water in mountain cut it through in the channels , & so on till elevation stopped the passage & river commenced. —

character of gravel is now quite altered, hardly any of St Julian yellow porphyry from great proportion of other rocks: —

[94] May 2^d [1834] Bad days tracking river tortuous. great blocks in stream of Feldspath & do slate. 1 & 2 yards square (also on lower plains) Heap of Guanaco dung 8 feet in diameter.

Cordilleras in view all day: The North shore trends away, with Lava much weathered into square masses & the patches below white. — The South shore shows no Lava white patches till far distant where the rock (Lava?) forms

[95] lofty abrupt escarpements — The valley has on each side a [illeg] weathered plain (as last measurement) with some [sea] pebbles of Slate & Porphyry:

(N B Saw some blocks of conglomerate ancient) —

There are still some low plains with white-washed gravel resting on a very pale, soft, finely laminated sandstone — This is the common stone

[96] soft Sandstone R

is this lower beds darker more argillaceous layers & concretions of hard-sandstone, & few pebbles of old rocks as in "old alluv". which it clearly belongs to. [illeg] resembling

E. coast of T del F. where such pass into each other current cleavage high dip —

curious pebbles   centre. layer hardish, decreasing on each side

[97]   Tierra del F East coast is same as this only here volcanic action has poured out the white
beds & Lava (which have not extended to the coast)
I see a gap in the mountains. The river is yet eating a deeper gut in the valley bounded
on each side by sandstone.
I do not believe the river could form a plain 20 feet high & many of low plains have
this elevation. —
Petrified wood only occurring in certain sites argument for its locality in bed:

[98]   Valley 10 or 12 miles wide
Lava from WNW Cone of Snow
No Lava pebbles in bed of river. —
May (3^{d}) [1834] signs of Indians yesterday skin feathers &c &c guanaco wild ostrich
feathers stick & skin
Little Hawk hovers all same Kestrel
at ship gravel cuts. marl
No Lava Blocks in this valley
Granite pebbles: other Block true: white: 5.3.2 dimensions Very few conglomerates
Pebble of strong clay with Turritellae

[99]
| Noon. Waters edge | 29.269 | A 42.5 | D 43.5 |
| Sunset | 29.280 | A. 41.5 | D 41 |

for measurement of River.
I could see horizontal lines in white beds beneath Lava on N. shore
Another modern Patella.
Valley though here 10 or 12 mile[s] wide, much lower & serpentine yet generally
in gut 20, 30, 40 or 50 feet of gravel, sometimes resting on sandstone: river
so clear; & from so frequently being within sides, too high to be now formed,
I believe it

[100] forms very few beds: very hard days tracking river so tortuous. — & some blocks: new
alpine plants. —
Brachinus
May 4th [1834]
| Sunrise | 29.507 | A 40.3 | D 40 |
| set | 29.713 | A 45° | |

Walked 7 or 8 miles up the valley, armed 15 men. Saw nothing: Lava to the
North broken by great valley — To the South the white horizontal lines may be
seen for 5 or 6

[101] miles further to the West. When inclined the strata of rock to the N W might be
seen. — Lava or what? —
Plain ~~150~~ 60 above river filling up valley & reaching to foot of Andes. —
old Basin of sea —
May (5th). [1834] —

- 10 feet

sun rise 29.813 A 44 D do?
Soft Sandstone more common than alluvium, lined like wood. — Lava field. — river.
generally remarkably in gut; especially in higher parts; scarcely any low land; very cold
5 & ½ days tracking 9. Condors. —

[102] May 6th. [1834] — Where I first ascended Lava: many slate pebbles but certainly not
like the higher parts much pale Porphyry. probably Andes rock: Valley here widens 10
or 12 miles, independent of in land Cliffs which I could not see: Before beds sometimes
striped white, sometimes real Calc stone: Beside the little knoll (which I measured) &
there are others on Lava, larger not same elevation:
Proportion of grand "noon cliffs" 1/7? —
A Little after 6th night from ship cliffs with calcareous, & green bands & concretions of
sandstone One Lava block as big

[103] as mans chest: bands not similar to "section." basin of calcareous matter
Stopped at night at "Resting place" (11th days in two). — pleasant day many guanaco
Slate pebbles — animals condensed — good many ostriches not having seen them
before proof of extreme wildness. —
May 7th [1834] (counting from ship. between 2^d & 3^d sleeping place.).

200

[valley plains] D C B A B C D
A River & its plain. L B lower & more irregular than R B. — (B degraded by River ? has
slate pebbles.)

[104] (C) common boundary of valley (D) inland <u>much worn</u> cliffs — Perhaps D here = (Z)
above. —
Slate pebbles perhaps only connected with straits: or = yellow St Julian Porphyry: —
Bed of river slate as yesterday. <u>no Lava</u>. — Faint white lines in cliffs. —
Carrancha soar. — Small chimango
May 7th 8 [1834] — Cliffs on N. Shore *[end]* as might be expected. <u>Marl</u> & beds at
ship pass into each other: Big

[105] Arcae: & 5 ribbed Pectens. faint white line

			Capt. — 30.157. —	
			My own — . 500: —	
			448.	48	46

May 3^d.		SS:[560]	448. 29.45	48 48.½	46 D do
	4{	SR	— 65.	47.	39:
		SS.	824.	50.	48
	5th-	SR	.912.	48:	44.½

Many oyster above & below. — Sandstone — arrived on board
Thursday 8. F[riday]. 9. S[aturday]. 10. S[u]n[day]. 11. M[onday]. 12
V Adelante

[106] May 12th [1834] PM 11. 12. /gr/ sand or rather minute shingle. — 2 & 3 small corallines (such as caught by Clams) & /shingle/ 3 or 4 lines in length. Porphyry pebbles. —
4 PM (7) 80 <6 68/.[561]
8 PM (8) 83
(3 AM) (9) (51)
PM

1.15	52	3
4	50	4
6	31	5
7	48	6
8	51	7
9 ½	49	8
11	52	9
5*	48	1
7	42	2

[107] Friday 9th
12th [May 1834] Went to **sea**. ~~good stock of provisions~~. shi**p** repaired. false keel masts up: good stock of provisions 10 guanaco bagged: took some long walks. killed a lion & curious wild cat[562] & 2 foxes & condors:
Bad blowing weather sea sick as usual & miserable: very extraordinary change in weather, frost one inch thick. sleet. clouds. gales. hunting for l'Aigle rock 120
miles to leeward, long beat up anchored in evening on the 16th close outside
C. Virgins: —

Inland Cliffs 6 miles back Noon cliffs 25th [May 1834] Stokes

---

560 SS = Sunset; SR = Sunrise.
561 ' 68/.' = sixty-eight fathoms with no bottom found.
562 Probably specimen not in spirits 2036 in *Zoology notes*, p. 401; listed as *Felis pajeros* in *Mammalia*, pp. 18–9, plate 9. See the same specimen in *Animal notes*, p. 17.

[108] P. Desire High Plains

30. 512
}75 A. —
29. 690.

+ N. /Capac/ ? rework /them/

What is cause of granite being to the West of Slate:

Slate wearing away, exposing greenstones.

Wollaston Island in line of Slate less altered: than others.

The WNW line of N part of /Clarence/ Island. Mineralogical change

Reread Pampas notes & copy out Gen observation

Color T. del F map[563] & read about direction of mineralogical change & M. Video notes about do:

[109] Coexistence in feeling in Polyclinum Clytia[564] flashes of light?? ~~Cellaria~~ /Crisia/[565] Flustra &c. —

/Thur/

With exception of Porph. pebbles & upper alluvium all quiet

no anticlinal lines:

High plains with gravel: low with earth: at time of upheaval. or Porph — pebbles:

Shells converted into silex & yellow Carb of Lime.

Ossiferous[566] gypsum ~~generally~~ entirely destitute of shells: Paris

Experimentation on shells

in Sicily. Blue clay with Gypsum without shells — Lyell Vol III P 64[567]

[110] Crystals of Selenite & <u>some</u> shells. base of Etna P 77

Similarly of Hypogenes[568] Volcanic metamorphic[569] — after Lava discussion

2 miles from P. St Anna close to 110 —

150/. ??

East Part of B. Channel 60 fathoms Christmas Sound Capt. Cook 130/. ?

/Immense/ irregularities in bottom.

As ancient alluvium, source of river, resembles in Tierra de F. /&/ does that at anchorage —

[111] Sandy earth contemp with <u>T del F</u> uego. ~~no~~ mem: Extent of main 840 range to the S of S. Anna

Block of lofty alluvium <u>angular</u>

---

563 Presumably CD refers to the hand-coloured map now in DAR 44.13.

564 '*Clytia* formerly included with bryozoans among the Sertularians, is a hydrozoan of order Leptothecata.' *Zoology notes*, p. xxvii note 48, see also note 54 and CD's notes on p. xvi.

565 Specimen 970 in *Zoology notes*, pp. 226–9.

566 Containing bones.

567 Lyell 1830–3, vol. 3.

568 Lyell's term for rocks formed at great depth (e.g. granite, gneiss). Lyell coined the term because he regarded these rocks as formed at the present day and therefore not primitive. See note 207.

569 Lyell's term (adopted from William Whewell) for stratified hypogene rocks such as gneiss formed by alteration of other rocks.

[112–236 blank]

[237–8 excised]

[239 blank]

[240] amygdaloid & Slate Pebbles

Basin shaped Strata facing East end of Lava. —

Inland highest cliffs Proportion of gravel at noon cliffs Height of lowest cliffs

Lava <u>columnar</u>. balls. (<u>upper surface</u>: cracks. vesicular *[2]* (& 2 other indistinct beds:

Hillock of gravel (or marl) line of springs & slips: denuded when inland cliffs formed &

covered partially with gravel which in the valley has again been denuded:

Hillock no lava pebbles: lime with pebbles & in Lava. same cause sonorous tough Lava.

no org: remains from & south cliffs & valley hig^r cliffs rather indistinct denudation

in other plains

Level surface of Lava prevents —

[241] Highest cliffs behind ship nature & height of ∠

Bones gravel

Voluta. oysters. old cliff; lowest small gravel bed of yesterday, some hundred feet above

river: — connection with Tierra del Alluvium.

Boulder 20 y^{ds} 6 feet above base? Old valley would not have been if old diluvium

was *[valley]* Lava & new diluvium would have been above sea. ∴ even if N & S Lava

streams had been distinct they would have blocked up valley: They lie in very head

of valley. yet within line of junction of Lava.

⁀ Old Strait ? Character of gravel much altered from coast

[242] 94.4

4806

468

[BACK COVER]

[INSIDE BACK COVER]

108
~~77~~
1
108
27
238
10
26

[A piece of paper (25 × 115 mm) holds down an added brown leather pencil holder. The following is written on the piece of paper:]

30.17 Therm 51°
30.108 54

23. $\begin{cases} 30.07 \\ 30.078.56 \end{cases}$

29.466 A 64 D 59
713
<u>280</u>
(433) 29.364

Textual notes for the *Banda Oriental notebook*

[2]   Charles Darwin] *ink.*
1.9] *Down House number, not transcribed.*
88202329] *English Heritage number, not transcribed.*
6] *added by Nora Barlow, pencil, not transcribed.*

   [5]   *12 ink marks on this page appear to be nib tests.*

 [14]   When Limestone … shells.] *ink over other entries.*

 [17]   *an ink mark after '*trouble*' appears to be a nib test.*

 [24]   concretion Uruguay] *ink.*

 [33]   Wheat Pubrilho] *written over the preceding paragraph.*

 [37]   *page in ink.*

 [38]   ∠ᴿ Blocks] *ink written over preceding paragraph.*

 [51]   *there is a brown fingerprint, possibly CD's, on the lower right margin of this page.*

 [73]   *page written perpendicular to the spine.*

 [78]   Curious … them:] *ink.*

 [90]   How … arrive] *added ink.*

[105]   Thursday … 12] *ink.*

[107]   Friday 9ᵗʰ] *ink.*

[108]   P. Desire … *[*Capac*]* ?] *ink.*

[241]   & → SH Oᴴ/O^{[u]} *not in CD's handwriting, not transcribed.*

THE *PORT DESIRE NOTEBOOK*

⤶

The *Port Desire notebook* is named after Port Desire [Puerto Deseado] in Patagonia, now Argentina. Port Desire was named by the privateer Thomas Cavendish in 1586 after his ship. The notebook has been rebound, the brown leather front cover is original. The front cover has a label of cream-coloured paper (74 × 55 mm) with 'Port Desire — Famine Wollaston Isl^d Navarin Is^d. E. Falkland Isl^d Measurements of curved hills East coast of Chiloe Boat excursion Chonos S. Carlos' written in ink. The notebook originally contained 91 leaves or 182 pages. Some pages bear the watermark John Morbey 1830. The entries were written in one sequence and cover January–April 1834 (pp. 1–55) and November–December 1834 (pp. 56–182).

The notebook is physically unlike any of the other notebooks. It has a narrow brown cover with the spine on the short side. It has the longest label of any of the notebooks. It is the only label to contain anything other than a place name. As discussed below, the 'Measurements of curved hills' on the label seems to refer to some calculations and diagrams made in the Falklands. The rest of the label can be summarized as being in two clusters, the first being Patagonia, the Falklands and Tierra del Fuego covering the period January to April 1834. The second is the East coast of Chiloe Boat excursion, in November, leading into various jottings from the Chonos archipelago and Chiloe, the period when some of the major Andean volcanoes erupted in January 1835.[570]

The *Port Desire notebook* should be read in the context of several of Darwin's theoretical essays on geological subjects. Firstly there is the c. March 1834 'Reflection' essay (DAR 42.93–6), then the c. May 1834 'Elevation' essay (DAR 34.40–60).[571] The end of the *Port Desire notebook* dates to around the time the 'Feb. 1835' note (DAR 42.97–9) was penned.[572]

---

570 Other notebooks used in the same time period as the *Port Desire notebook* are, in order, *B. Blanca*, *Banda Oriental* and *Valparaiso* for the southern winter of 1834 and *Santiago* which was in use from September 1834 at least until the end of the voyage (see the introduction to the *Santiago notebook*).

571 See the introduction to the *Banda Oriental notebook* for discussions of these two essays.

572 See below and introduction to the *St. Fe notebook*.

Port St Julian, January 1834

The notebook picks up shortly after the *Buenos Ayres notebook* left off at Christmas 1833 at Port Desire in Patagonia. There are various notes and calculations which seem to relate to the heights of plains in pencil and ink inside the front cover. A list of heights dated 2 January 1834 follows on p. 3. This was the date recorded in the *Beagle diary* when Darwin examined an Indian grave, but there is no mention of this in the notebook. The next few pages are dated 3 January and reflect Darwin's statement in the *Beagle diary* that he 'had some very long and pleasant walks'. Unusually he wrote the geological section perpendicular to the spine on p. 4, but returned to the normal writing parallel with the spine for the continuation of the notes. By comparing the rocks mentioned (e.g. 'white Feldspathic Rock') and the drawing of three little triangles on p. 10 these notes seem to link to the page from the geological diary for this time which is reproduced by Herbert 2005, p. 99, chosen to illustrate Darwin's technique for recording specimen numbers.

The next day the *Beagle*, striking a rock leaving Port Desire, moved down the coast to Port St Julian [Puerto San Julián].[573] Around 9 January FitzRoy landed Darwin, who wrote in the *Beagle diary* that he 'found some most interesting geological facts'. This was recorded in the notebook on p. 13. In a hand-drawn and coloured geological map of southern South America (DAR 41.13) Darwin showed how the geology around Port Desire down to about Port St Julian differs, in being predominantly of volcanic origin, from that of most of Patagonia.[574] On 10 January Darwin mentioned a 'Gecko, being kept for some days'. It is clear from the *Beagle diary* that on the 11th Darwin exhausted himself searching in vain for water and spent the next two days 'very feverish in bed'. On the 14th he recorded in the *Beagle diary* that they found a 'small wooden cross' which might have been a relic of Magellan or Drake.

Darwin recorded the geology of Port St Julian on the 16th on p. 13. The next day he described the cliff section with, at the top, a layer of mud containing mussels in an extremely good state of preservation '60 or 70 feet' above the present sea level, thus indicating fairly dramatic, geologically recent uplift. The mud in places filled channels in the underlying gravel bed and 'pumiceous mudstone', and in one of these channels on the 20th Darwin made his last but arguably most important fossil discovery. Unfortunately there is no mention of this discovery in the *Port Desire notebook*, but it is important in understanding how Darwin became convinced of descent with modification.

The fossil discovered at Port St Julian was the 'extinct Llama' which Darwin referred to on p. 129 of the *Red notebook* in January 1837, after Owen told him that the animal seemed to be a giant extinct llama, hence the new name *Macrauchenia*

---

573 See Keynes 2003, p. 188 for a map.
574 Reproduced in colour in Herbert 2005, plate 6, and as a colour fold-out facsimile in van Wyhe 2008a, p. 19.

patachonica, meaning 'big neck of Patagonia'. Darwin was unable to identify the fossil when he found it, but thought it might be a Mastodon, a type of extinct elephant. He was intrigued as to how this 'Mastodon' had died, as it was obvious from the completeness of the skeleton that the animal was found not far from where it had lived.[575]

It is evident from Darwin's geological diary (DAR 33.246) and his correspondence that he was not sure what the animal was when he found it, but he was certain that it was an extinct species. He continued to call it a 'Mastodon' for the rest of the voyage. The 'Feb. 1835' (DAR 42.97–9) essay, albeit apparently written more than a year after the find, is primarily a discussion of the reasons why this 'Mastodon' (i.e. *Macrauchenia*) was extinct. There was no evidence of it having been drowned by any 'debacle', indeed the bed of pebbles which might have been taken by some as evidence of a catastrophic flood was actually *below* the beds containing the fossil. Furthermore, there were no 'trees or stones' in the bed containing the fossil to indicate anything other than quiet estuarine conditions. The fossil itself was articulated, i.e. the bones were still arranged as they would have been in the living animal, and this can only occur when the dead animal is covered relatively gently.

This lack of evidence for a dramatic change in the 'Mastodon's' living conditions led Darwin to challenge Lyell's explanation for extinction, as Lyell would have insisted on a change of circumstances so that the species was no longer perfectly adapted to its surroundings ('station') and so became extinct. Lyell considered the possibility that a species might die out when its powers of self-replication became somehow exhausted, as posited by Giovanni Brocchi.[576] Lyell rejected that possibility on the grounds that it was 'hypothetical', by which he meant that it was not a *vera causa* (a testable 'true cause'). He had, however, left the door open to be proved wrong: 'if any animal had perished while the physical condition of the earth, and the number and force of its foes, with every other extrinsic cause, remained unaltered, then might we have some ground for suspecting that the infirmities of age creep on as naturally on species as upon individuals.'[577]

Darwin, having seen in July 1834 how apple trees grown from grafts on Chiloe sometimes all died simultaneously, was ready to believe a version of Brocchi's theory as a possible explanation for the extinction of the 'Mastodon'. Darwin held to this Brocchian position until 1838 and he mentioned it in *Journal of researches*, p. 212. Having thus broken with Lyell in February 1835 on the issue of species 'deaths', Darwin was ready subsequently to challenge Lyell on the issue of species 'births'. The

---

575 See Rachootin 1985 for historical discussion, *Red notebook*, Herbert 2005 for photographs of the fossils and Keynes 2003, p. 191, for an excellent diagram of the skeleton.
576 See the introduction to the *St. Fe notebook*.
577 Lyell 1830–3, 2: 129–30.

closest the field notebooks come to touching on this subject is the single reference, probably written on Chiloe in July 1834, to 'Apple trees'.[578]

When in Cape Town in June 1836, Darwin met the naturalist and explorer Dr Andrew Smith (1797–1872). Smith may have bolstered Darwin's belief in the Brocchian extinction theory, by convincing him that large mammals such as elephants could easily prosper on the vegetation now existing in South America, as they do in Africa. So lack of food would not explain the 'Mastodon's' extinction.

Once back in England, in January 1837, Owen told Darwin that his 'Mastodon' was related to llamas and guanacos, animals peculiar to South America.[579] This seemed to be another example, to add to the several Darwin already knew, of a 'past inhabitant' found only in South America being similar to a 'present inhabitant' found only in South America.

These speculations, together with Darwin's findings and experiences of geographical distribution, brought about a fundamental break with Lyell on the 'birth' of species. Darwin may have gradually begun to make this break as the *Beagle* returned to England from South Africa. This was probably what Darwin had in mind when he used the splendid phrase in his *Journal of researches*, p. 610, that such cases were 'more pregnant with interest' the more one reflected on them.

Macrauchenia is now classed not as a close relative of the guanaco, but as a litoptern, which was an order showing some similarities to elephants, such as the *Mastodon*, as Owen pointed out in *Fossil mammalia*, p. 35. Judging from the high position of their nostrils, for example, litopterns almost certainly had a trunk. They were also confined to South America and are now extinct.[580] Llamas and guanacos are species of camel. Darwin's section for Port St Julian appears with lengthy discussion of the *Macrauchenia* in *South America*, p. 95, fig. 16; p. 111, fig. 17.

Tierra del Fuego, January–February 1834

The *Beagle* returned to Port Desire where on 20 January Darwin recorded a section (p. 15) and mentioned 'Giant cliffs nearly 1000 feet high' at Cape Blanco [Cabo Blanco], p. 17. On the 22nd the ship ran down to the Straits of Magellan [Estrecho de Magallanes] and anchored at Gregory Bay on the 29th. Darwin may have geologized but there seems to be nothing in the notebook, which next mentions Port Famine on 3 February.

On p. 19 there is what appears to be a simple diagram of a 'Winters Bark' tree.[581] This is referred to in the *Beagle diary* for 7 February. On the 6th Darwin set out to climb Mount Tarn and he drew a diagram of a boulder. He recorded in the *Beagle*

---

578 *B. Blanca notebook*, p. 83b. The propagation of apple trees on Chiloe is described in *Zoology notes*, pp. 236–7.
579 *Journal of researches*, p. 209.
580 Benton 1990.
581 *Drimys winteri*. See *Beagle plants*, p. 177.

diary that he 'had the good luck to find some shells in the rocks near the summit' and on notebook p. 25 he referred to 'animals'. These are probably the fossils, including an extraordinarily large *Hamites* ammonite, which he described in *South America* on p. 152 (and appendix). *Hamites* is from the Cretaceous age, about 100 million years old.

Darwin's twenty-fifth birthday passed on 12 February. The notebook was set aside for a few weeks while the *Beagle* explored the east coast of Tierra del Fuego, until she anchored off Wollaston Island and Darwin went ashore on 25 February, p. 26. On p. 27 there is a rare example of specimen numbers from his specimen notebooks in a field notebook (geological specimens 1853–5). On p. 28 he recorded that the rocks on top of some hills were 'not Slate' and this comment found its way into the *Beagle diary*. This was the time when Jemmy Button was seen by the *Beagle* crew for the last time and the Fuegian Indians feature prominently in the *Beagle diary*, but there is no mention of them in the notebook. On 27 February the *Beagle* anchored off Navarin Island [Isla Navarino].

Falklands, March–April 1834

On p. 29 Darwin made a note about porphyry pebbles being dredged from the sea between Staten Island [Isla de los Estados] and the Falklands. He then deleted the note but the pebbles were mentioned in *South America*, p. 21. The *Beagle* arrived in Berkeley Sound on 10 March, providing Darwin with a second opportunity to examine East Falkland.[582]

There is no mention in Darwin's notebook of the horrendous events which occurred there before he arrived, although he gave a summary in the *Beagle diary*. It seems that about four months after the *Beagle*'s 1833 visit, in August, a gang of desperados murdered five of the men living in the Port Louis settlement, one of whom, the first British resident of the Falklands, Matthew Brisbane (c. 1787–1833), was mutilated. This was a bitter blow to FitzRoy especially, as he had come to know and like Brisbane when the *Beagle* made her first visit. It was FitzRoy who found Brisbane's partially buried body. The remaining thirteen men, three women and two children fled to a small island where they lived on what they could catch until H.M.S. *Challenger* arrived with marines to round up the desperados. On arrival of the *Beagle*, FitzRoy had the ringleader clapped in irons for transportation to England.

Darwin made an intriguing note on p. 30 that there are 'no dung beetles Falkland Islands'. As Darwin was unusually attentive to the ecological dependence of these beetles on large mammals, he may have concluded that the cattle on the Falklands were introduced. On the same page he asked himself why Tierra del Fuego seemed

---

582 Armstrong 1992 and Armstrong 2004 provide an exhaustive analysis of CD's two visits to the Falklands.

not to have been elevated in the same way that Patagonia had, and hinted that Falklands geology was more similar to that of Tierra del Fuego.

A 'crazy, very quick' Jackass Penguin (*Spheniscus magellanicus*) provided Darwin with an amazing human/bird interaction on p. 31.[583] On the next page Darwin inserted in ink a note that 'Rats & mice & foxes on small islands & [South] Georgia', showing that he was probably trying to work out which animals could reach oceanic islands by natural means.

The Falkland Fox or Warrah (*Dusicyon australis*) was described in *Mammalia*, p. 7, from *Beagle* specimens and although now extinct was common enough at the time of Darwin's visits. William Low, a highly knowledgeable Scottish sealer who helped the *Beagle* in various ways, told Darwin that specimens from West Falkland were smaller than those from East Falkland.[584] Darwin tended to spell the man's name as 'Lowe', but eventually corrected himself. The information about the foxes impressed Darwin immensely and he referred to it in his famous discussion of the Galapagos mockingbirds in his 'Ornithological notes' of June or July 1836. The Falkland Fox was the only case Darwin knew which was comparable to the tortoises of the Galapagos, which were said to differ on closely neighbouring islands. Faced with several slightly differing mockingbirds on different islands in the Galapagos, Darwin concluded that they were 'only varieties', presumably of an original species. Because the *Beagle* did not visit Ecuador, Darwin did not see the most closely allied mainland species at the same latitude. The idea that a species could adapt to a minor extent to suit new surroundings was commonplace and did not entail a belief in transmutation. As Darwin later recalled: 'When I was on board the *Beagle* I believed in the permanence of species, but, as far as I can remember, vague doubts occasionally flitted across my mind. On my return home in the autumn of 1836 I immediately began to prepare my journal for publication, and then saw how many facts indicated the common descent of species.'[585] It was experiences such as the fossil mammals from Patagonia, the Falkland Foxes and the different kinds of birds and tortoises on different islands of the Galapagos that stirred such doubts. Once John Gould pronounced the mockingbirds as separate species in February 1837, about four months after the *Beagle* reached Falmouth, the break with Lyell's view probably seemed inescapable to Darwin.[586]

Darwin's serious geological work in the Falklands resumed on 16 March 1834. He was intrigued by the ridge of quartzite hills which run east–west about 10 km (6 miles) south of Berkeley Sound, and these dominated his notes. There are also many references to natural history, such as notes on rabbits and foxes, both of which

583 See 'Ornithological notes', p. 50, and *Journal of researches*, p. 256.
584 See Armstrong 1992, p. 111.
585 CD to Otto Zacharias [24 February 1877], *More letters*, 1: 367.
586 See the introduction to the *Galapagos notebook*.

are now thought to have been introduced. Horses behaved very oddly and did not flourish on the islands as they did on the Pampas, hence they were 'very expensive 100 ps each', p. 36.

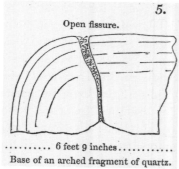

5.

Open fissure.

·········· 6 feet 9 inches ············

Base of an arched fragment of quartz.

Base of an arched fragment of quartz, East Falkland. Fig. 5 from Darwin 1846.

The 'measurements of curved hills' mentioned on the label of the notebook seems to refer to the Falklands. It is difficult to work out exactly what Darwin meant when he wrote the label but he several times used the word 'wave' when describing the quartzite ridges so he may have been thinking that the forces which formed the geological structures were in some way analogous to ocean waves, and this appears to be borne out by his discussion in the geological diary (DAR 33.198). In his 1846 paper on the geology of the Falklands Darwin made the startling comparison between some of the structures which he said were 'almost resembling those produced by the mingling together of two viscid liquids'.[587] Darwin thought that earthquakes might explain the mysterious 'stone runs' and he was open to catastrophic explanations for the contortions of the strata.[588] The diagram on notebook pp. 50–1 seems to be Darwin's attempt to compare the structure of the quartzite hills with the swell of the sea, and on p. 52 he called it a 'Wonderful scene of violence'.

The diagrams on pp. 43 and 44 are the prototypes for the woodcut ('fig. 5') published by Darwin in his 1846 paper and reproduced above. The one on p. 47 seems to be the prototype for 'fig. 1'.[589]

The last few pages from the Falklands in the notebook, pp. 53–5, are filled with fascinating ecological detail. Darwin spent time examining kelp and its role in supporting an extremely complex web of marine life. This, according to the *Beagle diary*, was on or about 6 April. The very last entry before the notebook skips forward eight months to November is simply 'no snakes'.

Chiloe, November–December 1834

The next dated entry is from Chiloe on 24 November, on p. 56. In the intervening period the *Beagle* sailed to the Santa Cruz river,[590] then down to Tierra del Fuego in May (*B. Blanca* and *Banda Oriental notebooks*), up the west coast to Chiloe in June

587 Darwin 1846, p. 274; *Shorter publications*, pp. 196–204.
588 See especially *Journal of researches*, p. 255.
589 Darwin 1846, pp. 271–2; *Shorter publications*, pp. 196–204.
590 See the *Banda Oriental notebook*.

and July and further up the coast to Valparaiso where Darwin took up residence on shore at the home of his old school friend Richard Corfield.

From Valparaiso Darwin commenced his first major expedition in Chile, where the *Valparaiso notebook* starts on 14 August. The *Santiago notebook* was opened at the beginning of September. At the end of that month Darwin fell seriously ill, being fit to travel again only in November. During these months Conrad Martens left the *Beagle* and FitzRoy was forced to sell the *Adventure* and abandon his hopes of perfecting the Tierra del Fuego charts. The alternative might conceivably have led to FitzRoy's own suicide.[591] It is frequently claimed that the Admiralty refused to pay FitzRoy for the *Adventure* because he was a Tory and the Admiralty Lords were largely Whigs. We are not aware of any evidence for this belief apart from Darwin's opinion expressed in a November 1834 letter to his sister Catherine: '... the cold manner the Admiralty (solely I believe because he is a Tory)'.[592] In fact FitzRoy had disobeyed his instructions on the matter of hiring another vessel, which would seem to be sufficient explanation.

On 21 November the *Beagle* entered the harbour of San Carlos [Ancud] on Chiloe and dropped off Darwin who 'took horse' to Chacao, p. 56, where he planned to join a surveying crew to commence the 'boat excursion' down the east coast of the island.[593] In Darwin's geological descriptions of the island in *South America*, he presented evidence showing how Chiloe was part of the great north–south Andean belt. His magnificent geological colour map of Chiloe is in DAR 35.306.[594]

From the first Darwin enjoyed Chiloe, now that it was summer and not always cold and wet as it was in July. He thought the indigenous population was superior to what he saw in the south 'not very like Fuegians', p. 56. He found the 'scenery exceedingly picturesque' with 'beautiful cleared spots & pretty enclosure. magnificent forest path like road to Castro'. He had a 'fine view of Straits of Chacao: Volcano beautiful many crosses, dangerous straits', p. 56. The great snow-capped volcano of Osorno was often admired by the men of the *Beagle* although they had no idea it would erupt eight weeks later.[595] Darwin was by now confident of his ability to detect crustal instability: 'everywhere signs of upheaval', p. 57. He noted the superstitions of the people 'Cheucau. making an odd noise people will not start',

---

591 This period from August to September 1834 is described in fascinating new detail by Simon Keynes 2004, pp. 159–68, partly on the basis of FitzRoy letters only recently acquired by Cambridge University Library. Several of Martens' sketches and watercolours from Chiloe, made before he left the ship, are reproduced for the first time by Richard Keynes 2003.

592 CCD1: 418.

593 This and all CD's time in Chiloe is described beautifully in Armstrong 2004, chapter 10.

594 The *Beagle*'s chart of Chiloe is reproduced as a colour fold-out facsimile in van Wyhe 2008a, p. 27. The manuscript version of the chart formed part of the collection of the late Quentin Keynes. This manuscript map is partially reproduced in Keynes 2004, p. 167.

595 See Herbert 2005, p. 138, for a fine photograph of Osorno.

p. 58, referred to their dread for the call of a bird known as a Chucao Tapaculo (*Scelorchilus rubecula*).

Darwin encountered various birds such as the Tapaculo which he grouped as 'Myothera', now classed as members of the family Rhinocryptidae, in various different habitats on mainland South America in 1834–5.[596] What makes these birds so important is that Darwin saw their *similarities*. This is perhaps unexpected, in view of the significance usually attributed to the *differences* Darwin noticed in island species, for example the Galapagos mockingbirds. If, as Lyell wrote, species should be adapted to their habitats, there is no reason *apart from heredity* for them to be similar if inhabiting different habitats.[597] In this sense the 'Myothera' were more important to Darwin than the rheas because, in the case of the rheas, there was no discernible difference in the habitats of the two species. It is, however, true that it was the rheas which Darwin mentioned several times in the *Red notebook* as replacing each other in space in a comparable way to the replacement in time of the extinct by the living species of 'llama', armadillo and other mammals. Darwin's recognition, albeit perhaps subconscious, that it was heredity linking the various 'Myothera', and that adaptation was relative, removed for him any logical objection to descent with modification. What he needed then was to find a mechanism that might cause the modification, and this he first found in the isolation of variants which occurred when barriers were imposed, such as the sea between the islands of the Galapagos. Thus Darwin's first tentative glimpse of descent with modification in living species was of a process which allowed change while retaining resemblances.

Chiloe at the time of the *Beagle*'s surveys had only been released from Spanish rule for about eight years and many people there were unhappy with the Chilean administration, to the extent that they hoped the *Beagle* had come to remove the Chileans. Thus Darwin noted on p. 58 how the now 'miserably poor' Governor asked him 'indifferently' if the English 'flag [on the mast of the yawl] would always fly' at Chacao? On the same page Darwin mentioned Charles Douglas, the surveyor, who gave Darwin much valuable information about Chiloe.[598]

On 25 November Darwin travelled by sea with the crew of the *Beagle*'s yawl and a whaleboat until 'foul wind drove us in' to a place called 'Huapilenous' [Haupilinae] on the northeast corner of the island. Darwin recorded 'Torrents of rain & in night Hurrah Chiloe', p. 59. He wrote '"Huapi" means Islands: yet now nearly all peninsulas. Proof of rise'. In his geological diary (DAR 35.297) he expanded on

---

596 In *Journal of researches* CD no longer used the name *Myothera* but instead *Pteroptochos*. See Steinheimer 2004, p. 304 (where the current names for all of CD's *Beagle* birds are given), note 11, 'Ornithological notes', p. 256 note 1, and *Red notebook*, note 154, for details of the various species.

597 We are grateful to Jonathan Hodge for generously sharing his unpublished research on this and other topics.

598 CCD1: 430–1.

this point which seemed to imply that in the time since South America was inhabited these islands had risen above sea level sufficiently to become joined up. The current view is that man only reached the continent about 11,000 years ago. If so, this rate of uplift would indeed be impressive.

On the 26th the boat crews saw 'Osorno immitting much smoke snow considerably melted, always in state of activity', p. 61. The locals 'expect earthquake, when no eruption during 1 to 3 years'. There is much fascinating geological detail concerning the Indians and how the Government kept them in poverty. Darwin was very interested in the history, languages and religions of the local tribes, much of his information probably coming from Douglas. The boats worked down the coast against a 'strong foul wind', p. 63, and Darwin made a note of 'Small Crustacea purple clouds of infinite numbers. pursued by flocks of P. Famine Petrel'.

By the 28th the party were at Quinchao 'splendid day & clear weather'. The next day Darwin noted 'many periaguas' which was a kind of boat, and 'hundreds & thousands' of petrels.[599] He was forming the impression that the east side of the island was covered in enormous quantities of gravel, and on p. 68 there is an intriguing diagram of this gravel resting in steps in the landscape. Darwin noted that the sea squirt (Ascidian) locally known as the Peures [Piure] had since 1825 'become abundant at Chacao were they formerly did not exist', p. 69. His 'Specimens in spirits of wine' list shows, against no. 1165, that they 'are good to eat & are much esteemed in Chiloe', as indeed they are today.[600] On p. 70 he recorded that 'The Nautilus is thrown up periodically in great numbers' along the coast, but gave no explanation. He mentioned that some of the Indians 'hold converse with the devil in a cave', a crime which would formerly have sent them to the Inquisition in Lima.

On 30 November Darwin and his party arrived at Castro, then the capital. Darwin admired the 'highly picturesque' wooden Jesuit church, p. 74. Darwin 'rode about in the neighbourhood to see geology' and saw how the conglomerate rested on mica slate. He also heard of granite occurring in the centre of the island. Pages 77–8 are excised but have been located in DAR 35.297A. Page 77 gives details, apparently from Douglas, of some recent earthquakes, while the reverse, p. 78, is a whole-page diagram (possibly not by Darwin) of the various levels in the cliffs at Castro.[601]

On p. 79 Darwin referred to a severe earthquake that occurred on Chiloe. The next day (1 December) the party left Castro 'early' in heavy rain for the nearby island of Lemuy, and Darwin made a note to remind himself to write a description of

599 See *Birds*, p. 137, and *Journal of researches*, p. 354.
600 See *Zoology notes*, p. 357.
601 This is the only excised page from a *Beagle* field notebook to be located.

the 'coralles', underwater hedges, exposed at low tide and used as natural fish traps. He found lignite (fossil wood used as low grade coal), p. 82, and a huge block of granite which he supposed came from the Cordilleras. He commented on the three different calls of the Chucao.[602] The next day Darwin found poorly preserved shells, including a *Cytherea* (clam), in the cliff but had expected to find more, so speculated that the absence was due to chemicals destroying the shells.

Darwin's party made slow progress and were impressed with the large numbers of 'purer and purer Indians', p. 88, who were 'excessively humble & civil'. Darwin found a beautifully preserved fossil tree apparently the same age as the fossil shells.[603] On p. 90 Darwin noted the extraordinary behaviour of some beetle species. On 3 December he described a splendid little frog and drew a sketch of it, p. 91. This was specimen in spirits of wine 1086. Its description was published in *Reptiles* as *Rhinoderma darwinii*. It was first named on the basis of Darwin's specimens by Gabriel Bibron (1806–48).

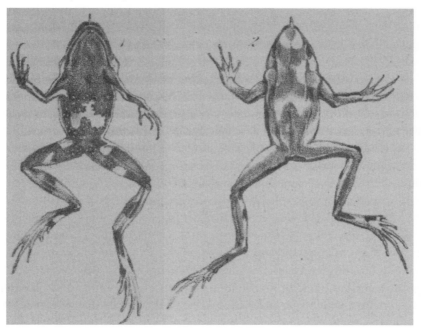

Rhinoderma darwinii from the island of Lemuy, part of plate 20 from *Reptiles*.

On 4 December Darwin continued to describe the geology and mentioned how Douglas, to protect the crew at night, simply told the locals that if they went out at night they might be shot. The party reached Quelen [Queilen] and on the

602 See *Beagle diary*, p. 271.
603 See Keynes 2003, p. 257.

5th Darwin found a slug-like mollusc, specimen 1092, *Onchidella marginata*.[604] They pressed on in 'squally weather' and Darwin told the story of a man who walked for three and a half days 'to receive value of an axe & a few fish'. On the 6th Darwin found a 'Doris' and an 'Eolis' (both sea slugs) crudely sketched. From *Zoology notes*, pp. 255–6 it is evident that these are both lumped as specimen 1091, *Anisodoris fontaini* and *Phidiana lottini* respectively. The party finally reached Caylen Island [Isla Cailín], 'el fin de Christianidad': the most southern point to which Christianity had reached in South America.

By the 7th they had to contend with the oceanic swell 'wild nasty day', but Darwin, while among rocks on the beach, was able to sneak up behind and kill 'rare fox with hammer', p. 106, while it curiously watched the officers with their surveying equipment. This was a new species (*Lycalopex fulvipes*) now known as Darwin's Fox or Darwin's Zorro, although it is closer to wolves than to foxes. Darwin suspected that it differed from the mainland species, and he noted in his *Animal notes* that it was a 'very uncommon animal' on Chiloe. Darwin later wrote the fox a fitting epitaph: 'This fox, more curious or more scientific, but less wise than the generality of his brethren, is now mounted in the Museum of the Zoological Society.'[605]

The boats arrived back at the island of San Pedro to rejoin the *Beagle* after an 'absence of fortnight'. The *Beagle* had 'failed in surveying outer coast', but had discovered that previous charts of Chiloe had overstated the length of the island by 25 per cent. On p. 107 Darwin described the dense forest on the island, and 'walk many feet above ground' was expanded in the *Beagle diary* for 8 December into a wonderful description of how Darwin and FitzRoy struggled to get to the top of the island by climbing over the dense mass of fallen trees.

From this point Darwin was back on the *Beagle* and the rest of the notebook contains notes and diagrams apparently relating to the Chonos archipelago.[606] The final pages of the notebook are, however, complicated by excised and blank pages and it is extremely difficult to date this part of the notebook. It is not even possible to be sure of the order in which these pages were written. On pp. 116–7 Darwin drew a 'ground plan showing the relation between veins and concretionary zones' which was published as Fig. 20 in *South America*. This diagram relates to the area around Castro on Chiloe, so it seems likely to have been drawn at the end of November. The simple sketch of a hollow concretion on p. 118 seems to be the prototype of the one which appears in the *Red notebook* on p. 160.

604 *Zoology notes*, p. 255.
605 *Journal of researches*, p. 352.
606 See *South America*, pp. 119–20.

A 'ground plan showing the relation between veins and concretionary zones' on Chiloe. Fig. 20 from *South America*, p. 124.

There follow pages of obscure calculations, and on p. 119 Darwin recorded the 'eruption of Osorno 26th of Jan^y.' This, however, is written within a pencilled box and seems almost certainly to be a later insertion, especially since on that date Darwin was making notes on Chiloe on what are now loose sheets, rather than in a notebook, in 'Chiloe Jan^r. 1835' (DAR 35.328).[607] This was during Darwin's overland expedition no. 6 across Chiloe which started, after the *Beagle* had sailed back up the west coast of Chiloe, between 19 and 28 January 1835.

On p. 121 Darwin recorded a 'valley of elevation' separating two escarpments tending north-south which he described in *South America*, p. 124, as on the peninsula of Lacuy [Lacui], near San Carlos. This would suggest a date around 24 November 1834 for this note, but it may have been made on 19 January 1835. On p. 122 Darwin seems to have calculated the height of the Corcovado mountain by triangulation, and this suggests a date around 26 November, as on p. 177 the wording about a mountain 'S of Corcovado not known to be volcanic' is similar to the reference in the *Beagle diary* for that date to mountains in the same direction 'not known to be active'.

Finally, on p. 177 Darwin measured a plant using his belt; this seems to be the Pangi plant, in the *Beagle diary* for 26 November. There are some ornithological notes of interest: 'is the brown vulture found at [South] Shetland?', p. 181, and 'Huitreu … comes near to a man if he is quiet', p. 178, which became, in *Journal of*

607 This document is transcribed in Chancellor *et al.* 2007.

researches, p. 352, 'let [a man] stand motionless, and the red-breasted little bird will approach within a few feet, in a most familiar manner'.

Thus ended one of the most varied and interesting of the field notebooks, which ranged across virtually all of Darwin's interests during the voyage, from zoology, botany and geology to ethnography, history, linguistics and anthropology. It also covered some of the most important episodes in the development of his thinking with respect to the 'birth and death of species'. Spanning the year 1834, it started on the east coast and ended on the west coast of South America.

[FRONT COVER]

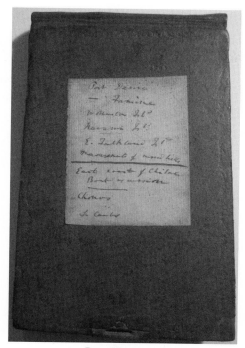

Port Desire
— Famine
Wollaston Isl^d
Navarin Is^{d.}
E. Falkland Isl^d
Measurements of curved hills
East coast of Chiloe
Boat excursion
Chonos
S. Carlos

[INSIDE FRONT COVER]

C. Darwin

(Capt. one dollar) Smith trousers

<Something was apparently glued to this page, which has been torn away,
thereby destroying words in ink>
< > *[* use*]* of < > T. del F*<uego>*
< >. exp< > Ponsonby. Sound: —
< > or N *[by]* W
Falkland Islands
[Gunner] [illeg] Head — 2 sorts
other *[species]* from *[ford]*
Pebbles
Height of N. Plain

South Barranca.
East Coast.
Structure of gravel at S. Barranca
/Gecko/ /Craters/
Shells low water
R. Chupat: 43° : 20'
Calyen 43.10'
492
667

300
285
503
1088
328
667
261   600
[1–2 excised]
[3]

	½		A	D
Sea		30.328	58.	— 56
7				
/Plain/ 9°		30.065	.58	55
12 ¾				
Hill		29.914	57	56
H. Plain		30.067	57	56 / 10 ¼
SW Hill		29.920		
6 ½				
First Plain		30.214	51.5	49.5
Sea 7 ½		30.492	51.5	49

(Jan 2nd) 1834
328
665
263
492
214
.298

SW	Hill	N: 64°
∠	NE Hill	14° 30'
∠	Penguin island	53°10'
∠	Mouth of Creek	84° 43'
	Guanaco Hill Island	122° 30'

[4] S̶t̶ Both 2 hills same sort of Porphyritic — Plain Barometrically of same height on both sides

(3ᵈ) [January 1834]

Plain above cliffs of same formation as generally — I see in many places, that the upper bed, is sand, calcareous, white, A: **the next more** clayey & yellowish B with salt flavour; this contains layers of fine gypsum — This plain slants down to the cliffs. —

Saw two breccia uncrystallized Porphyry bed dipping to

[5] the Westward // many of the pebbles on the hills white washed. —

Ferruginous Sandstone

Coarse as below

/like/ a specimen

White Mortar, silex particles D ?

red, white earthy

crystals, porphyry

dip to WNW

Coarse Ferruginous Sandstone

All dip to NW

Hard Calc white Sandstone F

Calcareo — Feldspathic ? E

red as below

white (D) Feldspathic rock

red jaspery porphyry

[6] (C) occurs amongst & forms great masses /amongst/ red Jasper Porph rock

Rocks generally resting on a compact hard earthy Feldspar

coloring matter of. —

Red rock seems to blend with white, making pink —

Dip NW

red color sometimes in lines in horizontal section in patches. circular & bands

[7] There are dykes only differing from surrounding rock /by figure/ in straight line 2 *[illeg]* [broad] In all pebbles

There is a bed of conglomerate of bright red colour, usual dip (about 10°) composed of pebbles of red highly crystalline porphyries. f̶r̶o̶ pebbles as large as your head to sand. — deposited in a current. There is a dyke tortuous of green. (sandstone!) running in SSE direction & a

[8] smaller one of same material crosses the ENE one. — This latter much varies in constitution passing into coarse uncrystalline substance. /Beds/ tortuous. —

The first must be only a crack filled up. —

The ENE one penetrates all beds excepting chalk with siliceous particles. —

Over the Jasper Porph bed (of which I brought specimen) there is a white (calc ?) bed with (G) w̶i̶t̶h̶ same

[9]   dip: I believe this ~~chalk~~ mortar with siliceous particles to have same dip generally with
other bed but not crossed by dyke — Hence period of uplifting strata at the beginning
of mortar Deposit — but anterior to Oyster bed. —
& of course subsequently covered by gravel when beneath water. —
(G) over the Jasper Porph same dip
(H) Dyke
(J) ENE Dyke
(M) Below (O)

[10] (N). do. There /it/ forms main mass of cliffs, especially this rather more coarse
(O) Above the two latter nearly as crystalline as the inferior bed. —
At West end of Bay. Mortar (?) & white Feldspathic Rock (P) lie over red earthy
porphyry & dip to W ½ N.

behind it (to S) [sketch of cliffs?] another of <u>fine red</u> Porphyry

(Q), which bed thickens out in usual dip & covers (T V)

[11]

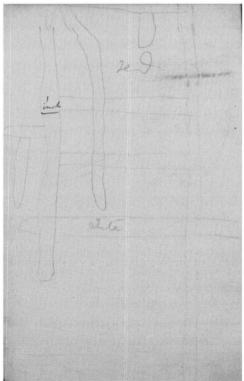

[section of dykes cutting dykes?]

white red inch

[12] Varieties of the earthy rock & with pebbles intermediate bed of Pitchstone (W) — Porphyry same as all hill external form &c. — Dip to W ½ N explained by form of Bay. —

Fels: rock, mica

Junction

coast broken many reefs breakers Barrancas like fortifications

[13] 9th. [January 1834] — Arrived Port St Julian. Went on shore with Captain. — Came in at night. —

10th [January 1834] Up harbor — mud-banks. late at night — Country rather better appearance:

Mud in St. Julian constant rolling of pebbles:

Gecko, being kept for some days colour uniform ~~colour~~ grey

Compare with Blowpipe Mytilus[608] with blue & shells from Barranca

16th [January 1834] Pebbles at top of woods Mount of usual ferruginous colour: plain of diff. height: oysters & Turritella high up: ~~Then a~~

[14] 17th [January 1834] Cliff of S Barranca, 60 or 70 feet high: ~~mun~~ after gravel time for deposit of mud from a stream. on surface many mytili with blue colour & & fragments of oysters:

There are many pebbles. some greenish; with quartz & Feldspar not very porphyritic; also some phonolite & fine black Basalt; Porphyries without crystals, light ferru: colour: —

      Port Desire

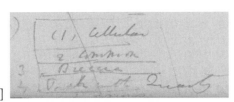

(20th —) [January 1834]

[Port Desire section] (1) cellular 2 common 3 Breccia 4 Porph with Quartz + common

[15] (2) is sometimes jaspery with very few crystals.

4 varies. — beds lower to the W by S (true)

The geology is only conformatory to that of the West Barranca Viz that whole Porphyry superincumbent to the Breccia — Variety one is curious appearing to be made of balls

Divisions in cliff strongly marked

[16] 14 miles South of Port St. Antonio cliffs. Shells

Cliffs Port Valdes

= do P: Cantor

Gravel line Point Castro [Latitude?]

Great inland Cliff (1500?)

608 A mussel.

Pt. Lotus /S.P./ /Dephine/

P. Union Tombo Atlas all rocky /rugged/ Points

Shells at Port St. Antonio

Behind Pinida Table land

[17]   About 30 miles South of Pinida great table formation, showing great alluvial action. —
This includes /Tilly/ road ~~in South~~ Table land not very even, — new made up of
parallel lines: Giant cliffs nearly 1000 feet high. St George

Cape Blanco /near/ ~~Cape three points~~

point Hard rock

[18]   Round Pt St Anna. on the of first little point: & other side 2^d little shells, on a stony
flat. — covered in gales: —

21st [January 1834] sea ½ &9 9.30 AM. 30.05

Attached 64.7 D. 62

11° 15' 29.778		A 71.5
		D 74.
12° 30'	29.690	A 78.5
		D 72
4.30 p.m.	29.992/68	A 77
		D 72

71
<u>23</u>
48
144
30.05
<u>1088</u>
30.012
71.23
3065
<u>29.686</u>
.36 °

[19]   8 A M —
10
12
2
9 ½
29.86
9 ½ .88
11 ½ .86
84
82

875
875
865
.020
855
835
.020
ne
Port Famine Feb 3[d] [1834]
bed-like cleavage N 70 W
Dip SW 56° —
do
— N.W. 68° D 29 S W
near R. Pant. bed like cleavage running 69° D NE 15°
a small space traversed by lines of cleavage NE by N & SW by S

[20]

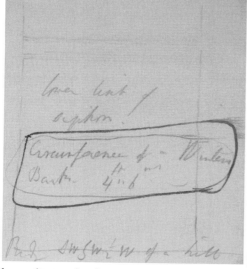 [diagram of Winter's bark tree]

lower limit of siphon
circumference of a winters Bark 4 [ft] .. 6[in]
Beds SW by W ½ W of a hill
& 17 from the central one Winter Bark
Feb. 6th [1834]
/Rough/ Point N 28 W. dip SW 25° —

[21] Highest point N 61 W

Dip 28° —

one from 10 to 20° dip W.

Hills NW & SE ? or NW by N

On Sea beach cleavage SW by S ½ S ?

Pyrites Iron stone — common ball: — Escarpement

[22] stone of Hornblendic Rock

~~On~~ 6 & ½ *[straps]* = 24 ft

4 ½ feet mean height

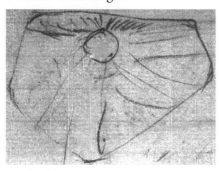

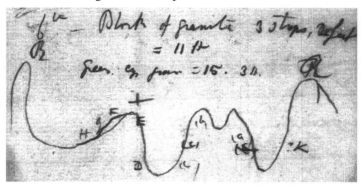

[gr SN or Hu] S 56 E for the more

southerly

6th — Block of granite 3 *[straps]* 2 feet = 11 feet

[green cy grain] = 15.3 ft R H G F E D (C) (C) (b) (a) (c) ?K a

[23] M. O'Hill: is under 1000 ft.?
     (a)   68 W   ∠25.
     (b)   70 W   ∠ 22   3.8
     (c)   55   ∠ 30   }irregular
     do   60   ∠ 38
     3.8
     6
     22.0
     2.1
     24
     (+ High peak S 46 E great /smi/ H S 54 E)
     (D)   no   strat
     (E)   dip   NW   ∠ 5   (?)
     F   81   —   11   north   D
     do   69   —   15
     G   75   —   17   do
     H   70   —   11
     (C)   70   25
     ┌──────────┐
     │ K 80-18 │
     └──────────┘

[24]

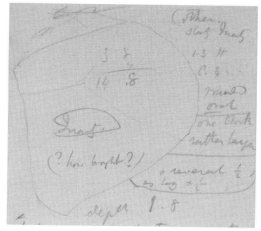

circum 4/.1/
(other slaty quartz 1.3 ft C.4
rounded <u>oval</u>.
(one block rather large)
& several ½ as long & ¼
depth 1.8
Quartz

3.8

$\underline{\quad 4}$

14.8

(? how brought?)

higher parts most stony: one Rock I should think 4 ft diameter —

curved mica slate —

pebbles generally few & small size apples to head. — angular — round some rounded, some angular. —

[25]   white with curved dark lines of small gravel. —

white more sandy & exactly like C Virgins. —

Was the oyster bed removed, or /prevented/ depositing: bed of alluvium continuing from animals & subsequently broken shells higher part

200 feet high? —

double elevation again convenient.

great boulders left in Shoal Harbor. —

[26]   P Rocky 26 N. Sarmin

        + 46 Peak E of S

Scarcely ¼ of mile from end of Point Rocky —

middle /of &/ snowy range 55

north side P R. 70. 45

Feb 25th [1834]

2^d Station A & B with common green rock — (B is globular concretions

[27]   Base S 30 E ∠ 18° —

great divisions: there were here several others at various /inclinat/ angles & quartz pebbles 1853

(D). Conchoidal basalt

Garnets 1854 Trachyte 1855 Variegated (vein like mass) green all together near /very/ /summit/ Porphyry

Ridge N 70 E = $\boxed{\text{E 3° S & W 3° N}}$

Vertical smooth planes generally common to tops

N 50 W = N 29 W & S 27 E first impure sort

Summit Porphyry & (D) (1853)

Quartz veins.

Cleavage S 50 E & low ridge

[28]   (W /Lakes/ most uneven: rugged — The breccia like rock generally lowest. — ~~little action of water over ledges uneven hill~~

At very summit an apparent South little Dip. —

Doubtful Breccia, some of the pieces were an inch large; — Read Wollaston Island. —

I do not think there is any slate & C. /Decert/ not Slate.

P 90

[29] Feb: 27th [1834] Slate dip 48° to exact South (true)

N: end of Navarin Isd

Cliff of Detritus — angular rounded stone. /no/ lines of deposition — small & few stones — earth color

W (Compass) 72 S D. —

W & E 59 S D

~~Porphyry pebbles between Falkland Islands~~: & ~~Staten Island~~.

[30] /Uni/ to /Front/ Slate run W ½ N sandstone dip W by S?

No dung beetles Falkland Islands

Why knowing that Patagonia has been elevated in some 1500 feet horizontally above sea (& probably 300 feet below) why not Tierra del Fuego? — Cleavage & stratification entirely distinct — F. Islands.

/S./ Usbourne pebbles outside coast.

What animals Toco Tosco. Aperea.

Rabbit Dung

Black Rabbit.

Read Falkland Geology

[31] Birds (: Penguin)

3 Pounds. Rabbit. megallanicus[609] ask Gregory Indians about do

Henslow importance of preserving labels

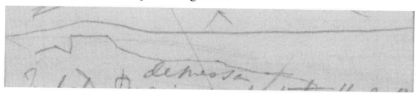

Jackass Penguin demersa:[610] perpetually rolls head side to side; as if he look out at anterior angle of eye: — noise exactly like Jackass at sea one deep note: — crazy, very quick, help of wings head stretched out — very brave diver — quickly moving wings very rapidly: fishing out at sea — cunning —

[32] The elevation of vertical cleavage in its exact direction would only be showing by a dip ~~outward~~ inward on both sides: Von Buch.[611] — M

Habits & geographical limits of brown vulture: Rats & mice & foxes on small islands & Georgia

Black rabbits breed. — are not found excepting where there are grey ones: there are white ones &c &c

---

609 Specimen 1885 in *Zoology notes*, p. 209. Waterhouse listed *Lepus magellanicus* as a black variety of the domesticated species of rabbit in *Mammalia*, p. 92.

610 See *Journal of researches*, pp. 256–7.

611 Possibly von Buch 1813.

Henslow cross means ~~Rabbit~~: Insects
Calcareous concretions in India —

[33]   Sunday — 16 [March 1834]
Hills W ½ N run —
Slate & sandstone ~~latter~~ both with cleavage
dipping N ½ ~~W~~ E 50–60
Hills of gneiss dip to within point of N —

= Flat topped very long oblong & upheaval S.   [curved strata]
great curved plates
(2) vertical plates
(1) small oblong
Slate (lower parts of country)
E & W run

[34]   The double Quartz range runs W by S & WSW
The Quartz is granular & very often contains mica specks no white /parallel/, but little

ferruginous the N. dip, the flat topped (castellated) &   [curved strata] &
the Southern curvature from anticlinal line: — in a /piece/ 4 or 5 feet long & above
3 thick curvature 1/8 of /circle/ — plain of divisions forming smooth domes: original
however there is much

[35]   fracture: — The 2^d Point of upheaval was at Southern base of great range — & directly
afterwards we had the vertical plates: —
Hail & wind, geese — Cattle country: — Sleep warm M
Monday [17 March 1834]
R. del Toro — Bone burn.[612] geld Bull — snow & Hail. — /Snipes/ geese Hawks
nest &c cry Sandstones ~~sometimes~~

[36]   generally more or less slaty running NW by W again appearing horizontal
very many valleys, parallel to range & ~~int~~ like furrows in the sea —
Horses very expensive 100 ps each. — out of 29 — 10 arrived safe & 4 now alive &
Bulls & cold. — no wild horses here. — only cattle. —
Sierra /larga/ — between /Physandria/ & /Inssiona/ crosses the Cochilla Grande. —

612 'the Gauchos soon found what to my surprise made nearly as hot a fire as coals, it was the bones of a bullock,
      lately killed but all the flesh picked off by the Vultures. They told me that in winter time they have often
      killed an animal, cleaned the flesh from the bones with their knives, & then with these very bones roasted
      the meat for their dinner. What curious resources will necessity put men to discover!' *Beagle diary*, p. 230.

[37] Where there are black rabbits there are others
Yellow legged hawks — females
Foxes have holes, generally silent excepting when in pairs
Hawks remain here whole year most in winter
Tuesday [18 March 1834]
Slate running NW by W Dip N
Granular quartz range dip WSW ½ S 40° ! /part/ of Sierra &boxed; & one point more S &boxed;
SE. by W. (SW. by W.)

[strata] 20 yards

[38] These quartz Hills run NW by W or ½ P more W. & from extreme point of Island
/At/ sleeping Place coarse slate D SSW
Wednesday 19th [March 1834]
returned home

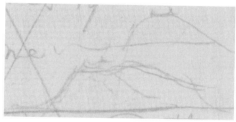

[sketch of hills]

Reread this after this
Carrancha kill Partridge let go again[613]
horses back — Chimango
(Caracara)[614]

[39] Tortuous cleavage connected with change in mineralogical change or mechanical. —
Barnacles in Tierra del F. from 20 to 50 fathoms
Degradation of rocks by snow ??
Sunday) [23? March 1834] DS ∠ 38° (true) (most general) SSW — ∠ 45° pure
~~crystalline~~ granular quartz Rock These subbordinate crests ran W & E
Valley of blocks ¼ of mile broad inclination 10° — from that to plain nothing like lava.
on crests detached blocks great earthquakes: Georgia (&c)

613 See *Zoology notes*, p. 212; listed as *Polyborus brasiliensis* in *Birds*, pp. 9–12.
614 Large carrion-feeding hawk of the genus *Polyborus*, but some other falcons may be called caracara.

[40] Flat plain or /Bog/ before last elevation. — When there were two islands: — a current eating out present arms: — /Sandra/ Orkney Shetland Georgia Desolation are all volcanic line of Elevation
line of greatest height runs perhaps WNW (½ N?) (contained a shell)
Intermediate rock, overlies & alternates with /stratif/ slate, itself becoming coarse — alternations or passages: passes into sandstone, suddenly assumes cleavage V Spec.^{n} W by W by S — very perfect.
or imperfect in proper form
Slate containing much sandstone mingled with

[41] the inclined laminae, & almost alternating, then suddenly the sandstone formed an inclined bed the extremities of which might be seen mingling with the laminae
Monday [31 March 1834] SSW 40° Quartz Crest same range as above
Quartz [Aten] R. /Quartz/. Alla R. — /Porphyry/-slate (Q A.R). suddenly slate
Southern limit of /Lievres/ between Port Desire & St Julian

[42]   **A**   8° 45' : 30.083
AT. 55.5.   D 55.5
11:45 29.04
A 55   D 55—
5. 50   30.07
A 54.   D 54
3.6 |55 = 12.7
= 12.7
55.5 = 13.
3.083
994
30.079
30.079
0.83
0.7
3 | 0.13
.004
~~30.083~~ 1013
30.087 7.004.
008
004
.079

[43]

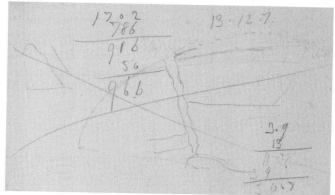

[base of an arched fragment of quartz]

1702	13.12.7
786	3.9
916	13
50	117
966	39
	5617

[44]

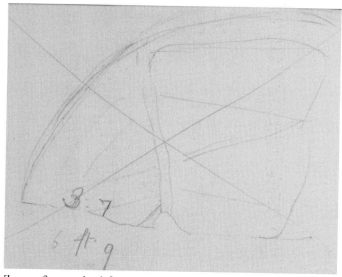

[base of an arched fragment of quartz] 3.7 6 ft 9

[45] & /traverse/ also ½ not a part
longitudinal veins with quartz on dome
Behind these broard crest dipping to SSW NNE or ~~S by W~~ N E — 50° — — 55° —

Crest running W by S.

Main chain do. — /height/ hills — N of main valley commence rather Northerly of this little range. —

[46] on very top. inverted arch: as /running/ of arches & /works/ from castle so these streams

9.4

120 s

14

3.8
‾‾‾‾
42.12

9.4
‾‾‾‾
51.4

[47]

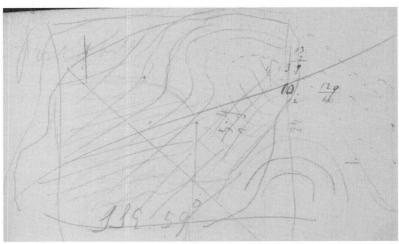

[hillock of quartz with summit of axis-plane thrown over to the south] Part of Hill SSE 59°

  13

   3
‾‾‾‾
4 |39

10

12

129
‾‾‾‾
 40

5.4

7.7

24

[48] Round Hill. +
next E Hill *[fort of on hill]*
S ½ W ∠ 70°
+ N by W ∠ 47°
Has Lowe[615] ever seen these big oysters of Patagonia?
Earth quakes. — F. Islas Gauchos

[49]

[cross section of the domed structure of South Berkeley Sound]

[50] Sunday ridge of greatest height W & E. — of wave WNW & ENE — Walking up a dome step by step from excisement: at foot due north a slight Southerly dip. evidently part of regular southerly crest, but more horizontal — points of upheaval not exactly in straight line in range like swell

[curved hills] ESE little dip WNW

615 Captain Lowe, a sealer. See *Beagle diary*, p. 148.

[51]

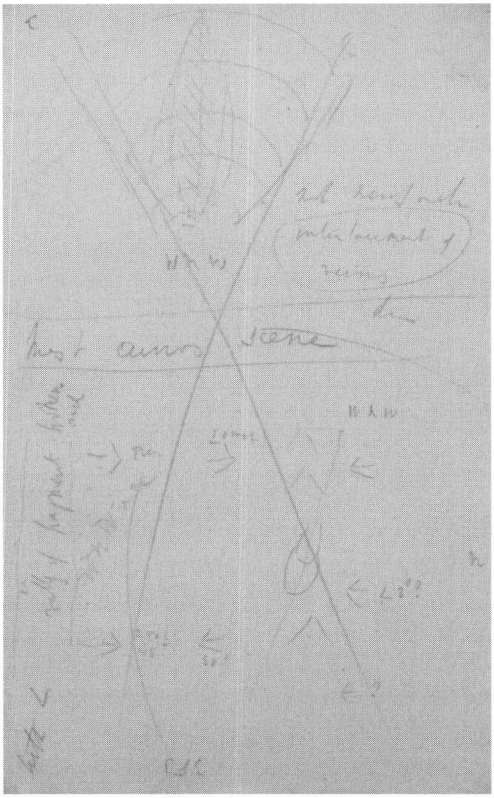

[diagram of zone of upheaval south of Berkeley Sound] *[a]* WNW not *[nearly]* *[such]*
interlacement of veins *[due]*
most curious scene S
valley of fragment — broken arch
WNW & & c *[green]* ∠ *[small]*
200 by 45 50° ESE WNW ∠ 3° ? N

[52] great hill long nearly as high as certain chain wave, more regularly anticlinal: in parts like
      crater. Wonderful scene of violence. Vegetation modern: — fragment as big as
      Churches: — valleys on each side a mile wide fragments: —
      line of view rather different from that of range in the line front hills —
      3 ranges between Berkely Sound & Sea. — W & E W by S the other end —

[53] Hills North of B. sound several *[orients]* (W & E (?)) centre of island more confusion —
      Thrush in flocks
      Experimentize with pods of Kelp. Nitre rapidly drying: —

Small Hawk iris "honey yellow" — legs — bright yellow . skin above blue beak do: male feeding on Carrion! — (1926)[616]

Blue leg Vulture. Male.

Super-abundance of inarticulate Corallines: Kelp Corallines

[54] Depth at which Kelp grows Limits on both sides of coast — all Southern Islands? — Kelp South Islands — Tristan da Acunha Sorrell?[617]

Exception of Kelp Fish in Falkland Islands[618]

Re-dissect Obelia[619]

Mr Stokes says kelp little to N of St Elena (about Lat 43°).

10 & 15 Fathoms Kelp grows generally

Degradation of land by snow Southern limit of Agouti.[620]

[55] Specimen of Kelp:

Pods of Kelp Experiment/ize about/ the Nitre. —

Barnacles very common. 30 to 50 Fathoms. Coast of Patagonia. Has Lowe seen big oysters? C. /Mammoth/ Gregory B. — Birds. Petrels: Puffins. where build?

Governor Andes *[illeg]* 11 14.89

Hares Viscachas at Coquimbo Molina Pitui

Silvago white tuff T del Muscicapa[621]

Red back shore finch builds in bushes

No snakes

[56] 24th. — [November 1834]

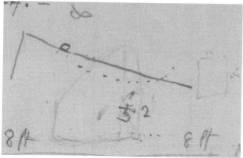

∞ 8 5 2 15 8 ft 8 ft

Nov. 1834

616 Specimen not in spirits 1926 in *Zoology notes*, p. 399, 'Ornithological notes', p. 233; listed as *Milvago leucurus* in *Birds*, pp. 15–8.

617 Possibly a reference to Thomas Sorrell the boatswain of the *Beagle*.

618 See *Beagle plants*, p. 211.

619 Specimens 597 and 1161 in *Zoology notes*, p. 147.

620 A rodent of the genus *Dasyprocta*, related to guinea pigs.

621 Compare with *Zoology notes*, p. 249; possibly the same species listed as *Xolmis pyrope* in *Birds*, p. 55.

Started, took horse in Chacao, followed the coast, winding in after every Bay: saw many civilised, pleasant to see in any case natives. small copper coloured men — not very like Fuegians: Scenery exceedingly picturesque, beautiful cleared spots & pretty enclosure. magnificent forest path like road to Castro fine view of Straits of Chacao: Volcano beautiful many crosses, dangerous straits

[57]   Cone of snow. — plains of different altitude. thickly wooded: higher ones soft earthy, sandstone with lines of pebbles, granite & greenstone & some few porphyries: lower plain chiefly ~~granite~~ pebbles: all showing curved lines of currents: perhaps different elevations: immense removal shown by reefs from points:
Granite increased as we approached Chacao: one nearly square, singularly little rounded, not more than fragment from mountain at base (very hard used for mills) 11. 1418. & 9 ft high: — every where signs of upheaval: one place apparently at [80] height of 200

[58]   200 ft — difficult to ascertain from people eating so much Marisco:[622] Said to occur in interior: Cordillera said to consist of several rows. —
? Are these plains contemp: with [Lecuy]? pebbles different? —
Cheucau.[623] making an odd noise people will not start
Governor old Spanish Colonel — Son without stocking — miserably poor indifferently asked whether flag would always fly: — — Cordilleras much more distinct peaks: the 2 or 3 first rows, coarse & fine Syenite & Granite (Douglas[624] — Volcan — very regular cone

[59]   Corcobado — Trachytic — (St Jago do?) — ~~Some~~ Many of the blocks ½ or ⅔ of the above size: —
25th [November 1834] Torrents of rain & in night Hurrah Chiloè
country plain from 100 to 200 ft thickets & low woods here & there patches of cultivated ground — Huapilenous, same general formation. many fragments chiefly granite. — some as ½ large as yesterday & angular: foul wind drove us in —
"Huapi" means Islands: yet now nearly all peninsulas. proof of rise. — ~~Big drops of rain~~

[60]   Expect a Ship at P. Castro Reason ?? =
(26.th) [November 1834] With a too strong SE breeze & a splendidly clear day made sail & reached Puerto Obscuro only good Harbor. — country same: Saw Indian party: one man exactly like — York Minster; others rather handsome. more color like Pampas — Manners <u>pleasant</u> dress &c &c like Spaniard — very poor. Cacique yet remaining but have little power since tributa ~~disb.~~ abolished — when any officer, as land surveyor come down, they accompany him with badge of office a silver cane: they talked their

622 Spanish name for sea food.
623 A bird known as a Chucao Tapaculo (*Scelorchilus rubecula*). See *Journal of researches*, pp. 351–2.
624 Charles Douglas, surveyor and pilot, long resident in Chiloe.

[61] Indian language & very little Spanish. = The Cordilleras very magnificent certainly much less lofty than in Chili <u>more regular than yesterday</u>. Osorno immitting much smoke snow considerably melted, always in state of activity. —
Saddle-backed, gentle cone, with immense crater emitting very little smoke — said to possess much sulphur till /some/ years since, was hardly known as Volcano. when great eruption made a plain concave —
Corcobado — 1 to 3 eruptions annually —
Inhab: of Castro expect earthquake, when <u>no eruption</u> during 1 to 3 years

[62] Some other mountains looked as if extinct Volcano. —
Geolog. Formation same, ~~any~~ proof that lava has not proceeded from Andes i.e ancestral crater — — few $\angle^r$ blocks of granite —
Presence of Indians owing to ~~their being~~ a large tribe being encouraged & land given them at time when Osorno was betrayed but these tribes remained constant. —
Puerto Obscuro beautiful little cove. several houses almost all grandchildren

[63] badly off for ground, but too idle to clear more only breed — Have to pay Government & surveyor, mentions price: & 3 auctioning settles it —
Chili government making retribution to Indians for their Tributa. Cacique 12 quadras wife half — Militia-man. — to wife ½: aged 4 quadras: during tributa Indians cleared land & were then driven away.
27th [November 1834] —Again fine day. strong foul wind — arrived at the cove of /Tuicari/. San Corcobado like same name at Rio.
Trachytic

[64] Blocks of granite. Saw at same time 3 volcanos & some most beautiful & <u>regular snow cones</u> far to the South: deception of semicircular arrangement: Table land from 2 to 300 ft high. — woods for fire Tropical from great intermixture & diff colored stems —
Formation of the Chaciques apparently the same as before —
Elmis beneath stones.[625] —
Small chimango eat Potatoe when planted & bread. —
Rapid tides amongst the

[65] Islands — // Small Crustacea[626] purple clouds of infinite numbers. pursued by flocks of P. Famine Petrel //
Natives eat much Marisco & Potatoes, — all Indians in the Chaciques & some other Islands. —
28th [November 1834] Chimango torments. swing swang the Carrancha[627] —

---

625 Specimen 2338 in *Darwin's insects*, p. 83.
626 Listed as specimen 1104 in Chancellor *et al.* 1988, p. 217. See *Journal of researches*, p. 355.
627 See specimens 1932–3 in 'Ornithological notes', pp. 233–4.

In the census of 1832, 42 thousand inhab: of all sorts in Chiloe islands. = granitic blocks rounder. & far fewer? —

Landed at P. Tenuan, <u>fine grained</u> clayed sandstone. with remarkably even laminae, the very

[66] same as at C. Virgins. greasy with water at sleeping place Quinchao (splendid day & clear weather) same substance with singularly <u>convoluted</u> thin layers in domes running NNE & SSW? All the Islands certainly once a plain from 200 to 300 ft. — Appears to dip to East. ∴ more interior higher land. — Channels in some places deep. — Met many ½ blood Indian families — when <u>picking up Marisco singularly like Fuegians</u>

[67] Something like St Julian Porph pebbles: ||| but coarser texture not Porphyritic + |||

29th — [November 1834] "*[Tey]*" Granitic blocks certainly rather rounder — country in all parts <u>more</u> inhabited many periaguas.[628] — Douglass sold 8 ½ square miles near Carlos for about 350 dollars. given by for Spanish bills. — nobody has an income here. — in a long life time industrious most rich person may gain 4 to 5000 dollars which is hoarded up. every family having his stow-holes. : Gomez, & wife **noblemen** of Spain came out

[68] to Lemuy by constant intermarriage with Indians present linear descendants can hardly be distinguished from Indians — whilst on other hand the present governor of Quinchao is very particular to keep his blood pure.

Name of saddle-shaped volcano — *[Serro]* de Vilcan ?? = Language of Chauwes [Chahues]

T. del Fuego Petrel: in a flock by hundreds & thousands flying in strength irregular number lines; occasionally uttering odd cry

[69]

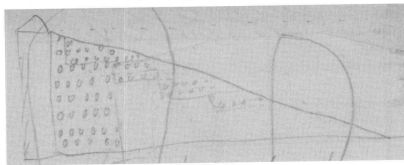

Explanation of gravel, during rise constant enormous accumulation: contrary facts the transportation of ∠^r boulders — & Julian pebbles —

Castro granitic greenstone & <u>many</u> Basaltic ones: Coarser horizontal strata of soft sandstone: land about 400 ft high: Reylan Point. —

628 Spanish name for small boats.

Formerly Peures[629] were brought in great abundance from Cabluco constant gathering have reduced in

[70] numbers but since 1825 they have become abundant at Chacao were they formerly did not exist: The Nautilus is thrown up periodically in great numbers at Juan Fernandez & coast of Chili. =

Layer of hard rock, composed of crystalline rock Volcanic crystals overlying horizontal slaty sandy clay — bed inclined, current inclination. — generally this degrades down into an earthy sandstone with black specks; the common formation

[71] seems as before —

Signs of steps in these forms [sketch of terraces?]

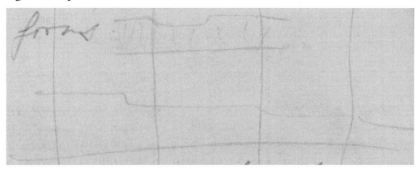

but very much alluvial action as probable

30.th [November 1834] —

Bybenes or Huyhuyenes Indians formerly inhabited Chiloe, some few Baptized probably driven away by consumption of Mariscoes: believed to have gone to city of Casares, perhaps those Indians seen by Bynoe[630] were /this/ tribe

[72] 1100 Indian surnames, ~~descended from as~~ all speak the same root Beliche language. — spring from some brought en commendo from the north supposed near Osorno & from Chauwes Chahues & Ragunos, /who/ remained at the taking of Osorno not more than one or two families of the Bybenes now remain in the Archipelago speak quite distinct language.

Douglas knows a family of Bybenes near S. Carlos. perhaps Chawes & Ragunos: Molina

Out of these 1100 some of the Indians are totally pure

[73] Spanish blood, others such as Gomez nearly pure Indian do not come in the list: yet perhaps a good approximate number —

En Commendo Douglass believes to be each Spanish family had a certain number of Indians to instruct in Christianity & obliged the Indians to work for so much time. —

629 A sea squirt (Ascidian), specimen in spirits 1165 in *Zoology notes*, p. 357.
630 Benjamin Bynoe (1804–65), Assistant and then Acting Surgeon on the *Beagle*, 1832–7.

All these Indians descended from but few Indians: A Cacique of the Ragunos, who had been placed after the taking of Osorno in the Islands of Calbuco, came with 12 periaguas to assist the Spaniards against the other Indians who had risen against them.

[74] Indians have few superstitions yet left hold converse with the devil in a cave: formerly sent, for this to Inquisition at Lima. The Bybenes now only inhabit Caylen. —
Arrived at Castro: never before saw a truly deserted city: roads thick pasture: a pleasing cultivated country fringing on the coast the dusky green woods. Jesuits church highly picturesque fine dome all built of plank: very poor people retired solitary spot. eating what they produce & hearing little of the

[75] rest of the World. —
(Chacao inhabitants driven away. in old Spanish time by burning their church) /fant/ numbers of chapels in all parts of country. Rode about in the neighborhood to see geology: all the lowest beds consists of Talcaceous (V. Spec) mica slate, with thick curved layers of quartz which sometimes is nearly pure laminae:
upon this I saw great bed of conglomerate, & near same locality the slaty clay — sandstone (heard certainly

[76] of Granite in the interior). In another locality, a mass of inclined columns of semi-

carious white Trachytic rock (V Spec) columns chiefly 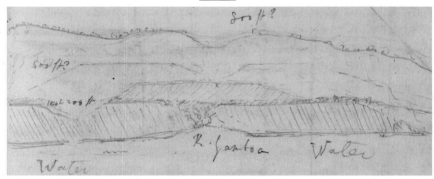 of this shape seem certainly to overlie the gneiss though did not actually see contact hence Chiloe & Cordillera of primitive rock from which melted matter has flowed & fringed by a modern bed. = There are inclined very distinct (3 sites at least) most manifest plains, with escarpements. The lowest from 100 to 200 ft. has a fringe about 30 to 40 feet of /comminated/ recent. The highest perhaps about 500

[77]   Castro

[sketch of terraces at Castro, Chiloe] 500 ft? 800 ft? 160 to 200 ft Water R. Yamboa Water

[78] Sea-shells — extensive formation: ?Perhaps gneiss dipped to NW small ∠°???
The dust of S. Carlos, came from the Corcovado, very thick, whole island, in 1827? & 14 miles to Seaward Douglass:
carefully reexamine Valparaiso shells

Point nat — Peninsula Lacuy shells high up Douglass

a doubt about Chacan Boulders

one strong shock & three days & dust band

Particulars & date of 1831 of small ones great Earthquake: shortly afterwards in activity Stone off S. Carlos. gradually was nearly laid dry & gradually the water regained: & *[3 words obscured by tape]*

[79] Cuy a rabbit in the Bichiche language. V. Falkner

The earthquake of Chiloe had a severe shock of ½ minute & three days small shocks. = The Valparaiso have forgotten their language Bichichas in Caylen Tapaculo[631]

December 1st [1834]

Left ~~St Ca~~ Castro early a large group watching us: Indian family with periagua — young man, in morning wet to skin from the torrents, with thin trowsers & drawers professed to be quite

[80] comfortable: heavy rain during day: proceeded but short distance to Lemuy. so thickly wooded to waters edge (for fish describe coralles) great difficulty in finding sleeping — large populations of ⅔ Indian blood, much surprised at our appearance. talking said the many parrots they had lately seen told them what to expect the "Cheucao" to beware. —

Many cellular Volcanic looking pebble near Castro

[81] Whole cliffs composed of this

Castro Creek — soft yellowish sandstone, passing with a finer grained brighter yellow sort of greenish clayey variety: globular concretions of grey hard sandstone — very like Lacuy

North W point of Lemuy composed nearly similarly but none uniformly soft sandstone. V spec: many ferruginous veins; few occasional line of pebbles: in very many parts patches & extensive horizontal layers

[82] of black Lignite, structure of wood yet very visible: is said to communicate much heat in furnace has been worked tried pieces all flattened; some layers several inches thick: probably from high land of Chiloe, previous to its elevation —

Summit of cliffs gravel

Many rounded blocks on beach of petrified wood same inexplicable occurrence as before.

One fine white

[83] granitic block. pentagonal mean height above gravel *[see]* 12 feet — one side however 16 feet — mean circumference at middle. 52 feet — like a house — Granite probably from Cordilleras:

Smaller ones high up

---

631 A bird of the genus *Sclerochilus* and the larger turca belong to the Rhinocryptidæ, peculiar to South America. The word tapaculo, derived from the Spanish, means 'cover your posterior' because of the absurd way the bird carries its tail. See *Journal of researches*, p. 270.

= ~~Fine~~ Beroe[632] fine ~~pu~~ up rose—

= purple with splendid /iriscident/ ciliæ

= Shells at Lemuy same elevat: as at Castro

~~Chucao.~~ I know 3 distinct noises[633] — nest ??

[84]　2nd [December 1834]

(A)　The common yellow <u>massive</u> soft Sandstone, in which the silicified wood is found

(B)　<u>soft</u> silic: wood from great branch in do

(C)　The wood coal showing the <u>leaves</u>, a laminated structure is the commonest, in which it appears just like wood. ? which has formed the whole mass? — Perhaps leaves may be applicable to ~~mar~~ much coal —

[85]　A little way above Yal. — I found in the usual sandstone (A). some <u>flat</u> concretionary & grey hard sandstone masses in which were horizontal seams of shells. These could be very imperfectly discovered, but from what I saw ~~seem~~ are certainly not such as now are heaped up on coast — a large common Cytherea.[634] taking the places of Concholepas[635] of Valparaiso — Before recent we may conclude — It is very strange, in such an extent of cliffs no trace of shells excepting in this one spot

[86]　where the rock was harder & rather different from general — Is not the presence or absence of shells owing to chemical nature of matrix? —

Mem: observ in E T. del Fuego: — ~~at the point of~~

The above Sandstone beds were capped as at Lemuy with thick bed of cemented gravel. —

At the point of Yal. Harbor the laminated sandstone. — passing in parts into a conglomerate alluvium like those to the north (round pebbles of all sizes) & capped as

[87]　before with great gravel bed. — plains 300 to 400 ft high.

Theory. where islands exist generally the softer sorts of sandstone abound, which probably are not "recent" beds. — they are capped with great bed of gravel, ~~is~~ are not the low gravel plains of the north of same nature as these & perhaps deposited & remodelled. as in Patagonia from the gravel of the higher series: near S. Carlos we also see the sandstone beds capped with gravel. — Mem: during depos Volcanoes existed? Pumice?

[88]　Where there are creeks this formation exists. =

Made little progress during the day owing to light weather: Continues thickly inhabited. coast more so than almost any place I have seen — Inhabitants purer & purer Indians: excessively humble & civil, most innocent good natured people (one most true Indian) wanting salt, Tobacco, Indigo & every one of these luxuries — speak little Spanish. —

632　A similar creature was described in *Journal of researches* 'The structure of the Beroe (a kind of jelly fish) is most extraordinary, with its rows of vibratory ciliæ, and complicated though irregular system of circulation.' p. 189.

633　See *Beagle diary*, p. 271.

634　A clam, possibly the species named *Cytherea sulculosa* in *South America*, p. 25, plate 2 fig. 14.

635　The so-called Chilean abalone, a species of large carnivorous sea snails.

~~Forme~~ I mentioned having found at Lemuy on beach very much silicified wood. — one piece penetrated by Teredo.[636] likewise

[89] I found many fragments on coast above Yal —

At <u>last</u> I found <u>in the</u> yellow sandstone (A), a great trunk (structure beautifully clear), throwing off branches: main stem much thicker than my body — & standing out from weathering 2 feet — central parts generally black & vascular & structure not visible — This tree coetaneous (near in position) with the shells of above: it is curious such a sandstone chemical action in sea holding such silex in solution vessel transparent quartz: This observation most important, as proof of general

[90] facts of petrified wood for here the inhabitants firmly believe the process is now going on —

Part of the great block was more sandy than siliceous (B):

The great Lucanius[637] makes when approached or molested a loud noise, which almost frighten a person

= <u>Inhabits. T. Firma.</u>

Carabus:[638] logs dark forest: most <u>powerful</u> disagreeable small & acrid juice

Bluish ~~T. del F~~. P. Famine Tyrannus[639] egg course large nest

[91] $\boxed{3^{rd}}$ [December 1834] shade green

[a frog] Under side[640]

Throat & breast & /cheeks/ rich chestnut brown with snow white marks thighs blackish with do. legs yellowish with do

Upper side

Pale rust color with posterior part of body, & thighs & anterior marks bright green: iris rust color pupil jet black. jumps dark forest point on nose.

other above blackish brown with narrow bright yellow medial line

[92] only proceeded to P. Detif owing to Calm: view of Corcobado (saddle-shaped Volcano ~~chitan~~ Lildeo near estero,

/Rengihue/ & 2 other snowy cones. over the inland sea, like glass, only rippled

---

636 A genus of bivalves that bore holes in wood.

637 A stag beetle. See specimens 2110–1 in *Darwin's insects*, p. 81. Also discussed in *Descent* chapter 10.

638 Specimen 2327 in *Darwin's insects*, p. 83.

639 Specimen 2081 in 'Ornithological notes', p. 249; listed as *Xolmis pyrope* in *Birds*, p. 55.

640 A frog, specimen 1086 in *Zoology notes*, p. 256; listed as *Rhinoderma darwinii* in *Reptiles*, p. 48, plate 20. fig. 1, 2.

here & there by a Porpoise or logger-headed duck seen from a cliff, through an avenue of various lofty evergreen trees, with fine white flowers & sweet smell, how mistaken in climate of Chiloe! The point is composed of

[93]   beautifully horizontally laminated clay-sandstone (specimen D), variation in color for every ¼ to ½ inch (beautiful basements to buildings & columns)
the junction of the [shapely] variations in color like carpentry; probably particle arranged after deposition capped to certain degree with aggregate gravel. Part 200 to 300 ft high.
Said to be Volcano far south of Corcobado.

[94]   4ᵗʰ [December 1834]
Isᵈ Sebastian, contains much very fine clay-stone, pale color. adhesion to tongue hard. — Douglass
                                        ⸮also Volcanic?
P. Famine Petrel very irregular in its migration suddenly appearing & disap. in countless numbers in certain parts of the Islands
Both night & day very squally (. Mem at Lemuy Douglas telling Sillador that by mistake we should certainly shoot anyone who walked about at night perfect humble acquiescence)

[95]   arrived at night beautiful harbor of Quelen: N B. where basins with narrow mouth geologically striking in number & must in upraised land give a curious lacustrine appearance. — Cause of this depression in bottom? —
Inhabitants certainly much less frequent, the whole line of Tanqui, without one cleared spot. —
At P. de S. [Aytui] cliffs from 100 to 200 ft composed of alternation of nearly loose gravel of different sizes & sand

[96]   The country bordering the coast. — Any all Tanqui would seem all to belong to one plain of low altitude upon which the higher series rests; extending to islands of Alao & Apiao but not reaching far to the South. —
5ᵗʰ [December 1834] —
Sea-snail.[641] body when partly crawling oval — post extrem. truncate & scooped out, with large Branch aperture always open: [ft] convex: when quite contracted a cone: above blueish black, with

[97]   pale white, projecting points & pale halo — edge with alternate spaces of narrow white & blue, the latter color being fimbriated. beneath white, excepting mouth: tentac short terminal black eye: beneath which a bifurcate membrane:
high up on rocks, near top of high water crawling on confervae nearly dry: —
Onchidium?[642] —

---

641 Listed as *Peronia*, specimen 1092 in *Zoology notes*, p. 255, identified by Keynes as the slug-like pulmonate *Onchidella marginata*.
642 An intertidal slug.

Two small Turbos same locality.

Pushed on to P. Chagua, same squally weather in several places on road, mass of alluvium hardened earth containing round &

[98] angular pebbles of granitic or <u>Volcanic?</u> (cellular) rocks (perhaps lower beds fine grained) ?? C͟h͟ Plains rather higher than before stated it is chiefly East end of Tanqui which is low. —

Very little cultivation: (a man travelled on foot round all the creeks from near Castro to Guildad 3 days & ½ foot journey, to receive value of an axe & a few fish)

[99] 6th. [December 1834] —

<u>Doris</u>[643] a͟b͟o͟v͟e͟ pale yoke of egg colour foot & mouth darkest colour mantle rough with rounded paps of two sizes. oblong. far surpassing foot on all sides — Anterior tentacula protruding through mantle. anteriorly pectinated with laminae: orifice on /rt/ side — Branchiae 8 Acanthus /head/ expansious Large same colour with mantle — dimensions when partly crawling

[a sea slug] common mantle surpassing inferior Tentac Onchidium bilabiate projecting /hard over mouth/

[100] <u>Eolis</u>[644]

[a sea slug] dimension when crawling

643 Specimen 1091 in *Zoology notes*, pp. 255–6, identified by Keynes as a cryptobranch doridacean, probably *Anisodoris fontaini*.

644 Listed as *Cavolina*, specimen 1091 (sic) in *Zoology notes*, p. 256, identified by Keynes as a aeolidacean nudibranch, *Phidiana lottini*.

general color "crimson red"? ~~with~~ (lilac with lead colour) under surface & mouth finer —
rose color — base of branchiae rather leader color in numerous transverse on each side of
clear space of back anterior & inferior Tentacule very long far apart tipped with white,
tapering — superior & posterior (arising between branchiae, some way back) blunted &
shorter: mantle not surpassing the inferior Tentac: body truncate. anterior. triangular:

[101] | Cheuqui,[645] bird like Chucau *[shore]* |
Reached Caylen (el fin de Christianidad), rather better inhabited: day same as before —
[In] wooded plains — bought a duck & cock for Tobacco value 3 ½ penny & 3 sheep
for ~~four~~ 3 shillings worth of goods. & great bundle of onions 4 handkerchief The
news had travelled, of the exact value we attached to the Tobacco (which was in the
relation of a shilling to *[1]* & ½ pennies)

[102]   The cliffs before entering St[s]. of Caylen were composed of mud with pebbles (& a
boulder) (Alluvium) which passed into consolidated mud (or laminated clay-stone)
contained small fragment of common Cytheraea.[646] Height 300 to 400 — ft
all composed of above sort with very few pebbles the lower plains were composed of
gravel which plainly had been deposited after *[wearing]*

[103]   the lower beds —
Beaches of present day <u>universally pebbles beach.</u> —
immense number of Granite pebbles
7th [December 1834]
dip WSW ∠ 22° mica slate Blackisk *[illeg deletion]* no small scale country plates (often
convoluted of quartz) ~~mica~~ (Feldspar ?). There are many vein like masses of quartz —
This first occurred extremity of I[d] Laytec from which it continued all the way to
S. Pedro

[104]   many lines of fracture
The dip was very constant: easily seen in the mass though not in very distinct laminae &
not above ½ a point on either side of WSW. the ∠ was generally small, but in one
place 60° — The coast here becomes abrupt. — there are some long saddle-shaped
hills — & the coast strictly resembles T. del Fuego —wild nasty day, long

[105]   *[beat]*. open sea: after leaving the Caylen stopped at a hovel, the last house about Lat
43° 10' This is the extreme point of S. American christendom a miserable hovel. —
People living on potatoes & mariscos = excessively fond of Tobacco =
Anecdote of gun & powder for feast. =
Features of S. America as regarding Vegetation

---

645   Specimen 2436 in *Zoology notes*, p. 408; listed as *Pteroptochus paradoxus* in *Birds*, pp. 73–4.
646   A genus of marine bivalve.

[106] killed rare fox with hammer[647] — Arrived late in the evening. S. Pedro after absence of
fort=night. — Found *Beagle* arrived yesterday — bad storm weather failed in surveying
outer coast: had visited Huafo & parts Chonos. is necessary to shorten Chiloe ¼ at
length! —
San Pedro — leafless trees = enormous

[107] trees of Winters Bark & Laurus Sassafras — delightful smell — T. del. Birch —
Alerce,[648] red Cedar Cones. — walk many feet above ground — like foxes at other
times — more moss than in T. del Fuego

[108] Is^d of Huafo & (Ypun Tertiary) horizontal strata composed of solid — hardish black
[grey] (Specimen 2358) earthy sandstone & the yellow sort characteristic of
Archipelago of Chiloe

[109] Same formation has silicified & carbonized vegetable substances Hence near a wooded
country as at *[illeg]*
Strap = 4. 8. ½

[110] (*[SE]*) Perhaps Island bears NW by S ¼ W

[section of concretions?] (a little more to SW) x

---

647 Specimen not in spirits 2431 in *Zoology notes*, p. 408; listed as *Canis fulvipes* in *Mammalia*, pp. 12–3. See
   Journal of researches, p. 341.
648 Spanish for larch.

[111]

[map of headlands?] 3 miles not so far E (calle length) NW by W ¼ n Compass

[112]

[section of concretions?]

[113]

[section of dykes?]

[114] S. Paleña

When did white Porphyry with much Silver in M^r /Budgton/ store come from?

6561.8	90: 46 ∴ 4320:y
2296.7	4320
8858.81	46
656.2	25920
196	17280.
852 20	9.0	198420
971 8	2208
	5
4320	2708
47	
30240	
17280	
9	2030 40	
2256	
500	
2756	

[115] 9

[section of concretions]
4820
46.5
24100
28920.
19280..
9 | 2241300
23

[116]

[ground-plan showing the relation between veins and concretionary zones in a mass of tuff, see p. 301.]

[117]

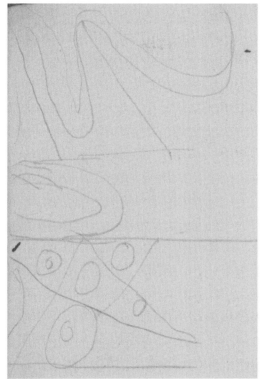

[continuation of previous diagram] 3 or 4 feet Fig. B

[118] Concret hollow — one /layer/ 3 — feet sand.

[a hollow concretion]

Dip W by N ½ N — 15°
Boulder 4 × 4 4
Syenite
Wood Agate Angular pebble
2700
26
2700

260
2440
6561.81
1312.41
131.28
8005.30

[119] │Eruption of Osorno 20th of Jany│
32808.6.9
4
131235.3.6
26246.10.3
157482.1.9
131.2.8
26.2.9
57.5.7
1575
13123.5.1
2624.8.2
15748.1.3
│157 68│
1/12 5/12 68
7/10+
12/1.00/0.08

[120] pebbles passage in /Cheverea's/ Creek —
East by dip —
West dip line of /Huapilenous/ shoal — horizontal

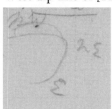 NW NE E
Extension East dip

[121]

[diagram of strata?]

The grand line runs (N 6° W) — valley of <u>elevation</u> separates the two escarpements 3 or 400 yards wide —
3 or 400 different regular variation in strata: —
= (dip in one place 25° — generally less in cliff 10° or 12°

[122]

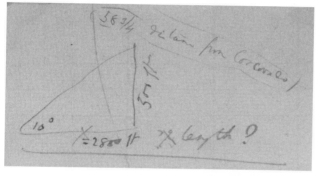

[triangulation of height of Corcovado] 583/4 distance from Corcovado 10°
300 ft = 2800 ft X length?
Mem concretions bordering line <u>of fissures</u>: yellow Carb of Lime, in veins: —

Thickening of strata in the West dip

[123] Osorno — 9005 ft
       Corcovado 7442 — Snow 3902. —
       Slate dip run W by N — NNW
       sediment dip W by S

Fig. 1 (A)

[124] Slate here dip
       S.W.
       Fig. 2

9 ft shells to 15 ft 9 in (B) NW to SE

[125] (C) dyke 18 inches wide — following strata, sudden bulge —
       D + D reticulated & [on] other part dyke 25 ft: irregular white inside (D). dip of Slate
       NE — dyke — /run/ NW by W compass

[126] E 2 dykes 10 to 20 ft inclined thus upwards
also ~~other~~ nearly the same E & W
SW compass /Test/ /Cony/ strata dip 30° —
NW by W compass dyke & cleavage

[127–74 blank]

[175–6 excised]

[177] Mellamoy name of four peaked snow mountain S of cone S of Corcovado not known to
be volcanic —
Huamlino name of Volcano in S. Pedro

[measurement of a plant by CD using his belt]
from Buckle to 3ᵈ regular hole[649]
from axilla of leav 1st + 4 inch
1 of & to the last hole from point
margin of leafe scalloped

[178] Different name in diff parts
Chiduco the good sign Huitreu — the bad sign of the Cheucau[650]
builds its nest in low bushes near the ground: erects its tail like Tapacola — but comes
near to a man if he is quiet. =
Height of /Desiertas Composition/

[179–80 excised]

---

649 Possibly the Pangi plant. See *Beagle diary*, p. 272: 'The leaf is much indented in its margin & is nearly
circular; the diameter of one was nearly 8 feet (giving a circumference of 24 feet!).'

650 'The cheucau is held in superstitious fear by the Chilotans, on account of its strange and varied cries.
There are three very distinct kinds, — one is called "chiduco" and is an omen of good; another,
"huitreu" which is extremely unfavourable; and a third, which I have forgotten. These words are
given in imitation of its cries, and the natives are in some things absolutely governed by them. The
Chilotans assuredly have chosen a most comical little creature for their prophet.' *Journal of researches*,
p. 352. See *Pteroptochos rubecula*, *Birds*, p. 73.

[181]   Is the brown Vulture found at Shetland?

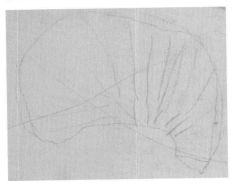

[182]   /Barmouth/

Textual notes for the *Port Desire notebook*

[IFC]   C. Darwin] *ink.*
1.8.] *Down House number, not transcribed.*
88202328] *English Heritage number, not transcribed.*
< > /use/ … Sound:—] *ink.*
Pebbles … Coast] *ink.*
R. Chupat … 20'] *ink, written perpendicular to the spine.*
Calyen 43.10'] *written perpendicular to the spine.*
7] *added by Nora Barlow, pencil, not transcribed.*
several ink marks appear to be nib tests.

[3]   *an ink mark appears to be a nib test.*
1834] *ink.*

[4]   Plain] *ink.*
(3ᵈ)] *ink.*

[5]   Ferruginous … jaspery porphyry] *written perpendicular to the spine.*

[11]   inch] *ink.*

[18]   144] *ink.*

[19]   ne] *ink.*

[20]   circumference … 6ⁱⁿ] *circled in pencil and ink.*

[23]   M. O'Hill: is under 1000 ft.?] *ink.*

[29]   ~~Porphyry~~ … ~~Island.~~] *ink.*

[32]   Rats … Georgia] *ink.*

[37]   SE ….W.)] *ink.*

[38]   *two ink marks after* 'island' *appear to be nib tests.*

[39]   Tortuous … mechanical. —] *ink.*

[42]   'N' *of* 'AN' *in ink, second and third* 'D' *in ink.*

[43]   *page written perpendicular to the spine.*

[45–6] *pages written perpendicular to the spine.*

 [46] 9.4 ... 51.4] *written parallel to the spine.*

 [47] *page written perpendicular to the spine.*

   13 ... 40] *ink.*

   5.4 ... 24] *written parallel to the spine.*

 [49] *ink sketch written perpendicular to the spine.*

 [53] <u>Blue leg Vulture. Male.</u>] *underscored in ink.*

 [55] Specimen ... build?] *ink.*

 [56] Nov. 1834] *not in CD's handwriting?*

[77–8] *leaf excised, now in DAR 35.297A. Diagrams not by CD?*

[101] /shore/] *ink.*

[111] *sketch perpendicular to the spine.*

[112] *sketch perpendicular to the spine.*

[114] When ... from?] *ink.*

   6561.8 ... 2708] *calculation written perpendicular to the spine.*

[115] *page written perpendicular to the spine.*

[116–7] *sketch perpendicular to the spine.*

[117] 3 or 4 feet] *ink.*

   Fig. B] *ink.*

[118] 2700 ... 8005.30] *calculations written perpendicular to the spine.*

[119] Eruption ... Jan^y] *ink.*

   32808.6.9 ... 12/1.00/0.08] *calculations written perpendicular to the spine.*

[122]   = 2800 ft] *in ink.*

[123] NNW] *in ink.*

   Fig. 1] *in ink.*

[124] Fig. 2] *in ink.*

[177] from Buckle ... 4 inch ...] *written parallel to the spine.*

[182] *page heavily soiled – indicating where original back cover was once missing. The back cover has been replaced since microfilming. /Barmouth/] written perpendicular to the spine.*

THE *VALPARAISO NOTEBOOK*

⌒

The *Valparaiso notebook* takes its name from the port city of Valparaiso (Valle Paraíso 'Paradise Valley') in central Chile. The notebook is bound in black leather with the border blind embossed: the brass clasp is intact. The front cover has a label of cream-coloured paper (73 × 26 mm) with 'Valparaiso up Aconcagua to St Jago' written in ink. The notebook has 52 leaves or 104 yellow edged pages. The text is written in two sequences, pp. 1a–98a and pp. 1b–6b. The notebook covers 14–30 August 1834. It is the first of six of the more or less square-shaped Type 5 (Velvet paper) notebooks which Darwin used in the field during the latter stages of the *Beagle* voyage.

The notebook was the first used exclusively on the west side of South America. It was first used in Valparaiso, covering the trek up the coast to Quintero, then via Quillota up the Aconcagua valley and a night camped on Bell Mountain, to San Felipe. Darwin then spent five days at the copper mine at Jajuel [Jahuel], thence, via the plain of Guitro, to Santiago, where he stayed for five days and the notebook ends. It thus covers the first half of Darwin's first inland expedition on the west coast and is mainly concerned with his study of the geology of the coast ranges of the Andes.

The first three months or so of 1834 saw Darwin make his spectacular discovery of the fossil *Macrauchenia* in Patagonia, followed by the *Beagle*'s final lengthy spell of surveying around Tierra del Fuego, and her second visit to the Falklands. This was recorded in the *Port Desire notebook*. April to June saw the Santa Cruz river expedition and the sail up the west coast for the first visit to Chiloe, whence the *Beagle* left for Valparaiso where she arrived on 22 July. There is no notebook coverage for about two months, from early June until the opening of the *Valparaiso notebook*.

Valparaiso to Jajuel to Santiago, August 1834

Darwin took up residence with his old school friend Richard Corfield in Valparaiso on 2 August. Darwin recorded in the *Beagle diary* that in the first week in Valparaiso he took 'several long walks in the country' and that he immediately found evidence of dramatic elevation of the coast. He was struck by the low diversity of animals and he conjectured that this was 'owing to none having been created since this country was raised from the sea'.[651]

Unusually, the first entry in the notebook is a precise literature reference, on the inside front cover, to Vargas y Ponce's 1788 account of his 'Ultimo Viage al Estrecho de Magellanes'. This could well be the book Darwin told FitzRoy to expect to receive, via Alexander Caldcleugh, in Darwin's last known voyage letter to FitzRoy of 28 August 1834.[652] The notebook entry itself is impossible to date with certainty.

On p. 1a Darwin wrote the first date, 14 [August 1834] and recorded that he and his guide, most probably Mariano Gonzales,[653] after a 'very picturesque' ride, arrived 'benighted' and spent the night at the Hacienda de Quintero, formerly owned by the British sailor Lord Thomas Cochrane (1775–1860). Cochrane was a great local hero, who had successfully been Commander of the Chilean Navy 1818–22, before moving on to command the Brazilian Navy.

The next day Darwin wrote that the Tapacolas (Tapaculo birds) 'are very numerous and active', and he immediately noted that the granite of the coast range was giving way to greenstone, a lightly metamorphosed basic igneous rock.[654] He found the Quillota valley to be 'A delightful smiling pastoral count[r]y', p. 7a, and 'the picture of fertility; all the land irrigated, beautiful oranges'. He slept at 'a most perfect' hacienda and noted how the 'very tame & abundant rat, lives chiefly in hedges, curls its tail', p. 9a.

On 16 August he thought the country was 'like Wales' and he was disappointed by the Chilean version of gauchos (huassos) who 'do not look as if born on a horse' like their Pampas counterparts, p. 13a. Darwin began to get his first close look at the Andes as he climbed the Bell or Campana Mountain at 1,880 meters (c. 6,200 feet), now part of the Parque Nacional la Campana. There were 'magnificent views' and Darwin wrote in the *Beagle diary* that from the summit Chile was 'as in a Map'. That evening he noted the 'Setting sun, ruby's points against red, sun, black valleys!' and he spent the night on the summit: 'oh for the camp', p. 15a.

The next day Darwin recorded the detailed section and noted that the 'Night Jar emits shrill plaintive cry', p. 22a. The greenstone was in places 'extraordinarily shattered' and 'appear as just broken; the ruins on the greatest scale & evidently the effect of Earthquakes', p. 24a, to the extent that Darwin recorded in the *Beagle diary* that he 'was inclined to hurry from beneath every pile of the loose masses'. He 'Staid whole day up mountain, very pleasant, understand Cordilleras … most interesting are such views, when connected with the reflection how formed', p. 30a.

651 *Beagle diary*, p. 250.
652 CCD1: 407.
653 See the introduction to the *St. Fe notebook*.
654 Described in *South America*, pp. 162, 169–75.

On 18 August they descended back to Quillota and there follow pages of detailed geological notes, including the remarkable observation that heat has 'different action on different layers of sedimentary rocks', p. 37a, and the dramatic conclusion that if 'the gneiss [was] the lowest & most affected rock' the entire range 'has all been once covered!', p. 39a. Darwin referred to the Bell of Quillota in a note alongside a diagram of the Cordillera on p. 150 of the *Red notebook*.

On 19 August they started up the Aconcagua valley to San Felipe. There were 'peace-blossom, orange trees & date palms', p. 41a, and Darwin 'Saw large Kingfisher — long-billed Furnarius — a black Icterus, with orange head', p. 43a. The next day they crossed the San Felipe valley to the copper mine 'superintended by Englishman', p. 46a, at Jajuel, where they stayed five days before heading south for Santiago. Darwin was amused by the 'simple Cornish miner', (one of a number of Cornish minder emigres in Chile) in the *Beagle diary*, where he related that on being told that 'George Rex' (King George IV) was dead, the miner asked Darwin 'how many of the family of Rexes were yet alive'. Darwin added in the *Beagle diary*, that 'this Rex certainly is a relation of Finis who wrote all the books'. Darwin was shocked to see the conditions under which the Chilean miners slaved all day for '5 dollars a month', p. 46a. He gave a full account in the *Beagle diary* and explained that the ore was shipped to Swansea for smelting. He noted that lions (pumas) sometimes 'kill men' and the Tapaculo was 'well adapted tail erect, hop, very fast, large one most ridiculous', pp. 47a–48a.

On 22 August Darwin measured a large cactus before recording the geology for many pages. On p. 59a there is a circled note 'like fragments Cape Town' which is almost certainly a later insertion. Darwin was almost convinced that the valleys were once under the sea: 'The mist well represents the sea in the basin & showed probability', p. 61a. He recounted that 'It is said Lion if he covers his prey returns, if not, not', p. 63a, and that concerning a lake in the mountains the locals 'say it is arm of sea'. Darwin noted that 'Biscatcha shrill repeated noise Stony place connected with habit of collecting sticks and stones', p. 64a.

The next day Darwin described a snake, then drew a remarkable geological diagram, pp. 66a–67a, which as he explained on pp. 68a–72a depicts a dyke cutting across 'jaspery rocks and breccia-conglomerate' and also apparently cutting across a complex of smaller dykes or veins. Darwin often spelled the word 'breccia' with only one 'c' in this notebook. On p. 73a he noted how the locals obtained huge quantities of sap by carefully felling a palm tree. On the 24th his geologizing was restricted by snow and on the 25th after making a list of representative rock specimens he 'got into a waste of snow — up horses belly — difficulty of returning, clouds threatening day began to snow heavily should have been shut up'. The travelling was probably very unpleasant.

They left Jajuel the next day to cross the Cerro de Talguen. The clear air and the fresh snow created a 'Most magnificently splendid' view of Aconcagua, which, at

6,960 meters (22,840 feet), is the highest mountain outside the Himalayas. The rocks dipped dramatically 40° to the southwest, indicating great upheaval. They slept at a 'small Rancho', where the servant was 'very humble' and would not eat with Darwin. On the 27th they reached the Llanos of Guitròn, where Darwin found 'immense quantity of petrified wood'. There were sandstones and limestones, as well as the usual porphyries, and Darwin noted the acacias and mimosas growing in the 'beautiful' valley on the approach to Santiago. He thought the entrance to the city 'very splendid', p. 85a.

Darwin stayed in Santiago until 5 September, before setting off southwards to complete a circuit back to Valparaiso. While in Santiago he recorded the geology around what is now the old part of the City, including the Cerro of St Lucía, at the foot of which Pedro de Valdivia founded Santiago on 12 February 1541. These notes eventually formed the basis of the description in *South America*, p. 59. On pp. 92a–93a Darwin declared that 'The general form of country convinced me that these hills & places have all been formed by water.'

On 30 August Darwin rode to the 'half leather & chain' bridge over the Maypo, p. 96a, but it seems from the *Beagle diary* that he may not have actually crossed this 'miserable affair' until 5 September, which is covered in the *Santiago notebook*. There are a few notes at the back of the notebook, including what seems to be a list of local names of mammals such as 'Guanque' against 'Mus cyanus', but sadly the top of the list has been torn off, p. 1b. There are two more pages listing rocks which allow a tentative dating of the notes to 18 or 19 August as Darwin's last comment is that they 'correspond to what I saw yesterday'.

[FRONT COVER]

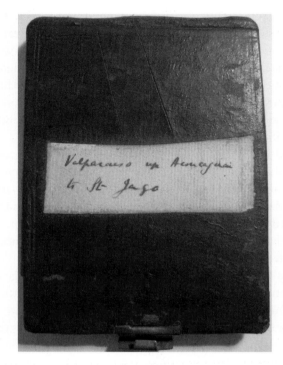

Valparaiso up Aconcagua to St Jago

[INSIDE FRONT COVER]

Ultimo Viage al Estrecho de Magellanes. S. Maria de la Cabeza 1785–86.
Extracto de todos los anteriores & MSS. A.D. 1788.[655]
6 inch in diameter

[1a] 14th [August 1834] — Started, slept at Quintero; pretty very green lawn — scenery some of the views very picturesque, benighted at last we reached the house. — The gneiss seemed to run in a NW & NE line but very irregular. The beds of shells are

[2a] at but a small elevation; they chiefly consist in Monodon.
Rusty Red & Black Muscicapa.[656]
T del F Creeper.[657]
(15th) [August 1834]
The Tapacolas are very numerous & active, their crys incessant & varying run quickly from bush to bush. —

655 Vargas y Ponce 1788.
656 Specimen not in spirits 2208 in *Zoology notes*, p. 406.
657 See *Zoology notes*, p. 279.

[3a] In approaching the Sierra de Chilicauquen, the granite rock disappears & in its place greenstone rocks appear. — These consists generally of a compact slate coloured bases with slaty structure like (A*) which sometimes contains

[4a] acicular crystals of glassy feldspar, at othertimes common feldspar. — The bases in many places become granular & crystalline then earthy without crystals so that having a slaty structure, which is common to much of these

[5a] rocks, it appears like siliceous green altered clay—slate — I found in two places the cleavage running SSE & NNW — Perhaps the most porphyritic varieties were at the

[6a] upper parts of ridge; this Sierra runs E & W. — As far as I can see there is nothing to lead me to suppose this mountain has been formed differently from the Cuestas near

[7a] Valparaiso, ∴ by sea. —

A delightful smiling pastoral county, cottages scattered in all parts; in the valleys beautiful evergreen forest trees. — The vally of Quillota, the picture of

[8a] fertility; all the land irrigated, beautiful oranges. Slept at a most perfect Hacienda[658] of English Merchant.

the course of the River a remarkable scene of violence from the water but apparently only rolling over stones when snows

[9a] melt. — A plain of shingle.

The Rat is very tame & abundant, lives chiefly in hedges, curls its tail.

(16^(th)) [August 1834]

In ascending, first Granitic, Quartz & White feldspar. — Altered Slates & Porphyritic greenstone, alternating many times, Slate with

[10a] metal, greywacke with water lines, Brecia, & red Limestone, particles blending together, angular cleavage: then directly, fine porphyritic sonorous greenstone, a reddish grewwacke passing into porphyry.

[11a] (Red Limestone with contained bed of compact clay slate)

generally red & some green porphyry, red porphyry sometimes with apparent fragments as at P Desire. At the base of El Morro the mountain

[12a] appears stratified, beds of altered slate dipping to E by N — therefore under the rough summit of Mountain.

Town of Quillota each house with Garden, send produce to ~~Chiloe~~ Port. Country

[13a] like Wales with broard flat valleys: — Gauchos not like true men — can walk & climb, are not gentlemen, do not look as if born on a horse, & eat bread & potatoes. enormous stirrups, spurs

[14a] with many rowels, diameter length of this page. — Have not Chilipa. — worsted boots untidy laso, no bolas![659]

Ascended mountain 4000 feet, very bad road, magnificent views, brush wood on North side

---

658 Spanish for ranch, estate.
659 Balls attached to leather thongs which the gauchos use with a whirling motion for catching game or cattle.

[15a] Bamboos on South Palms 3500! Setting sun, ruby's points against red, sun, black
      valleys! Fire amongst Bamboo a little harbor very pretty — most comfortable evening,
      oh for the camp.

[16a] Bamboos 15 feet. — Could see ships. Condors ~~have~~ sleep in cliffs.
      In Baldiva, <u>petrified</u> shells, — white stone on river bank
      Cancagua: Agua del Obispo, the name of place

[17a] 17th [August 1834]
      (A) Breccia, often containing large fragments here, small very compact, red calcareous?
      basis
      (B) — Red Limestone, do. Situation, low in the mountains
      (C) Porphyritic, sonorous greenstone, lying close to above *[rocks]*

[18a] (E) & (D) — Varieties of what I call altered slate alternating with various porphyritic
      greenstones.
      (F) Breccia altered & becoming Porphyritic
      (G) Tolerably perfect Porphyry *[more]* crystalline varieties, compose great part of
      base of

[19a] mountain
      (H) — a white Porphyry with copper
      (J J) Greenstones composing whole summit of the Bell
      (K) do with vein of green mineral, this is so abundant as to form masses —

[20a] (L) A mineral in a small mine very summit of mountain
      (M & N) — the great slate, (much altered, rather sonorous & little, which underlies the
      greenstone summit.

[21a] Bearings from 2d N Point of Campana — North highest hummock of the Patillos N 32
      E Aconcagua — N 64 E A The North high point of several peaks N 88 E South extreme
      of do distance from each other S 78 E (Between these bearings came S de Palmas Small
      Group of Point S 65 E

[22a] Large Julus,[660] emits yellow fluid which smells like mustard. —
      Night Jar emits shrill plaintive cry —
      our resting place is called Agua de Guanaco.
      The whole top of mountain is fine greenstone, very steep & precipitous & lies

[23a] directly on the altered slate which dips under it —
      Greenstone with much green mineral some solid masses; the surface is extraordinary
      shattered like the T del Fuego Mountains but with this wide difference, that very many
      large

[24a] *[Masse]* <u>appear as just broken</u>; the ruins on the greatest scale & evidently the effect of
      Earthquakes, must wonderfully lower a mountain. I do not understand the
      superposition of the greenstone

---

660 A genus of millipede. See specimen in spirits 1058 in *Zoology notes*, p. 353.

[25a] on Slate. —

This slate extends whole foot of Mountain. All lower parts of mountain a mingled mass of greenstone Porphyries & altered slates: Brecias:

One of the highest mountains. The Grand Campana runs N by W & S by E; the small do. Connected by small oblique ridge; to south not continued, perhaps

[26a] are to North across valley of Quillota. —

Sierra de las Palmas (deserves name) runs same direction, (as does Cordillera approaching coast), to South is continued in mass, connected by high transverse range, & which I suppose from Zapata & Prado — Chilicauqua

[27a] is all small, perhaps N & S chain, much cut up by valleys. —

But seeing such immense flat valleys & many of them as Aconcagua & the side valleys making principal hills like stars, it is impossible to say what was original form of this land, it really

[28a] might have been table land —

(Sierra Palmas to the East higher than Campana) Cordilleras, remarkably level topped, parallel to lower line of snow, over which road must pass, does not look like roof of house

[29a] Some greenstone dykes?

but a small serrated plain, with here & there (4 sets) groups or parts of one Volcano. These hills by no means appear branches of the Andes —

Staid whole day up mountain, very pleasant, understand Cordilleras, mine — most interesting

[30a] are such views, when connected with the reflection how formed — Warm — snow — different vegetation from England. — Too dry for insects (Greenstone summit 800 ft thick)

(18th) [August 1834]

Descending in the most direct line, a very steep side there is much of what I call the altered

[31a] Slate of which (O) is another specimen — this rock has a distinct cleavage, & coloured bands all parallel — I was surprised a parallel layer, with a gradual passage of the two rocks of greenstone (P) Again altered slate do with many parallel layers of the green

[32a] mineral so common at the summit:

Above many 100 feet thick.

There is a great fragmentary of a red color & more or less crystalline. (2) & 1 represents the ~~finer~~ less coarse sorts, in the more crystalline varieties (R) Much green mineral is arranged in layers — the imbedded fragments themselves

[33a] seem to have been altered becoming porphyritic

Again altered slate (all these beds same cleavage), soft porphyritic greenstone & green mineral.

We have a great mass of dark purple porphyry. Specimen (S) is not so crystalline as some, ~~nor shows there~~ & than other parts.

This forms the main

[34a] basis of mountain, has entirely lost all cleavage passes into & contains much pale Porph (T) & pale purple Porphyry (V) ~~Beneath~~ There was also a fine greenish porphyritic greenstone, but at this point the section finishes, for the road followed a winding as descending course, amongst

[35a] the hills at base of mountain, generally however it continued Porphyry — At the very basal parts some greenstone & a sort of gneiss, decomposed, without mica. —
It is clear all these Porphyry, slates, are the same, which appeared still more so in the undulating & ascending

[36a] road — in my descent I did not meet with so much Porphy — greenstone as in ascending
This descent was at right angle to direction, towards Escarpment. —

N — greenstone — S
I have to remark in the lower part no where have I seen such glassy greenstone as at summit —

[37a] I feel no doubt that the appearance of whole above mass of rocks is owing to heat having different action on different layers of sedimentary rocks — (I omitted to state that much of the lower porphyry, consists in the basis alone of harsh texture, like what

[38a] I called Limestone in which there have been worked Gold mines) — how else can we account for the layer of greenstone: Is all the Porphyry grewwacke altered, I feel no doubt it is in this case, — Is the upper greenstone flowed

[39a] out through N & S — Rent, tilted rocks on each side, hence a vertical cleavage would dip towards the hills: Old puzzle comes into play — Is the gneiss the lowest & most affected rock? If so has all been once covered! But not all metamorphic, hence *[&]* moved only to where greenstone has burst
With respect to great

[40a] valleys, Mem Pat — pebble bed. — Perhaps in Pacific if seen, wonder would be reversed. —
in our return we met with many pretty spots, & beautiful clear brooks & fine trees. the little clear spots were not infrequently the vestiges of some old mine, the whole country may be

[41a] said to be *[drilled]*. — Slept at same Hacienda.
(19th.) [August 1834]
Proceeded up the valley — remarkable agriculture, aquaducts, peach-blossom, orange trees & date-palms.
the rocks we saw were chiefly much altered slates, red Porphyries, in some places with apparent breccia — much very

[42a] fine & largely porphyritic greenstone —

Also there was a good deal of gneiss without mica which alternated with an imperfect greenstone; some of the high hills, from color porphyry — (There must be a transverse ridge to the E. of the S. de Palmas although I could make but

[43a]

little of them.) // — Saw large Kingfisher — long-billed Furnarius[661] — a black Icterus,[662] with orange head. — //

Late in the evening found a house to receive us.

(20.th—) [August 1834]

The side of one of the mountains to the

[44a] West of the Basin of Aconcagua, is composed of nearly horizontal beds (or slight dip to NW) of fragmentary rocks with particles of Crystal Lime (old shells?) & partly porphyritic (much coarsely brecciated). (W) fine variety

[45a] with a large bed, intermediate of a greenstone (X) —

The basin of Aconcagua most clearly marine with Islands —

Esteros: The town of S. Felipe very large straggling like Quillota about 5 leagues up

[46a] in a crack in the very Andes — arrived at a Copper mine. — superintended by Englishman.

21st [August 1834] —

Wages of men 5 dollars a month; work from light to light summer & winter hard work, carrying stones, very short time at meals — Food given them 2 loaves & 16 figs from breakfast

[47a] Beans for dinner, broken wheat grain for supper. —

Veins run to E & W — Petorce Coquimbo — or rather to N of E (?) —

Lions more savage kill men

Little tufted bird, nest deep simple full of feathers

Tapacola well adapted tail erect, hop, very

[48a] fast, large one most ridiculous — —

Mines here, so quiet, not like England — No engine. — Miners live Entirely in mines

Mine laws; any one by paying 10 Rials can open a mine. May try for 20 days anywhere

Was about the family of Rexes, simple Cornish miner —

[49a] The hill North of houses, composed of Feldspathic altered slates — fine grey greenstone, both of which become coarsely porphyritic with Feldspar.

These rocks contain many large veins of copper ore & iron /plane/, adjoining rocks appear always more or less altered, there is

661 See specimen 1467 in *Zoology notes*, p. 247; listed as *Opetiorhynchus patagonicus* in *Birds*, p. 67.

662 Specimen 1784 in *Zoology notes*, p. 247; listed as *Agelaius chopi* in *Birds*, p. 107.

[50a]   on the adjoining hill much red fragmentary rock & some do porphyry these are said never to contain veins. To the South of houses an abundance of <u>large</u> 4 or 5 inches & most decided breccia, all the included pieces porphyritic, & some Conglomerate, there is also more jaspery &

[51a]   rocks such as <u>Limestones</u> of Campan[a] — arranged in nearly horizontal strata (pebbles of quartz) the lines, seen in vertical face incline to North. (& W ?) (Mention specimens): these rocks have angular cleavage & sonorous & pass in every stage from decided breccia to true Porphyry — (& the red Porphyries pass into the

[52a]   Porph – greenstone in the Campana). — Here the following succession is seen — Red compact — Fragment — Porph – greenstone fine — — Red fragmen[t] & red Porph — Porph greenston[e]. Red — Porph — & some very fine sonorous greenston[e] lying on a Porphyritic altered Slate

[53a]   (<u>Mem</u>: at Campana I do not know that the <u>beds</u> of Red Frag. dipped with slate). They had sort of cleavage.
at Petorca much conglomerate.
(22^d) [August 1834]
6 4 — H — 1.7
3.2 12 15
Cactus depressed globular circumference 6^{ft} " 4^{in} height .1.9
Common size of cylindrical sort 3.7 — height 12 — 15

[54a]   Campana S ~~28 E~~ 35 W
~~N 54 E~~
Volcan de Aconcagua or N, part of great irregular mass S 62 E
Great Volcano N 18 E
The spot intermediate between high peak & Volcano
The hill is <u>NW</u> of mine, higher than Campana, it lies a little to the South of line from Campana to Great Volcano

[55a]   I can hear of 3 volcanos in these parts — in front of Juapa — Patos where there is a pass: Aconcagua — Another at Bunsters Mine.[663] Another other side near Mendoza, ashes. —
14 leagues N. of R. Juapa, hill of Teniente, land locked harbor vessel lost near it —

[56a]   Good water in Pechi languel, 1 degree North of Valparaiso — Bunster murdered
In ascending above mountain, much altered slate, with needle of glassy feldspar, some pale porphyries with acicular Hornblende no mica (this is rare) pale greenstone, porphyries & much fragmentary &

[57a]   puddingstone rock. — At very summit, these breccia were closely joined by vertical suture, (which imparted color) to altered Slate (or rather a very fine greenstone). This I thought had burst up & hardened breccia, found however, itself penetrated by narrow tapering veins of Breccia, (

663   Humphrey Bunster, a Cornish émigré, had mines at St. Pedro Nolasco.

[58a] (much porphyritic) which proceeded from a mass which also appeared like dyke, in other cases the two mingled together, each in turn looked as if melted & protruded both must have been softened & jammed together. —

[59a] Found altered slates running N by W̶ E to N N W̶ E agreeing with some ridges. — (Cordilleras clearly trend to West) All the lower parts of neighbouring mountains & great <u>Cordilleras</u> may be seen to be divided or stratified these often appear nearly horizontal. N B. **B**aked greenstone porphyritic like fragments Cape Town

[60a] or dipping somewhat to NW — I found one bed lower at 10° to EW by W but this is too much

? 80 3300

The geological construction of hills all sides of Basin of Aconcagua same with this same inclination. — Has the

[61a] N. dip any connection with the Volcano? —
The mist well represents the sea in the basin & showed probability
It is remarkable the Breccia & Slate having here vertical junction in this one place; must be owing to disturbance during baking

[62a] The N & S ridge were visible, the transverse chain of Chacabuco shows connection with m̶o̶u̶n̶t̶a̶i̶n̶s̶ Andes — This mountain forms a part of do —
Broad flat valleys seem only to have disrupted & found the N & S ridges. —
There was some pretty perfect slate without any crystals!!

[63a] It is said Lion if he covers his prey returns, if not, not. — In one Hacienda in one year killed 800 young a̶n̶i̶m̶a̶l̶ cows
Ancient Indian buildings in mountains
Where did all fragments come from in red & globular Breccia ??? old range of Andes.

[64a] Lake, people did not like to open it. Padre say it is arm of sea
Molinas statement, Culpeu[664] not Falkland Isl Fox

---

664 Listed as *Canis magellanicus* in *Mammalia* p. 10, plate 5: '[This animal] is mentioned by Molina in his account of the animals of Chile, under the name of Culpeu, which he supposes to be derived from the Indian word "culpem," signifying madness; for this animal, when it sees a man, runs towards him, and standing at the distance of a few yards, looks at him attentively. He adds, although great numbers are killed, they do not leave off this habit. Molina states that he has repeatedly been a witness of this, and I received nearly similar accounts from several of the inhabitants of Chile: yet I must observe, that the people of the farm-house, where my specimen was killed (after it, together with its female, had destroyed nearly two hundred fowls) bitterly complained of its craftiness. From this bold curiosity in the disposition of the Culpeu, Molina thought that it was the same animal as that described by Byron at the Falkland Islands, but we now know that they are different.' *Mammalia*, pp. 11–2.

Biscatcha shrill repeated noise stony place connected with habit of collecting sticks & stones

(23ᵈ) [August 1834]

N & S line of igneous action shown in T. del Fuego: Height of plains at S. Cruz. horizontal elevation & Mendoza

Ref. hor: elevat: going S. Ward. — for /deposits/ look /more earlier/

Central band greenish amber /br./ edged with yellowish do. /shading/[665]

[65a] into grey — belly blueish

Snake. — foot of Cordilleras

Mʳ. J. Murray, has given notice to Ro Soc. respecting Luminous property of glow-worms[666] — — same result as me

Barranca with shells — Las Vacas — gold mine — Conchilee (near Petorcia?) road to Coquimbo

[66a]

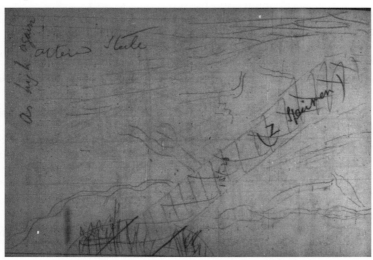

[section, continued on following page] As high again Altered slate (Z specimen)

---

665 Possibly the snake listed as specimen in spirits 1054 in *Zoology notes*, p. 353.

666 John Murray read a paper entitled 'Experiments and observations on the light and luminous matter of the *Lampyris noctiluca*, or glow-worm' before the Linnean Society (not the Royal Society as indicated by CD) on 18 November and 2 December 1823 which was widely reported in scientific periodicals. Presumably CD refers to one of these reports. The paper was later published as Murray 1826. This line was copied into CD's *Edinburgh notebook*, Barrett *et al.* 1987, p. 477. This notebook is fully transcribed by Rookmaaker on *Darwin Online*.

[67a]

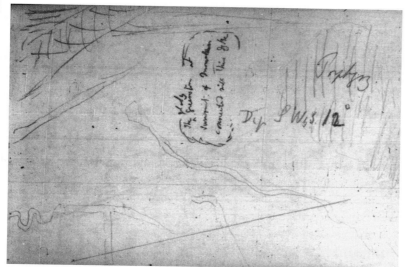

[section continued from previous page] The slaty greenstone at summit of mountain connected with this dyke
Porphyry Dip SW by 12°

[68a] This drawing [pp. 66a–67a] represents a mass of bright red jaspery rocks & breccia — conglomerate dipping at ∠ 12°–15° to /about/ SW by S. (The rock is fragmentary in patches more than in planes) This is penetrated in every direction by dykes, zig-zag — strata-shaped or of every sort of grey, slightly porphyric greenstone: the main dyke is from 15 to 20 ft wide, & rises from a pap of similar

rock — in the smaller veins,   such sides prove it is crack opened & filled, edge tinged purple, but quite distinct & fine; numerous projections show much melted matter has not

[69a] flowed through it — the drawing does not represent nearly all or the intricacy of these dykes; rock not particularly altered: These beds reach 3 or 400 ft high

above them is a mass of equal height of rock like the dykes, in every respect, excepting in having few crystals of feldspar — this is a conformable mass to the strata of breccia: the junction of the two rocks is quite gradual in colour & nature: the main dyke I believe seems to penetrate it, equally

[70a] with the Brecia but I could not prove it: till I had noticed these facts I believed it to have been melted mass which had flowed through the cracks; but the above reasons show it to be altered bed. in same degree as Breccia is altered. (Quartz pebbles might be seen

in Breccia) at no great distance <u>down</u> the stream I found the Breccia reposing on conformable bed of rock like the <u>dykes</u>? whether dyke or bed: I believe latter? — On Rt Hand there is a mass of pale purple porphyry, with fine

[71a]   semi = cryst. of Feldspar, which is similarly is covered by the altered rock — ~~Perhaps~~ In this the Breccia structure could only in one place be distinguished. Perhaps this is protruded mass of <u>melted</u> inferior Brecia, breaking through the semi-porphyritic brecia — as the melted plates have cut through both breccia & altered slates. (We must recollect the vein of Brecia yesterday). — I see no way of distinguishing greenstone produced in situ & ones that have flowed

[72a]   perhaps this may hold good to Porphyries. — (it is likely, that some of the <u>beds</u> of greenstone have been dykes): — The Brecia being traversed by greenstone as well as Gneiss interesting: — (a mass of Breccias & Slates to a thickness of <u>at least</u> 5000 ft have been baked & altered. Mem: yesterday): This West (+ N or S) dip caused by the protrusion of dykes, same force which has long been sent forth Lavas in Andes. — (This is same hill. East base, which I ascende[d] first day on South side: on other

[73a]   side of ravine, on Cordilleras corresponding altered rock, contains very much copper — (The breccia is not <u>particularly</u> altered <u>close</u> to dykes)
to make Mel. Palm tree cut down in August. must fall upwards — slice cut off every. from beneath the truncate head — flows for many Month, juice concentrated by boiling one good tree will give 90 gallons /~~annually~~/ of sap. when sun hot most
one estate has counted many hundred thousand trees on it near Petorca. — nuts & /roofs/. —

[74a]   the great use of English money has been reducing by air furnaces the /Bronce of/, ores, & cleaning the scoriae. — Air furnaces & stamping More lines. —
(24th) [August 1834]
In the hill facing yesterday section, the part corresponding to the altered slate is a greenstone or rather porphyry. contains great veins of largely crystallised Carb of Lime parts of this mass become breccia with Copper

[75a]   & little gold. — the Lime is in lines as if deposited by water —
unfortunate full of snow threatening weather for some time — weather here is very regular — 25th [August 1834] —
Series of specimens from this places
(1). for <u>small</u> spec. well illustrate Brecia structure, also much porphyritic
(2 & 3) Very common sorts, where

[76a]   Breccia structure is yet evident but otherwise considerably perfect Porphyry
(4) — Pale red porphyry, with crystals of quartz, <u>common</u> here but otherwise uncommon
(5) a curious sort of Porphyry not abundant (saw one hill)
(6) The least crystalline rock I have ever met with in the mountains. bed near Breccias
(7) The "Altered Slate"
(8) Fine grained greenstone, often becoming Porphyritic, is very

[77a] abundant in all parts —

But the main rock is the Breccia — I see in the Main chain this dip to the NW & contains much crystals of quartz. I see the hill in front of the "section" is streaked with Trap dykes. —

Tried to reach Laguna — got into a waste of snow — up to horses belly — difficulty in returning, cloudy threatening day began to snow heavily

should have been shut up.

[78a] 26th [August 1834] —

Charles /account/ of 15 leagues of /fire/ Crater south of Aconcagua

Took my departure from the mine. Snow almost down to house. — Most magnificently splendid the view of the mountains: I see that one of the most isolated islands is composed of Breccia which dips to SW, the usual method (will apply to origin of Basins) which helps to show that the greater part of this

[79a] space (inclining to seaward) perhaps 10 miles across has been cut by the sea.

Crossed the Sierra Talguen, saw a mass of beds dipping at about ∠ of 40° to SW. — this great dip shows that the upheaval has not always been so regular. This ridge seems

[80a] continuation of East & West Chacabuco range — could see from it, about 5 parallel N & S ranges (NB the lower beds in Aconcagua basin seem all Breccia). — Could trace the form of one of two Volcanos to S of Aconcagua)

Slept at very small Rancho — people much more different grades of

[81a] Life. servant not eat with me — pay every-where. man very humble about his country, some see with both eyes — &c &c —[667]

(27th) [August 1834]

The road lay over a more rounded undulating country (with patches of soft coarse sandstone) & the rocks /there being/ not of so determinate a character, there was (no^s) 1 & 2 — Common varieties of greenstone

[82a] however some Breccia — Porphyry & the high hills evidently composed of it. —

We arrived at Llanos of Guitron, near Porparco. In a valley ~~flanking~~ leading into it, immense quantity of petrified wood like at Pisada. — enough to fill Cart, mixed with scattered ∠^r fragments of Porphyries &c

[83a] Llano — dips to Seawards very level, covered with Espina & Algarroba, two species of Mimosa's now without leafs: Saw Limestone. Black colour conchoidal fracture, crystalline

(N. 3)

beds dipping to about West about, about ∠ 10° — This plain is separated

[84a] by the ridges composed Limestone, associated with usual porphyries from the great St Jago plain: perhaps former rather lower but not much — view very striking large lake. — forest of Acacias. — green turf isolated hills — & magnificent Cordilleras —

667 'We crossed the Cerro del Talguen, & slept at a little Rancho. The host, talking about the state of Chili as compared to other countries, was very humble; "Some see with two eyes & some with one, but for his part he did not think that Chili saw with any"'. *Beagle diary*, p. 256.

[85a] with various lines of Clouds beneath summits (plain of land more beautiful than sea. —

All this country is a sort of valley between Andes & great range of which Prado is termination? — St Jago plain dips seawards. — Entrance of city very splendid.

[86a] 28th & 29.th [August 1834] —

The Fort of St Lucia is a hillock composed of sonorous, conch. fracture. Slate colored porphyry (?)

Specimen 4 . it is roughly divided into circular masses, which are partly subdivided into columns these dip about $\angle$ 45° to SE W~~ obliquely to to the horizon.

[87a] on the Eastern side the column ~~finest~~ pass into by globular concretions of a coarser greenstone This knoll has a singulary abrupt appearance in the plain; in the same line however with its ridge there is a low ridge or peninsula in former

[88a] sea, which runs about NNE ½ E & SSW ½ W — the greater part of which is composed of the same rock, on the part however the columns dip in directly opposite point to the St Lucia & in another

[89a] place rather toward this hill. —

At the foot there are some globular concretionary decomposing mass of iron greenstone (N^{r}5) which I suppose is directly connected with the Porphyry — on the West side there is

[90a] some of the pale purple porphyry. — the rock which so many reasons lead me to suppose has been an altered Brecccia in site I imagine from a crack has issued a mass of greenstone through which a grewwacke — which

[91a] in the case of St Lucia has been entirely removed when its present insulated form was given to it — Also near the base where the [Quartz] exists there is some of the same porphyry (Spec: 6) this has formed an island & is a remnant of general covering

[92a] of breccia — porphyry — (NB. — I see in the building in the town a rock precisely similar to that at P Desire — where a sort of Porphyry has water lines. —)

The general form of country convinced

[93a] me that these hills & places have all been formed by water. I was pleased to find on the top of St Lucia & St Chrystophal, a mass of Breccia (& at St Lucia, a variety of rock which is not found on summit) united

[94a] & coated by layers of white, friable light Calcareous Matter which (spec 7) evidently has

been deposited by water St Christophal ▨ feet above plain. The Quartz Hill, Cerro blanco, is partly composed of a

[95a] white Feldspath base with crystals of do — (Spec 8) of which the Cathedral is built cemented by similar, Calcareous Matter — this rounded hill of Brecia, must once have formed part of a large mass — before it was rounded & modeled

[96a] to its present form

(30th) [August 1834]

Rode to bridge over Maypo: half leather & chain, curious evidence of negligence. — plain not really level, but some undulation corresponding to great valleys in Andes might be expected: on approaching Maypo appearance of old river lines would if taken alone make one suppose it all formed by river this part might have been remodeled. Barranca there consists of nothing but large rounded stones: mixed with *[illeg]*

[97a] blocks & some lines of coarse sand. origin dubious — no organic remains could be expected, & from mountains in basin such a bottom would be formed. — On the plain on which *[only]* stands road to Valp. inequalities great part composed of reddish sandy clay, firmly aggregated containing many pebbles in parts & in others passing into white sand: in all parts much pumice. *[Tupungato!]*

[98a] Specimens of porphyritic greenstone, from near St Jago. First range of Cordilleras near here. Fragmentary rocks not so plainly seen as in some other parts alternating with *[great]* dip WSW $\angle$ 15: could see many dips, & one apparently towards mountain on large scale some high up rather horizontal: many greenstones & N^r 5 colina — mountain plains. (F, Fish, Rapel) St Lucia; section of this plain — Limestone at St Christophal — Big Animal Note Book & whipp *[illeg]*

[BACK COVER]

[INSIDE BACK COVER]

C. Darwin
H. M. S. *Beagle*

Gypsum
Biscatcha
Mice
Big Bones
Shells in plain of town
Earthquake 22
do Green & Brec pass into each other
Petorce
Las Vacas Conchilee
Cuy. Lepus minimus[668]
Puda. — Chiloe deer ?[669]
Guemul — Equus bilsulcus.[670]

[1b]   Gato del Mar[671]
Chinchimen. c<*excised*>
Guillino. Castor
Coypu — Nutria[672] ?

668 See *Port Desire*, p. 79.
669 The wild deer of Chiloe, *Puda puda*.
670 The heumul or southern Andean deer (*Hippocamelus bisulcus*) was first described and named *Equus bisulcus* by Molina 1794–5.
671 A marine otter, specimen 2529 in *Zoology notes*, p. 280; listed as *Lutra chilensis* in *Mammalia*, pp. 22–4.
672 The coypu or nutria is a large, herbivorous, semi-aquatic rodent, see specimen 2530 in *Zoology notes*, p. 280; listed as *Myopotamus coypus* in *Mammalia*, pp. 78–9.

Chingue. Vivena, Zorilla ?

Cuja & Quiqui. Mustela[673]

Porcupine: Culpeu, large Fox

Guigria & Colo Colo. Cats

Guanque. Mus cyanus

Degu. Sciurus[674] [list concludes with three further lines on IBC, above]

[2b]   Crater (/Maypo/ in Aconcagua

     *<excised>*phyry

[3b–4b excised]

[5b]   (Spec)

Great mass of altered Slate. band of Greenstone — (Spec) Slate

Greenstone slate & green mineral —

Red fragmentary mass (Spec) Lower parts with green mineral in lines, & more crystalline (Spec?)

Slate soft Porph greenstone & green mineral. —

[6b]   Darker red fragmentary mass becoming very crystalline

Spec is half way between most crystal & most fragment

Such crystalline varieties with some pale porphyries Form whole base of mountain —

These clearly correspond to what I saw yesterday

Textual notes to the *Valparaiso notebook*

[IFC]   1.15.] *Down House number, not transcribed.*

      88202335] *English Heritage number, not transcribed.*

      7 8] *added by Nora Barlow, pencil, not transcribed.*

      Aug. 14, 1844] *added, by Nora Barlow? '1844' is a mistake for 1834, not transcribed.*

[3a–13a] *pages written perpendicular to the spine.*

[18a] *there is a dark brown stain on this page, obscuring '[more]', and mirrored on the following (facing) page.*

[59a] N B.] *ink.*

[64a] Ref .... /more earlier/] *ink.*

[66a] *page written perpendicular to the spine.*

[67a] *sketch drawn perpendicular to the spine.*

[97a] *The bottom right hand corner of the page has been excised.*

---

673 Mammal names copied from Molina 1794–5. Cuja: the lesser grison (*Mustela cuja*). Quiqui (*Mustela quiqui*) a species of weasel. Mustela: *Felina plantis.*

674 A genus of squirrels.

[98a]   colina … *[illeg]] upside down from other entries on page.*
        The top right hand corner of the page has been excised.
[IBC]   Las Vacas Conchilee] *written perpendicular to the spine.*
        Cuy. Lepus … bilsulcus.] *upside down from other entries on page, continuation of a*
        list on the opposite page 1b.
[1b–2b] *lower half of leaf excised.*

THE *SANTIAGO NOTEBOOK*

⌒

The *Santiago notebook* takes its name from the city of Santiago (Saint James or 'St Jago'), the capital of Chile. The notebook has black paper covers, with black leather spine, and originally contained 69 leaves or 138 pages. No watermarks have been found. Both covers have c. 18 × 45 mm cream-coloured paper labels with 'Santiago Book' written in ink. There are 89 originally unnumbered pages (pp. 1–89), followed by 34 numbered pages (pp. 90–124), and a different sequence, several excised, at the end (pp. 125–38). The dates in the notebook range from 2 September 1834 (p. 2) to 27 September 1834 (p. 67) and some dates from February 1835 ending April or May 1836.[675] This is probably the most closely studied of all the *Beagle* field notebooks. It has been quoted from by many scholars in connection with the famous passage which confirms that Darwin thought out his coral reef theory before seeing a coral island, as he claimed in the *Autobiography*, p. 98. Partly for this reason, the *Santiago notebook* has been displayed in recent years in some of the world's great museums.

The earlier theoretical pages of the *Santiago notebook* record Darwin's transition from a view of crustal mobility dominated by his observations of elevation of the continental mainland, to a literally more balanced view, in which he came to see the oceans as predominantly areas of subsidence. By the time Darwin sailed across the Pacific and Indian Oceans, and was re-crossing the Atlantic, he was equating elevation with volcanoes and subsidence with coral reefs. These he linked respectively with the Tertiary and Secondary eras of geology of Europe, and was beginning to see how the 'Geology of the whole world will turn out simple'.[676]

David Stoddart (1976) reviewed the background to the scientific theories of coral reef formation prevalent at the time the *Beagle*'s orders were drafted, and, fortunately for Darwin, these theories were ably summarized for him in the second volume of

---

675 Sulloway 1983 dated the first use of the *Santiago notebook* to about 16 August on the basis of a mention of the noises made by Biscatches. However, this note could have been made in September, at the start of which month the *Santiago notebook* was in use. There is a reference to the noises made by these animals in the *Beagle diary* for 16 August, but this has its origin in an almost identical notebook entry for 23 August on p. 64a of the *Valparaiso notebook*.

676 *Red notebook*, p. 72; see Rhodes 1991.

Lyell's *Principles of geology*.[677] The development of Darwin's brilliant theory for the origin of coral reefs, starting with the passages written in the *Santiago notebook* before the *Beagle* departed South American shores, has been analysed in some detail by Sulloway 1983. Darwin mentioned his theory in a letter to Caroline Darwin dated 29 April 1836.[678] The *Santiago notebook* coral passage is analysed in an appendix to the first volume of the *Correspondence*, and by Herbert 2005. Armstrong 2004 and Herbert 2005, in particular, have also shown how Darwin, having seen some reefs in December 1835 and from extensive reading, worked up his theory into perhaps the first of the synthetic essays from the voyage to be almost ready for publication. This essay, headed 'Coral islands' was first published in Stoddart 1962. The essay is now in DAR 41.1–23, and formed the basis for one of Darwin's first scientific papers, which in turn developed into his classic book *Coral reefs*.[679] Stoddart 1976 also reviewed contemporary reactions to Darwin's theory and Rosen 1982 took this further to place the theory in the general context of global tectonics.

The *Santiago notebook*, as the lone example of Type 6, is physically unlike all the other field notebooks. The manufacturer's label shows that the notebook was made in France. Darwin may have purchased the notebook in Santiago. This is suggested by the fact that Darwin had already filled the *Valparaiso notebook* by the time he reached Santiago and would have needed another notebook for the rest of his circuit back to Valparaiso.

There is one other feature of the *Santiago notebook* which sets it apart physically from the other field notebooks. All of them have a paper label gummed to the cover on which Darwin wrote in ink the names of the places described in the notebook. In the case of the *Santiago notebook* there are two labels, one on each side, each bearing the name 'Santiago Book'. Darwin seems to have added identical labels to both sides of notebooks used after the voyage. Evidence is presented below to argue that Darwin used the *Santiago notebook* not only while in Chile, but in fact for several years after the voyage, by which time he was using his theoretical notebooks.

These differences between the *Santiago notebook* and the other *Beagle* notebooks seem to tally with its unique position in being transitional between a field notebook (i.e. one used in the field to record observations on a day to day basis) and one used to record theoretical speculations. The notebook was used in the field in September 1834, pp. 1–67, but this includes a very long retrospective entry for 18 September, written in ink. There are then about five pages of 'theory' which may result from Darwin's forced immobility due to illness during October, although they could have been written any time between September 1834 and February 1835. There is

---

677 Stoddart 1976.
678 CCD1: 494–7 and note.
679 Darwin 1837a; *Shorter publications*, pp. 40–5.

sometimes a subtle shift in Darwin's notebook style when he was more theoretical and this shift can be detected in these pages. Darwin's field notes are always written in the first person (for example, 'I rode to' or 'we ascended'), whereas his theoretical style tends to be in the third person.

The *Santiago notebook* was next used in February 1835, pp. 73–87, although true field notes do not begin again until p. 79. This was around the time Darwin also made notes in the *B. Blanca notebook*, pp. 82b–end, and also just after he wrote his short but highly important essay on the extinction of the 'Mastodon' fossil which may have been written between Chiloe and Valdivia around 7 February.[680] The 'Mastodon' notes record Darwin's first theoretical break with Lyell on the issue of extinction and it was during the following few months that Darwin broke from Lyell more forcefully on the issue of coral reefs.[681]

For the recording of field notes, Darwin then switched from the *Santiago* to the *Galapagos notebook* for a few pages dated mid March but mainly to the *St. Fe notebook* for his March–April expedition across the Andes.[682]

Darwin continued, however, to use the *Santiago notebook* from March 1835 for more speculative or theoretical notes and the remainder of the notebook, pp. 88–124, plus excised pages, which have not been located, became the precursor of the *Red notebook*, as Sulloway 1983 was the first to show. The *Red notebook*, which was the first notebook dedicated to theory, was apparently opened in May 1836[683] and used until it was supplanted in July 1837 by the *A and B notebooks* which started the sequence of notebooks in which Darwin developed some of his greatest scientific theories.[684]

Historians have concluded that it was the completion of the *Santiago notebook* which caused Darwin to open the *Red notebook*, or at any rate that the *Santiago notebook* was not used after May 1836. However, it is arguably the case that the *Santiago notebook* was only partially filled by May 1836 and may have been used in parallel with *notebooks A, B* and *C*, and with the *St Helena Model notebook* at least until the summer of 1838. There may well be other important links between the *Santiago notebook* and Darwin's post-voyage researches. This chronology, if correct, will require some reappraisal of Darwin's scientific activities in the months leading up to his discovery of natural selection.[685]

---

680 See the introduction to the *Port Desire notebook*.

681 See Herbert 1982.

682 The view of Barlow 1945, p. 231, that there is a missing notebook, dedicated to Concepción (late February to early March), seems unlikely.

683 Armstrong 1985 dated the first use of the *Red notebook* to just after the *Beagle* quit Australia in March 1836, but that dating is at odds with the evidence for a post-Mauritius first use.

684 Barrett *et al.* 1987.

685 For example, Rudwick 1982; Browne 1995.

Santiago to San Fernando to Valparaiso, September 1834

The *Santiago notebook* was apparently first used at the beginning of September 1834. The notes record Darwin's setting out southwards from Santiago, where he had spent a very enjoyable week relaxing after his two week northern circuit from Valparaiso, recorded in the *Valparaiso notebook*. Darwin's friend Corfield came up from Valparaiso to meet him in Santiago to 'admire the beauties of nature, in the form of Signoritas'. The two men stayed at an English hotel, the Fonda Inglese.[686]

From Santiago Darwin went more or less south to Rancagua and San Fernando, then down the Rapel valley to Navedad on the coast and from there back up to Valparaiso. The whole August–September expedition itself gave Darwin an excellent geological overview of the outer ranges of the Andes. These he described in *South America* (mainly pp. 169–75), but he also published several papers dealing with the tectonics of the Chilean coast in which he used information collected on this expedition. The outer ranges are composed primarily of gneiss and granite, as Darwin showed, but today it is understood that they are the result of melting of the edge of the South American continental plate as it overrides the Nazca oceanic plate. Herbert 2005, p. 224, reproduces a small diagram of the coast range from Darwin's geological diary, DAR 36.438, which gives a sense of how he saw the basic structure of the country.

Darwin became ill after leaving Santiago and by the time he was back in Valparaiso on 27 September he was in no state to travel any further.[687] He was sufficiently ill to require the attentions of the *Beagle*'s surgeon, Benjamin Bynoe (1804–65). Darwin attributed his stomach disorder, in a letter to his sister Caroline, to drinking chichi, the local Indian aguardiente, but historians including Keynes 2003 generally blame typhoid.[688] Darwin drank the chichi at a gold mine around 16 September and seems to have become ill within 24 hours.

While it is obvious that Darwin had a miserable time getting back to Valparaiso in the last week of September his handwriting at this time was no more 'wild and straggling' than usual, as Barlow claimed, or the 'straggling hand of a sick and exhausted man', according to Huxley and Kettlewell.[689]

The notebook opens on p. 1 with a complex sequence of jottings almost certainly written in Santiago, starting with 'Biscatchas making a noise'. Darwin described the behaviour of this rabbit-like animal (*Lagostomus trichodactylus*), with great care in *Journal of researches*, p. 143. Several names appear on p. 1: Corfield and Darwin's servant Covington. 'Mariano 7£' is no doubt a fee to Darwin's guide Mariano

---

686 CD to FitzRoy 28 August 1834, CCD1: 406.
687 See the introduction to the *Port Desire notebook*.
688 CCD1: 410. There seems to be some confusion about chichi. Parodiz 1981, p. 120, claims it is made from corn, whereas Covington (Young 1995, note 151) claims it was made from cider. CD himself also says it was made from the fruit of a Bromelia (*Journal of researches*, p. 361).
689 Barlow 1945, p. 228; Huxley and Kettlewell 1965, p. 40.

Gonzales but one can only guess why Darwin gave £12 to a man called Ramirez, perhaps for horses. Alexander Caldcleugh of Santiago is mentioned in connection with a book and with the 'geology of Concepcion'.[690] A Mr Conrad seems to have provided some evidence of the elevation of North America, which would have been highly interesting corroboration to Darwin of his emerging views on South America. Conrad's information is referred to in Darwin's 'February 1835' essay on the extinction of his 'Mastodon' (DAR 42.97–9), and Conrad later provided Darwin with observations concerning erratic boulders in Alabama.[691] John Wickham (1798–1864), First Lieutenant of the *Beagle*, is mentioned without explanation, although he had taken some charts to Santiago in August.[692] Finally there is Juan Ignacio Molina's name in connection with the behaviour of the Carrancha, a bird of prey. Darwin's copy of Molina 1794–5 is inscribed 'Valparaiso 1834'.

The field notes proper start on 2 September on p. 2, with Darwin riding to a place northwest of Santiago beyond what is now Padahuel (which Darwin called 'Padaguel' in the notebook and 'Podaguel' in *South America*, p. 59) 'to see Limestone'. In the published version Darwin described this as a 'calcareous tuff' overlying sandstone with 'water lines' (that is, cross-bedding, caused by current action), all of which he was certain were of marine origin. In *South America* Darwin wrote that he found a similar rock on the summits of the San Lucía and San Cristóbal hills in Santiago, and mentioned that the barometric height of 2,690 feet (c. 900 meters) of San Cristóbal provided by Frederick Andrew Eck indicated dramatic elevation of the coast range.[693]

The visit to Padahuel seems to have been a day excursion, and it is unclear what else Darwin did until he left Santiago on 5 September: 'started gloomy day' and 'crossed bridge of Hide'.[694] That night they stayed at a 'nice Hacienda' where there were 'Signoritas', which became 'several very pretty Signoritas' with 'charming eyes' in the *Beagle diary*. In the *Beagle diary* Darwin recounted how he assured the Catholic signoritas that he was 'a sort of Christian', but in the notebook he wrote that they showed 'astonish-ment at clergyman marrying' and could not accept that Darwin's God was the same as theirs! There seems to be more detail about this or a similar conversation on p. 18.

On 6 September, p. 4, Darwin continued to Rancagua on an 'interesting ride' across the 'great plain', p. 6, which was incised by rivers flowing rapidly from the Andes. Darwin noted the nest building habits of various birds and seemed to record that some 'talk' while others 'sing'. The habits of the 'Tapaculo and Turco' were

---

690 See the introduction to the *St. Fe notebook*.
691 See *Journal of researches*, p. 614.
692 Keynes 2004, p. 161.
693 The measurements are to be found on a list in the geological diary DAR 35.232, see CCD1: 407. They are dated September 1834 in CD's handwriting.
694 See the introduction to the *Valparaiso notebook*.

eventually described at some length in *Journal of researches*, p. 329. Darwin also added an *aide mémoire* to check which names Molina 1794–5 gave to these birds and on the 7th he recorded the behaviour of a 'Goosander .... runs quick. Very active in the rapids'. Darwin's note on the nest building of the 'Furnarius' on p. 9 is referred to in *Natural selection*, p. 504.

Darwin's party turned up the Cachapol [Cachapoal] valley to visit the hot baths at Cauquenes [Termas de Cauquenes]. Since the bridges were taken down every winter Darwin's party had to ride across the torrents and one 'could not tell whether horse moved or not'. Darwin was hugely impressed by the erosive power of the torrents which in summer could move 'fragments as big as cottages', p. 10. Because of heavy rain, the last he was to see for seven and a half months, Darwin stayed at the baths for five days, until 13 September.

Darwin later expressed his view emphatically that the Cachapoal valley was sculpted by the sea[695] and remarked in the notebook on the 'decided fringe ... composed of pebbles', p. 9, marks of a 'fossil beach' which was once at sea level. Darwin recalled that to his mind: 'This has been one of the most important conclusions to which my observations on the geology of South America have led me; for we thus learn that one of the grandest and most symmetrical mountain-chains in the world, with its several parallel lines, have been together uplifted in mass between 7,000 and 9,000 feet, in the same gradual manner as have the eastern and western coasts within the recent period.'[696] Darwin's recognition of former sea beaches was reinforced by those he studied in 1835 near Coquimbo.[697] As is well known, he overstretched this explanation in 1839 by applying it to the Parallel Roads of Glen Roy in Scotland.[698] These horizontal ledges were afterwards shown to be glacial lake beaches, although the methodology he used at Glen Roy was scientifically impeccable.[699]

The baths as Darwin found them were just 'miserable huts' and it was 'necessary to bring everything' to sample them, but there was a 'pretty view', p. 8. Today they are regarded as one of the most beautiful hot spring resorts in Chile, with marble baths and a mid nineteenth century sala de baños and water around 45° C. Darwin noted that the springs were hotter in summer 'Some people can enter hottest', and he reported that 'M. Gay' (1800–73), the 'zealous and able' French geologist,[700] 'say Muriat of Lime Spring flow through the pebbles cemented by Carb of Lime hence Gas', p. 11. Darwin met Claude Gay in Santiago, where he was Professor of Physics and Chemistry, but his information that the springs were partly hydrochloric acid

---

695 *South America*, pp. 66–7.
696 *South America*, p. 67.
697 See the introduction to the *Coquimbo notebook*.
698 Darwin 1839; *Shorter publications*, pp. 50–91.
699 See Rudwick 1974 and Rudwick 2008.
700 See *Journal of researches*, p. 323.

obviously intrigued Darwin, who was especially keen that Henslow should get his sample of 'water and gaz' analysed.[701]

On 8 September, Darwin noted that the 'strata may be seen gradually becoming more inclined till they are vertical', p. 14. On p. 15 he drew a sketch map showing Rancagua to the west and a southwest-trending spur of mountains coming off the Andes. The view was 'singular', with 'little wind — blue sky & clear — very few birds excepting an odd note from Tapacola or Turco — wonderful few insects, dryness of soil', p. 17. On the 9th Darwin took a ride up the valley to a deep ravine where he could see for the first time close up the 'real Cordilleras', p. 19, and scrambled up 'a very lofty mountain' at least 6,000 feet (2,000 meters) high. On p. 25 Darwin jotted 'Pinchero Pass' and in the *Beagle diary* he explained that it was through this pass that the 'Renegade Spaniard' (*Beagle diary*, p. 258) Pinchero once ravaged Chile. Darwin noticed a block of granite 'several hundred feet above bed of river' which he thought was 'placed in present place by former sea', p. 20. He later noted that blocks such as this one could not normally be moved by the sea.[702] Instead he suggested that they may have been moved by catastrophic floods unleashed by the damming of rivers during earthquakes.

Thursday 11 September was a 'Miserable wet day', p. 30. On the 12th 'Have seen but few Condors — yet this morning 20 together soaring above' and a man told Darwin that this meant there was a 'Lion' (i.e. puma) guarding its prey. Since there was a bounty on pumas, the condors usually triggered a hunt with dogs. Darwin was impressed to hear that the pumas were as cunning as foxes and could escape 'by artifice of returning close to former track', p. 31. Pumas on this side of the mountains were said to be fiercer than on the east 'killed a woman & child lately & man at Jahuel!!', p. 31.

On p. 32 Darwin noted that the hot springs had stopped in 1822 (when there was a great earthquake) but were gradually returning and getting hotter again, as 'known by feathers coming from a fowl'. Darwin recounted in *Journal of researches*, p. 321, how the locals at the baths estimated the temperature of the water by the ease with which they could pluck fowls which were scalded in the water.

On the 13th they 'At last escaped from our foodless prison — threatening day snow low down' and 'reached the Rio Claro by night', p. 33. The next day they rode on to San Fernando across a plain: 'to give an idea of its extent, the distant Cordillera only showed their snowy part, as over the sea', p. 35. Darwin here made the extraordinary speculation that this is how Tierra del Fuego might 'appear if elevated'. Furthermore he recalled, presumably from Humboldt, that Quito in Ecuador at 3,000 meters (9,000 feet) looked like this, so 'elevation decrease to South?', p. 36.

---

701 CCD1: 401.
702 *South America*, p. 66.

The next date in the notebook is 18 September and the gap is presumably the four days ('during two of which I was unwell'[703]) when Darwin stayed at the gold mine at 'Yaquil near Rancagua' owned by an American named Nixon. As Darwin drank the chichi at a gold mine, this means that he became ill about the 16th. It seems, therefore, that the descriptions of the mines in the notebook are retrospective, which tallies with the fact that they are written in ink.[704]

The literary style of the ink notes is more discursive than is usual in Darwin's field notebooks, as witnessed by the fact that the whole entry dated 18 September runs to fifteen pages, before pencil is used again on 19 September, p. 52. According to the *Beagle diary* Darwin 'took leave of Yaquil' on 19 September so it seems almost certain that the ink section dated the 18th represents the first use of the *Santiago notebook* as a medium for recording reflections as well as observations.

On the first page dated 18 September Darwin described a 'Lizard with blackish tail basking on stones in sun … half of body brilliant ~~greenish spot~~ blue scales', p. 37. This lizard was probably the one listed as *Proctotretus tenuis* in *Reptiles*. He mentioned fungus on the 'Roble or oak tree', p. 50, reminding him of England. There follow several pages of geology, then on pp. 40–1 Darwin 'Saw the Descabozado & heard certainly of Peterra', two mountains to the south. A series of pages follow in which Darwin recorded information about the geology of the Durazas and Cruzero mines which seems likely to be information direct from Nixon.

Proctotretus tenuis collected in Chile. Plate 3 from *Reptiles*.

On p. 50 Darwin 'Rode up to the Mine' and 'Saw lake of Tagua-tagua' which he noted in the *Beagle diary* was described by Gay. Darwin was shocked to see the conditions endured by the miners who had to carry incredible loads up the crudest of ladders, with 'nothing but beans and bread' to eat, all for '6 & 7 dollars a month'. In 1880 Darwin was asked for his opinion on vegetarianism. With typical

---

703 *Beagle diary*, p. 260.
704 Sulloway 1983, p. 373, suggests that CD may have used a travelling writing kit to write this section.

candour Darwin replied: 'I have always been astonished at the fact that the most extraordinary workers I ever saw, viz., the laborers in the mines of Chili, live exclusively on vegetable food, which includes many seeds of the leguminous plants. On the other hand, the Gauchos are a very active people, and live almost entirely on flesh.'[705]

One can only guess why Darwin left the sentence at the end of p. 51 unfinished. On p. 52 he wrote 'feudal system of men' immediately followed by the word 'slaves'.

Darwin met the German merchant and naturalist Juan Renous at Nixon's house, and on p. 52 he recorded what an 'Old Spanish lawyer' thought when Renous asked him for his views on Darwin's activities. After long consideration the old man declared 'no man so rich that he sends naturalist for pleasure — there is a "cat shut up here", do you think King of England allows man to explore stones & reptiles'. Renous related how he was arrested by the superstitious people in San Fernando for 'having snakes [and] feeding Caterpillars'! The ink section ends with Darwin's surprise at the strength of that hybrid beast the mule: 'wonderful so slim an animal', p. 53, and he expanded on this in the *Beagle diary* with 'One fancies art has here out-mastered Nature'.

Darwin returned to writing in pencil on 19 September on p. 53 as his party began the descent down the Tenderirica [Tinguiririca] Valley. This eventually became the Rapel, towards San Antonio and the ocean (there is a small ink sketch of part of the valley on p. 71). No doubt Darwin was feeling rather ill and the 'miserable place where we slept (nothing but beans & milk)' cannot have helped. In the *Beagle diary* he noted that he was ill from 20 September to the end of October.

They crossed the plains west of Rancagua which Darwin concluded were 'without doubt', from their 'levelness & great size', p. 55, and 'barrancas upon Barrancas', p. 57, episodically upraised sea bed. He noted the 'Bishops Cave' which was a consecrated cavern in which Indian remains were found. There were 'no trees' and the landscape reminded Darwin of the Pampas, p. 58. The 21st was 'Rainy day & I very unwell stopped at kind Chilotan schoolmaster'. On the 22nd they 'Rode on to Navidad' which had a front 'very like Valparaiso — but longer', p. 60. The *Beagle diary* records that they stayed at the house near the sea of 'a rich Haciendero' and that on the 23rd Darwin was 'very unwell' but 'managed to collect many marine remains from beds of the tertiary formation'. These fossils were of great importance, and included plant material and 'Fishes teeth', p. 64. Many of them were figured by Sowerby in *South America*. Darwin wrote, 'though suffering from illness' he collected thirty-one species, all extinct and of genera no longer (i.e. in 1834) found living so far south.[706]

705 Darwin 1880; *Shorter publications*, p. 428.
706 *South America*, p. 127.

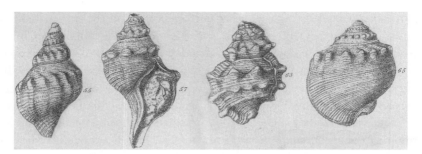

Fossil shells collected by Darwin in Navidad, Chile: *Fusus* figs. 55, 57, *Triton* 63; *Cassis* 65 from plate 4 of *South America*.

On the 24th and 25th they pushed on to Casa Blanca where they arrived on the 26th. Here Darwin had Covington arrange for a biloche (a kind of wagon)[707] to take him back to Corfield's House in the Almendral area of Valparaiso. After the expedition there are a few pages difficult to date, pp. 68–72, which may represent the beginnings of Darwin's theorizing, perhaps the result of being ill and not able to get about observing and collecting. The *Zoology notes* written at this time show that Darwin was not entirely incapacitated, and he was clearly still intrigued to hear of evidence of sea level change: 'It is said that the sea 70 years ago reached foot of Dr Stiles house', p. 71. These are the last entries in the *Santiago notebook* which probably date from 1834. The notebook was then put aside for the southern summer, while Darwin went south, for the last time as it turned out.[708] The notebook was not used again until February 1835.

Valdivia, February 1835

Darwin next used *Santiago* as a field notebook on his way back north from Chiloe, around 10 February 1835. This part of the *Santiago notebook* covers the time in Valdivia on 20 February when he experienced a massive earthquake, and this may be the reason why a week or so later he started to record his theoretical speculations on crustal instability in the notebook. It has been estimated that the 1835 earthquake was of magnitude 8.5 on the Richter scale, although this is dwarfed by the Chilean earthquake of 1960 which was 9.5.[709]

It seems that Darwin spent the next two to three weeks examining the aftermath of the earthquake and the ensuing tsunamis at Talcahuano (the port of Concepción), but the details of this are not recorded in any extant notebook. Darwin often spelled Valdivia 'Baldivia' at this time.

Darwin arrived near Valdivia in Chile's Lake District on 8 February, but he seems not to have landed until the 11th. The first entry in the notebook which is certainly

---

707 See Young 1995, note 155.
708 See the introduction to the *Port Desire notebook*.
709 For more detail about the earthquake see Keynes 2003, p. 274, and for geological background see the introduction to the *St. Fe notebook*.

from Valdivia is in ink and dated the 10th. It records sunset at 7.15 p.m. and sunrise at 5.15 a.m., p. 73. On p. 75 there commences a long list of Andean volcanoes, two of which are noted on p. 76 as having been added to a map 'reduced from one of Dalbe's Republic of Chili surveyed &c &c in 1819'. Page 77 has a piece of tracing paper gummed to it on which is traced a map of the rivers flowing down to a section of the coast about 25 km (15.5 miles) long, with Valdivia at the top.

The 11th February is recorded on p. 79 'Went in Yawl to Valdivia', whence according to the *Beagle diary* he went for a ride. Darwin obtained a room from 'obliging Governor' then 'after crossing in a canoe a small river entered forest'. He mentioned the poison made by the Indians and noted the 'Upright Bamboos', p. 80. He recorded the easterly dipping mica slate overlain by clay with quartz pebbles 'as in Chiloe'.[710] On p. 81 Darwin noted the 'Bright red soil' and referred to Febres's 1765 'Spanish & Chilian & Peruvian' dictionary, which FitzRoy mentioned as very useful.[711]

'After travelling some hours through forest (tormented by innumerable flee bites, pigs, dogs, & cats)' the country opened out: 'the curious fact of plains banishing trees', p. 82. Darwin explained in the *Beagle diary* how the Governor's house was so flea-infested that Darwin actually preferred to sleep in the open. He thought the Llanos 'very pretty' and recorded having 'Heard great guns in the morning', presumably from one of the coastal forts, p. 83. On 12 February (his twenty-sixth birthday) Darwin stopped at the Mission of Cudico, where the 'Discontented padres [sic]' was teaching the 'very industrious' Indian children. Darwin expanded greatly on these ethnographic observations in the *Beagle diary*, where he expressed his opinion that the Padre was wasting his life for want of something more stimulating to do. There follow some jottings about mines and earthquakes on p. 84. Darwin's activities in the ten days or so up to leaving Valdivia on 22 February are covered in the *Beagle diary*. On notebook p. 87 Darwin mentions the Padre who was distressed by the extermination of the Indians, then there is a note about a 'Fleet of whalers' and he noted 'There are Whalers cruising for fish' in the *Beagle diary* for 24 February, which is the date immediately following in the notebook. His experience of the earthquake of 20 February while 'on shore and lying down' is, therefore, not recorded in the notebook but given a memorable account in the *Beagle diary*.

Other authors, including FitzRoy in his *Narrative*, wrote detailed accounts of the earthquake. FitzRoy's initial report was 'communicated to the President' of the Geological Society of London (i.e. Lyell) by Captain Beaufort and was read on 18 November, immediately before Professor Sedgwick's reading from Darwin's letters to Henslow.[712]

---

710 Discussed in *South America*, p. 160.
711 *Narrative* 2: 397.
712 See the introduction to *Cape de Verds notebook*.

The month before, FitzRoy's report to the Court Martial 'on Capt. Seymour and his Officers for the loss of His Majesty's Frigate *Challenger*' had been read at Portsmouth. The wreck of H.M.S. *Challenger* was an important part of the *Beagle* voyage. While Darwin was trekking from Coquimbo to Copiapò, FitzRoy was engaged in a successful mission to rescue the crew of the *Challenger*, which was wrecked on the treacherous Dormido Shoal, near Araucho, *c*. 30 km (19 miles) south of Concepción.[713] So while Darwin was in Copiapò the *Beagle* was under the command of Wickham. He collected Darwin on 5 July and took him to Iquique and Callao, where FitzRoy rejoined the *Beagle* on 9 August.

The earthquake of 20 February came between 11 a.m. and noon.[714] It lasted about two minutes and the main shock which destroyed much of Concepción took just six seconds, and there were aftershocks for several weeks. There were three tsunamis at Talcahuano, each one larger than the last, starting about half an hour after the earthquake. The island of Santa Maria was raised an average of about 3 meters, but seems to have subsided somewhat, later in the year. The fact that the coast appeared to have been raised more than the Cordillera was a bit awkward for Darwin as he had come to expect the Andes to be the central axis of uplift.

Darwin arrived at the island of Mocha [Isla Mocha] on 24 February en route to Concepción, 'having sailed 22d' from Valdivia. This is the last dated entry in the notebook as on the next page, p. 88, in an entry written before reaching Concepción, Darwin embarked on the longest series of theoretical speculations in any of the field notebooks, a series which may have continued until the notebook was full several years later.

The *Beagle* spent a week at Mocha (and returned once, without Darwin, on 27 March). Darwin later wrote that he never landed on Mocha but the officers collected some fossils for him.[715] While there the ship felt a strange 'jerk' of the sea which Darwin noted in the *Beagle diary* snapped an anchor cable in two and must have been caused by an aftershock. FitzRoy resolved to head north as they were now dangerously short of anchors. They reached Concepción on 4 March where Darwin saw the terrible destruction from two weeks before: 'the most awful yet interesting spectacle I ever beheld'.[716]

The *Beagle* sailed for Valparaiso on 7 March, arriving on the 11th. Darwin stayed at the house of his friend Corfield. The ship, with new anchors, headed back to Valdivia on the 17th and Darwin did not see her again until she returned on 23 April, by which time he had traversed the Andes, and was ready for another trek,

---

713 FitzRoy's account of his daring mission is in *Narrative* 2: 428–80 and the whole episode is covered by Keynes 2004, chapter 24.
714 Accounts differ regarding the exact time; CD refers to this in the *Galapagos notebook*, p. 8b and in the geological diary DAR 36.424.
715 *South America*, p. 124.
716 *Beagle diary*, p. 296.

this time northwards to Coquimbo where he experienced another earthquake, on 20 May.[717]

As a field notebook, the *Santiago notebook* was replaced by the *St. Fe notebook* on 12 March, but Darwin also made notes dated 14 March on his trip from Valparaiso to Santiago in the *Galapagos notebook*. The entry in the *Galapagos notebook* on p. 4a is 'Started for St Jago in a Birloches or gig' which appears almost verbatim in the *Beagle diary*. There are also notes which are obviously of a similar date at the back of the *Galapagos notebook*. As recorded in the *St. Fe notebook* Darwin made his Andean preparations while staying for a few days in mid March with Caldcleugh in Santiago. He also stayed with him for about four days when he returned to Santiago on 10 April.[718]

The fact that Darwin had at least two field notebooks (*St. Fe* and *Galapagos*) with him on his Andean traverse suggests that he may, for reasons of document security, have preferred to leave the *Santiago notebook* on the *Beagle* around 11 March. If so the pages after about p. 91 were probably not started until 23 April. If, on the other hand, he had time, perhaps while at sea around the 7th to the 11th of March, to write up all his field notes from the *Santiago notebook*, he might have decided to take it with him exclusively for theorizing. This seems unlikely, in view of the relatively short time available. Darwin wrote to Henslow on 18 April 1835 that he had sent 'With the Specimens ... a bundle of old Papers & Note Books'.[719] These may have included the field notebooks that were no longer used.

The 'theoretical' pages (April 1835–September 1838?)

Pages 88–124 are completely undated discussion, with references to at least nineteen authors and notes of conversations with at least eight different people. There were in addition ten pages excised from the back of the notebook. Darwin wrote most of the pages, but not all, in ink and, uniquely in the field notebooks, added ink page numbers from p. 90 onwards. Darwin's page numbers are referred to by several scholars, but since page numbers have been assigned to all the pages of the notebook in the present edition, these numbers are used rather than Darwin's.

The theoretical pages discuss a great range of geological issues and they nearly all relate to one another in complex ways. The unifying theme is vertical crustal mobility. Among other topics, they deal with the effects of earthquakes and the sea on the transportal of gravels and erratic boulders, pp. 88–94. This section includes the only diagram in this part of the notebook, showing the effects of

---

717 See the *Coquimbo notebook*.
718 See the introductions to the *St. Fe* and *Galapagos notebooks*.
719 CCD1: 444.

elevation on the coast at Chiloe. There is some correspondence between the subjects touched on in these pages and those in the *Galapagos notebook*, around p. 32a.

These earlier numbered pages in the *Santiago notebook* contain the famous 'coral passage', in which Darwin sketched out the essentials of his coral reef theory, pp. 95–7. They also present Darwin's novel and completely unstudied analysis of the geological history of England in stratigraphical order, that is from older to more recent rocks, pp. 97–104. Darwin never published on this subject and his comparative ignorance of English geology was one of the main reasons he was reluctant, in 1837, to become Secretary of the Geological Society of London.

There follows a mixed discussion, partly in pencil and partly in ink, of conglomerates and related sediments, and of the geology of Rio de Janeiro. The conjecture 'I suspect that Granite heated at bottom of ocean', p. 107, is declared 'Very important' on p. 109. A slip of paper in DAR 5.B73-B74 refers to 'p. 18' of 'Santiago note book'. The note refers to an issue of the *Madras Journal* dated 13 October 1836. There is a subtle shift to even more general theorizing, such as the following, which is surely the basis for Darwin's great 1838 paper on crustal mobility, p. 108.

> Amplify on importance of proving extent & recency: of upheavals & manner over whole America. — Explaining generally Continental upheaval. so important in understanding valleys, diluvium escarpments & successive lines of formations &c. — Showing that they are not mere Local effects, as so many authors suppose.

Some passages belie a highly sophisticated, almost adversarial, approach to the marshalling of evidence according to strict logical principles. This approach first appeared in the *B. Blanca notebook* in September 1833. Examples include the layering of evidence 'that they have moved … that they have travelled … that this took place … that they thus travelled', pp. 90–1. Darwin was to apply this technique in print, such as when arguing for the elevation of Patagonia in *South America*, p. 18. There are other, more conversational, examples in the notebook where Darwin expressed himself as if trying to persuade his peers, using such phrases as 'we may suppose..' and 'we know that', p. 104. There is also some rhetoric: '꞉ Do not the successive terraces', p. 109, 'Is not the presence', p. 110. Darwin knew when to emphasise a point: 'It must be re re urged', p. 118, and he did not want to waste his powers of persuasion by bad presentation: 'State this with clearness', p. 109.

Biological topics begin to appear around p. 110 with mention of 'Hydrophobia' (rabies) and 'If the Pacifick Isl^{ds} have subsided there ought to be a peculiar vegetation', p. 110. This last conjecture bears comparison with the almost opposite entry in the *Red notebook*, p. 127, written around March 1837: 'my idea of Vol: islands. elevated. then peculiar plants created'[720] and to Darwin's first published reference to

---

720 See *Red notebook*.

species origins in his 1837 paper on the oceans, also drafted in the spring of 1837: 'That some degree of light might thus be thrown on the question, whether certain groups of living beings peculiar to small spots are the remnants of a former large population, or a new one springing into existence'.[721]

There is an increasing number of references to the opinions of other people and a return to the effects of earthquakes, p. 112, and various new topics such as concretions, p. 115, and cleavage. On p. 116 there is another mixture of pencil and ink jottings on many subjects, some of which were certainly taken up in Darwin's theoretical notebooks. The discussions range far and wide and although centred on the Andes include several references to the Falklands and even a reference to the North Pole, p. 118.

The last extant geological entry, p. 124, is in pencil: 'S. America fundamentally & systematically different S. America; shows that the geology of the world cannot be taken from Europe.' This entry, which seems rather obscure, is so similar to one comparing the geology of America with Europe in the *Red notebook*, p. 18,[722] that it is likely that in the *Santiago notebook* entry Darwin meant 'Europe' when he wrote either the first or the second 'S. America'.

Finally, there are some field notes on the inside back cover which may date from the *Beagle*'s visit to Mauritius in late April and early May 1836, as Darwin in the note calls it Isle of France as he sometimes did around the time of the visit.[723] The reference to 'Stokes a quire & a half of foolscap' (i.e. 36 sheets) may relate to Darwin's visit with John Lort Stokes to the house of the surveyor general and the purchase of paper on Mauritius.[724] 'I have about 900 pages' is another tantalising clue to dating of these notes, as this is almost exactly the number of pages Darwin had written in his geological diary at that point in the voyage.[725]

Sulloway 1983 discussed the dating of the 'theoretical pages' in the *Santiago notebook* in great detail in a pioneering attempt to apply his chronology of Darwin's spelling habits during the voyage.[726] Taking all available evidence into consideration, approximate dates can be assigned to some pages.

Page 89: 'ask at Valparaiso and Concepcion', must have been written before 4 March 1835.

Pages 88–91: these pages seem from subject matter (gravels and boulders), literature references (mainly to De la Beche but also to Playfair) and locutions, 'they have travelled on each side from Cordilleras', p. 88, 'they have travelled in opposite directions', p. 90, 'they thus travelled at two very different periods', p. 91, to have been written at the same time, that is before 4 March.

---

721 Darwin 1837b, p. 554; *Shorter publications*, pp. 35–7.
722 See *Red notebook*, p. 36.
723 See Armstrong 2004 and the introductions to the *Despoblado* and *Sydney notebooks*.
724 See below in connection with the dating of p. 116.
725 DAR 38.
726 Sulloway 1983.

Page 95: Sulloway suggested that the reference to Caldcleugh having seen a guanaco 'near Cordovean range', p. 95, was a personal communication to Darwin, therefore dating that page to March or April 1835.[727] Sulloway supported this dating with orthographical evidence: on the same page Darwin used 'Pacific', predating his switch to 'Pacifick' in July 1835, and 'corall' predating his switch to 'coral' in February 1836. Sulloway was careful to point out, however, that Darwin used 'Pacific' once later, in October 1835.[728] The dating of this page and others up to p. 104 is also discussed in the *Correspondence* where it is concluded that the pages predate Darwin's September 1835 sailing to the Galapagos.[729]

Page 105: 'If the circle of change has always gone' is similar to 'if the crust of the world goes on changing in a Circle', written to Henslow on 12 August.[730] 'Alison's notes' probably refers to the letter from R. E. Alison received by Darwin after 19 July.[731]

Page 111: 'a swelling of the fluid nucleus of the globe, to form the statical equili [brium] destroyed as Sir J. Herschel suggests', as for p. 110 (see note 720), although a related discussion occurs on p. 114 of *Notebook A*.[732] That page was probably written in August 1838. The page in *Santiago* may, therefore, have been written around the time of *Notebook A* or may have inspired the passage of the latter notebook.

Page 113: 'I observed in K: George's Sound' postdates the *Beagle*'s visit there in March 1836.[733]

Page 116: 'Introduce in cleavage discussion' may refer, as Sulloway and Herbert suggested, to the 'cleavage paper' (DAR 41) Darwin had apparently already begun.[734] It was written on paper which Sulloway deduced was purchased in Mauritius in the beginning of May 1836. This is not the only possibility, however, as on p. 62 of *Notebook A* Darwin wrote 'In Cleavage discussion, state', an entry perhaps dating to early 1838. The 'Mem: to tell Lyell about crocodile in South Sea Island' suggests that Darwin may have already met Lyell, which would date this page to at least late October 1836. The book Darwin referred to, Mariner 1817, is not known to have been in the *Beagle*'s library. It is certainly true that the subject matter here is picked up at several points in the transmutation notebooks, especially *Notebook C*, in passages dating to the spring and summer of 1838. The page in *Santiago* may, therefore, have been written around the same time or may have informed the transmutation notebooks.

727 Sulloway 1983, note 10.
728 DAR 37. 791.
729 CCD1: 567–71.
730 CCD1: 461.
731 Now in DAR 36.427. See CCD1: 450–3.
732 See Barrett *et al.* 1987.
733 Not February as stated by Sulloway 1983; see Armstrong 1985.
734 See Herbert 2005.

Page 117: 'N.B. in general discussion introduce diluvium generally submarine' is similar to some entries in the *Red notebook* and so seems to have been written very late in or after the voyage.

Page 118: 'Introduce in cleavage paper', as for p. 116.

Pages 118–19: 'In discussion on Porph: Breccia, I should state to gain confidence, that it was sometime before I fully comprehended origin' and 'Introduce into my general discussion the gneiss & Quartz of central Patagonia, on coast' both imply that Darwin was preparing material for publication. While it is true that Darwin began this process as early as 1834 the general tone is also consistent with a post-voyage date.

Page 120: 'coral' in 'The Coral theory' suggests post April 1836.

Page 123: 'Before concluding the Cleavage paper. Consult the VI Vol. of [Humboldt's] Pers[onal] Narra.[tive]'. Again, this seems to link to the paper in DAR 41, but it is undated so, as for p. 116, an 1838 date is possible.[735]

Page 130: 'Wild dog of Australia copulates freely with the tame ones' is similar to an entry on p. 26 of the *St Helena Model notebook*, which has been dated to August or early September 1838.[736] A similar reference appears on p. 4 of Darwin's *Edinburgh notebook*: 'In Australia I was assured wild dog copulates freely with tame', which was certainly written no earlier than 1837.[737]

In summary, the theoretical pages of the *Santiago notebook*, left undated but paginated by Darwin, from context and by comparison with other Darwin manuscripts of more or less established date, can be dated with perhaps decreasing degrees of probability. The earliest pages were written in March 1835, some eighteen months before the end of the voyage. The later pages, however, were written at various points towards the end of the voyage and began to overlap with entries in the *Red notebook*, the first notebook dedicated solely to theory.

It seems possible, however, that the *Santiago notebook* was still being used after the *Red notebook* was replaced with *Notebook A* devoted to geology, and *Notebooks B* and *C* devoted to species. More accurate dating may be possible from chemical analysis of the various inks used in these notebooks. Unfortunately, until the excised pages are located, it is impossible to know exactly when the *Santiago notebook* was finally put aside.

The *Santiago notebook* is one of the most interesting and problematic of all Darwin's notebooks and surely the only one in which we can trace in some depth the intellectual development of Darwin, from field observer to published author. The precise dating of the theoretical passages has challenged scholars for decades and seems likely to remain elusive. We believe, however, that the analysis presented here will provide a firm foundation for further study of this remarkable document.

---

735 Herbert 2005, p. 403 note 132, makes clear that this volume of Humboldt was central to the argument of the cleavage paper.

736 Chancellor 1990.

737 See Barrett *et al.* 1987, p. 478. A complete transcription of the *Edinburgh notebook* is on *Darwin Online*.

[FRONT COVER]

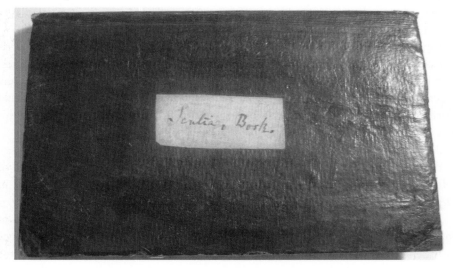

Santiago Book

[INSIDE FRONT COVER]

> DUFOUR
> CHABROL
> M^d Papetier
> Rue S^t. Martin,
> N°132,
> A PARIS

    Corfield

[1]   Corfield. 3^oz (3 ½ dollar Covington)
    Biscatchas making a noise
    Carrancha's throwing back their head — Molina
    Culpeu not Falkland Fox
    Agreed with Ramirez for the 2^d day of month gave 12 ££
    Mariano 7 £ —
    N. America elevated M^r. Conrad
    Caldcleugh geology of Concepcion
    There is heel of Megatherium at Mendoza. English doctor.
    Return Book Caldcleugh
    Wickham

[2]   2nd [September 1834]
    Rode to beyond Padaguel to see Limestone: which formed part of the plain: it
    consisted in a layer overlying sand — with water lines & parts semi-crystalline
    <u>exactly</u> like Tosca rock; other parts <u>white</u> friable, something like that in chinks of
    rock: a most indisputably formed beneath water, no wonder no shells Mem:

B. Ayres: plain has been subject to considerable alluvial action: this Limestone probably passes into

[3] the agglutinated sandy earth. —

5th [September 1834] — started gloomy day — plain — nice Hacienda sleeping place — Signoritas attempts at conversion, astonish-ment at clergyman marrying — ∴ not same god because clergy marry — sisters names not saints: Crossed bridge of Hide Palm at Concepcion

/Rana/ eats buds

[4] 6th [September 1834] —

The plain of St Jago contracts & ascend a little towards the Angostura where a rapid river remarks the declivity it cuts across a low chain with branches from Andes, the valley expands & rises till it forms a great plain with water flowing to South: no horizon to South. — but some

[5] islands. — a river flows to the West also. in the middle of plain two islands which dip to E. small ∠. probable that part of outer chain once existed in plain. (the oblique range South of Aculeo I believe is pieces of N & S lines. At head of Angostasa mass of earth & pebbles covered by great bed of

[6] fine white sand & pumice apparently water-deposit:

All the rock porphyry & some Breccias

interesting ride great plain

Tapacola & Turco buil[d] deep straight hole in ground

Sternus ruber[738] on ground

Black Icterus[739] talks in bushes

yellow spot Icterus[740] in water reeds — Molinas names

[7] Thrush[741] talks a little nest lined with mud

When in holes, nest not covered. nest of Black Furnarius[742]

great long prickly nest little pointed tail creeper

Callandra[743] best singer, plain round nest: only one time of year. Mem: R. Negro.

(7th) [September 1834] Goosander in river back white. — belly brown — breast black. top of head black, beneath white bill, plain cry — runs quick. very active in the rapids —

[8] /Plenty/ of Granite pebbles

Arrived at Baths. road with pretty valleys like to the north with fine trees. Baths square of miserable huts — necessary to bring everything — pretty view

---

738 Specimen not in spirits 1146 in *Zoology notes*, p. 386; listed as *Sturnus militaris* in *Birds*, p. 110.

739 Specimen 1784 in *Zoology notes*, p. 247; listed as *Agelaius chopi* in *Birds*, 'It can be taught to speak, and is sometimes kept in cages', p. 107.

740 Specimen 2186 in *Zoology notes*, p. 246; listed as *Xanthornus chrysopterus* in *Birds*, p. 106.

741 Specimen 2125 in *Zoology notes*, p. 247; listed as *Turdus falklandicus* in *Birds*, p. 59.

742 Specimen 1823 in *Zoology notes*, p. 247; listed as *Opetiorhynchus nigrofumosus* in *Birds*, pp. 68–9.

743 Specimens 2169–70 in *Zoology notes*, p. 244; listed as *Mimus thenca* in *Birds*, p. 61.

unpleasant passing river, could hardly tell whether horse moved or not. R. Cachapol.

I think *<1 word destroyed?>* local circumstance *<3 words destroyed>* formed under [water] but *<1 word destroyed>* at this spot where this valley is so narrow, there is a

[9] decided fringe of level plain, through which the river has cut. Thi[s] is composed of pebbles, sometimes arranged in planes, in a white marl, which coats & white-washes the stones.

other places almost entirely of Lime crystallized (12) in stars. ~~This~~ These gives me the idea it has been formed beneath water. (The Lime & Marl is too abundant to have been subsequently infiltered), (It is not easy to give any idea of the degrading power

[10] of these torrents, fragments as big as cottages are moved & I am told that fragments from a yard & ½ square move rapidly in <u>winter</u> rains — (big rivers flood in summer): the rock 50 or 60 feet above level was a freshly scraped as in bed of river All this cut through rock

The Springs consist of 4 or 5 springs of diff. temp: from diffe proportion of cold pure

[11] water: it is positively stated that water is more abundant & much hotter in summer — (source in <u>Cordilleras</u>) in all parts bubbles appear to escape not only where water appears M. Gay say Muriat of Lime Spring flow through the pebbles cemented by <u>Carb</u> of Lime hence Gas. — <u>Some</u> people can enter hottest — Spring further up <u>simply</u> acid,[744] said so.

Rocks close to well, the usual red Breccia or Conglomerate, one mass

[12] I found where the coarsest Breccia passed & alternated with a dubious looking green (10) sort & fine jasper porphyry — the strata dipped to West (Compass) ∠ 37° — there were fine parallel water lines & alternations of Breccia. The whole has been so much baked, that it assumes a slightly globular structure placed in lines, ready to form columns — If I was not certain that the Breccia & Conglomerate was

[13] really so: some of the pebbles would have startled me.

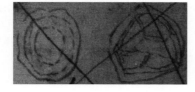

[concretions][745]

744 The normal chemical meaning CD used, for example when he wrote of 'muriatic' (=hydrochloric) acid meaning an acid which reacts with limestone. Also, the petrological meaning for igneous rocks with relatively high amounts of silica (e.g. granite).

745 See the fair copy of this sketch in the geological diary DAR 35.396.

Red bands traced external forms. = forming also regular figures inside

On road saw whole hill composed of nearly vertical beds, & many alternations of greenstones & red Breccias —

I should have stated that above mass of Breccias is traversed by dyke of Greenstone (11)

[14] 8ᵗʰ) [September 1834] The patch of Breccia in bed of river with ∠ 37° dip is continued up in a high hill to South of Bath. here I found the dip Ɇ W ∠ 67. following the line, or crest of Breccia, on for a mile or two a parallel crest is seen where the strata may be seen gradually becoming more inclined till they are vertical, this part forms lofty hill with snow. —

a parallel valley lies on East on the opposite side

the rock appear to dip East & it is

[15]

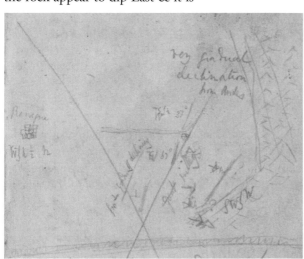

[map of Rancagua]

Rancagua W by N ½ N front gradually declining W by N 37° W 67° great valleys very gradual declination from Andes (A) W SW by W

to be observed where the strata are vertical the Andes have suddenly sent off great ridge (or bent) in a SW by W line; the end of this which faces the ∠° dips towards the mass —

a parallel case to that seen near St Jago

[16] to the East of the crest (A). there is a great mixture of various porphyries, porphyritic greenstones the curious sort of Jajuel — bases of the Porphyries & decomposing greenstones with much iron — remote from the crest these present pure & without Breccia & perhaps have been ejected & caused dip. — NB Much of the Porphyry is amygdaloid with Agate. —

[17] There is a group of Peaks like Volcano here not known to be active

General appearance of the hills singular peaks & ~~ridges~~ are much connected together by ridges — the whole valleys sides & /gullies/ sprinkled or dotted with evergreen trees &

bushes, the intervals are nearly bare. — very little pasture. — there is little wind — blue sky & clear — very few birds excepting an

[18]   odd note from Tapacola or Turco — wonderful few insects, dryness of soil — Plains old bays, now irrigated dry summer

Signoritas conversation, eyes — pious horror bewitching. at my having entered a church to look about me. Why not become Christian — our religion is certain — we are one sort of Christian. Would not hear it — Do not your Priests even bishops marry — what reasoning — attempts at /spelling/. ~~Ta~~

[19]   10. — Breccia near wells

11. — Dyke in do

12. — Calcareous conglom. Baths

14. Porphyry. From high peak

15. — do. South of Baths. the latter very abundant

16. — curious cellular porphyritic rock (amygdaloid with ?)

horizontal bed

9th [September 1834] Rode to the last house in the valley, short distance from where the valley of the Cypressos & Yeso divides — real Cordilleras scenery — the size of the valley most striking especially geological —

[20]   crossed the torrent.

On the road noticed block of granite several hundred feet above bed of river in depression where present causes could not have placed them. Yet I think it probable that Granite /exists/ in valley del Yeso, & placed in present place by former sea. I noticed about 200 feet above present brook pieces of the Conglomerate

[21]   as at baths, & a band of enormous Rocks of a cellular porphyry which, does not occur immediately on the spot. I believe this to have been a line of former sea beach

I ascended a very lofty mountain. I should think ~~Mon~~ nearly 6000 ft. at its base I saw a mass of white rock, with its edge entangled with

[22]   the porphyry. — the main mass of the hill is porphyry — in many parts of which ~~it assumes~~ the breccia structure was clearly visible. — even in the Pitchstone Porphyry near summit I could see it. The Porphyry is often red like brick, with white spot — near Summit columnar. Very summit was capped with white rock

[23]   singularly entangled with the Porphyry. — I suppose some igneous rock —

with the Porphyries & Breccia there was some iron greenstone & varieties of cellular porphyry, much of this I saw in the form of large dykes others were massive pieces, the origin of which it was difficult to say — the

[24]   most remarkable fact. V. X. /as/ < > the Cordilleras. < > nearly horizontal arrange=ment < > the Porphyry & the lower probable /line/ < > as /numerous/ < > strata: the < > dips did < > very regular there < > have

[25]   been one of some degrees, without my perceiving < > Particular points /where/ the white rock was < > the strata were more disturbed

Pinchero pass — few Condors

Rande

17 — Pitchstone Porph. near summit — columnar

18 — Bed beneath conformable to < > Breccia

19 & 20 — the Paradoxical bed, most < > /having/

[26] flowed in dykes

10th. [September 1834] —

N$^{r.}$ (21 & 22) ~~The~~ A rock abundant in the Peak South of Baths, the relation of which I am quite uncertain whether to consider injected & contemp. with Porphyry it is immediately connected with it. —

also amygdaloid with Agate

N$^{r.}$ 22 The **upper part &** cellular of 23

[27] which is same with 16. — These rocks form a nearly horizontal bed between two lines of hills with dip. — the upper surface is highly cellular I am almost inclined to think that it has burst forth & formed a plain. Saw some broard dykes. —

There was some little <u>altered</u> Slate with the Breccia Porphyry. —

[28] X

the nearly horizontal arrangement of the great beds of Porphyry & immense valleys shown by them — there might have been small one towards me — great dips only occasional & as well without as within the Cordilleras.

[29] 17. Pitchstone from near summit. Columnar.

18 — Bed beneath, but conformable to Breccia — Porphyry

19, 20 — the ambiguous Dyke rocks

[30] 11th [September 1834] Miserable wet day

12th [September 1834] Have seen but few Condors — yet this morning 20 together soaring about. Man said at once probably a Lion if ~~afte~~ a Vaccano see the Condors alight & suddenly all fly up, they know the Lion

Audubons theory—

[31] is watching the Animal which it has <u>covered</u> with branches & is watching. — A Lion once hunted never covers up his prey, but eats it & retires to a distance: escape like fox by artifice of returning close to former track. — single dog a breed will kill one — if up tree with lazo [lasso] or stones

[32] makes noise when hungry — — killed a woman & child lately & man at Jajuel!! central line of Granite in T del Fuego distinct from Porphyries.

In 1822 Spring stopped for one year, returned little by little — but never so much water or so hot — known by feathers coming

[33] from a fowl — same proof for summer & winter. —

~~12~~ 13th [September 1834] At last escaped from our foodless prison — threatening day snow low down: We reached the Rio Claro by night: the plain seems certainly much affected by the river—

~~13~~ 14th [September 1834] Soon after leaving the town pass an Angostura,[746] to which the plain comes up level & the road crosses a round basin 30 or 50

[34] feet lower coming to another Angostura ~~pass~~ which gradually rises with a river exactly in same manner as at the 1st August: drains ~~gr~~ north part of great plain of S. Fernando. The city itself stands at foot or large hill insulated by the Quillota = like plain which runs to the sea & the great Southern plain which extends to

[35] Talca (?) — appears perfectly horizontal with <u>few</u> islands; to give an idea of its extent, the distant Cordillera only showed their snowy part, as over the sea. — Scenery very interesting — Had often wondered how Tierra del F. would appear if elevated: — no shells — Mem.: East St. of Magellan: exact

[36] identity between here & South of C. 3 Montes:
Valparaiso to Mendoza 3 to 50 chain of hills
Is not this country same as a low Quito: elevation decrease to South?
Pebbles of similar granite flesh colored — feldspar in small Quantities & black mica also here & Cauquenes a white syenite abundant
plains of S. Fernando

[37] saw during the day much of the regular Breccia
small dips —
18.th [September 1834] Lizard[747] above blackish tail ~~very~~ basking on stones in sun half of body brilliant ~~greenish spot~~ blue scales. — anterior greenish. —
colors shade down till some individuals are simply brownish black with transverse black bars, & the [formost] scale in head colored
white breasted Creeper of Chiloe
The Fungi[748] on the Roble or oak tree footstalk longer. — shape more irregular. color paler — cups inside much darker color: fewer of them. — in young state large cavity: — occasionally eaten grow to large size 3 or 4 times any of my specimens.

[38] The mass of hills behind. Yaquil which lie between the ~~valley~~ plain of the river & S. Fernando & that of the Cachapol. is fronted on the East as we have seen by the Breccias & Porphyries: As far as form is concerned they resemble heap of mud washed by water with no observable direction. —
The mountains are composed of crystall. — feldspar & quartz & more or less of green semi-cryst hornblende? (N^r 23). they become so fine grained in a few places as to be Eurite. — the rock is generally

[39] very much decomposed & soft to a great depth. it is ordinarily of a very uniform character: strikingly resembles in many respects gneiss of Valparaiso. Hornblende replacing mica. — it <u>often</u> contains. where most decomposed large round concretions of more crystall: varieties, is traversed by many small veins, many of which run parallel. — is auriferous as

---

746 Spanish for a narrow pass.
747 Specimen in spirits 1063 in *Zoology notes*, p. 354; listed as *Proctotretus tenuis* in *Reptiles*, pp. 7–8, plate 3.
748 Specimen 1065 in *Zoology notes*, p. 252, and *Beagle plants*, p. 228.

at Valparaiso. — many of the main veins runs about N & S — I saw specimens of a fine grained

[40] micaceous sandstone with regular cleavage. — passes into it? — I could see very little cleavage & nothing of stratification. — Know nothing of origin. — An enormous plain to NW. — said to extend to Melipilli. & from there to Casa Blanca. — From the /crooked/ inland bays. — even the flat valley of river could not have been levelled by river. —

Saw the Descabozado &

[41] heard certainly of Peterra.

The whole of these mountains well known from very numerous gold mines & some lavaderous.[749] — It occurs with pure — quartz rock They may be generally considered as auriferous pyrites. This is Bajos iron, in the Durazas copper Pyrites.

(there are many veins of Iron ore — micaceous iron ores &c. &c.) — where this occurs the pyrite does not occur gold is not found & if pyrites become pure in a mine it is at once better to give it up. — the better ones sorts are soft

[42] & smally crystallized. those of less value contains numerous large crystals of quartz. Carb. of Lime. — But in Durazas there are the /richest/

Pyrites confusedly blended & masses of copper (the usual cementing substance is a silvery grey fine scaled Talc. — (There are is also a black mineral crystallised in stars N^r 24). —

All the best mines are here formed by a cruzero on a that the line of intersection

[43] of a large stony vein or veta with contains small subordinate metallic ones gias (or veins). — with one or more larger small (but yet richer than the contained ones) gias metallic gias. — The whole stony veins with its included gias become infinitely more metallic with the Auriferous pyrites. — In the case N^r 2 where 2 good gias cross at veins? different ∠ &

[44] & directions — there whole included mass is metalliferous following the shape direction of the gias. — it may be observed that the influence extends beyond the wall to the section of the gias. —

In this case the included gias themselves become infinitely more rich. — These included gias are in reality much the same as those which cross the Veta

[45] as they only run parallel to it for a short distance, leaving the vein at the same ∠ at which they entered it — on the number & quality of these gias. & when when crossed by the larger gias the richness of Cruzero depends. — In those cases, where the metallic vein or gias is in its whole course parallel to the vein it

[46] is called Veta Real. & these veins almost always run nearly N & S. —

749 Lavaderous = washing gold ore.

(NB. — Cleavage) Where veins cross each other at great $\angle^{s\cdot}$ — one is often shifted by the other or fault is formed. — the veins at a short distance bends back to old line. — these faults seem to bear no relation

[47] to size of veins — take place more generally where veins intersect each other at great $\angle^{s\cdot}$ — (NB some of the gold occurs in quartz rock) The gias run in every direction. the main veins perhaps N & S. — The <u>metallic</u> veins seem to attract each — other The better gold veins seem superficial. 300 ~~yard~~ Varas is the greatest depth to which any good metal has been known

[48] to extend. —
Copper Pyrites does not seem to follow the same laws of increase as the Gold. — The Gold ores are ground washed & amalgamated the same substance after repeated washings continues to give a little gold. the sediments contains much copper — much Sulph – Copper effloresces. — it is very curious to observe that this substance divides itself into a curious

[49] (R. Quinderidica)
(S. Fernando) *<c. 7 lines excised>*
sort of Breccia. — the angular polymorphous fragments, (few inches each way) are nearly black. imbedded in a yellow soft substance have a regular

[50] cleavage can only be broken by a hammer. — were *<c. 7 lines excised>*
gold $\therefore$ /necessary/ than the iron
Rode up to the Mine. Chilean oaks. <u>Roble</u> reminded me of English oak. from picturesque manner of growth.

[51] Saw lake of Tagua-tagua. miners pale. nothing but beans & bread. — like latter best. but not so strong for work. — wives once in three weeks — carry up on back one quintal = 104 Lbs on back. perpendicular depth 150 $y^{s\cdot}$ — miserable poles with steps. — do not appear muscular beardless boys 6 & 7. dollars a month. support family. — in Hacienda men have nothing to eat but beans (Curious how every place become penetrated with

[52] 30 dollars from Court yard. — washing interesting operation — feudal system of men slaves. — Old Spanish lawyer. — considered for long time. no man so rich that he sends naturalist for pleasure. — there is a "cat shut up here", do you think King of England allows man to explore stones & reptiles. —
Don Pedros. Heresy. having snakes & feeding Caterpillars.[750]
Mules on level road 4 quintals. — 6 mules

[53] to each muleteer. — wonderful so slim an animal. —
14
12

---

750 An anecdote related by Juan Renous, a German merchant and naturalist resident in Chile from 1825, see *Beagle diary*, p. 261: 'two or three years ago [Renous] left some Caterpillars in a house in S. Fernando under charge of a girl to turn into Butterflies. This was talked about in the town, at last the Padres & the Governor consulted together & agreed it must be some Heresy, & accordingly Renous when he returned was arrested.'

168 4

~~Volcano of Peleroa said~~

19th. [September 1834] Passed Colchagua, & followed the valley. — examined one round island. — composed of quartz. — this valley must run about NNW — the whole plain is formed of pale reddish non-Calcareous Tosca rock: in some places conglomerate. — At the miserable place where we slept (nothing but beans & milk)

[54] the plain is uneven to the NW a great Bay runs & to /N/ the rivers bend up & then must again turn to W, flowing through the plains which form the whole of that country 20th [September 1834] We crossed the Bay & the ridges which formed its Northern side. These are formed principally red feldspar & quartz alone occasionally with mica — in many places with Hornblende

[55] These were traversed by fine-grained, porphyritic iron-greenstone — running about N & S. = The levelness & great size of plain is not perceived till it is approached. on account of the very numerous & broard valleys which intersect, the plain seems to dip to East & valleys run in same direction, to the great inland sea 20 or 30

[56] miles to the hills behind Rancagua & N & S of it. — These Present streams totally inadequate to form these valleys. I have no doubt the sea — — some steps in the ∠^r edge of wall, were visible marking without doubt the residence

[57] of different elevation — (Islands of granitic rock). — These walls & flat valleys — & barrancas upon Barrancas — have rather a strange appearance. — Bishops Cave irregular — smooth, passages pebbles, none near, formed by sea used gold Indian burial place — The upper beds generally coars Breccia — Conglom

[58] beneath this (perhaps generally) hard, softer hard, softer earthy indurated sandstone ranging in fineness in horizontal layers containing lines of pebbles. — General appearance quite Pampas Climate & soil similar, blacker & damper — no trees. & green plains

the numerous barranca

[59] give it another character — very numerous sheep

~~no~~

21. [September 1834] — Rainy day & I very unwell stopped at kind Chilotan schoolmaster — in Rozario. —

22nd [September 1834] Rode on to Navidad the very point house Breakers lying off the Point about one league or more seaward. mouth of Rapel descended from great plain must be very high 1000 to

[60] 2000 — front very like Valparaiso — but longer — plain ascends gradually from the cliffs, but I think must be distinct from the highest one. —

No visible chain of hills between coast & Yaquil. —

Cordilleras appear much higher — Planchon

[61] & North of it very low.

(23^d) [September 1834] The Barrancas here are about 100 to 150 feet high — I saw some Concholepas & Fissurella lying on the surface. —

Descending from the great plain I believe all the beds are similar — consisting of pale yellowish earthy fine sandstone, with numerous ferruginous thin veins & concretions

[62] shells, fruit of tree wood. —

Barrancas, consist entirely of above, varying horizontally but to a very slight degree in texture, excepting numerous <u>thin</u> beds & lines of flattened concretions (general law of forming beds) of grey crystalline calcareous rock — in one

[63] part there was much small conglomerate, porphyries & quartz — which passed into a very coarse one — in midst of one fine bed, great boulder, yards square of greenstone. — mixed with numberless fragments of shells, which do not occur very regularly in any ~~place~~ any one bed

[64] but abound in different places. — The concretions contain the largest — Pectunculus Oliva. & most abundant Turritella also, & Fusus — much wood charred in layers, some petrified with Serpula Fishes teeth. —

Most hospitable house unwell yet —

[65] Barrancha above high-water mark —

Great line of escarpment shortly after starting

 Shells in parts high up in layers, in <u>ravines</u> & plain, Bucalemu in Quebrada Onda[751] having passed Maypo

[66] (Maypo shells) all monodons

~~3 or 4~~ 2 or 3 feet above present brook. All north of Maypo Gneiss

SSE ½ S — decomposing

[67] 24 & 25th [September 1834] to C. Blanca

26th [September 1834] at C. Blanca

27th [September 1834] Returned

The Basin of Casa Blanca connected by Porte-suele with other basin — This /by/ other Porte-suele with whole granitic country North of Maypo

[68] Tufted Bird nest ~~middle~~ end of ~~Augu~~ August —

Black Furnarius young end of September

Wren — middle of October

F. Diuca[752]

751 Quebrada Honda is a mountain region in Chile.

752 Specimen 2172 in *Zoology notes*, p. 249; listed as *Fringilla diuca* in *Birds*, p. 93.

Contest of opinion in mountains believe torrents can do anything — in plains front of
Carpana — sea must have in modern time been up there: in

[69] present state rivers do little. sand flows down big pebbles remain pretty stationary till
floods, when they are replaced in few places touch the side of valley. but /causing/ to hus
band man the widest devestation in valley —
fall of 1700 ft in 50? miles, hence rapidity of river

[70] The Action of sea in continuing during its depression these valleys owing to greater
extent of beach — certain action tides. & streams ~~great clime~~. — & exposure to West
swell: Rivers subsequently modify plain —
Captain of Port Rise of Land

[71] It is said that the sea 70 years ago reached foot of Dr. Stiles[753] House?
17 years ago up to cellar of Mr Alison house?
The /nucleus/ chain, cause of diff formations on *[illeg]* *[diff]* chains of mountains which
have sometimes been observed. —

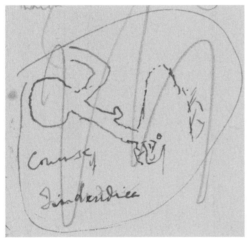

course of [river] Tinderidica

[72] Comparison of Patagonia. Pampas & Mendoza with the plains starting with
supposition of horizontal upheaving:

16 / 5000 (313
  48
  /120/
  /112/
  46

14
20

---

753 John Stiles, American botanist residing in Valparaiso.

280
56
336
Miers states each mule in mining district carries 312 Lb:

[73] Feb. 10ᵗʰ. [February 1835] —
Sun sets. ¼ before 7
Rising … 5 + ¼ —

[74] 30.450
30.084
AT. 67°
D 63°

[75] Volcan. Copiapo
V Coquimbo
V Limari
Pates ?
V. Aconcagua *
Tupungata
V. Maypo — *
V. Curico (same with Peteroa ?)
V̶ El Descabezado
V. Chillan
V. Tucapel o. Antuco
V. Callaqui
V. Chinal
V. Villarica
V Hotucoto. (to the E).
V. Chigual — *
V. Rauco
V. Guanahaca o. Rananahuca

[76] V. Osorno
V. Quieh
V. Puranaqui (our Osorno.)
V. Mine himadiva. (—flat top)
El. ! Corcovado
Mellamoy. —

The * signifies, that these volcanoes have been super-added to the map: —
This map is reduced from one of Dalbe's Republic of Chili surveyed &c &c in 1819. —[754]

---

754 This work has not been identified. Joseph Albert Bacler d'Albe (1789–1824), a French officer and engineer who fought during the wars of independence in Argentina, Chile and Peru; Principal engineer of San Martin, Chile in 1818. See Puigmal 2006.

[77]

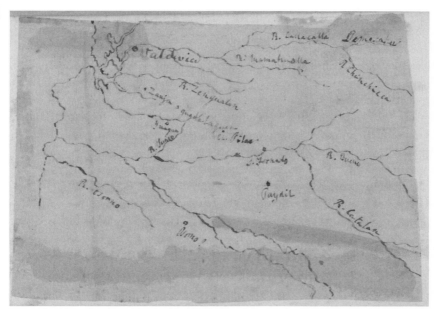

[traced map of region around Valdivia]

[78 blank]

[79] (11ᵗʰ—) [February 1835] Went in Yawl to Valdivia Stopped on road; found old man who made spirit wine — Cyder & treacle from apples. = Olives sometimes bear fruit = & grapes. — Soon procured, from obliging Governor — after crossing in a canoe a small river entered forest: 2ᵈ cleared spot slept at: forest rather different. — diff: proportion of trees: — Indians parties; the 3 Caciques who had just arrived. —

[80] Nearly all the road true Mica Slate dip very various. some SE by S — but generally NE by E. — (Compass). — Covered as in Chiloe by Clay & round pebbles of quartz. — Country very undulating, rises in the interior, no inhabitants. =
Meloe poison & Caustic[755]
Upright Bamboos

[81] Lima Junio: 14: 1765 *[Por]* Andres Febres.[756] Spanish & Chilian & Peruvian Volcanos none but Calbuco SE (Easterly) Compass. – & Villarica
On Road, Saw (Granular) Gneiss & Mica Slate with regular dips, but inclined with very little or no regularity: Bright red soil. —
After travelling some hours through forest (tormented

---

755 A blister beetle. '[Specimen] 2546. Meloe, common. crawling about grass and flying about, Cudico, S. of Valdivia. The Padre told me, that the Indians use this as a poison, and likewise apply it as a caustic or Blister.' *Darwin's insects*, p. 87.
756 Febres 1765.

[82]   by innumerable flee bites, pigs, dogs, & cats) — began to open a little park forest
       scenery & become more level
       the curious fact of plains banishing trees
       ~~Near~~ At Cudico, pale coarse-grained earthy (that is particles of different degrees of
       fineness. — Sandstone which is said to extend over whole plains South of Osorno

[83]   View of Llanos very pretty, intermediate mamillated district: (Heard great guns in the
       morning) corn & brown grass & great plain in horizon & cottages of Indians. —
       Discontented padres, is easy to teach the Indians very industrious. —
       Appearance of Indians curious universal desire of wearing their long hair
       broad faces; browner more expression — far less civil —

[84]   Speak same language at Chiloe & Here
       Many less Earthquakes here than in Valparaiso
       Gold mine in mountains of Plains. — José Xavier Gruzman — El Chileno
       Instruido
       Valdivia — gneiss — Auriferous
       Quicksilver mine — Quillota

[85–6 excised]

[87]   *[Anchor swell]*
       Indians
       Padre *[extreme Form of Country]*
       Mist for rain
       (Fleet of whalers)
       (24th [February 1835] — Arrived at Mocha) Having sailed 22^{d} [February 1835]

[88]   It is important to remark the great transportal of gravel in Patagonia is <u>unaccompanied</u>
       with Faults — as also in Chiloe. — In Contradiction to M^r De la Beches argument
       P 160[757] — Nor any signs of sudden debacles only where channels existed. —
       Was there a fault at Port Desire S. Barranca.
       They have travelled on each side from Cordilleras
       (Waves from Earthquakes

[89]   generally poco à poco, ask at Valparaiso. & Concepcion —
       It appears to me that the two sorts of upheaval are so distinct, that the effects on water
       (& hence debacles) cannot be judged from analogs of one to the other. —

---

757 De la Beche 1831, section 3, argued that there was a positive correlation between the degree of
    faulting observed in the bedrock in many areas and the occurrence of gravel deposits. He drew
    attention to the fact that rivers often exploit faults as lines of reduced resistance to erosion and that
    these valleys often contain gravels. He also suggested that faulting indicated disturbance which he
    speculated might link to increased agitation of water masses and hence increased gravel deposition.
    CD observed that the thick gravels of Patagonia and Chiloe are neither clearly associated with
    faulting nor with 'any signs of sudden debacles'.

Not many /blocks/ in Chili?

Even many of the facts in de la Beche can be accounted for beaches

[90] (1)

Every conclusion is of consequence with respect to Erratic Blocks —

That they have moved beneath the sea subsequently to existence of chain of mountains like the Andes. — mixed with gravel — that they have travelled in opposite directions from ridge of Cordilleras; that this took place during upheavel of land. quite gradual only small steps. — That there is no particular evidence of faults. = $\angle^d$ nature & analogs of Boulder in true Tertiary strata of Navedad.

[91] (2)

That they thus travelled at two very different periods in Patagonia. —

At S. Cruz high up, where Lava blocks disappear, I should think they were buried. & that the latest epoch of the /presens/ of the sea, carried the Primitive block to that locality. — Anyhow the ~~obs~~ alterations of kind of these blocks proves that they are not progessing in one direction from present cause. —

The vertical movement described by Playfair[758] assisted by earthquakes, down inclined smooth surface. —

[92] (3)

I think the ~~gr~~ step-like upheaval of the land, assisted by earthquakes ~~would~~ is the index of explanation. —

V. De la Beche In Europe the N & S. lines may be explained by channels. — the accumulation at base of escarpment. — the lines of same height are all /presenations/ of same fact: of gradual marine ~~deposi~~ transportal: — In the case of being behind Mountain ranges. —

We may imagine tidal action of surf occasionally to have same effect of rescuing from the main line of depression as eddies are supposed to have done in N & S Debacle. —

[93] (4)

The almost established fact in La Plata — that the great quadrupeds are there marine deposits, renders it not improbable that many of the Diluvial deposits in S. America have thus been formed. — I much suspect this to be the case. —

Re-state the <u>numbers</u> of shells found on plains of Patagonia:

In ~~An~~ drawing in De la Beche is exactly that of the figure of Many of the blocks.

The transportal has taken place where there has been no narrow currents ~~as Lyell?~~ ~~supposes~~: not probable glaciers—

758 Playfair 1802.

[94] 5

Between Chiloe & Mainland —
bottom is (V. soundings)

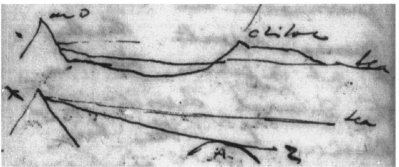

[section of Chiloe and Andes] Andes — Chiloe — Sea X A Z sea
A. Hard nucleus of Chiloe, subsequent elevations might cut inclined line XZ into a
curve — ??
Depth of Channel too great ? —

[95] 6

Mr Caldcleugh saw Guanaco near Cordovean range
Again I think M: Hermoso bed coeval with Pampas diff. mineral: constitution owing to
proximity of S. Ventana ?
As in Pacific a Corall bed forming as land sunk. would abound with those genera which
live near the surface. (mixed with those of deep water) & what would more easily be told
the Lamelliform Corall forming Coralls. —

[96] 7

I should conceive in Pacific wear & tear of Reefs must form strata of mixed.
broken sorts & perfect deep-water shells (& Milleporae). — Parts of reefs
themselves would remain amidst these deposits, & filled up with infiltrated
calcareous matter. — Does such appearance correspond to any of the great
calcareous formations of Europe. —
Is there a <u>large</u> proportion of these Coralls which only live near

[97] 8

surface. — If so we & may suppose the land sinking: I believe much conglomerate on
the other land is an index of ~~land~~ bottom coming near the surface.
If so Red Sandstone Epoch of England will point out this: Mountain limestone[759] the
epoch of depression. — Do these numerous alternations of these two grand classes of
rock point out a corresponding opposite & repeated /motions/ of the surface of that
part of the globe.

759 The Carboniferous Limestone of the English Pennines.

[98] 9

Do not the Sandstone (generally ~~sat~~ containing subordinate beds of pebbles) have imbedded wood (or other land remains??) which at least renders probable the proximity of land or uprise of grands oceans bottom: Consolidated gravel occur in Transition series. in the old Red Sandstone Millstone grit & coal. Sandstone. in lower part of R. Sandstone. — in sand strata beneath chalk. in gravel beds associated with plastic clay. & of course diluvium Phillips. introduct.:[760] ℞ vegetable remains with do — P. XV

[99] 10

Entire absence of fossil large quadrupeds in Chili conformable to what we suppose the original form of S. America continent. —
The /comminuted/[761] bits at Lows Harbor in <u>state</u> & kind I certainly believe are recent Height at Red ~~Marl~~ Sandstone P 299. (300–400 ft).
No. organic remains consult Daubisson.[762] — De la Beche.! —
<u>Conglomerate separates the Magnesian Limestone from Red sandstone. — & the latter from the Coal Measure</u>
It has been remarked that beds of conglomerates or gravel, generally separate different formations, now.

[100] 11

the very distinctness of formation implies, that an interval time or change of surrounding circumstances has taken place in the regular successions of deposits. —
— The different formations as seen in England are not of that different height as to allow a supposition that ~~when~~ the present area of /any/ formation was elevated into dry land. before all the superior formations were deposited. —Yet the step-like succession seen in /E/—W section would to certain extent point out such a

[101] 12

conclusion. — May we not imagine each band of conglomerates marks an epoch when that part of the ocean's bottom was near to a continent or shoal water; or that having again being depressed. calcareous fine sediments were deposited. (if under circumstance to allow of corall reefs. such would be very abundant). — In a long series of such /near/=undulations: the thickness of the deposition would cause the most ancient ~~finally~~ first to form. the present dry land. —

[102] 13

It will well explain immense wear & tear. —
Calcareous depositions in the deep ~~water~~ (& ∴ distant) seas owing to facility of solution. — For instance I suppose, that ~~when~~ the lowest conglomerate of new Red

---

760 Phillips 1816.
761 Broken into small fragments.
762 Aubuisson de Voisins 1819.

Sandstone,[763] or those upper beds of Magnesian Conglomerate containing fragments of Mountain Limestone were deposited in a shoal ocean ~~at near~~ at foot to the hills of older rocks. — that depression then commenced. sands were deposited. (Red color owing

[103] 14

old Red Sandstone & reddish Transition Limestone?) that as the ocean became deeper & land less in quantity & more remote. Lias took its place, ~~as fine sediment~~ & all the succeeding fine depositions took place. . —
perhaps in middle divisions ~~land rose & allowed of numerous corall reef-forming coralls.~~ — there were alternations, certainly Corall reefs. —
at Period of Iron Sand with so much wood. we may certainly suppose that the bottom had been raised. — (It ~~m~~ cannot be raised solely by deposition

[104] 15

otherwise there would be no room for subsequent deposits) We may suppose the ocean sunk to rise again at the period of plastic Clay. We know that in this formation Pebbles of flint are found which could hardly happen without a coast line of chalk. — at this period outlying chalk islands existed: The Test of depression on in strata is where great thickness has shallow coralls growing in situ: this could only happen. when bottom of ocean was subsiding:

[105] 16

(NB Terebratula not always deep water)
Peat Bogs. — Maccullock Edinburgh. Phil. Journ. 1820[764]
If the circle of change has always gone, those unaltered conglomerates ~~ought~~ which immediately overlie the metamorphic strata, ought to contain other pebbles besides the ~~Granite~~ Metamorphic
In Alisons notes: mentions great fall of Mercury at store in Barometer of 1822[765] —
Is it mentioned in the account.

[106] 17

This is very good. — In transporting boulders. What would be the affect of depressing any great valley or plain with blocks at the bottom. Would it not sea send them up opposite slopes. — not applicable on account of such taking place in slope of Patagonia. —
That is that well worth considering
With respect to utter confusion of pebbles in Europe state after Hypothetical case of subsidence rounding escarpments.

---

763 The red sandstones of the English Midlands thought to have been laid down in arid conditions in early Secondary times after the Carboniferous.
764 MacCulloch 1820.
765 Alison to CD 25 June 1835, CCD1: 450.

[107] $\boxed{18}$

Mention Barnacle above level of /water/ at /Repel/. returned Cacique *[illeg]* — in like irreversible passes masses of cellular Porphyry. —

Mem at Rio. I suspect that Granite heated at bottom of ocean. — Was Granite ever covered? Lithomarge /appears/ to contain diff fossils. from harder rock to certain extent is it not more Auriferous? Crystals in Lithomarge fractured by admission of water to heated mass. —

Lithomarge found at mines & to the South toward S. Paulo — Prince Maximil road to Bahia

D^r. Forchhammers[766] fact about Brazil

[108] 19)

Amplify on importance of proving extent & recency: of upheavals & manner over whole America. — Explaining generally Continental upheaval. so important in understanding valleys, diluvium escarpments & successive lines of formations &c. — Showing that they are not mere Local effects, as so many authors suppose. —

[109] (20

The conjecture of Brazilian Granites being heated in deep ocean, is a Very important conjecture. —

？ Do not the successive terraces point out a gradual rise, whilst circumstances changing determine. which shall be a slope & which a cliff. — opposed by the plain of S. Cruz. & the equality of height in different plain. —

A more sudden & greater upheaval would probably determine this change of circumstances & so agree with all the facts. State this with clearness. —

[110] 21

Australia no Hydrophobia[767] & Burchell[768] says during 5 years he never heard of a case in the C. Colony

Tosca would be Limestone if deposited in a deep ocean

If the Pacifick Isl^{ds} have subsided there ought to be a peculiar vegetation

Is not the presence of Flint with Limestone a very common occurrence? Does flint frequently occur in any other strata? —

[111] $\boxed{22}$

The uprising of East Coast plain side of America, owing

The horizontal rise of America, a swelling of the fluid nucleus of the globe, to form the statical equilibrium. destroyed as Sir J. Herschel[769] suggests by the gradual, wear & tear of surfaces in other parts

Galapagos from the cont. 480.

---

766 Johan Georg Forchhammer (1794–1865), Danish geologist and chemist.
767 Rabies.
768 Burchell 1822–4, 2: 524.
769 Herschel 1833.

Juan Fernand (Mem: erupt): 315

Mem. Silicified wood so frequently found on surface, in proportion to within strata: (from greater duration) Humboldt gives instances. Vl. P 626[770]

It is a fact of some interest, as being different from Europe, that in Chili, the secondary rocks repose <u>conformably</u> & <u>pass</u> into the oldest Granite & Mica slate

formations &c &c. —

[112] 23

Forster P 156.[771] states that Capt. Davis in the year 1687 being 450 leagues from the main of America felt an Earthquake very strongly, when at the same time its most violent effects were observed at Lima & Callao. — In 1692 a Davies accompanied Cavendish

Investigate this case <u>well</u>. Mem: double occurrence at Juan Fernandez: nearly 400 miles (British) from Coast

This must be the same Davis, who gives name to Davis's Land:[772] & at Galapagos: account written by Wafer.[773] —

[113] 24

I observed in K: George's Sound on a sand bank, which never drys. being 6–8 ft below surface the same ripple marks as on tidal beach: it may help to explain waving structure in hard rocks. which could hardly have been formed on a beach.

Mem: in General observations. it is known. in Chili Earthquake that from mill courses, that the /ruin/ was greater inland. it would appear that the disturbance could not have been far to sea wards.? V. Journal of Science

Write Alison. to enquire whether the Millers at. Concepcion

[114] 25)

observed. whether the level of these dams were affected ~~from the~~ in their level. from the Sea inland or vice versa. —

Capt Fitz Roy states, that the older writers say, that the water in the river Cauten, Tolten & Bueno ~~rivers~~ & Valdivia rivers. — was deep enough for large ships to enter.

Vide Molina[774] — date of Molina at Guatemala I have read of similar case & Feuille & Frezier.[775]— Capt. Fitz Roy. doubts the old writers from position of river mouths to swell.

770 Humboldt 1814–29, vol. 6.
771 Forster 1778.
772 John Davis (1550?–1605), explorer who reported sighting an island in the Pacific 500 leagues from Copiapò (in 27° 30 S) and 600 leagues from the Galapagos Islands. Later explorers could not find 'Davis's Land' although it was sometimes listed on maps.
773 Lionel Wafer (1640–1705), Welsh explorer and privateer. Wafer 1699.
774 Molina 1794–5.
775 Frézier 1717.

Do fragments of shells in Tosca rock at Coquimbo blend into the solid Limestone, like rocks of Bermuda described by Capt Vetch. Fitton App. P 588:[776]

[115] 26)

Silex is well known to form globular concretions in Limestone rocks; Perhaps Limestone in siliceous stones may not be so common: It would appear however that it is the Calcareous matter, which determines the great spheres in the recent Sandstone beds: & the curiously shaped bodies in the Clay Slates of Port Famine. — M. Peron. at Benier I^{sd} on W. Coast of Australia. has described calcareous balls, formed of concentric layers. imbedded & intimately united with the surrounding "terra sablouneuse ocracèe" — Peron Voyag. Aux terres Aust. Vol I. 204.[777]

There are concretions of Clay Ironstone. Iron Pyrites & great Clay Shale balls. as at Bahia

Fitton App: P 619.

Introduce this somewhere S. Cruz ??

[116] 27)

Introduce in Cleavage discussion fact mentioned by Lyell that Modern Limestones elevated 1500 ft to SW ∠ 45°. :[778] NW strike in St Eustacia Maclure Lyell III V. 132 Officer of Isle of France says that it is said dogs never grow mad there. Vol I. P 248.[779] mentioned by M. Lesson[780]

Mem: to tell Lyell about crocodile in South Sea island. Mariner's Tonga:[781]

Carl Thuenberg.[782] says mountain in whole Cape. NW & SE. C of G. Hope

Mem. to make a note about cleavage. being owing to currents, as sands /act/ in lines on sand-dunes: also Herschels.

[117] 28

idea of equilibrium. acting on the nucleus of the globe. —

There is much resemblance. between Pann.nama West Indies. T del Fuego mark this symetry. —

& Sumatra: the Aleutians turn other way

In 1764 April & Febr. earth quake felt in Atlantic. on Equator 25° W.

---

776 William Henry Fitton was the author of the geological appendix to King 1827, which cited James Vetch (1789–1869): 'The cemented shells of Bermuda, described by Captain Vetch, which pass gradually into a compact lime-stone, differ only in colour from the Guadeloupe stone; and agree with it, and with the calcareous breccia of Dirk Hartog's Island, in the gradual melting down of the cement into the included portions, which is one of the most remarkable features of that rock.' pp. 587–8.

777 Péron 1807.

778 Lyell 1830–3, 3: 133.

779 Bernardin de Saint Pierre 1773, 1: 248.

780 Duperrey 1826–30.

781 Mariner 1817, 1: 334ff.

782 Thunberg 1795–6. The edition used by CD is not known.

Officer of Isle of France. P 98 I Vol:[783]

N.B. in general discussion introduce diluvium generally submarine

Examine old note Books. about Tufa. Mr Lambert[784] statement that Granite hills capped with horizontal volcanic rocks. in favor of subaqueous deposit of Tufa. — Used

[118] 28

It must be re re urged that what ever has caused cleavage, has in primarized districts, divided mineralogic substances.

Mem: depth of mud in Dr Mulgrave sounding at N Pole: — (Playfair ?)[785]

introduce in cleavage paper[786] when S. Cruz balls are discussed in aid of separation of ingredients

> Mem: state that in the Falklands the layers of Slate & Sandstone are nearly 20 ft high: | ∴ capable of movement in the depth

Ought I not to state that my metamorphic ideas obtained from Lyell. III Vol. might be put in a note?

In discussion on Porph: Breccia, I

[119] 29

should state to gain confidence, that it was sometime before I fully comprehended origin

Introduce into my general discussion the gneiss & Quartz of central Patagonia, on coast

The composition of Mica Slate must be owing to "hot" metamorphic action acting in same planes as those of the Clay-Slate, which were determined, in a liquid shortly after period, of deposition. — This is valuable argument.

In Falkland paper. nothing is said about, uplifting laminae not affecting vertical strata

[120] 30

The Coral theory rests on the supposition of depressions being very slow & at small intervals

Can it be believed that any heat, as for instance from a dike, ever convert a shale or slate into Mica Slate: is there any tendency even to the separation of the different minerals into laminae — no —

Then is it not

[121] (31

probable that the patch which formed rocks like mica slate was accompanied by other principle Electrical current

783 Bernardin de Saint Pierre 1773, 1: 98.
784 Charles Lambert (1793–1876), a Franco-British entrepreneur who made a fortune with copper and silver mining and smelting in Chile.
785 Playfair 1802, p. 415.
786 Possibly DAR 41.59–77.

The laminae of feldspar are in certain lavas places in one direction.

In the Cordilleras dikes have not commonly affected the nature of the strata?

[122]   32

Discussion on presence of Salt at Lima: read Humboldts account of grand salt formation. — Arica

The tilted alluvium in the plain of Uspallata proof that central chain there elevated before the flank one.

I do not see certainly that the Uspallata formations are unconformable to the Gypseous. although probably so: certainly from Alluvium elevated

[123]   33

posteriorly

Sierra Parince. & partial chains. extend in N 85° W line: (not introduced into my paper). Humbol. VI. P503 & 519 & 527[787]

Before concluding the Cleavage paper. consult the VI Vol. of Pers. Narra.[788]

Ascertain nature & position of all the strata in all the basins, on summits of Andes — New Spain essay: Superposition: Dessalines D. Orbigny. Titicaca.[789] —

[124]   34

S. America fundamentally & systematically different from S. America; shows that the geology of the world cannot be taken from Europe

[125–6 excised]

[127–9 blank]

[130]   Wild g dog of Australia copulates freely with the tame ones near of the houses.

—Note very different, from the domestic dog

[131–8 excised]

[INSIDE BACK COVER]

M. Desjardin[790] was the man, who found Dodo bones at Isle of France

De la Beche P 142

Stokes a quire & a half of foolscap

In Falkland paper 2 pages contains 2200 syllables in one page — 2500 of Beecheys /J/[791]

I have about 900 pages

The Captains papers are in quarter quires or 6 sheets.

787 Humboldt 1814–29.
788 Ibid.
789 Orbigny [1834]–47.
790 Desjardin 1832.
791 Possibly a reference to Beechey 1831.

[BACK COVER]

Santiago Book.

Textual notes to the *Santiago notebook*

[FC]   50] *in pencil, not transcribed.*

[IFC] DUFOUR … PARIS] *printed label.*

          With 31 pp at end with draft for geol. papers & good remarks. Santiago.] *a white paper label pasted in, in pencil, not in CD's handwriting, not transcribed.*

          88202338] *English Heritage number, not transcribed.*

          1.18] *Down House number, not transcribed.*

          9] *added by Nora Barlow, pencil, not transcribed.*

[24–5] *the pages were stuck together, then parted badly, so that much is difficult or impossible to read.*

[32] distinct … Porphyries.] added *ink.*

[33] '13th' '14th'] *dates corrected in ink.*

[37] day] *added ink.*

          18.th … specimens.] *ink. From here to p. 57 (19 September) the text is in ink.*

[41] This is … Pyrites.] *interlined.*

[42] But … [richest/]] *interlined.*

[49] *a 72 × 88 mm section is excised leaving stubs at the top and bottom.*

[50] *page partly excised.*

[53] to each … ~~said~~] *ink.*

[67] Casa] 'asa' *added ink.*

[71] It is … House.] '?' *written over these lines.*

          sketch and caption in ink.

[72]   Comparison ... upheaving:] *ink.*

[73]   *page in ink.*

[74]   *a fragment of gummed paper suggests that something was once glued to the page.*

[75–6]   *pages in ink.*

[77]   *map on tracing paper glued to page perpendicular to the spine in greyish ink.*

[81]   Compass. -] *added in ink.*

[90–124] *numbered on the top corners by CD as (1) – (34). These numbers are in ink, even if the page is written in pencil.*
That they ... Navedad.] *in ink.*

[91–104] *pages in ink.*

[105]   (NB ... water)] *ink.*

[106]   This is very good. —] *added pencil.*
In transporting ... Patagonia. —] *ink.*
or plain] *added pencil.*

[107]   Mention ... Porphyry. —] *overwritten by added ink paragraphs* 'Mem ... to contain'.
Mem ... road to Bahia] ink.
D^r. Forchhammer ... Brazil] *added ink.*

[108–9] *pages in ink.*

[110]   Is not ... strata? —] *ink.*

[111]   Mem ... &c &c.] *ink.*

[112]   *page in ink.*
In 1692 ... Cavendish] *pencil.*

[112]   *page in ink.*

[114]   observed ... versa. —] *ink.*
at Guatemala ... case] *added ink.*
Do fragment ... P 588:] *ink.*

[115]   *page in ink.*

[116]   Mem: ... Herschels.] *ink,* 'says' *and* 'C of G. Hope' *added pencil.*

[117]   idea ... other way] *ink.*
nama] *added pencil.*
Examine ... Used] *ink.*

[122–3] *pages in ink.*

[123]   Before ... Narra.] *pencil,* 'concluding' *underscored in ink.*

[130]   Wild ... houses.] *ink.*

[BC]   /50 P/] *pencil, not transcribed.*

THE *GALAPAGOS NOTEBOOK*

꙳

The *Galapagos notebook* has 50 leaves or 100 pages and is bound in red leather with the border blind embossed: the brass clasp is intact. The front of the notebook has a paper label with 'Galapagos. Otaheite Lima' written in ink.[792] It was written in two sequences, pp. 1a–34a and pp. 1b–66b, primarily between August and November 1835. It is currently lost and is, therefore, available only as black and white microfilm images taken in 1969.[793] About one third of the notebook deals with Darwin's visit to the Galapagos in September–October 1835.

Nora Barlow, in her study of the field notebooks, stated that 'there is disappointingly little in the Galapagos pocket-books of interest'.[794] Yet when the *Galapagos notebook* was first published in full on *Darwin Online* in October 2006 there was enormous world-wide public interest. The number of visits to the texts of the field notebooks on *Darwin Online* in April 2009 showed a massive level of interest in the *Galapagos notebook*. Working transcriptions of all the notebooks were available online since October 2006, and although there were unexpected differences in the numbers of visits to particular notebooks, few were accessed more than four thousand times, and four of the notebooks were accessed less than two thousand times. By dramatic contrast, the *Galapagos notebook* was visited 62,000 times.

This great and growing interest in the notebook is mainly because Darwin, evolution and the Galapagos are so closely and so famously associated. In many popular accounts the voyage of the *Beagle* is almost synonymous with visiting the Galapagos Islands. Despite Frank Sulloway's decisive refutation in 1982, it is still widely believed that Darwin 'discovered' evolution on the Galapagos.[795] This belief is constantly reinforced in TV programmes, travel brochures and so forth.

The belief that the Galapagos are where Darwin became an evolutionist is a mid twentieth-century version of Darwin's life story. If anywhere can be said to be the

---

792 In his *Beagle* writings CD used the name Otaheite less often than Tahiti, which is the name he used in the *Beagle diary*, zoological diary (DAR 31.345) and geological diary (DAR 37.798–801). CD sometimes used 'Tahiti (Otaheite)', as for example in *Coral reefs*, p. 6, and *Volcanic islands*, p. 25.
793 It has been missing from Down House since at least 1985 (CCD 1: 545).
794 Barlow 1945, p. 247.
795 Sulloway 1982b.

geographical source of Darwin's doubting the fixity of species it is the South American mainland, where he was struck by the key relationships between fossil and recent mammals and between the various 'representative' species of living mammals and land birds. These crucial observations, combined with his study of Lyell's *Principles of geology*, prepared Darwin for a break with received wisdom on species origins, as he himself stated in the opening lines of the *Origin of species* and as he explained with great clarity in the introduction to *Variation* 1: 9ff.

There were, however, many other observations made by Darwin during the voyage which contributed to his eventual development of his theory of evolution, including, for example, the mammals of the Falklands and Ascension Island, the absence (with a few minor exceptions) of indigenous mammals on New Zealand, the antlion of Australia and the dramatic differences between Englishmen and Fuegians.[796]

It remains the case, however, that Darwin's famous retrospective note on the 'stability of species' (discussed below), written in June or July 1836 in his 'Ornithological notes', was a result of considering the distribution of different kinds of Galapagos mockingbirds and tortoises. It is also true that Darwin's retrospective Journal (DAR 158) entry for July 1837 attributed great importance to the facts of the Galapagos (though not to the experience of his visit): 'Had been greatly struck from about month of previous March on character of S. American fossils — & species on Galapagos Archipelago. — These facts origin (especially latter) of all my views.'[797] In his *Autobiography* Darwin mentioned his accomplishments on the *Beagle* voyage, and listed the Galapagos thus: 'Nor must I pass over the discovery of the singular relations of the animals and plants inhabiting the several islands of the Galapagos archipelago, and of all of them to the inhabitants of South America.' And later in the *Autobiography*, p. 118, he listed the kinds of evidence that first convinced him of evolution.

> During the voyage of the *Beagle* I had been deeply impressed by discovering in the Pampean formation great fossil animals covered with armour like that on the existing armadillos; secondly, by the manner in which closely allied animals replace one another in proceeding southwards over the Continent; and thirdly, by the South American character of most of the productions of the Galapagos archipelago, and more especially by the manner in which they differ slightly on each island of the group; none of the islands appearing to be very ancient in a geological sense.
>
> It was evident that such facts as these, as well as many others, could only be explained on the supposition that species gradually become modified; and the subject haunted me.

---

796 See the introductions to the *Rio, B. Blanca, Santiago* and *Port Desire notebooks*. Many scholars have discussed these and other key lines of evidence, for example Browne 1995, Herbert 2005, Hodge 1983, Keynes 2004, Kohn 1980 and Sloan 1985.

797 Journal, p. 13r.

This passage names fossils, geographical distribution, and the organisms of the Galapagos as suggesting the mutability of species. Here and elsewhere Darwin consistently described the Galapagos as one of three important influences.

Darwin was strongly associated with the Galapagos long before he became famous for his theory of evolution by natural selection. From 1839 to 1859 Darwin was known to the general public primarily as a scientific travel writer. It was Darwin who introduced most readers to the exotic Galapagos. His *Journal of researches* remained popular well into the 20th century. Only decades after his death did the story of a eureka-like discovery of evolution while on the Galapagos emerge. After so many decades of association between the exotic islands and the birth of the theory of evolution, it seems the connection will endure for the foreseeable future.

In 1980 John Chancellor (1925–84) selected the Galapagos as the setting for his first of two paintings of the *Beagle*.[798] Several authors have reconstructed Darwin's excursions in the archipelago in far more detail than has been attempted for any other part of the voyage.[799]

As Armstrong 2004 stressed, in terms of Darwin's intellectual development it was fortuitous that he visited the Galapagos having just left the South American main-land. By February 1835 he had become convinced that species had 'life spans' and so had already made a break from Lyell on the issue of the 'death' of species. By the time the *Beagle* sailed for the Galapagos in September Darwin had not only rejected Lyell's explanation for coral atolls but was also ready to challenge the older man's rejection of any natural 'birth' of species and his prediction that islands would be '*foci* of creation'.[800]

The fact that varieties or species of organisms with apparently close links – not only to each other but also to mainland species – appeared to be confined to particular islands in the Galapagos, was a great puzzle for Darwin. He could not fit this into the theistic view of perfect adaptation to environment which he had imbued from his teachers such as Henslow and Lyell. He was also struck by the absence of insects which would certainly have flourished in the well-vegetated uplands, further undermining his faith in Lyell's view that species were created where the conditions would suit them. Darwin had already observed the striking similarities between the different 'allied' mainland species such as 'Myothera', occupying differing habitats, for example on either side of the Andes. It was not until he visited the Galapagos, however, that he saw how isolation might provide a mechanism for forcing *differences* between varieties or species in *similar* habitats.

Eight or nine months after the *Beagle*'s visit to the Islands, probably just after his sojourn at St Helena, Darwin drafted his 'Ornithological notes' and began to see

798 See Chancellor 2008.
799 See Armstrong 2004, Estes *et al.* 2000, Herbert 2005, Sulloway 1984.
800 Lyell 1830–3, 2: 126.

how the Galapagos land birds had actually gone one step further than diverging from their presumed mainland ancestors. They had, it seemed, started to diverge from each other within the archipelago itself, because the chances of three varieties of mockingbirds migrating exclusively to three separate islands were extremely small. John Gould's confirmation, in early 1837, of the American character of the birds and of their specific differences, struck Darwin profoundly. Here at last, along with the facts he learned from Owen about his South American fossils, was proof for Darwin that adaptation was relative and that descent with modification was the best explanation for the observed patterns.

In the *Origin of species* Darwin ended his discussion of the biogeography of oceanic islands with one of his key points about evolution by natural selection.

> The relations just discussed … [including] the very close relation of the distinct species which inhabit the islets of the same archipelago, and especially the striking relation of the inhabitants of each whole archipelago or island to those of the nearest mainland, are, I think, utterly inexplicable on the ordinary view of the independent creation of each species, but are explicable on the view of colonisation from the nearest and readiest source, together with the subsequent modification and better adaptation of the colonists to their new homes.[801]

At first glance there appears to be almost nothing in the *Galapagos notebook* which can be interpreted as a glimmer of understanding of speciation in the Galapagos. There are, however, clear indications that Darwin did quickly perceive the American character of the land birds and immediately asked himself if the plants of the archipelago might, if he only knew enough botany, reveal the same links with the mainland.[802] The notebook also contains Darwin's first reference to the finches which were named after him during the centenary of his visit to the islands in 1935.[803] The finches' beaks are still of great scientific interest today and are routinely cited as a beautiful example of adaptive radiation.[804] There are many other clear examples of evolution in action to be seen in the Galapagos, several of which were used by Darwin himself as evidence for his theory.

So the *Galapagos notebook* does not provide any clear evidence of Darwin's 'conversion', but this actually brings one closer to understanding how Darwin came to a new view of nature. Gradually during the voyage, and not by any 'eureka' revelation, Darwin synthesized all that he saw with what he had read and discussed with others and out of this mixture began to see a natural explanation for the otherwise obscure or mysterious 'birth' or appearance of new species. In the years following the voyage he was able to use his discoveries to build new theories for a

---

801 *Origin of species*, p. 406.
802 In *Variation* 1: 9 CD apparently implied that he perceived the 'American-ness' of the Galapagos cactus flora.
803 See Lowe 1936.
804 Grant 1984.

whole range of geological phenomena, of which the origin of coral reefs is by far the most elegant and ambitious. On this foundation he went on to construct a new theory for the origin of species, and it is far from a co-incidence that he thought of calling his first attempt at a diagram of evolution 'the coral of life'.[805] If Darwin ever did have 'eureka' moments they were his reading of Malthus in September 1838, which made him realize the effects on reproductive success of mostly destroyed superfecundity, thus revealing a driver for natural selection, and 'the very spot in the road' in the mid 1850s when the solution to the problem of divergence suddenly occurred to him.[806]

Summary of the notebook

The March 1835 entries are on the first seven front pages and perhaps the first ten back pages, although none of these early back pages is dated. The March entries relate to Darwin's trip from Valparaiso to Santiago in advance of his Andean traverse. These merge into field notes from Peru and some rather more theoretical notes which suggest an interesting relationship between this notebook and the *Santiago notebook*, which Darwin used at about this time and which became the main repository for his theoretical musings. As is obvious from the 'Birloches' note,[807] Darwin had the *Galapagos notebook* with him on his Andean traverse, unless he left it with Caldcleugh for safe keeping in Santiago and picked it up on his way back around 8 April. Therefore Darwin used the *St. Fe notebook* as his main notebook for the traverse but may also have carried the *Santiago* and *Galapagos notebooks* with him.

The next few front pages date from late July and are again partly of a theoretical nature. 'Reached Lima' on p. 17a is unambiguously from 29 July, so there is a gap in use of the intervening pages of about three months. The next back pages are almost impossible to date although they too seem to relate to Peru in July. The Lima front pages extend to about p. 28a at which point Galapagos island notes commence and continue until four pages from the end.

The Galapagos field entries continue on the back pages after various probably updateable notes. The Galapagos notes continue through the second half of September and the first half of October until the final note registers Halley's Comet on 18 October on p. 50b.

The last four front pages seem to be concerned with theoretical topics of the kind discussed in the *Santiago notebook*. The entries seem to relate to South America and the Pacific and they date from the *Beagle*'s passage from the Galapagos to the Low Islands [Tuamotus] in late October or early November. The next dated note is 18 November in Tahiti and Darwin's visit to this Island ends the notebook on p. 66b.

---

805 *Notebook B*, p. 25 in Barrett *et al.* 1987.
806 *Autobiography*, pp. 120–1. See Browne 1995.
807 See below and the introduction to the *Santiago notebook*.

What follows is a detailed analysis of the notebook, divided into 'Chile', 'Lima', 'Galapagos' and 'Tahiti' sections, with the proviso that these divisions are not always clearly defined in the notebook. On the inside front cover of the notebook 'Lima August 4th 1835' is written perpendicular to the spine and there are a few jottings including 'Amblyrhyncus', the famous Galapagos iguanas. 'Benchuca', was the bug which bit Darwin on 26 March. Sadly the first two front pages are excised. The inside back cover has 'Anchored 30th at Blonde Cove' and refers to Banks' [Tagus] Cove on Albemarle [Isabela] Island in the Galapagos where the *Beagle* anchored on 30 September.

Chile, March 1835

The back pages commence on p. 1b with names of places perhaps to be visited on the coast between Valparaiso and Coquimbo. These are followed by jottings of provisions to take on an expedition and on p. 4b of names of persons, equipment and pieces of intelligence. The note 'Letters home' followed by at least eight names suggests Darwin reminded himself to attend to his correspondence. He wrote to several of these people at roughly this time.[808] The lists, which are very difficult to date, continue on the back pages up to p. 17b after which entries clearly relate to the Galapagos. There is a general correlation in the subject matter between these earlier front and back pages and since the former can be dated it is best to consider them in more detail, noting where the same subject crops up in both front and back pages.

The surviving front pages begin on p. 3a with more jottings including the question 'Aconcagua active. When? Before Birkbeck ??' apparently answered by '19th of January [1835]'. The same occurrence is mentioned on p. 15b followed on the next page by a remarkable note of the eruption of 'Cosiguina [that is Coseguina in Nicaragua] at 6 & ½ in the morning of 20th'. Darwin apparently heard that Aconcagua, the highest mountain in the world outside the Himalayas, erupted on 20 January 1835, and he referred to this in his 1840 paper on volcanoes and earthquakes as an example of how three Andean volcanoes all erupted at more or less the same hour one month before the great Concepción earthquake.[809] FitzRoy, who measured the height of Aconcagua, stated in his short paper about the mountain in 1837 that it was 'a volcano in the Cordillera of the Andes' and that it was active 'at intervals'.[810] However, the Smithsonian Institution does not recognize Aconcagua as a volcano.[811] A careful reading of Darwin's 1840 paper shows that he was uncertain

---

808 CD wrote to Fox on 7–11 March, Henslow on 10–3 March and 18 April, and to his sisters Caroline on 10–3 March and Susan on 23 April. Perhaps the last two are the 'Women' mentioned on p. 4b? See CCD1.

809 Darwin 1840, p. 610, in *Shorter publications*, pp. 97–124.

810 FitzRoy 1837b.

811 We are grateful to John Woram for the suggestion (personal communication) that Birkbeck is CD's misspelling of Byerbache, the name of 'a resident merchant' in Valparaiso, who CD recorded in the

whether the reports of Aconcagua's eruption could be relied on and in his private papers he noted that it was not always easy to identify these mountains when viewed from afar.[812]

A related pair of entries on p. 6b and p. 8b refer to a Mr Croft's clock stopping at about fifteen minutes to noon, which apparently corresponded to ten minutes to noon at Santiago. This was no doubt the time of the Concepción earthquake on 20 February. Mr Croft is also credited on p. 4b in connection with a 'back bone of Cetacea'. There is a reference on p. 3a to Henry Cood, whose name also crops up on p. 6b with a note about a 'letter to Iquique', and citation of a 1826 Peruvian map. This is followed on p. 4a by the first true field notes, dated 14 March 1835, when Darwin 'Started for St Jago [from Valparaiso] — in a Birloches or gig'. There is a brief account of the geology then 'Slept at the foot of the Prado & reached St Jago by 9 oclock on Sunday morning (15th)', p. 7a, implying that Darwin had made an early start. In the *Beagle diary* he called the mountain the Rado.

It is at this point that the notebook has been put aside until late July, although some of the back pages may date from the gap months. In these months Darwin completed his classic circuit to Mendoza and back, for which he used the *St. Fe notebook*, then the *Coquimbo notebook* for May, the *Copiapò notebook* for June and the *Despoblado notebook* for July, with the *Santiago notebook* for theoretical work.

Lima, July–August 1835

The reference on p. 7a to 'Guancavelica' is to Huancavelica, about 300 km (186 miles) southeast of Lima, and 'Caneta' is Cañete. The place is also mentioned on p. 6b supporting the view that Darwin was using both ends of the notebook at this point. It is impossible to date the entry but it was probably written in Callao, where the *Beagle* arrived on 19 July, remaining there for seven weeks. Due to the unstable political climate Darwin spent most of August on board writing up the *Beagle diary* and 'Geological notes about Chili', including presumably his 'Recapitulation and Concluding Remarks'.[813] FitzRoy, however, after returning aboard H.M.S. *Blonde* on 9 August, resided in Lima.

On p. 8a Darwin recorded a lizard and a frog but apparently did not collect specimens of these. On a 'Wednesday' on p. 8a Darwin recorded the barometric altitude of some shells, and it is a reasonable guess that the date was 21 July.[814] The

1840 paper informed him 'that sailing out of the harbour one night very late, he was awakened by the captain to see the volcano of Aconcagua in activity' (Darwin 1840, p. 611; *Shorter publications*, pp. 97–124). Woram also suggested that Byerbache could be Edward Beyerbache, United States Consul in Talcahuano. We are, furthermore, grateful to Sergio Zagier for suggesting that Byerbache misinformed CD and that it was more likely to have been the nearby true volcano Tupungato which erupted.

812 DAR 36.442–3.
813 DAR 41.23–39.
814 This is slightly at variance with the dating of Barlow 1945, p. 244, as 19 July.

notes relate to Darwin's fieldwork on the island of San Lorenzo, 'the only secure walk' according to the *Beagle diary* that he was able to take in that vicinity. These notes follow on from p. 45b in the *Despoblado notebook*. Page 9a in the *Galapagos notebook* is dated 'Thursday' which must then be 23 July. The next three pages, which relate to volcanoes, earthquakes and tsunamis are impossible to date more precisely than before 29 July. The note at the top of p. 12a that 'The *Beagle* called in on the 23 April. Valparaiso' is retrospective. That was when Darwin told FitzRoy of his promotion to Post Captain.

The entry on p. 13a starts with 'Casma, Huaraz', which are two places reported to Darwin by a civil engineer in Lima called Mr Gill, between which this gentleman had found evidence of changes in irrigation having forced the abandonment of a settlement.[815] The entry continues with a field note, which must date to around 21 July, from San Lorenzo and the reference on p. 14a to salt in the sandstone on the island occurs in a footnote in *South America*, p. 48. The entry 'Petrified wood, Gypsum & Salt' on p. 10b seems to refer to this, providing the first reasonably solid dating for the back pages.

The question 'Was Bellavista destroyed in 1846' on p. 15a is a mistake referring to the terrible Lima earthquake of 1746. This is also mentioned on p. 12b. The 'Cruikshank' noted on p. 15a is obviously the 'Mr. A. Cruckshanks' reported by Lyell to have found sea pebbles 700 feet above sea level at Callao.[816]

Pages 12b–17b could date to almost any time from late July to mid September. In some ways perhaps the most important note is 'Mem Dessaline D'Orbigny excellent memoir' on p. 14b as one of only three field notebook references to the great French naturalist who subsequently made a major contribution to Darwin's *South America*. The note shows that Darwin had seen Blainville 1834, which contained the early results of d'Orbigny's research in South America and this is confirmed by Darwin's letter to Henslow of 12 August.[817]

On p. 16a Darwin made an unusual reference to an article in 'No 32 Magazine of Natural History' which is not listed in the *Correspondence* (vol. 1) as on the *Beagle*. It is possible that someone, perhaps the author of the article Charles Waterton (1782–1865), sent him a copy. The article and others in the same volume discussed the smelling powers of birds and this was a subject of great interest to Darwin.[818] At a garden in Lima he carried out an experiment, described in his 'Ornithological notes', p. 244, and published in *Birds*, pp. 5–6, which seemed to show that Condors found

---

815 See *Beagle diary*, p. 319.

816 Lyell 1830–3, 3: 130. CD referred to Cruckshanks in *South America*, p. 51, as providing yet more excellent evidence of dramatic elevation, some of it in historical times. See the introduction to the *Copiapò notebook*.

817 CCD1: 462.

818 Waterton 1833.

carrion by sight rather than smell.[819] After the voyage Darwin entertained the Fellows of Christ's College, Cambridge with this story.[820]

Field notes from Lima resume on p. 17a with 'Reached Lima. Wednesday morning by coach'. In the *Beagle diary* Darwin recorded that this was on 29 July and that he 'spent five very pleasant days' at Lima which was about 12 km (7 miles) from Callao. Lima, founded in 1533 by Pizzarro after conquering the Incas, is likely to have been a great attraction for Darwin as being the only place he visited which was also visited by Humboldt, who visited in 1802. The notebook entry is worth quoting in full:

> road, uninteresting. not like Tropical country — many ruined houses, owing to long state of anarchy: Lima, passed gate; wretched filthy, tropical smell, ill paved — splendid looking town, from number of churches painted, like stone, cane as are upper stories. But every thing exceeded by ladies, like mermaids, could not keep eyes away from them: — remarkably mongrel population.

There is a hiatus in Syms Covington's *Journal*[821] for the period February to November 1835, but he made eight watercolours of the Limenian ladies.[822]

Darwin 'Rode to Merchants home to sleep; nice garden & large house only 20 £ per annum!!', p. 18a. Presumably the next day (the 30th) 'Before breakfast' Darwin 'rode to neighbouring hills' but experienced the disadvantage of visiting Peru during the *garua* season: 'I dare say view would be very pretty, but ceaseless mist & gloom: Have no idea of merits of view; excepting from day, from S Lorenzo — Clouds clearing away leaving strata, always give a majestic air to landscape', p. 19a.

That evening Darwin 'Dined with Consul General' Belford Wilson (1804–58) and had a 'very interesting evening'. Wilson was Simon Bolivar's (1783–1830) *aid de camp* and Darwin recorded in the *Beagle diary* that Wilson knew South America 'right well'. He had suffered the indignity, a few years previously, of having all his clothes stolen by robbers who cried out patriotically 'Vive la patria off with your Jacket', p. 26a. Presumably the geology notes which follow on pp. 20a–23a were made the following day (the 31st), and the 'Peruvian geologist' mentioned on p. 20a is the 'M. Riveiro' on p. 22a. As usual Darwin noted any shells at high altitudes, p. 24a, then on p. 25a 'August 3rd returned' to the *Beagle*. He observed the 'Condors flight, close wing — remarkable motion of head & body' which became the beautiful description published in *Birds*, p. 6.

Galapagos, September–October 1835

In several letters home written at this time Darwin expressed his eagerness to see the Galapagos, largely, perhaps, because they were almost new scientific territory. To his

---

819 This observation has since been confirmed, see Steinheimer 2004, pp. 307–8.
820 See the recollection of G. E. Paget from DAR 112.A86–A91, transcribed on *Darwin Online*.
821 Young 1995. FitzRoy also kept a diary (location unknown) which formed the basis of *Narrative* 2.
822 For colour reproductions of one of these see Keynes 1979, p. 289, and Taylor 2008, p. 84.

sister Caroline, for example, he wrote with eager prescience: 'I am very anxious for the Galapagos Islands, — I think both the Geology & Zoology cannot fail to be very interesting'.[823] Darwin visited four of the islands: Chatham [San Cristóbal] starting 16 September, Charles [Floreana] starting 24 September, Albemarle [Isabela] starting 1 October and James [Santiago] starting 8 October.[824]

The first entry on p. 18b is a description of the eel *Muraena lentiginosa* collected off Chatham Island.[825] This was probably on 17 September when the *Beagle diary* records that Darwin collected many plants and animals. The notebook had not, therefore, been used since 3 August, the *Beagle* having 'Sailed from Lima 7th September'.[826]

The next entry is dated 'Saturday', that is 19 September 1835, and reads 'left our anchorage & stood out to outside of Island, did not anchor'. Darwin made no more entries until the following day: 'Continued to beat to windward high side of island rather greener waterfalls of Water!', pp. 18b–19b. This was Freshwater Bay, Chatham.[827]

The next day, 21 September, 'A boat being sent to some distance, landed me & servant, 6 miles from the ship. Where we slept — I immediately started to examine a Black — Volcanic district deserving name of Craterized', pp. 19b–20b. The past tense means that the field notes from the 21st are retrospective.[828] Darwin 'Met 2 immense Turpin'. He noted that these giant tortoises (*Geochelone elephantopus*) 'took little notice of me' and one was 'Eating a Prickly Pear — which is well known to contain much liquid', pp. 20b–21b.

Estes *et al.* 2000 give a full account of Darwin's activities on the 21st and attempt to identify where exactly in the 'Craterized district' he made his detailed observations on the lavas there, which Darwin noted had 'been compared to most boisterous frozen ocean' , p. 21b. He drew two small sketches here, the first showing a wave-like fold in the lava, the other showing hexagonal cooling joints. He was trying to identify the boundary between two lava streams. Lavas on oceanic islands are generally of 'basic' composition, that is they have low silica and are not as viscous as so-called 'acid' continental lavas. Thus Galapagos lavas tend to flow freely rather than cause explosive eruptions like their Andean counterparts.[829] In his geological diary, DAR 37.768, Darwin discussed the apparently quite fresh lavas in the Banks' Cove area where he was on 1 October, and compared them with the Chatham lavas described on p. 21b. In Darwin's classic account of the Galapagos in chapter 5 of *Volcanic islands*, p. 105, he refined the language used in the geological diary entry so that the

---

823 CCD1: 458.

824 See Thalia Grant's map in Keynes 2004, pp. 308–9, and Estes *et al.* 2000 for details.

825 Specimen in spirits 1286 in *Zoology notes*, p. 360; listed as *Muraena lentiginosa* in *Fish*, pp. 143–4.

826 *Despoblado notebook*, p. 43b.

827 The notebook entry supports the statement by Estes *et al.* 2000, p. 346, that they could find no evidence that CD went ashore there.

828 Sulloway 1984, p. 34, provides a fine photograph of the district CD visited.

829 For an excellent account of Galapagos geology see Simkin 1984.

final version became one of the most poetic verbal comparisons he ever published: 'At Chatham Island, some streams, containing much glassy albite and some olivine, are so rugged, that they may be compared to a sea frozen during a storm; whilst the great stream at Albemarle Island is almost as smooth as a lake when ruffled by a breeze.'

Darwin recorded 'little less than 100' 'very small' craters on p. 23b, and tried to estimate their average distance apart, p. 24b. He sketched a crater, p. 25b, describing the different types of lava. Today the type he called 'black & glossy' would be given the Polynesian term *pahoehoe* and the more 'slaggy' type *aa*. On p. 26b he drew what may be a spatter cone, judging from the size and description on pp. 27b–28b and he noted collapsed lava tubes on pp. 28b–29b.

There follows, on p. 30b, one of the most important entries in all the notebooks: 'The Thenca very tame & curious in these Isl^ds. I certainly recognise S. America in ornithology, would a botanist?' This note contains three key elements. Firstly, there is the observation on the tameness of the 'Thenca', the Spanish name for the mockingbirds. The tameness applied to all the Galapagos animals but was extreme in the case of the 'Thenca', as indeed it still is today. This tameness, which he attributed to the absence of carnivorous mammals on the islands, suggested to Darwin that animals must acquire instinctive fear by a process of hereditary learning. Since at this point he may or may not have realized that the Thenca had migrated from the mainland, where mockingbirds are less tame than the Galapagos birds, we cannot know whether he was thinking that they had actually become more tame than their ancestors. Secondly, there is the obvious resemblance between the Chatham (and Charles) 'Thenca' and 'Thencas' and 'Callandras' from the mainland of which he had extensive knowledge, albeit from further south on the mainland. The resemblance could be the result of some mysterious 'halo of creative force' but could also be explained by the birds migrating to the islands from the mainland. Thirdly, there is the question: did this potentially historical link apply to the plants, as from the presence of cacti such as the prickly pear (*Opuntia*) it certainly seemed to? Would 'a botanist' be able to confirm this link? It is obvious from Darwin's letter to Henslow written four months later that it was his Cambridge mentor who he hoped would confirm this point.[830] As he put it in the less private confines of the *Beagle diary* entry for 26–7 September: 'It will be very interesting to find from future comparison to what district or "centre of creation" the organised beings of this archipelago must be attached.'

It is of course desirable to date the 'Thenca' entry, which, judging from the reference to 'Finger Point' two pages further on, seems to set it firmly on Chatham on 21–22 September.[831] It is, however, followed by '¾ of Plants in flower', p. 31b, which is almost

830 CCD1: 485; see also Porter 1985.
831 Estes *et al.* 2000, p. 347.

identical to wording which occurs in the next island entry in the front pages: 'Big tree — Misseltoe tree on various other kinds – ¾ plants in flower', p. 28a. The 'Misseltoe' may be the *Phoradendron henslovii*,[832] from Charles Island, indicating that the 'Thenca' note may have been written on Charles, where Darwin was on the 25th, when he met the Vice Governor Nicholas Lawson who said he could tell which island a tortoise came from by the shape of its shell. It was believed at the time of the *Beagle*'s visit that the tortoises were introduced by sailors from the Indian Ocean within the previous few centuries. Darwin was impressed, on later reflection, by the rapid divergence of the tortoise shells implied by Lawson's claim. Apart from the Falkland foxes, this was the only case he knew of different varieties on different islands in an archipelago.[833]

In attempting to date the 'Thenca' entry, it appears likely that the tell-tale phrase is 'in these Islds'. Darwin would probably not have used that phrase if he had only visited one island, so it seems that Darwin wrote the note on Charles and that he assumed the 'Thenca' was the same bird on both islands.[834] In *Zoology notes*, p. 341, written after leaving the Galapagos in late October, Darwin reflected on the notebook entry and made the following observations concerning the mockingbirds: 'This birds [sic] which is so closely allied to the Thenca of Chili (Callandra of B. Ayres) is singular from existing as varieties or distinct species in the different Is^{ds}. — I have four specimens from as many Is^{ds}. — These will be found to be 2 or 3 varieties. — Each variety is constant in its own Island.....This is a parallel fact to the one mentioned about the Tortoises.' This entry shows clearly that Darwin was now confident about the link between the islands' land birds and those of the mainland which he knew so well. He was now seeing, however, something which had not occurred to him before: that the mockingbirds might be even more divergent than the tortoises, with varieties *or distinct species* apparently confined to particular islands. For the moment this was an anomaly Darwin could not take further. He would have to ponder it until he had another chance to lay out his specimens, which in the event was around the time he was at St Helena in July 1836, when he wrote the famous entry in the 'Ornithological notes', pp. 73–4:

> In each Isd each kind is *exclusively* found: habits of all are indistinguishable ... When I see these islands in sight of each other, and possessed of but a scanty stock of animals, tenanted by these [mocking] birds, but slightly differing in structure & filling the same place in Nature, I must suspect they are only varieties ... If there is the slightest foundation for these remarks the zoology of Archipelagos will be well worth examining; for such facts undermine the stability of Species.

832 Specimen 3244 in *Beagle plants*, p. 182, identified by Porter as *Phoradendron henslovii*.
833 Nicholas Lawson's claim is somewhat exaggerated but it is the case that the subspecies of tortoise vary in the shapes of their carapaces. 'Saddlebacks' are from islands where the tortoises obtain their food by reaching up. 'Domebacks' are from islands where there is sufficient low level vegetation.
834 Herbert 2005, p. 313, places the entry on Chatham.

Darwin later inserted the word 'would' so that the line reads 'such facts would undermine the stability of Species'. Obviously Darwin needed to wait for an expert ornithologist to determine whether his 'Thencas' were varieties of one species or not. Gould responded by describing three new species from the *Beagle* specimens. The 'Thencas' of Chatham and Charles are today usually but not always regarded as separate species, *Mimus melanotis* on Chatham only and *M. trifasciatus* on Charles only. Sadly *M. trifasciatus* is now extinct on Charles itself but survives on two nearby islets. The other species described by Gould is *M. parvulus*, on the other large islands, including Albemarle and James.[835] Gould's assessment in 1837 that the types from different islands were distinct species was of immense importance to Darwin as this did indeed 'undermine the stability of species'. Darwin responded in *Birds*, p. 64, by stating firmly his belief that *M. fasciatus* and *M. parvulus* are species 'as distinct as any that can be named in one restricted genus'.[836]

Mimus melanotis from Chatham and James's Islands, Galapagos. Plate 16 from *Birds*.

Mimus trifasciatus from Charles Island, Galapagos. Plate 16 from *Birds*.

Mimus parvulus from Albemarle Island, Galapagos. Plate 18 from *Birds*.

Three species of Galapagos mockingbirds collected by CD.

The field notes are not conclusive with respect to location but the note 'I now understand St Jago Lava' on p. 32b perhaps indicates that he was looking for signs of 'upheaval' and the reference to 'King Landscape' is suggestive of Charles Island.[837] Darwin's comparison between the Galapagos lava fields and craters and the 'Phlegrean fields' and craters of Vesuvius and Etna, which he knew so well from Lyell's accounts, is memorable.

835 See Grant 1984.
836 Sulloway 1982c, p. 350, provides photographs of CD's specimens of *Mimus*. All of the colour plates from *Birds* are reproduced on *Darwin Online*.
837 Philip Gidley King painted a watercolour on Charles Island which is reproduced in Keynes 1979, p. 301, and in Taylor 2008, p. 136.

The next page, p. 34b, is also highly important. There is the first reference to 'Lizards' which may be iguanas or lava lizards, followed by 'Black Mud & parasites — Brazil without big trees. Feast. Robinson Crusoe Gross-beakes'. The mention of 'Gross-beakes' is the first known explicit recording by Darwin of Galapagos finches. This notebook entry also provides previously unknown evidence that Darwin collected the exceptionally thick-billed large ground finch, named by Gould in 1837 as *Geospiza magnirostris*, on Charles Island.[838] The reference to black mud, Brazil and Robinson Crusoe all tally exactly with Darwin's *Beagle diary* entry for 25 September. The 'Feast' was probably the meal provided by Mr Lawson on that day (*Narrative* 2; p. 491).[839]

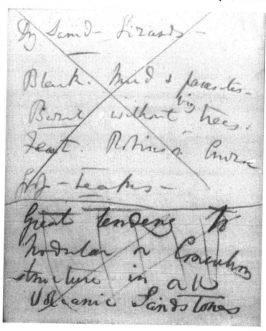

Page 34b of the *Galapagos notebook* with Darwin's first record of Galapagos finches.

The birds listed on pp. 28a–29a 'Duck male Bitterns 2 female Heron female' may be next in date sequence since they precede 'Craterized Point — Perfect cones — subsequent to streams' which seem to fit exactly with the *Beagle diary* report for 26–7 September. The duck is almost certainly *Pœcilonitta bahamensis*, *Birds*, p. 135, the bitterns *Nycticorax violaceus*, *Birds*, p. 128, and the heron *Ardea herodias*, *Birds*, p. 128.[840]

The next entry on p. 29a is clearly for 29 September as the *Beagle* rounded the southern end of Albemarle and Darwin had his first view of Narborough

838 See Sulloway 1982a.

839 We are grateful to Thalia Grant and Greg Estes for this suggestion (personal communication) about the 'Feast'.

840 The listing of these specimens as from James Island in *Zoology notes* is incorrect.

[Fernandina] which he declared 'Desolate'. He likened Albemarle to a 'continent built of big old Volcanos' and it is clear from *Volcanic islands* that he was deeply impressed by the northwest–southeast alignment of the volcanoes on this island.

Geospiza magnirostris from Charles and Chatham Islands, Galapagos. Plate 36 from *Birds*.

Zenaida galapagoensis from Galapagos. Plate 46 from *Birds*

After sunset on 30 September the *Beagle* 'Reached Blonde Cove', then usually called Banks' and today called Tagus Cove. This event which is also mentioned on the inside back cover seems to be the last Galapagos entry in the front pages and so there is an apparent gap in the notes until 12 October. Estes *et al.* 2000, p. 352, describe Darwin's work at Banks' Cove in great detail and point out that one-third of his geological notes from the islands refer to the area around the cove, an illustration of which appears in DAR 44.32 and was eventually published in *Volcanic islands*, p. 107 (reproduced below).

A sectional sketch of the headlands forming BANKS' COVE, showing the diverging crateriform strata, and the converging stratified talus. The highest point of these hills is 817 feet above the sea.

Banks' Cove, Galapagos. Fig. 13 from *Volcanic islands*, p. 107.

The bottoms of pp. 34b–35b at first seem difficult to date or locate precisely other than that they must predate 12 October, the date of the next page. An important element is, however, the observation on p. 34b of a 'Great tendency to nodular or Concretionary structure in all Volcanic Sandstones' which resonates with Darwin's description of structures now called accretionary lapilli at the *Beagle* Crater near Banks' Cove, which he explored on 1 October.[841] It therefore seems reasonable to consider p. 35b as Darwin's field notes from Albemarle, but it is certainly possible that they relate to his first few days on James. In the final element of p. 35b Darwin estimated the area of the archipelago and equated this to the area of 'Sicily & Lipari Isds' and compared the craters to those figured in Scrope's 1825 *Considerations on volcanoes.*

The *Beagle* stood round the north end of Albemarle on the 3rd and was therefore in the northern hemisphere for the first time in three and a half years. She then struggled eastwards against the trade winds and currents via Tower on the 7th before Darwin was put ashore at James Bay on the 8th, in company with Covington, Benjamin Bynoe, Harry Fuller and one or two other men. They remained on James until the 17th. On 9 October Darwin set out for the interior, visiting the hovels where a party of men was employed hunting tortoises. On the 12th Darwin went there again 'Walked up to the Houses — Slept there Eating Tortoise meat By the way delicious in Soup', p. 36b. According to the *Beagle diary* he 'enjoyed two days collecting' in this 'very green & pleasant' area, p. 36b, which Estes *et al.* 2000, p. 357, identify as Jaboncillo and which in Darwin's time was thickly covered in endemic *Scalesia* trees. It was here that Darwin made his classic observations of giant tortoise behaviour, pp. 37b–39b.

> Extraordinary numbers of Turpin — When drinking bury head above eyes — Will drink when a person is within 2 yards of them about 10 gulps minute. noise during cohabitation [copulation] & length of time certain — Eggs covered by sand soil from 4 to 5 in number — require a long time before they are hatched. — Quickness of travelling certain — now said come every three days for Water — Eat Cacti in the dry Islands

All of these entries were expanded in *Journal of researches.*

The next entry on p. 39b shows that he dissected a 'Yellow Iguana' (*Conolophus subcristatus*), whose intestines were 'full of Guyavitas & some large leaves' and who laid its 'eggs in a hole'. He observed that the 'Caracara habits like Carrancha round Slaughter house <u>kill</u> chickens — <u>run like</u> a cock', pp. 40b–41b. This was the Galapagos hawk (*Buteo galapagoensis*). The final entry is 'Thenca eat bits of meat', p. 40b.

On the morning of 13 October Darwin 'descended highest Crater' in which he collected trachyte lava specimens. For years these rocks, which were described using

---

841 See Estes *et al.* 2000, p. 354.

techniques not available to Darwin by Richardson 1933 and discussed at length by Herbert 2005, pp. 121–6, plate 4, have unsettled geologists because they are of an apparently anomalously high silica content. In the standard book-length geological description of the Galapagos, McBirney and Williams 1969 went so far as to suggest that Darwin's specimen 3268 might not actually be from the Galapagos. Recent work by Pearson 1996 shows that on the contrary Darwin's specimens are of the highest value as substantiating his then radical approach in *Volcanic islands* to explain the fractionation of lavas. Darwin's theory bears comparison with natural selection as a mechanism for sorting variants, with gravity supplying the force but without of course any analogue of the self-replication process found in living things.[842]

On returning to Buccaneer Cove Darwin found the freshwater spoiled by the surf, and would have suffered considerably if it had not been for the 'extraordinary kindness of Yankeys' from the whaler anchored in the bay who 'gave us Water', p. 41b. On 14 October Darwin 'Wandered about Bird collecting' and observing with quite exceptional empathy the iguanas' behaviour: 'shakes head vertically; sea — one no = dozes, hind legs stretched out walks very slowly — sleeps — closes eyes — Eats much Cactus: Mr Bynoe saw one run walking from two other carrying it in mouth — Eats very deliberately, without chewing — Small Finc[h] picking from same piece after alights on back', pp. 42b–43b. This is the second of only two references to the finches in the notebook. On pp. 44b–45b Darwin recorded the temperature: 77° Fahrenheit in the trade wind, 108° 'On rock out of wind'. On the 16th, Darwin's last full day in the islands, there was no wind at noon and his thermometer placed in the sand shot to 137°, p. 45b.

The remaining field notes from the island must date from the afternoon of the 16th. They may indicate one final trek into the interior: 'at <u>highest</u> central Crater'. Darwin noted that one side of the 'perfect' crater was lower, due to ash piling up more on the lee side, and that it was '⅓ of mile diameter', p. 46b. He included a section on the striking asymmetry of the Galapagos craters in *Volcanic islands*, pp. 113–4. The description in the notebook matches that in the geological diary (DAR 37.770). On pp. 48b–50b Darwin made his observations of the land iguana:

> Iguana digs tail motion slow — appear = same stupid from low facial ∠ = Very fond of Cactus run away like dogs from one another with pieces — Excavate Burrow shallow — first on one side & then on other — two or three time throw dirt with one arm & kick it out with well adapted hind leg then on other side

The very last entry actually written on James is the single word 'Comet', p. 50b. This was Halley's Comet, which was at its minimum distance from Earth on

---

842 Pearson 1996, p. 57, provides photographs of some of CD's Galapagos lava specimens.

13 October 1835. Darwin would have been able to see the comet with the naked eye soon after sunset on any clear evening at about that time. He probably had a telescope with him so would have had an exceptionally good view of the comet.[843]

The *Beagle* returned from her nine day circuit via Chatham, Hood [Española] and Charles to collect water, tortoises, wood, potatoes, pigs and mail, sent in boats at 2.30pm on the 17th to retrieve Darwin's party: 'Ship came', p. 50b. This is the moment depicted in John Chancellor's painting of the *Beagle* in the Galapagos.

HMS *Beagle* in the Galapagos, 17 October 1835 2.15 p.m., by John Chancellor.

Darwin was back on board by 5 p.m. and recorded that on the 18th they 'Ran along Albermarle Isd' and there was one last geological note: 'One of the mounds in Albermarle most covered with bare Lava — Mounds 4000ft high', p. 51b. This was probably Volcán Darwin. The final notes from the Galapagos are calculations on the speed of tortoises, in ink: '30 yard in 5 minutes 360 — in 1 hour' multiplied by twenty-four hours gives 4 ⅓ miles (6.8 km) per day. On 19 October the *Beagle* sailed north to pick up Mr Chaffers in the yawl at Abingdon [Pinta], so briefly entering the northern hemisphere for a second time. On the 20th FitzRoy set a course west-south-west and the *Beagle* commenced her month-long 5,000 kilometre (3,000 mile) passage across the Pacific to Tahiti.

---

843 Materials in the David Stanbury archive at Christ's College, Cambridge, seem to show that Stanbury was the first to realize that this entry referred to Halley's Comet.

Tahiti, November 1835

The last four front pages following the Galapagos field notes apparently date from the passage across the Pacific in late October or early November. This dating is not secure but is based on consideration of the topics discussed which seem to relate to South America and the Pacific. Darwin noted that William Ellis in his *Polynesian researches* 'states that the Austral Is[ds] have only lately been inhabited are they low Is[ds]? Corall rapidly growing in the Low Is[ds] ?', p. 30a.[844] Darwin was perhaps reading up in advance of reaching the Low or Dangerous Archipelago [Tuamotus] which the *Beagle* sailed through starting on 9 November, giving him his first close views of coral atolls. Admiral Krusenstern's chart of the Low Archipelago and the Society Islands, including 'Otaheite or Tahiti', showing the *Beagle*'s track through the islands, is included in the appendix to *Narrative* 2.

The following two pages refer to four other books, all in the *Beagle* library.[845] On p. 31a Darwin's thoughts appear to have returned to boulders 'being carried on plains of Patagonia by any violent motion excepting by one beneath the sea'. He then wondered in response to reading Daubeny's 1826 book on volcanoes 'Is not Olivine present?' perhaps in one of his Galapagos trachyte specimens. Ellis is mentioned again on p. 32a in connection with Hawaii, then Darwin noted 'Strong Earth quake useful to Geologist, can believe an amount of violence has taken place on earth's surface & crust'.

On the next page Darwin jumped with no obvious connection from 'Mad dogs. Copiapò' to 'S. Cruz — Glaciers', then observed 'On the Atlantic side [of South America?] my proof of recent rise become more abundant at the very point where on the other [Pacific?] side they fall'. This passage is unclear although if 'proof' and 'fall' are opposites then Darwin was testing the effectiveness of his argument for continental uplift. The Santa Cruz entry may mean that the passage refers specifically to the fact of Andean glaciers mainly flowing towards the Pacific. Finally, on p. 34a he noted to himself 'Collect all the data concerning recent rise of Continent' and the Tahitian river name 'Tia auru [Tuauru] Piho'.

Switching now to the back pages, a note on p. 52b 'Fresh water fish' surely indicates that Darwin had started to explore Tahiti on foot. This entry must date from just before the first date 18 November further down the page, in which case it was the day when Darwin saw the beautiful view of Eimeo [Moorea] Island in the distance. As Armstrong explained in his excellent account of Darwin's stay on the island, Tahitian dates were a day ahead of those the *Beagle* brought with her at the start of her ten day stay because of the international date line.[846] The

---

844 Ellis 1829.
845 See CCD1, Appendix VI.
846 Armstrong 2004, p. 139.

references to Bynoe's information on 'direction of slate in Obstruction Sound' in southern Chile and 'angular fragment of Granite on Icebergs' seem oddly out of place. Bynoe is mentioned in connection with Obstruction Sound in the *Red notebook*, p. 141e. 18 November commenced with 'travelled up valley'; this is the Tia-auru valley on the northwest side of Tahiti which was mentioned on p. 34a. The Tahitian entries are intensely atmospheric: 'at first beautiful view over cocoa nut trees — 2 <u>fine</u> men, take no cloths or food — Higher up valley very profound — most dangerous pass — ropes — Kotzebue — The one with pole, ropes, dogs & luggage. — Wonderful view — Cordilleras nothing at all like it', pp. 52b–53b. Darwin's small party 'Ascended a Lava slope' which was 'excessively steep', pp. 53b–54b. With a vertical sun it was 'Steaming hot' with waterfalls and 'enormous precipices' all around. The vegetation was lush with 'Bananas & trees', perhaps reminding Darwin of Brazil. After climbing the 'fern hill' he threw himself into the shade of 'thick trees surrounded by sugar', pp. 54b–55b. There were edible plants everywhere: 'Yams, Taro — Sweet root like sugar' and the rivers were full of 'fish & Prawn'. His two companions caught these creatures 'diving gracefully amphibious', p. 55b. It appears that the 'Darwin' prawns belonging to the genus *Macrobrachium* which are today preserved at the Oxford University Museum were among those caught that day.[847]

On the next page Darwin thought 'only famine & <u>murders</u>' could have induced people to explore the 'really most fearful road'. He noted that his guides had learnt 'a little English' and that their way of making fire by rubbing a 'Carpenters tool' reminded him of the gauchos, p. 56b. The valleys were like 'mere crevices'. The rocks were 'grey base, <u>with nests of olivine</u>', together with conglomerates, clays and sandstones, pp. 57b–58b. In *Volcanic islands*, echoing the notebook, Darwin recorded that he 'picked up some specimens, with much glassy feldspar, approaching in character to trachyte', p. 26. The men who were guiding Darwin told him 'not to tell Missionary' after accepting a drink from his flask of spirits, p. 58b. They bivouacked at a 'cool stream where we bathed buried in peaks'; their shelter was thatched with banana leaves which ensured a 'dry bed' even though that night there was 'much rain', pp. 58b–60b. They enjoyed a 'supper baked in stones' of 'fine vegetables' and could hear 'another cataract of 200 ft'. The men said their 'prayers & grace [with] no compulsion'. Darwin declared the evening 'sublime', p. 59b. On 19 November Darwin 'returned by other road' so 'avoiding cascades', p. 61b. They walked along a 'knifes edge' in one place using ropes. He noted that the men were very strong and tattooed with 'flowers round head'. They 'caught some fine Eel', presumably by hand as Darwin noted this as '<u>Evidence how man</u> can live by hand',

847 See Chancellor *et al.* 1988, p. 222. The cliff of columnar lava on the Tuara river where CD ate his lunch is beautifully photographed in Catling 2009, fig. 5.

p. 62b. He was deeply impressed by the steepness of the ravines and the ridges separating them which were 'about same angle as a ladder'. Somehow the 'vegetation clings on … up to the highest peak', that is over 2,000 meters (6,500 feet). By breakfast he had an 'enormous appetite', p. 63b and ate a 'mass of Banana!!', but it was 'fatiguing travelling so far poising each step with greatest care', p. 64b.

That night they 'slept under [a] ledge of rock' and watched the stars. The next day, Friday, they returned and met a 'party of noble athletic figures travelling for Tayo'. Darwin found the *Beagle* had moved to Papawa so 'walked round' to meet her. On the 21st the 'Ship returned' to Point Venus, where Captain Cook had observed the transit of Venus in 1769. On Sunday 22 November Darwin 'went down to Papiete in boat' to hear the 'Tahitian service'. He thought the congregation did not pay much attention but that they looked 'respectable'. There was 'good singing' although the sound was 'not Euphonious'. Darwin judged the missionaries to be 'good' and that one could never 'believe what is heard', p. 65b. This favourable view of the missionaries fed into Darwin and FitzRoy's joint letter to the *South African Christian Recorder*.[848]

The single word 'reef' for his canoe trip on 23 November must be the shortest notebook entry of all, and gives no indication that probably within a week or so Darwin would be drafting his 'Coral Islands' manuscript.[849] On the 25th Darwin recorded having breakfast with the missionary Charles Wilson at Papiete and attending Queen Pomare's party. Finally on the 26th he noted the 'great Parliament' called by the Queen to decide how to respond to FitzRoy's 'request' for settlement of an unpaid compensation fee to the British Government. This appears to be the last notebook entry Darwin made in 1835.

The *Beagle* headed out of Matavai Bay on the evening of 26 November 1835 and resumed her west-south-west bearing for New Zealand. Darwin stowed his fish and other Tahitian specimens safely away for the month-long voyage, started to write up his diaries, putting the *Galapagos notebook* aside with the others. It seems he looked over the *Galapagos notebook* when writing *Journal of researches* and the *Geology of the Beagle* after which, as far as is known, he never used it again. Certainly Darwin would not for a moment have believed that in the distant future tens of thousands of people from all around the world would want to see his jottings in the notebook he carried in his pocket on the Galapagos for a few short weeks in 1835.

---

848 FitzRoy and Darwin 1836; *Shorter publications*, pp. 15–31. A 29 June 1836 letter from FitzRoy to Lady Herschel reveals that a note on the subject of missionaries was proposed by Lady Herschel and that FitzRoy sent the copy from St Helena. (Royal Society, Herschel Letters, vol. 7 (E–F)). We are grateful to Simon Keynes for calling our attention to this letter and for generously sharing his research with us.

849 See the introduction to the *Santiago notebook*.

[FRONT COVER][850]

Galapagos.
Otaheite
Lima

[INSIDE FRONT COVER]
    Freyrina
    Benchuca
    Limestone & Salt all parts of Island
    Amblyrhyncus[851]
    Covington *[7 R]*
    Lima August 4th 1835
[1a–2a excised]
[3a] Bones of Megatherium at Mendoza. —
    R. San Francisco—
    Paraguay. —
    Aconcagua active. When? Before Birkbeck ??

850 Photograph from Dobson 1959.
851 The iguanas of the Galapagos. See *Zoology notes*, pp. 293–7. CD found the name in Byron 1826.

19th of January [1835]

Mr Cood[852] Mapa Fisico & Politico Alto & Bajo — Peru 1826[853]

[4a]   14th [March 1835] Started for St Jago — in a Birloche or gig: Hills all soft & worn into that sort of bifurcating ridges which is peculiar to degradation of soft matter; in all the ranges which I saw were composed of very soft gneiss or protogene, which is interlaced by veins of same figure as

[5a]   quartz of a siliceo-Feldspath nature or Trapoidal or semiporphyritic, intersect each other, thin out downwards; & pass in places into surrounding rock yet some localities have air of true dykes: Having passed Prado true Porphyry

[6a]   The plain of St Jago it is only on the upper parts or border of basin where the sand or red clay & Tosca & Volcanic ashes are found, the lower part either excavated by rivers or retreat of ocean presents the true & pure

[7a]   gravel. —
Slept at the foot of the Prado & reached St Jago by 9 oclock on Sunday morning (15th): [March 1835]
Two roads to Guancavelica one by the coast, turning off a little above Caneta another passing by Guarochin

[8a]   Viviparous Lizard near Leposoma of Spix[854] — Frog with pointed nose. Rhinella of Fitzhinger[855]
Wednesday [22 July 1835]
Level of Sea 29.952
Shells 29.877
Thermometer 72°
Comminuted & large shells a considerable patch

[9a]   Thursday [23 July 1835]
Higher level 29.941
~~Thermometer~~ Lower level **30.175**
+ 5 ft.
Thermometer– 67
29.580
Ther 62+

[10a] Great blocks of Granite other side of Ancud Bay Chiloe, like the Geneva block[856]
Salt mine near mouth of R Guyaquil
Roe — stone of Copiapò

---

852 Henry Cood, an English merchant residing in Valparaiso.
853 Hacq 1826.
854 Spix and Martius 1824. *Leposoma* is a genus of lizard.
855 Fitzinger 1826. *Rhinella* is a genus of toad.
856 This is a reference to the famous erratic boulder known as the Pierre à Bot above Neuchâtel in Switzerland. See Rudwick 2008, fig. 6.4.

[11a] Letter paper

Volcanoes — Sea waves retire — St Jago or Valparaiso most affected Direction of cracks

Earthquake — Coquimbo Copiapò

[12a] The *Beagle* called in on the 23rd of April. — Valparaiso

Lima = Is^d of Fronton & Lorenzo said to be united? Foxes & Mice & Rats. — passage said to be much smaller —

V Freziers[857] Chart

[13a] Casma, Huaraz

But river course which formerly irrigated a tract, covered with Indian ruins = river course 50 yards wide & about 8 ft deep in solid rock — sand & shingle, after ascending some distance suddenly descended —

[14a] As the Sandstone in Island contains Salt: some veins nearly 2 inches thick — Lignite. black *[glossy]* coal — parts vegetable structure. —

Difficulty of understanding Salt deposit in such open districts —

[15a] Was Bellavista destroyed in 1846

Cliffs of Banos del Pujio about 200 ft above sea — 200 above Riman

Amancares 200 ft above sea

2 rivers unite Islands

Cannot understand, A. Cruikshank[858] some pebble at about that elevation on some of the hills

[16a] High white powder

30.230 — T 68 | 70 AT—

30.350 — T 68 | 80

30.436   do do

Smelling properties discussed of Carrion Hawk Crows Hawks N° 32 Magazine of Natural History. —[859]

[17a] Reached Lima. Wednesday morning by coach — road, uninteresting. not like Tropical country — many ruined houses, owing to long state of anarchy: Lima, passed gate; wretched filthy, tropical smell, ill paved — splendid looking town, from number of churches painted, like stone, cane as are upper stories.

---

857 Amédée-François Frézier (1682–1773), French engineer and explorer. Frézier 1717.

858 'Mr. A. Cruckshanks' reported by Lyell 1830–3, 3:130 to have found sea pebbles 200 meters above sea level at Callao. CD referred to Cruckshanks in *South America*, p. 51, as providing yet more excellent evidence of dramatic elevation, some of it in historical times (see the introduction to the *Copiapò notebook*).

859 Carrion crows = Waterton 1833. A letter by P. Hunter in the same volume disputed that vultures found their food by scent: 'The means by which the vulture (Vultur Aura L.) traces its food', pp. 83–4 and extracts, pp. 84–8, of an article by John James Audubon from *Jameson's Edinburgh New Philosophical Journal*, which argued that the Turkey Buzzard did not find its food by scent. Waterton replied, pp. 162–71, disputing Audubon's claims. See *Journal of researches*, 2d ed., p. 184.

[18a] But every thing exceeded by ladies, like mermaids, could not keep eyes away from them: — remarkably mongrel population. =
Rode to Merchants home to sleep; nice garden & large house only 20 £ per annum!! — not fashionable — Before breakfast rode to neighbouring hills — I dare say view would be

[19a] very pretty, but ceaseless mist & gloom: Have no idea of merits of view; excepting from day, from S Lorenzo — Clouds clearing away leaving strata, always give a majestic air to landscape. But even that day air not clear — Indians very different from Puelche: — Dined with Consul General[860]

[20a] very interesting evening. Bolivar's[861] aid du camp: Peruvian geologist walk to Alameda=
Plain of Lima great extensive shingle, Islands on a large scale — They do not think but am not sure about bits of brick being peruvian: I examined hills near

[21a] Limekilns — Main rocks are calcareous & shaly, some tolerably pure black Limestone with thinner strata, intervening as to the South, others rather siliceous & grey coloured; — general dip, Easterly — injected & fronted by hills of green feldspathic rock passing into a greenstone There is much of the

[22a] black impurer Limestone, as to the South. —
M. Riveiro, considers that there lie in Porphyry sometimes conglomerate structure, beneath which are Syenites & Granites, above the Limestone there comes Sandstone as of Lorenzo, with coal of Pasco: &

[23a] probably Gypsum of other side of Cordilleras — Sandstone often capped by conglomerates —
The Salt & Coal belong to one formation —
mines in sandstone? Salt there? Shells from there & from coast

[24a] Piura, shelly, Tosca — Payta — many /appear/ // above 14 leagues Inland Chira — Amatape Dr Lynen Berlin[862]
Shells 1000 ft above sea, ~~several~~ 15–24 leagues from Payta

[25a] July 29th [1835] Went up to Lima — August 3^d returned
Condors flight, close wing — remarkable motion of head & body — Lima numerous churches seemed to be upwards of 80. — large houses
plain /green/ structure with Islands —
Anarchical state

---

860 Belford Hinton Wilson (1804–58), British Consul in Lima, 1832–7; Chargé d'Affaires, 1837–41; Consul General in Venezuela, 1842–52.

861 Simon Bolivar (1783–1830), general and patriot leader who was instrumental in independence movements in Venezuela, Colombia, Ecuador, Peru and Bolivia. B. H. Wilson was Bolivar's aid de camp.

862 Not identified. This may be a phonetic spelling for Leinen.

[26a] black flag — Robbers Vive la patria off with your Jacket. — Few carriages or Carts, mules & water, donkeys;

Zapata, dust shot

Limestone of Mr Gay[863] distinct to mine

[27a] Channels back of Chilie penetrate to East of highest peaks — V valley of Cordilleras of Chili =

Mountain plants — Sea side do — The commonest leafless

[28a] /coton/ tree. Big tree — Misseltoe[864] tree on various other kinds —

¾ plants in flower

Duck[865] male Bitterns[866] 2 Female

[29a] Heron[867] female

Craterized Point — Perfect cones — subsequent to streams. Desolate Narborough very beach — Crater: Albermale continent built of big old Volcanos

30th [September 1835] Reached Blonde Cove

[30a] Soap — Shoes

~~Does~~ Ellis[868] states that the Austral Is[ds] have only lately been inhabited are they low Is[ds]? Corall rapidly growing in the low Is[ds]?

M. Gay Limestone different from mine at Porpico

[31a] Murray grammar[869]

The argument of alternate bands of boulders — prevents ideas of boulder being carried on plains of Patagonia by any violent motion excepting by one beneath the sea —

Does Miers[870] talk of loose blocks near S. Luis —

Mem: Galapagos. Is not Olivine present? in one of the Trachytes P 93. Daubeny.[871] never

[32a] Ellis[872] great /grey/ fissure in Hawaii

Strong Earth quake useful to Geologist, can believe an amount of violence has taken place on earth's surface & crust

In Chiloe I notice accumulation of Rocks at foot of escarpments

863 Claude Gay (1800–73), French naturalist and traveller. Professor of physics and chemistry in Santiago, 1828–42.

864 Possibly specimen 3244 (*Phoradendron henslovii*) in *Beagle plants*, p. 182.

865 Specimen not in spirits 3299 in *Zoology notes*, p. 413; listed as *Pœcilonitta bahamensis* in *Birds*, p. 135.

866 Specimen not in spirits 3300 in *Zoology notes*, p. 413; listed as *Nycticorax violaceus* in *Birds*, p. 128.

867 Specimen not in spirits 3296 in *Zoology notes*, pp. 300, 413; listed as *Ardea herodias* in *Birds*, p. 128.

868 William Ellis (1794–1872), missionary in the South Seas. Ellis 1829.

869 Murray 1824.

870 Meirs 1826.

871 'olivine rarely, if ever, occurs [in Trachyte], and therefore appears to be the only mineral which has any claim to be considered as peculiar to basalt', Daubeny 1826, p. 93.

872 Ellis 1829.

[33a] Mad dogs. Copiapò
        S. Cruz — Glaciers
        On the Atlantic side my proof of recent ~~side~~ rise become more abundant at the very
        point where on the other side they fail —

[34a] Collect all the data concerning recent rise of Continent
        [Tia auru Piho]

[BACK COVER not microfilmed]

[INSIDE BACK COVER]

Anchored 30th [September 1835] at Blonde Cove

> Soap
> [Drying]
> Letter
> Paper
> Chaffers—
> Keg of
> Spirits
> Captain

[1b]   Paguado
        Pichidongue in two places
        Conchalee
        Maytencillo (round Basin) to the ~~South~~ North of Amalanas & South Talinay— ~~& the~~
        Tonguy & Tonguy cilla: to the North

[2b]   P de la Vaca —
        Mauli —
        Calle
        Bombalee
        Mattee

[3b] Chiffles
Tin Mattee: ~~Bom~~
Saddle for mule
Provisions
Cigaritos
Silla fitted for Pistol &c
Turin a dulce
Made in Concepcion
2 dollars worth

[4b] Mr Green[873] — localities of Shells — Letters home = Henslow = Eyton[874]
Women — Journal — Mr Lumb[875] — Fox[876] Arrange Specimens — Insects
Microscope = Caldcleugh — Letter for Iquique — Coquimbo Books — Budge
Money for Coquimbo Alison[877] — age of St. Domingo — Fort which came in view —
Mr Croft back bone of Cetacea
Compass for Padre Pistols
Quick silver mine Waddington

[5b] Boots & Shirts Jacket Corfield[878] — letters to home
Books — Therm — Candles Bread Meat Chocolate = Tea = Cigars Passport Agree
with Mariano[879]
Caldcleughs message Miers — Mr White[880] — (Haines)
Capt Fitzroy — Shoal of Ulloa (M^r *[illeg]* charts) (Arica soundings Mummy
161 HD St Augustin, *[illeg]* from of:

[6b] Covington Instructions —
Write Bill
Mule-shoe
[illeg] = Mr Croft time of clock stopping = 12–370 ft elevation of Guancavelica
letter from Duncan for Corfield
Chissel Snuff
Hat clean

---

873 Mr Green, a ship owner in Valparaiso. See *Narrative* 2: 559.
874 Thomas Campbell Eyton (1809–80), Shropshire naturalist and contemporary of CD's at Cambridge.
875 Edward Lumb (d. 1872), British merchant in Buenos Ayres.
876 William Darwin Fox (1805–80), clergyman naturalist, CD's second cousin and contemporary at Christ's College, Cambridge.
877 Robert Edward Alison, English author who wrote on South American affairs and resident of Valparaiso and later managing director of a Chilean mining company.
878 Richard Henry Corfield (1804–97), English merchant living in Valparaiso and Shrewsbury school-fellow of CD's. CD stayed several times at his house in the Almendral, including his illness in September and October 1834.
879 Mariano Gonzales, CD's hired guide in Chile.
880 Possibly Nicholas White (b. 1806), Second 'Master' on the *Beagle*.

Date of old Sea Wall

Mr Cood letter to Iquique

[7b]   Eck[881] do

Miers[882] map & Passport

Bracers Tooth Brush Stirrups Letter paper

Mr Millers. Tea

/Sharp/ Pencil

Gillman to Arica Sugar

Letter to Henslow.

Mr Jewel. —

[8b]   Alison Barometer correct: & Bivalves — Flustra.

Medicin

Mr Crofts clock stopped 17° — 15° to 12° oclock mean time: St Jago nearly at same

time 10 minutes to — corresponding

[9b]   hour

Money — small gold

Ship 100 miles of Bubbles of Air

Mr White

Aqua – diente.

Don Pedro Abadia Fossil shells—

[10b] Shells on surface on other parts or Islands

Best river course ~~Limestone~~

Rocks of Lima, same as Isl^d.

Petrified wood, Gypsum & Salt:

Shell-fish Dredgers

Blue Beads &c &c for Indians. —

[11b] Glass — Saucer — Books — Mr Thomas = Fossil shells — M. Barometer Stags —

Horns — Museum Razors — Medicin

Sweet smelling oil. —

Night caps — Stockings

Black Ribbon

[12b] Was Bellavista destroyed in Earthquake 46 — No — /Mr Maclean/

Pill Boxes

Callejon la Maravilla

Mr Cunningham, thinks Carb of Lime as in Mortar is produced, is turned into

Nitrate & this

---

881 Frederick Andrew Eck (1806 or 1807–84), Swiss-born banker and collector of mineral specimens in
     South America, 1825–52. See CCD1: 407.
882 Meirs 1826.

[13b] decomposes the Murate of Soda ???
For fourty leagues, salt-petre worked, plain 14 leagues wide, — 420 miles long. —
Extend just within margin of plain.

[14b] Mem
Dessalines D'Orbigny excellent memoir[883] — *[Cobija]* elevated — 300 ft. —
NB. The Syenite & purple Sandstone &

[15b] purple Porphyrie — all appear sparingly found here.
Great eruption at New Orleans on the 7th? of February
Rialeja. —
Osorro —
Aconcagua 19th of January

[16b] ~~Volcan~~ Cerro
Cosiguina[884] at 6 & ½ in morning of 20° of Enero: lasted 20° to 23 — most *[violence]* 4
first days
Ria Leje latitude
Cholutega & Nacaome & Viejo

[17b] Costa de Leon
Letter Paper to pay to Chaffers
Falkland fossils Tropical?
[Ell]

[18b] Eel dark reddish purplish brown with pale or whitish brown spots. Eyes Bluish[885]
Saturday [19 September 1835]: left our anchorage & stood out to outside of
Island, did not anchor
(Sunday.) [20 September 1835] Continued to beat to windward high

[19b] side of island rather greener waterfalls of Water! — Came to an anchor in harbour
where whaler was. — —
Monday [21 September 1835] A boat being sent to some distance, landed me &
servant, 6 miles from the ship. Where we slept — I immediately started

[20b] to examine a Black — Volcanic district deserving name of Craterized —
Met a 2 immense Turpin:[886] *[hiss]*, took little notice of me. — They well match the
rugged Lava. — Eating a Prickly Pear — which is well known to

---

883 Blainville 1834.
884 Coseguina in Nicaragua erupted on 22 January 1835 for four days.
885 Specimen in spirits 1286 in *Zoology notes*, p. 360, collected in September 1835 on Chatham Island,
Galapagos; listed as *Muraena lentiginosa* in *Fish*, pp. 143–4.
886 Galapagos tortoise. CD recorded in the *Beagle diary* entry for 21 September 1835: 'In my walk
I met two very large Tortoises (circumference of shell about 7 ft). One was eating a Cactus &
then quietly walked away. — The other gave a deep & loud hiss & then drew back his head.'
p. 354.

[21b]   contain much liquid —

<u>Craterized district</u>

Lavas, of two ages, one rugged [image] [curved strata] little cemented
bits — has been well compared to most boisterous frozen ocean; but with wide
cracks — other apparently has had the outer crust weathered

[22b]   away a more solid rock now remains [image] [crystals] prismatic, very uneven; but cracks
filled up: & covered with low trees — These characters, define the very line of junction
<u>otherwise not</u> distinguishable — the old ~~cone~~

[23b]   (from the Sandstone crater?) & newer kind proceeded from same district ~~of~~ covered
with Craters. — in space of few miles little less than 100 — Examined several both
old & new — All very small ~~50–75~~ 30-15. yards in diameter

[24b]   50–80 ft above a tolerably even district sloping a little from each of these points —
Crater nearly as deep as the plain —
~~Average~~ Average distance of Crater from Crater, ⅓ of mile — some more distant,
2 quite distinct very small ones

[25b]   30 yards from rim to rim — not possible to distinguish Stream from Stream — if Craters

removed appear one uneven coule: [image]   [crater]
this form consisting of ~~ft~~ of circular rim of cemented fragments —

[26b]   where whole surrounded by very porous
laminated clinkstone as if in flat surface
one st[r]eam had /b/ been poured out
on all sides —

[crater]
Other irregular piles, Crater more or less
distinct of black & glossy. /stalach/.

[27b]   from slaggy lava, from the base of which the Lava appear to have flowed or rather the
/pile/ formed directly subsequently to Stream: In the streams near all, Craters, covered
gutters tolerably smooth side. 2.4

[28b]   ft deep. — well arched — At base of very many crater
large Circular masses have subsided evidently from crust of /caverns/ being lowered.
These vary from 30 to 60 ft

[29b]   deep — 3 of these surround the Crater in one spot. —
Also parts of the Stream appear upheaved in arch — or down by escape of gas & have
large irregular holes

[30b] in consequence. —

These circular depression pits at first look like Craters. — || The Thenca[887] very tame & curious in these Isl[ds]. I certainly recognise S. America in ornithology, would a botanist? — ||

[31b] ¾ of Plants in flower — age of freshest Lava not great

Pumice grey on beach

Near edge of Lava generally /found/ by crack parallel to the border —

I now understand St Jago Lava — 50 years in the sea —

[32b] would remove the Crater & upper surface of Lava. Who could tell points of upheaval — no direction of points — ("King Landscape) These are Phlegrean fields to /Etna/ — Finger Point — & Vesuvius Crater Hill =

[33b] These other such district on opposite side of Hill —

So does Hoods Is[d] appear to be

Clinking plates of "iron Lava

[34b] Dry sand — Lizards —

Black Mud & parasites —

Brazil without big trees.

Feast. Robinson Crusoe —

Gross-beakes[888] —

Great tendency to nodular or Concretionary structure in all Volcanic Sandstones

[35b] Galapagos Lava generally $\angle^r$ cellular: therefore little fluid — agrees with little Volcanic Ashes /&c/ — But columnar, built up or concretionary —

Covers about 160 nautical miles square: in space = to Sicily & Lipari Is[ds]

View of Craters like in Scrope;[889] & Lyell /3/

[36b] 12th — Monday [October 1835]

Walked up to the Houses — Slept there Eating Tortoise meat By the way delicious in Soup. —

Followed down the ravine with water — soon drys: very green & pleasant

[37b] Extraordinary numbers of Turpin —

When drinking bury head above eyes — Will drink when a person is within 2 yards of them about 10 gulps in minute. noise during cohabitation

[38b] & length of time certain — Eggs covered by sand soil from 4 to 5 in number — require a long time before they are hatched. — Quickness of travelling certain — now said come every

---

887 Spanish name for mockingbirds.

888 Galapagos finches. This is the first known explicit recording by CD of the finches and here presumably referring to the thick-billed large ground finch listed as *Geospiza magnirostris*, *Birds*, p. 100, plate 36; see specimen 3331 in *Zoology notes*, p. 297.

889 Scrope 1825.

[39b] three days for Water — Eat Cacti in the dry Islands
Yellow Iguana[890] intestine full of Guyavitas & some large leaves
eggs in a hole —
Caracara[891] habits

[40b] like Carrancha sound Slaughter house <u>kill</u> chickens — <u>run like</u> a cock

Thenca eat bits of meat
|13th| Tuesday [October 1835] Returned — Cloud

[41b] all morning descended highest Crater — Glassy Feldspar — red glossy scoriæ:
Whaler gave us Water — extraordinary kindness of Yankeys

[42b] |14th| [October 1835] Wandered about Bird collecting —
Iguana[892] — shakes head vertically; sea — one no = dozes, hind legs stretched out
walks very slowly — sleeps — closes eyes — Eats much Cactus: Mr Bynoe[893] saw one

[43b] run walking from two other carrying it in mouth — Eats very deliberately, without
chewing — Small Finc[h] picking from same piece after alights on back —

[44b] generally in the Tent generally 85–80° —
Trade wind & sun 77° or 78 —
On Rock out of wind 108° — —

[45b] |Friday|. [16 October 1835] No wind 12 oclock in Tent. 93° on sand above 137° In
little wind & sun } 85°

[46b] at <u>highest</u> central Crater; marks where streams have flowed — Crater perfect one side
lower — red glossy scoriæ — ¼ ⅓ of mile diameter. —
Near Bivouac on Coast to Eastward —

[47b] many <u>thin streams</u> of Trachyte all with much glassy Feldspar. Some very old Craters
The covering of Volcanic Sandstone on the hill which has been particularly described
is marked in /channel/

[48b] or arches — hence not dust — since land assumed present height or there would have
been a cliff —
Iguana digs tail motion slow — appear =

[49b] same stupid from low facial ∠ = Very fond of Cactus run away like dogs from one
another with pieces — Excavate Burrow shallow — first on one side & then on other —
two or three time

---

890 Specimen 1315 in *Zoology notes*, p. 294; listed as *Amblyrhynchus demarlii* in *Reptiles*, p. 22, plate 12, current name *Conolophus subcristatus*. Today sadly extinct on James Island.

891 The Galapagos hawk, specimens 3297–8 in *Zoology notes*, p. 299; listed as *Craxirex galapagoensis* in *Birds*, p. 22, plate 2, current name *Buteo galapagoensis*.

892 Galapagos land iguana, listed as *Amblyrhynchus demarlii* in *Reptiles*, p. 22; see *Zoology notes*, pp. 295–7. It is curious that there seems to be no mention of the famous marine iguana in the notebook.

893 Benjamin Bynoe (1803–65), assistant Surgeon on the *Beagle*.

[50b] throw dirt with one arm & kick it out with well adapted hind leg
then other side

(Comet)[894]

Saturday ~~18~~ 17th [October 1835] Ship came —
Sunday ~~19~~ 18th [October 1835] Ran

[51b] along Albermarle Is^d
One of the mounds in Albermarle most covered with bare Lava — Mounds 4000 ft high
30 yard in 5 minutes[895]

360 — in 1	6	3	1
hour	4	2	6
24	0	1	5
1440	2	0	0
720.	0	0	
8640	0		
4. ⅓			

[52b] Fresh water Fish
Bynoe direction of Slate in Obstruction Sound
Bynoe says they were quite angular the fragment[s] of Granite on Icebergs —
18th Wednesday: [November 1835] travelled up valley — at first beautiful view over cocoa nut

[53b] trees — 2 <u>fine</u> men, take no cloths or food — Higher up valley very profound —
most dangerous pass — ropes — Kotzebue[896] — Then one with pole, ropes, dogs & luggage. —
Wonderful view — Cordilleras nothing at all like it. Ascended a lava slope

[54b] excessively steep — middle of day vertical sun.
steaming hot — cascades in all parts, enormous precipices — columnar — covered with Lilies — Bananas & trees — after ascending fern hill threw myself in shade of thick trees surrounded

[55b] by Sugar — Bananas — — Food so abundant — Yams, Taro — Sweet root like Sugar, size of /log/ & forest of shady Bananas. different /tribe/ Rivers fish & Prawn — catch them diving gracefully amphibious. Ellis horse story[897]

[56b] this track so wonderful only famine & murders /ever/ could have induced people to have discovered them — Men speak a little English — breakfast — make fire rubbing. — Gauchos like Carpenters tool — really most fearful road 7000 ft

894 Halley's Comet.
895 Calculations on the speed of tortoises.
896 Otto von Kotzebue (1787–1846), Russian explorer and navigator. Kotzebue 1821.
897 Ellis 1829 recounted how a horse being landed at Tahiti in 1817 fell overboard and the natives on the ship 'plunged into the water, and followed [the horse] like a shoal of sharks or porpoises'. CD cited this in *Journal of researches*, p. 486, with the observation 'The Tahitians have the dexterity of amphibious animals in the water.'

[57b] mountain split by mere crevices to the very base valley three sources, we took South one —

Geology — rocks chiefly grey base, with nests of olivine vesicular in bands, some hard conglomerates — fragments of Trachyte & [do] large Crysts of glassy Feldspar & Hornblende

[58b] volcanic clay & Sandstones. Great pile of Lavas eat through by rivers. —

Ava not drink — Spirits. Men tell me not to tell Missionary. — Bivouac — cool stream where we bathed buried in peaks — house of Banana profusion of fruit — green leaves, thatch, dry

[59b] bed. 20–25 ft high shade as dark as noon —

After supper baked in stones fine vegetables. walked up to a little way in Banana grove by Valley. heard another cataract of 200 ft — Peaks infront ∠ 45° — Evening sublime —

[60b] Ava, brook shaded by knotted, deadly ava — acrid, poisonous stimulative taster — eat small bit, fear missionary.

Say prayers & grace no compulsion

Valley without breath of wind, unbroken Banana leaf — much rain

[61b] during night [pretty] dry

19ᵗʰ [November 1835] — returned by other road — Bananas, avoiding cascades sweeping round hill side, entered valley high lower down — only one place where rope was required — knifes edge, enormous precipices on each hand. — Men very

[62b] strong Tatooed, recall picture of S American forest — flowers round head, in the gloom of Bananas — caught some fine Eel. — Banana 3–4 ft circumference.

Evidence how man can live by hand.

Geology conglomerates certainly marine. —

Specimens will show the

[63b] kinds — Valley crevice very little wider at base & some way up than twice bed of torrent —

Knife edges, about same angle as a ladder — vegetation clings on a ∠ surface of decomposing rock — up to the highest peak — enormous appetite at breakfast

[64b] mass of Banana!! — fatiguing travelling so far poising each step with greatest care. —

20ᵗʰ [November 1835] — Slept under ledge of rock, very dark valley, starlight. — returned, met party of noble athletic figures travelling for Tayo. —

Found ship moved Papawa, walked round — Friday

[65b] 21ˢᵗ [November 1835] Ship returned Point Venus.

22ᵈ [November 1835] Sunday went down to Papiete in boat — Tahitian service — not much attention, appearance respectable, good singing. — not Euphonious sound — good missionaries

walked back in rain

Never can believe what is heard. —

[66b] ~~24~~ 23^{d} [November 1835] reef —

25th [November 1835] Breakfast Mr Wilson[898] — Papiete — Queen party

26th [November 1835] great Parliament. —

Textual notes to the *Galapagos notebook*

[IFC] 1.17] *Down House number, not transcribed.*

Lima August 4th 1835] *written perpendicular to the spine.*

13] *added by Nora Barlow, pencil, not transcribed.*

[10a] Lower level+ 5 ft.] *ink.*

+ 5 ft.] *ink.*

[IBC] Anchored … Cove] *written perpendicular to the spine.*

[1b] in two places] *ink. Numerous ink marks appear to be nib tests.*

[12b] Callejon la Maravilla] *not in CD's handwriting?*

[13b–14b] *lower third of leaf excised.*

[31b] age … great] *added ink.*

[51b] 30 … 50] *ink.*

[52b] *small ink marks appear to be nib tests.*

---

898 Charles Wilson (1770–1857), missionary at Matavai in Tahiti since 1797. See FitzRoy and Darwin 1836; *Shorter publications*, pp. 15–31.

THE *COQUIMBO NOTEBOOK*

⌐

The *Coquimbo notebook* takes its name from the port town of Coquimbo in Chile. The notebook is bound in red leather with the border blind embossed: the brass clasp is intact. The back cover has a label of cream-coloured paper (60 × 35 mm) with 'Valparaiso to Coquimbo. Coquimbo valley' written in ink. There are 68 leaves or 136 yellow-edged pages. The text was written in one sequence, pp. 1–132. Pages 133–6 are excised. The notebook is written entirely in pencil and contains about twenty geological diagrams. Some of these are drawn with exceptional neatness and some are unusually complex, such as the one on p. 27. This notebook was one of only two of the field notebooks which were lent by Francis Darwin for the 1909 centenary exhibition at the Natural History Museum, the other being the *Santiago notebook*.[899]

The notebook begins where the *St. Fe notebook* left off in late April 1835, when Darwin was in Valparaiso about to trek northwards up the coast to Coquimbo, the port city in Chile. This was a few days after Darwin wrote to his sister Susan.[900] In that letter Darwin forcefully expressed how he felt about the great traverse of the Andes he had just completed: 'I cannot express the delight which I felt at such a famous winding up of all my Geology in S. America. I literally could hardly sleep at nights thinking over my day's work.'

The *Coquimbo notebook* was used continuously from 27 April to 26 May after which it was not used again. The *Copiapò notebook* replaced the *Coquimbo notebook*, after a few days break, for June 1835, then the *Despoblado notebook* was used for July, with the *Galapagos notebook* for August to November when the *Beagle*, after three and a half years, finally left South America to cross the Pacific and complete her circumnavigation.

The notebook traces the first part of Darwin's expedition no. 8, from Valparaiso 600 km (370 miles) northwards through Chile to Copiapò. The second part of the expedition is covered in the *Copiapò notebook*. The *Coquimbo notebook* is the only notebook to provide a direct report of an earthquake, the one of 22 May recorded on

---

899 See Harmer and Ridewood 1910, p. 11.
900 CCD1: 445.

p. 99, the time of which Darwin jotted in the inside back cover as 8.00 to 8.15 p.m. Darwin himself did not feel much motion, but the noise caused great consternation at the house where he was dining at the time.

A very high proportion of the *Coquimbo notebook* is devoted to geology. Indeed, apart from recording occasional botanical observations and mentioning his horses, Darwin barely refers to a single living organism in the *Coquimbo notebook*, the zoology being confined to shells uplifted from the sea. There is general interest however, for descriptions of the mines and haciendas Darwin visited and the only mention in all the notebooks, on p. 132, of his favourite poet John Milton.

There is an excellent description of Darwin's mode of travelling on this expedition in a letter to his sister Catherine on 31 May from Coquimbo, just before setting off for Copiapò:

> I am tired of this eternal rambling, without any rest. — Oh what a delightful reflection it is that we are now on our road to England. — My method of travelling is very independent & in this respect as pleasant as possible. I take my bed & a Kettle, & a pot, a plate & basin. We buy food and cook for ourselves, always bivouacking in the open air, at some little distance from the house, where we buy Corn & grass for the horses. — It is impossible to sleep in the houses, on account of the fleas.[901]

The notebook contains detailed observations of the famous Coquimbo terraces, visited by Darwin on 19 May. These were discussed by Lyell in the third volume of his *Principles of geology* and Darwin must have been eager to see them since reading Lyell's book the previous year. Lyell doubted the explanation for the terraces given by Hall 1824, vol. 2, p. 9, who saw them as the shores of lakes, and Lyell suggested that uplifted sea beaches seemed more likely.[902]

Darwin's discoveries of marine shells demonstrated to him that the terraces were indeed uplifted sea beaches, this being clear proof of his by now mature view of South America as a continent rising from the ocean. Darwin discussed these terraces at length in *South America*, pp. 39–44, where he several times referred to his own identical 1839 explanation for the Parallel Roads of Glen Roy in Scotland. Darwin's work at Glen Roy was probably the most important piece of geological fieldwork that he carried out after the voyage and he was very reluctant to give up his sea beach theory for their origin, even after most geologists had accepted a glacier lake theory.

Darwin's basic problem with Glen Roy was that it had never occurred to him in 1839 that the lakes which formed the Parallel Roads might have been dammed by ice. As he wrote of his 1839 paper in his *Autobiography*, p. 84, in the same year

901 CCD1: 449.
902 Lyell 1830–3, 3: 131–2.

(1876) that the second edition of *South America* appeared: 'This paper was a great failure, and I am ashamed of it. Having been deeply impressed with what I had seen of the elevation of the land in S. America, I attributed the parallel lines to the action of the sea; but I had to give up this view when Agassiz propounded his glacier-lake theory. Because no other explanation was possible under our then state of knowledge, I argued in favour of sea-action; and my error has been a good lesson to me never to trust in science to the principle of exclusion.'

By 1861 the Scottish case was shown conclusively to have nothing to do with Darwin's theory of crustal elevation, but ironically much more to do with Hall's explanation for the Coquimbo case.[903] Darwin did not take the opportunity in the second edition of *South America* to correct the statements he had made about Glen Roy in the first edition of 1846.

Valparaiso to Coquimbo, April–May 1835

The *Coquimbo notebook* opens on Monday 27 April 1835 with 'Started from Valparaiso: farewell look — smokeless — (perfumed) red — bare hills — picturesque', p. 1. It is clear from the *Beagle diary* that Darwin was travelling with his guide Mariano Gonzales, four horses and two mules. All the animals cost just £25.[904] The country was 'so brown & bare' that Darwin compared it to the Cape de Verds Islands he visited three years previously. The following day Darwin 'passed Limanche & Umiri' and took a detour to the foot of the Bell Mountain [Cerro La Campana] which he climbed in 1834. He immediately plunged into detailed descriptions of the granites, gneisses, porphyries and other rocks which he eventually published in *South America*, pp. 169–75. He was especially interested in the metalliferous rocks which were the big attraction of Chile to European mining engineers and adventurers. He 'Lodged at Chacia', p. 3, and the next day had a short ride to Quillota where he was struck by the high price of corn and the poverty of the natives.

On 29 April there was 'very light shower during the last night', p. 4, the first rain Darwin had experienced since being pinned down at Cauquenes seven and a half months before.[905] On the following day he thought the new snow on the distant Cordillera might have been 'perhaps most beautiful view I have seen in Chili' and he noted the 'immense quantities of broken shells' on the hills, p. 5. On Friday 1 May Darwin noted the valley terraces of pebbles and shells 'said by Guasso (in all parts of valley) to be "Diluvium"' but the quotation marks show how Darwin had by now abandoned this term for such obviously marine deposits, even though they were '1 & ½ league from sea', p. 7.[906]

903 See Rudwick 1974, p. 114 *et seq.*, and the introduction to the *Santiago notebook*.
904 *Journal of researches*, p. 416.
905 See the introduction to the *Santiago notebook*.
906 These deposits were described in *South America*, p. 35.

The next day Darwin examined the slates and sandstones exposed on the beach where there were 'roaring Breakers' and the trees were replaced by a Yucca-like plant, p. 11. Geologizing continued on the 3rd, on the road from 'Quilimar to Conchalee'. The 'immense quantities of large Concholepas = Venus & Donax' mollusc shells put 'question beyond doubt if there had been any: great sea in Chili', p. 15. The climate was 'much drier here than in Valparaiso' and Darwin was amused that everyone had heard of the *Beagle* and 'all think' she must be a smuggler!

The expedition continued on to the copper mining town of Illapel which was about the half way point to Coquimbo. On the way Darwin was struck by the stark contrast between the green 'Alfarfa fields' and the brown hills above, and on arriving in Illapel he found the Governor to be an 'extraordinary arbitrary character', p. 18. They 'Staid at Illapel whole day [to] rest animals' and just as at home in May summer would be starting they had to bivouac against a 'cloudy gloomy windy day — leaves falling. Winter — Oh the difference with England!', p. 20.

From Illapel Darwin struck inland on the 6th to Los Hornos to investigate the mining district. He crossed the north–south line separating the granite from the porphyritic breccia derived from it, and came upon a group of miners carrying a dead companion for burial. Each miner wore a 'small Scarlet cap — extraordinary wide trousers. — & officers broad parti-coloured belt'. Four men ran carrying the body while the other four galloped ahead on horseback so they would be fresh to carry the body when the runners caught up. 'Such a funeral' wrote Darwin. These miners, he noted, were 'remarkably brown' and they were 'like Sailors in extravagance', pp. 24–5.

By 7 May Darwin's highly varied rock collection for this expedition was growing steadily, reaching specimen number '(24) Gold ore', p. 26. The whole of the next page is taken up with a complex map showing how the mines seemed to be concentrated in a NW–SE trending band of Tertiary 'gypseous formation' of immense thickness, with breccia on both sides. This was followed by page after page of geological description, including notes of large, silicified, metalliferous fossil trees, p. 36, which Robert Brown subsequently identified as coniferous.[907] Some of the hills were so 'drilled through' with mines as to seem 'like bank with Rabbits', p. 45.

This was a red letter day of 'interesting Geology' and 'Very hot the sun although now winter'. That night Darwin 'dined in little Rancho mining hut — good dinner, pumpkin bean …. pepper grease, onions — roasted Pumpkin', p. 47. He continued collecting the next day but as he approached the 'pretty looking town' of Combarbalá he was surprised to find that 'there are no mines!!', p. 55. This is the narrowest part of Chile, the border with Argentina being only 80 km (50 miles) from the sea.

907 *South America*, p. 209.

On 9 May Darwin continued on his rather wearisome way to Punitague [Punitaquí]. This was a 'Stupid day — barren like turnpike road — wheatfields in valley — curiously stony', p. 59. The 10th continued in a similar vein, but on the 11th Darwin was able to visit Alexander Caldcleugh's copper mines at Panuncillo. The notebooks do not suggest that Darwin actually went far inside any mines in Chile. He did, however, witness the extraordinary feats of labour performed in the mines by the 'Apires' with their 'breath like fire; odd noises stream of perspiration of breast'. They had 'no meat, no diseases' and laughed at Darwin's puny '25 pound hammer', p. 65. Darwin could barely lift off the ground the 200 lb (c. 90 kg) loads these men carried up near vertical notched poles twelve times each day.

The Mayor-domo of the mines was Don Joachin Edwards, described by Darwin on p. 75 as 'young ½ Chilean & English'. He recalled to Darwin that as a boy in Coquimbo he was taken to see a visiting English sea captain and such was the dread of the English buccaneers that he was terrified of 'contamination sticking to him', p. 76. Darwin stayed at the mines for three days, leaving on the 14th for 'Herradura bay South of Coquimbo', where he found all hands from the *Beagle* living in tents while the ship was refitting for her long passage across the Pacific. Darwin was able to find partially silicified whale bones and large quantities of shells, p. 85. He collected at least fifty-eight geological specimens and was no doubt relieved to offload these onto the *Beagle*. Darwin spent 15 May on board, then on the 16th 'with Capt. FitzRoy' (as is stated in the *Beagle diary*) he took lodgings in Coquimbo.[908]

On the 18th Darwin examined the plain behind the town and drew a sketch map of the bay with two ships at anchor, one of which was presumably the *Beagle*. He found numerous marine shells at various heights and was certain the plains were marine terraces. Darwin ridiculed Basil Hall's lake hypothesis: 'How could B Hall fail to see the plains curling out to seaward?', p. 93.

Darwin was told by '"Honest man — he is <u>old Spaniard</u>"', p. 94, that the first light shower in Coquimbo was on 14 May, and that some Apires could carry 300 lb loads up mines. On Wednesday the 20th Darwin made barometric height measurements of the terraces and drew a detailed cross section of the valley on p. 96 showing the heights on either side. Darwin made a diagram of the terraces in the margin of the *Beagle diary* once he was back on the *Beagle*.[909] The section on notebook p. 96

908 Keynes 1979, pp. 280–1, reproduced a remarkably atmospheric watercolour of Coquimbo attributed to FitzRoy, dated 'May 25th 1835', also in Taylor 2008, pp. 166–7, who notices that the watercolour could easily pass as the work of P. G. King.

909 Herbert 2005, p. 221, reproduced two of CD's worked-up diagrams which formed the basis of the published versions. These are both now in DAR 44.30.

became the lower of the two in DAR 44.30 which itself became diagram fig. 10 in *South America* (reproduced below).

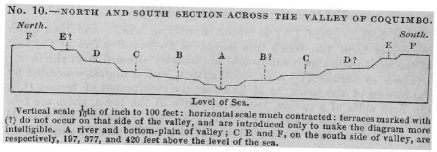

No. 10.—NORTH AND SOUTH SECTION ACROSS THE VALLEY OF COQUIMBO.

North.
F   E?   D   C   B   A   B?   C   D?   *South.*
                                                                    E   F

Level of Sea.

Vertical scale 1/10th of inch to 100 feet: horizontal scale much contracted: terraces marked with (?) do not occur on that side of the valley, and are introduced only to make the diagram more intelligible. A river and bottom-plain of valley; C E and F, on the south side of valley, are respectively, 197, 377, and 420 feet above the level of the sea.

An idealized north–south section through the terraces in the Coquimbo Valley, approximately 5 km from the sea. The section is approximately 1.7 km across and shows the five main terraces of shingle and sand, of which B, C and F are the most conspicuous. Fig. 10 from *South America*, p. 40.

According to the *Beagle diary* on the evening of the 18th Darwin dined with 'Mr Edwards', but it is unclear whether this was the father Joaquim or his son José Maria Edwards. Darwin visited the former's Arqueros silver mine a few days later with José Maria. It was during this dinner that Darwin experienced the 'smart shock of an Earthquake'. This does not, however, tally with the notebook where the earthquake is described at the end of the entry for the 20th: 'Earthquake, screams of women great roar — little motion — called severe — appalled the men', p. 99. This entry, taken together with the dated entry on the inside back cover, seems to demonstrate that the 20th is the correct date for the earthquake and that the *Beagle diary* is incorrect.

Coquimbo to Rio Claro and return, May 1835

The next day (the 21st) Darwin set off across 'mountainous grand alpine country', p. 101, with Don José and arrived at the mine after dark. Darwin recorded in the *Beagle diary* the pleasure of sleeping that night derived from the absence of fleas above 1,000 meters. He was, however, puzzled at this absence as the small temperature difference did not seem an adequate explanation. The 22nd was an opportunity to collect specimens and to record the geology in detail, much of which was published in the final chapter of *South America*.

On p. 102 Darwin recorded that at the S. Rosa mine he found some red rocks and green rocks. In the green rocks he found a specimen which deserves to be regarded as one of the most important he ever collected, as it is one of the only geological specimens ever described by Darwin in one of his publications. The specimen, which survives today at the Sedgwick Museum, is recorded thus: '(64) in one locality were near red beds (65) green bed with Silver near vein' and on p. 105 this was 'Considered most rare instance'. Once Darwin was back on the *Beagle* he entered this specimen in his list as 2940 and described it as 'Mixture of paler green mineral & Carb of Lime as (2937.) interstices with Silver & Muriate of Silver'.[910] He next mentioned the specimen in his geological diary (DAR 36.575–6) as follows: 'The green rock has in one case to the distance of a yard from the side of the argentiferous vein been had its particles cemented by native or the Horn Silver — the variety is exceedingly beautiful & the circumstance considered rare.' In *South America*, pp. 211–2, Darwin singled out 2940 for special mention: 'I have a specimen of one of these green rocks, with the usual granules of white calcareous spar and red oxide of iron, abounding with disseminated particles of glittering native and muriate of silver, yet taken at the distance of one yard from any vein, — a circumstance, as I was assured, of very rare occurrence.' Darwin mentioned the specimen again in the special section on 'Metalliferous Veins' in the concluding chapter of *South America*, p. 236: 'I have also described the singular and rare case of numerous particles of native silver and of the chloride being disseminated in the green rock at the distance of a yard from a vein.' Alfred Harker took a thin section of 2940 for petrographic analysis. He numbered the section as 47194 and labelled it 'Calcareous tuff impregnated with ore'.[911] So 2940 is a remarkable instance of a specimen for which we have the original note and field specimen number, made by Darwin at the site of collection, in Chile on 22 May 1835, then his specimen list entry and geological diary description, followed ultimately by his published account drafted about ten years after collection. The specimen is now receiving the highest quality of curation at Darwin's *alma mater*, a short distance from where he unpacked his specimens at the end of the voyage in Cambridge.

The notebook contains mention of many other important specimens, such as '(70) part of Coral reef', p. 102, and on p. 107 Darwin noted the presence in the limestone of 'immense quantities of **large oysters** & Gryphites' and other fossils. There is some confusion over dates on p. 109 but it seems that Darwin reached the Hacienda of the Marqueros by the night of the 22nd and left there on the 23rd. He continued to fill the notebook with an extraordinarily dense discussion of the rocks and their complex field relationships, largely as seen from the saddle. On p. 110 he drew a section showing purple porphyritic breccia cut by a fault, and by p. 119 he felt ready to attempt an 'eye-sketch section' showing the breccia, limestone, hornstone, syenite, gypsum, conglomerates, clays and lavas, most of which are described in *South America*.[912]

On the 24th Darwin related how, at the Hacienda of Gualliguaca, the 'Signorita not 17 mere child mother of two children' was apparently pregnant again. He adds that she was 'very pretty — dressed like lady' but that she was living in 'Some second rate farm house', p. 120. The scenery, conversely, was 'first rate' and reminded him

---

910 We are grateful to Francis Neary and Katherine Antoniw, of the Sedgwick Museum of Earth Sciences, for this information and colour photographs of the specimen.

911 See photographs of Harker's manuscript catalogue from the Sedgwick Museum on *Darwin Online*.

912 CD's fair copies of the two combined notebook sections are now in DAR 44.21.

of pictures of the 'Alps in the Annuals'. The next day Darwin parted from Don José and continued up the valley to where the Rios Claro and Elque met, his specimen tally there reaching 100. He found 'very great quantities of small white quartz pebbles in the white sandstone ([specimen number] 97); remarkable quantity, first time I ever saw such a thing, ground strewed with Beans — common Belief', p. 123. He also found some 'Secondary' (i.e. Jurassic-Cretaceous) fossils. The numbers of some of these were 'quite wonderful', p. 127.[913] One of these fossils was the new species of brachiopod named and described with an illustration by Edward Forbes as *Spirifer linguiferoides* in *South America*, p. 267 (below).

Spirifer linguiferoides, from the Rio Claro. Plate 5 figs. 17–8 from *South America*.

On 26 May Darwin 'having seen what I wanted', as he wrote in the *Beagle diary*, 'returned to Coquimbo', p. 128. On the next page there is an obscure sketch or map which may show the terraces mentioned on p. 130. On the 27th Darwin and Don José were back in Coquimbo where the rain of two weeks before produced 'thin scattered delicate grass. One inch long like hairs.' There follow four names of key ex-patriots (Douglas, Corfield, Alison, Caldcleugh) and two of Chileans who probably had important information and should be sought out in Copiapò.

On p. 131 Darwin reminded himself to repay FitzRoy '2 dollars & Rial'. There are some more names, such as Charles Lambert (1793–1876) who provided Darwin with much useful information[914] and played a prominent role in the Welsh copper industry.[915]

The final two leaves are excised, and the inside back cover has jottings relating to the 20 May earthquake; '5 more were felt up in the country same night' and the time of the shock at Coquimbo noted as '8 PM to 8.15 (20th) — Wednesday'. 'What

913 Discussed in *South America*, pp. 210–3.
914 See for example CCD1: 455.
915 See Alban 1990.

money has been got out of Arkeross' relates no doubt to the discussion of mine revenues in the *Beagle diary* for 22 May. The final decipherable jottings are 'Mar[iano] G[onzales] is paid up to the 15th of May + 20 Rials in Quillota for spurs'.

The next notebook, *Copiapò*, continues from the *Coquimbo notebook* on 31 May almost without break and the two notebooks are perhaps closer in content than any other two. Almost the first jotting in the *Copiapò notebook* is '20 Riales in Quillota', almost identical to the last in the *Coquimbo notebook*. The geological notes simply continue into the *Copiapò notebook* and then into the first forty-two back pages of the *Despoblado notebook*, where the final rock specimen number reached 185. In this way *Coquimbo*, *Copiapò* and *Despoblado* are a continuous trilogy of notebooks which took Darwin up to the point where he sailed from Copiapò on 6 July 1835.

[FRONT COVER]

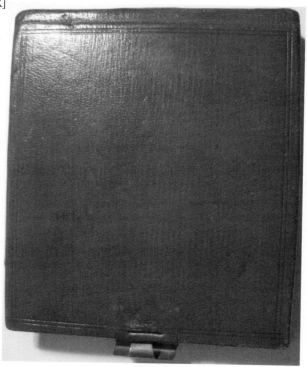

[INSIDE FRONT COVER]

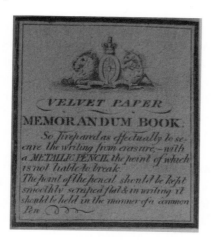

Charles Darwin
H. M. S. *Beagle*

Oliva
2 /Penus/
Barnacles
Pecten
/mien/
/Granitic/ blocks
/Trigona/
inclined strata
Bones
no /petrel/ drift
pebbles high up.
NE — /lan/ NW dip SW

25
 8
200
14/ 179 / 14
   14
   57
   56
   18
      112
       20
      2240
12
 200
2400

[1]   27th — (Monday) [April 1835] Started from Valparaiso: farewell look —
      smokeless — (perfumed) red — bare hills — picturesque: passed through highly
      auriferous country to near Limache — supporting or the gold near all the streams a
      scattered people: Like Plazilla. but more earthy sandstones pebbles — certain layers
      appear auriferous — so that sort of mines are excavated: Country like the C. Verds so
      brown & bare: — in great want of rain: — I saw 2 paps very small of granite &
      syenite: —
      28th — Tuesday [April 1835] — passed Limache & Umiri to south foot of Bell

[2]   found strata of amn crystal; without distinct crystals of green rock with green patches,
      alternating with some of the regular purple PB & dipping South high ∠ covered by
      considerable mass of altered slate like summit chiefly pale without trace of crystal & part
      of which has a brecciated blended structure: skirts of very foot composed of harsh
      whitish, Euritic, but chiefly quartz rock (with large veins of do). gre resembling parts

of Cuesta[916] on road to St Jago; & certain aspect of being part of granitic formation although not seen in Cordill: I have no doubt the Por. Br. overlies the

[3] gneiss — auriferous: what connection? Copper mines on SE slope of Bell: Due South a considerable hill perhaps running E & W of Syenitic Granite with some Mica: in which has been a Silver mine — Mem: Gold in Prado. Caldcleugh[917] — Loadstone & Quicksilver mine —
Lodged at Chacia — ugly people as most Chilenos — produce of garden & Chacia support Inhabitants a ~~great number~~ of small freeholders?
29th Wednesday [April 1835] — from beyond Umire — (quicksilver mine is at Caleo) to Quillota short day — passed over — decomposing hills of granite — gneiss with trappean veins — which manifestly fringe

[4] the foot of the West Bell: as beyond Umire — Corn much dearer than at Valparaiso — from the entire want of capital agree to sell their corn for so much before crop, to buy

necessaries &c &c — first rain very light shower during last night. —
Thursday 30th [April 1835] — Passed Chilicauquen. perhaps most beautiful view I have seen in Chili — Cordilleras Winter covering — granitic platform of Valparaiso — Chilicupu E & W range — perhaps melted rocks — country to Plazilla ugly low, brown undulating —

[5] nice village rested there — Shortly before decomposed granitic rock — Just beyond village form of Bay — immense quantities of broken shells — close to Catapilco again in very many places. hill tops thickly covered with layers of broken shells — many Donaxes;[918] in <u>road finely comminuted</u> — very sandy soil. — <u>not blown by wind</u> because packed in layers — in many places I should think certainly 300 ft

[6] sides of these hills — or valley — these layers could in quantities like Iniaquin
May 1st Friday [1835] — Catapilco to valley of Lingua — much granitic rock generally — decomposed — some — syenitic — much quartzose Granular fine rock with green spots
Valley of Lingua fringed by exceedingly also obscure 2d plain levels plain of some height — form like Valdivia apparently composed of pebbles, extends (with broad flat, numerous Patagonian formed valleys) like well come to Longotomo
Hence Valley of this little rivulet

[7] grand broad, formed on South side by the /gravel/ plain — yet bottom & North side could with immense quantities of comminuted shells — said by Guasso[919] (in all parts of valley) to be "Diluvium" — Here valley marine — 1 & ½ league from sea —

916 Spanish for slope, escarpment.
917 Caldcleugh 1825, 2: 43.
918 Marine clams of the genus *Donax*.
919 Chilean equivalent of Argentinean or Uruguayan gaucho.

Barren ride dotted — trees scarcer replaced by the Bromelias moderately few small inhabited valleys — At (Tilla/ped/.) Granitic hills apparently capped by stratified (P. Breccia ?

[8] Saturday 2^d or 3? [May 1835] On the North Slope of valley of Longotomo to the height of I should think 200 ft — ~~plenty~~ immense quantities of shells excepting in road not much broken, chiefly Donax; showing that even side ravines are formed by sea; after passing over some more granitic we come to green crystalline rock even in one case porphyritic with Feldspar: altered slate. we then come to true Blue

[9] Slate — Beyond Guachen — on the Beach is seen to alternate with thin layers of sandstone & a pale rotten slate — some compact purple slate — auriferous quartz veins — dip to E by N about 40° — also a cross sort of cleavage dipping North. — The Slate is smoothed over into a low plain with pebbles which abuts against a less regular one — & behind round mountains — The Slate is continued whole way to

[10] Quilimar, an irregular platform. studded over with curious abrupt cones & broken paps — These consist of that kind of Greywacke where fragments almost blend in each other — fragments occasionally of a large size very compact — other paps of a slate almost become a greenstone: All very hard rocks — origin of this structure common in granitic countries is explained by the equally numerous peaks

[11] on coast surrounding by roaring Breakers — spaces between paps very level, like little Bays or coves — universal East dip — excepting in these massive paps: — Scenery rather curiously broken, with Bromelias — stems, dead like brooms, like Agave or Yucca — Shells in valley of Quilimar, but this is not above a mile from the sea — some of the Clay-Slate glossy — is connected with quartzose rocks — Dyke N & S

[12] Sunday 3^d [May 1835] (1 Brecciated Slate — South foot of Bell of 2 (2 & 3) — altered green slates 3̶ 4 Breccia like P.= B = (5) — do compact jaspery
Quilimar to Conchalee. — On the road there was less true Clay-Slate the paps not quite so numerous. composed either altered crystalline green slate (now Trappean) (2 & 3) which I saw, in one place becoming truly Porphyritic — & many of a purple very compact. conch: fract., brecciated rock (4) structure same as what I have called green blending

[13] Greywacke: fragments generally small & ~~some~~ some rocks almost jaspery in nature of same purplish color (5). also one quartzose pap. — Paps probably upheaved & rounded by Breakers on Beach — At a point of Conchalee, we find true granitic formation, but forming a most singular intricate mass of dykes & distinct masses (junctions violent) of different Granite
1^s most common is a soft decomposing syenite — with much green mineral sometimes true /Quinca/ & sometimes

[14] Hornblende. — structure like Gneiss of Valparaiso but no cleavage: a white compact (Euritic) quartzose granite, smaller dykes & proceeding from masses. — & black almost pure Hornblendic rock inter/leaved/ with thin granitic veins. — These rocks appear to have burst through a Porph greenstone Base like the altered Slates. — In a point to the North we again have the Slate, such as (2 & 3). — Hence altered Slates explained = The above

[15] rocks are just like the Porph= Breccia & greenstone of <u>Cordilleras</u> excepting less
      Porph. —
      At Close before Conchalee, two plains most distinct. lower one 50–80 ft above sea —
      also a lower one a little above sea. — The 50–80 one covered by <u>immense</u> quantities of
      <u>large</u> Concholepas = Venus & Donax — This puts question beyond doubt if there had
      been any: great sea in Chili. 1831 — much dryer
[16] here than in Valparaiso — hardly expect any rain till end of May — All think *Beagle*
      Smuggler — complained of want of confidence — features of country same. — No
      cattle. —
      Monday 4th [May 1835] Conchalee to Illapel. various Granitic & Syenitic in the
      lower ravines beautiful Granite, fine black scales of Mica: a miner described the mine
      of Las Vacas as occurring in this rock — Cuesta of Las Vacas lower parts chiefly
      Feldspar & Quartz, & much harsh rocks — Euritic & Quartzose, as in the Prado:
      higher up a small
[17] crystallised Syenite (which also happens in the Prado} Feldspar singularly apt to
      decompose.) = This Cuesta has a considerable elevation, probably 2000 ft runs about
      ENE & WSW. this seems rather a common direction there are some also N & S. —
      All irregular = on the N=Eastern slope many dykes of blackish-greenish close-grained
      crystalline rocks, just like Prado — all much decomposed. Hill certainly of a
      syenitic character: There was some little of the Porphyry with plates = crystals
      of Feldspar. But a whole road to Illapel, in bottom
[18] of all the valleys. we have fine grand /masse/ of well developed Syenite — & those
      white porphyries which we have seen associated in Andes — In C of Las Vacas
      copper mines. = Valleys of Chuapa B. of Illapel, with very high plains of gravel,
      on each side of rivers. = Crossed to day considerable stream of Chuapa — glimpse
      of Cordilleras = I believe trees like Chili — soil peak of Chuapa = Valley of Illapel,
      town pretty extraordinary contrast of the point — like green of Alfarfa fields &
      turnpike road hill on each side. a hedge separating the colors: Governor of Illapel
      formerly /days/ extraordinary arbitrary character — honest people. honestly
      distributed in district
[19] Tuesday 5th [May 1835] Ascended hills NW of Illapel All Syenitic, sometimes granitic
      accordingly as black mica preponderated over fine crystals of Hornblende with quartz —
      varying suddenly from coarse to very fine grained = feldspar white almost powdery in
      some cases — <u>much</u> Hornblende or Chlorite with Feldspar without quartz (6). Will
      such be called Syenite. — Generally little quartz in all the Syenites excepting in some
      parts of C. of Vacas = Dykes of close-grained dark greenstone = Metalliferous & grand
      copper vein running N by W & S by E; more than one parallel — containing much
      micaceous iron ore (7): — The syenite near vein differed suddenly in the degree of
      fineness — In some parts
[20] of hills. Hornblende black & glossy — generally green = Staid at Illapel whole day rest
      animals — bivouaced at cottage making a screen with horse — rugs & purchasing fire

wood & provisions. cloudy gloomy windy day — leaves falling. Winter —
Oh the difference with England! —

Wednesday 6th [May 1835] Illapel to Los Hornos; we proceeded at first rather
NEasterly & met a good deal of true Granite: after this very much Syenite, & those
mixtures of white Feldspar & Hornblende; saw ~~many~~ several Copper Mines. — We
proceeded thus far some way till at last we came to a crystalline stratified greenish
rock — /after/ this purple

[21] Porphyries, & then the Road ascending a hill we had true coarse Porph. Breccia in
abundance. Saw some of the white rock brecciated soft rock. Such as in the Valley of
Maypo = The stream in this part is highly auriferous. = The strata dipped to about
NE at a small angle. — we therefore (Travelling N & S) crossing the line obliquely. —
The green rocks were certainly the lowest. —
As the strata here in the

[22] first escarpement were of a considerable thickness. the denudation from Syenite to
west must have been very great: — After passing for some time through a broken
(instead of as before rounded country) of the P. Breccia we came to an undulating
open piece of ground where I found hillocks much water /worn/ of a pale friable
light porous earth, containing in layers of Silenite patches & mamillated

[23] clay centres: thin layers of Calcareous tufa — metallic patches. — Salitra patches.
Hence water bad. — thinly stratified reposing on, conformably & dipping $\angle$ ~~30~~–20 to
NE, apparently covered by P. Breccia — immense thickness, which appears to vary —
associated with siliceous sandstone & ~~purplish crystalline Limestone~~ ? —
Appear to S traversing the mountains, which

[24] are lofty we crossed this NW & SE band in a NNW & SSE direction. —
Mine Earth — Cordilleras road — miners — front coloured base — small Scarlet
cap — extraordinary wide trousers. — & officers broad parti-coloured belt — party all
on horseback — four men carried body for about 200 yard as hard as they could run,
four on horseback darted in front change in ~~min~~ second, very quick travelling, uncouth
shouts — such a funeral — remarkably brown

[25] men — the miners — like Sailors in extravagance. =
The gypseous earth, full of pores, friable in the fingers. —

Thursday 7th [May 1835] — (7) White softish aluminous stone, which passes by
every shade into (8) & (9) latter exceptionally abundant laminated — translucent
like Uspallata (10) — still more altered (11) black siliceous rock (basis of Breccia
Pitchstone) (12–13) other varieties of do (14. 15) Compact & laminated light
green indurated & luminous rock, like Uspallata (16) fine grained brown siliceous
rock —

[26] (17) Extraneous particles quartz /formed/ by white aluminous base
(18) do ferruginous base
(19) white, hard, (non cryst) earth Calcareous stone Tosca
(20) Green extran particles? crystalline structure

(21) Green Porphyry
(22) Syenite — like do — associated with latter
(23) silenite & copper ore
(24) Gold ore. —

[27]

Porph Breccia Mines Tertiary Porph Gypsum Gypsum High Hills NW by
W Porph Breccia

[28] Again examined junction of Gypseous formation & Porph-Breccia — find that they are decidedly conformable some where about dip 20° — 30 — The gypseous formation, or rather it — recognised by a broad band of whitish soft stone, is properly contained in a ~~whitish~~ white aluminous stone, soft & easily decomposed into a friable pale earth — such as (7) — This generally in parts passes into a laminated rock such as (8 & 9) in every degree of softness: Such beds ~~bound~~ join on to the P. Breccia.

[29] This latter near junction is only a compact purple, or <u>greenish</u> sandstone, precisely like those at Incas bridge High up. — I believe from fragments there are some strata of coarse conglomerates. These are thinly stratified: they rest on such varieties as seen at Incas bridge of partially crystalline Sandstone & beneath /it/ to the West we have the true Porph-Breccia of the Gypseous formation, amongst the lowest beds. I noticed some of the green rock (14) which

[30] will be /mentioned/ & some harsh hard siliceous grey rocks of same structure as a very common Rock one to be mentioned, mingled in ~~masses~~ veins with calcareous tufa; but the main rock is this aluminous softish Slate. —

The P. Breccia close to junction traversed by dyke — I should state that the Gypseous form crosses country NW by W & SE by E —

When well within the Gypseous formation appearance has been described —

[31] specimen (23) Is not Characteristic too compact — Copper ore probably from some metallic vein. —

All the waters just like those in Uspallata range —

I saw one place where this substance was traversed & curiously contorted by a /pap/ of green base Porphyry white Feldspar crystals. — here all the Gypsum was white fine grained non transparant like in Cordilleras: many pure layers some inches thick — worked — (Mem in Cordilleras frequent association of Gypsum & Alumen) Is this an accident

[32] or rather the effect of heat? I found in another place a small anticlinal line, produced by a white Feldspathic dyke on the North Boundary of Gypsum in the aluminous slate: — On the NE boundary we find the associated aluminous Slate or stone dipping still generally to NE by N, right beneath a mass of strata which I consider Tertiary or anyhow as same as Uspallata — The Gypseous district from soft /nature/ now removed from great basin, but

[33] to N & S is seen /conformably/ in the high mountains of P Breccia — In the Gypseous form we see particularly in NW corner of it the effects of the Porph = injected Hills — & the Strata accordingly show much irregularities & opposite dips — but on a large scale especially in SE part the NE dip is the characteristic one:

So return above the Aluminous Slate we have a great mass of thinly stratified rocks, which almost at first inspection

[34] I recognised a not very easily explicable resemblance. —

These strata are most wonderfully torn with ~~Hills~~ Paps of Porphyry & more so with a network of Metallic veins — All this occurs chiefly in NW corner where I most examined strata. — The rocks vary much — ~~Next to the~~ & are here much altered — Next to the hard Aluminous Slates thinness of strata alternate — which even become such as (10) & remind one of Uspallata we have a jet black very siliceous rock such as (11)

[35] This is thinly stratified & very common contains masses & traversed together with Alum slate with large veins of Crystall Carb of Lime, such as in Uspall. also some little agate (12 & 13 other specimens) — & quartz — I saw in one instance layers of calcareous tufa — partially like Tosca rock. Alternate with it — Is remarkably by in a few spots passing into Pitchstone & containing precisely as in Uspall brecciated Pitchstone (This first made me suspect similarity of origin) — identity complete: —

[36] Also clear impression of short pieces of horizontal branches of trees: — I doubt whether any part of the world can produce such immense quantities of silicified blackened, often metalliferous wood as here lies scattered, many of the pieces appear in form like the root = stumps — being many feet in diameter — All so very much altered as to require lens to recognise character — excepting when external shape is preserved. The specimen like necklace

[37] structure was common. —

Mem. quantities of Silic = wood — unaltered near St Jago & Plazilla = patches of Sandstone — Excepting where igneous action has hardened Tertiary strata, they do not remain.

There is a good deal of a compact green rock, sometimes laminated, which reminded me of Uspallata (14–15)

(16) brown siliceous Sandstone — others where the distinct grains of quartz are very visible —

(17 + 18) Extraneous Particles are cemented together by Aluminous & ferruginous particles

[38] These are abundant — Some little (19) white calcareous earthy stone = other layers assumed a partially crystalline structure, but are really sedimentary (20) — The trees = Brecc Pitch = Green-stones — some of the Aluminous Slates — & those with Extraneous particles, but chiefly a general structure recalled Uspallata to my mind. — These beds (near to the injected Porph Hills) contain an infinite number of metallic veins chiefly <u>Copper</u> & a little Gold — both similar to

[39] what occurs in Uspall. — The best veins run from N by W to NW. — They are equally worked (but I think in less number) in the injected Porph: The Copper ore is associated with immense quantities of most brilliant micaceous black Iron ore — ((used as flux) miners black as in coal) just like vein in Syenite of Illapel — This substance traverses all the lateral crevices & sides of veins — Thus we see here & in Uspallata, strata most

[40] worked on the more patches which adhere to the injected rocks.

I believe these strata do not alternate with Lava —

Specimen of Gold ore (24)

As may be seen in map, the NW end of band of Gypsum & Tertiary Strata is almost broken off by Hills of Porphyry, these strata sewing up more /area/ between them, this also occurs to the NE or /point/ dip — at this point apparently beyond the lines of Porphyry, there

[41] come a /first/ lofty escarpement of Porphy — Brec dipping to NE — The Porphyries are closely connected with Syenite, are greenish with white Cryst of Feldspar & green specks of Hornblende such as (21), & these pass into what I consider Syenite (22), Although Quartz does not seem present — black angular fragments, plates of Epidote. — straight or largely conch — fracture — These rocks form paps & some dykes in all parts of the NW corner & cause the strata to dip in every direction; large masses

[42] with an undulating outline having burst through the lower strata, lie & contort the black rock with trees. — Although the main dip is directly toward the line of hills to the NE, yet at their foot in places, an anticlinal (or NW) dip will be seen. — The strata being softer, valley are formed between paps of Porph = which are thus only fringed with Tertiary strata & in

[43] such are richest mines — One of the best marked included line of Paps runs NNW
      to SSE. — ~~one pap~~ Much of the Porphyry deviates from the character of specimen,
      which perfect sorts only occur in ~~main~~ larger paps
      I suspect from fragments, & form, Porph, Brecc caps parts of NE line of hills. M^r
      Caldcleugh remarked similarity of Gold ore of Uspallata & Coquimbo
      All the mines this side of Cueli de los Hornos called Los Hornos

[44] To return to Porph: some of the paps — are composed of a purple Claystone Porph
      (like P. Brec & that near houses of Uspall) also a compact green base, with Epidote
      white & pink crystals of Feldspar, very pretty — this had a coarse brecciated structure,
      like some Lava — interstices chiefly filled up by Crystall Carb of Lime —
      Must nearly have burst out of Volcanos — if not actually so — These Porph paps
      indiscriminately have burst through

[45] the Gypseous & Tertiary. —
      Some of the hillocks of Porph & Tertiary are drilled through & /every/ direction by
      chiefly parallel mines — like bank with Rabbits
      The Gypseous & Aluminous soft slates of immense thickness seen to South
      I am certain of age of P. Breccia & of Tertiary — I am certain Gypseous is
      conformable (seen in range of transverse mountain to S of Basin)
      I feel little doubt Gypseous by degrees passes into lower beds

[46] of Tertiary — I strongly suspect that Gypseous corresponds to highest beds of Andes —
      Consequence is evident — that here there has been no interruption in deposition —
      there then has been — It also proves what I suspected, posteriority (Mem: identity
      of Copper vein in Syenite) posteriority of outer lines of mountains) It is manifest
      that the NE dip, is the regular one — the other only depending on Paps &

[47] the NE chain, over which has had no general effect —
      Staid whole day — interesting Geology dined in little Rancho mining hut — good
      dinner, pumpkin bean — /kian/ pepper grease, onions — roasted Pumpkin —
      Remarkably brown color of the miners (& often curious expression of eyes?) —
      Very Hot the sun although now winter
      The black siliceous rock contains much Iron — It is certain that /Gap/ is between
      the soft A Slates, that these alternate with Tertiary & overlie confor the red
      sandstones

[48] These Tert-strata being without old pebbles — worthy of remark & easily
      explicable. —
      Friday 8th [May 1835] — (25) White Aluminous stone various fractured cleavage
      (26). Variety abundant of the Syenite
      (27) Almost jaspery, small brecciated rock — now porphyritic —
      (28). Sandstone or rather base of P.B purple with specks of almost Crystall
      Limestone I noticed that much of the white aluminous rocks such as (25) &c &c has
      a shattered fracture, & is exceedingly like some of the rock of Uspallata; especially
      where near a

[49] Veta it is traversed by network of ferruginous threads, as well as micaceous iron ore. —
The arch (A.) of Tert formation runs along way up to the North, & **full** of mines,
it is a mere strip, rises to some height up the Cuesta — in a similar manner as to
the west another creek was yesterday seen on other side of group of Porph-Hills —
I see the main Porph Pap (B) has on its top & East side a quantity of very
Porph. Brecc, of a greenish color, patches stained green with Epidote, & purple
with lines

[50] of fragments parallel — It is probable that these have been carried up on the top of
Porph — through Gypsum & Tertiary strata — In all this mineral — although many
mines are worked in the Porphyries the greater number are in the Tertiary strata: —
The Cuesta at head of last creek (of Hornos) is formed by (E & W) mass of the greenish
Porphyries already described. The tops of which are in places capped with (as I supposed
with Porph Breccia:

[51] The road pursuing a North course passes through a grand long district of
these Poprphyries & mountains & Syenite such as specimens — there is
much white quartzose??? Syenite as (26) — We then ~~come see on Paps~~ to P.B
formation dipping NE. & afterwards, road bending a little Easterly enters same
formation dipping Easterly; hence I imagine a North & South range of Syenite
bends at

[52] the Cuesta de los Hornos E & W & afterwards continues its course about
NW & SE

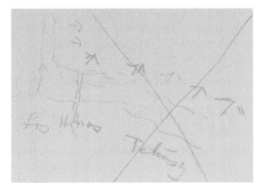

Los Hornos Tertiary

The formation of Porph Breccia was very extensive of which we now entered — near
Combarbala, the Red Sandstone

[53] becomes thinly stratified & contains agate, much & cryst Carb of Lime: they are
jaspery & very <u>red</u>: — Close South of the town, I examined strata. Found a ~~br~~ <u>red</u>
compact, conch: fract almost jaspery rock (27) with small /n/ brecciated structure &
/Basis/ of the Claystone Porphyry sonorous & brittle: above there came a very
red sandstone with white specks (28) almost crystall of Lime, & white nodules of
do; this rock had the appearance to an extraordinary degree of mass of $\angle^r$ & round
pebbles

[54] basis same as pebbles, at first looking like the common Porph: The angular pebbles & nodules of opake Limestone made me suspect it was not true Conglomerate — Above this came great mass of white often carious, opake cherty rock — (in some spots interlined with red sandstone): All these beds dipped to S a partial disturbance, apparently owing to a pap of the same greenish Porphyry, which

[55] I found at base of escarpement: yet here there are no mines!! These beds come high in the series of P. B. formation. — Remarkable for their harshness hardness — redness & want of Porph-structure. — Is the East great Escarpement of stratified rocks under which these appeared to dip. & beyond them other Escarpement — (thus all face W) which by color &c appeared to contain much syenite

[56] These varieties of P.B formation differ a little but not essentially from any I have seen. —
Pebbles of red cellular rock & Black Basaltic, probably Volcanic
Including the Escarpement W of the town, probably three coast-wards
Mountainous, barren country, dog, describe method bivouacing, small & here always pretty looking town *[illeg]* In the creek I found some of the strata dipping at an $\angle°$ of about 60° —

[57] Saturday 9ᵗʰ [May 1935] About half way to Punitague: in first part (we steered about NW, so forming a zigzag section) we found much hills of green porphyry with plates of white Feldspar (curious sort) which seem to have certainly been injected: passed the Jaspery & cherty rocks — occasional dykes of greenstone & Paps — In Cordilleras apparently granite beneath the P. Breccia. — dips various, none (in any part) high — dome-shaped strata:

[58] In Paral, in P Brecc form Silver Mine — ~~metal~~ ore in mixture of quartz & white decomposed Feldspar, Very like S. Pedro de Nolasko (29) The road for several miles P.B — nothing particular: till before near junction of the hills of Porphyry & Syenite, there were several small copper mines —
Syenites, more truly Porphyry, generally darker coloured, but same structure: during day much little altered P. Breccia or rather Sandstone: Mem: our course NW explains:

[59] the occurrence of syenite: — Stupid day — barren like turnpike road — wheatfields in valley — curiously stony: —
Sunday 10ᵗʰ [May 1835] — The Syenite district may be described as occurring a little to the East of Quillay — the road then continues amongst small rocks — till we come to a considerable line of mountains apparently running SW & NE & called Los Hornos — I have no doubt part of same mass as former so called hills. — Rock, true syenite

[60] (& before) with also much black mica — much however without the Quartz. — The road during this time had much *[Westing]*. — at the Mineral de Puntaque of S side of valley innumerable Copper mines & some gold — much black micaceous Iron ore (used as *[Liga]*) they run as miners say, about NNW, some Gold — absolutely in the Syenite; one part of the circumscribed mass of metalliferous ~~rock~~ rock

[61] was grains of quartz & white Feldspar tending to decompose: On the N. side very many poor Quicksilver mines — I examined two, they were in soft white feldspar, few quartz grains — Scales of Talcaceous mica & decomposed spots of Hornblende — rock very soft: Quicksilver ore spoils in sun (30). affects the quartz (31), the specks seen often to occur in minute drusy cavities lined with quartz: This side of /far/ valley — the granitic

[62] rock is capped, & succeeded, a little Westerly by a very black Hornblendic Porphyry (32): which I have not seen before —
I believe the Mercury mines are also worked in this: There are but very few Copper mines in same side with Mercury: — I believe these veins identical with Los Hornos — the Granite seems to stretch far to West — Much Quartz paps in one Part of Syenite. Cuesta — & Trappean dykes: —

[63] Beyond Punitaque, road turns to N — the Granites are succeeded by the abundant black Porphyry which is blended with very much of the plate /grey/ Feld in pale green base — & other varieties; in no other place clear stratified or brecciated structure, many /miles/ broad this formation — one variety almost composed of large white Cryst of Feldspar just united together = The road lay through low hills bordering on, a great plain, like St Jago, these hills

[64] islands &c &c — Coast lines of hills — Section at the river Tuqui or Ovalle two sets of plains seem to be composed of gravel — upper parts coarsest — Plains Traversia. = — In the above Porphyries, near Punitaque, some few scattered Copper mines. = In the Porphyries, there was a small ridge of the syenitic Porphyry. —
Monday 11th [May 1835] Rode over plain to Panuncillo before ascending hills

[65] in which mines are seated I believe the whole country is Porphyry From the mine to the West I saw an escarpment of Porph-Breccia which seems blended with unstratified hills. — Saw miners at work — Apires[920] light load-stone (8 arrotas — 3 pound) breath like fire; odd noises stream of perspiration of breast: only a custom they have of blowing: — no meat, no diseases — legs bent: 80 perpendicular yards — no right to breathe or halt for 200 yds — /Banelesos/ 25 pound hammer laughed at Englishman — Apires carry load 12 times a day.

[66] 33: 34: +35 Varieties of cryst. granular. siliceous rocks in which mines are worked
36) do slaty less common —
37: 38 low down. in mountain
NB 35 is too much crystall or carious for characters
39 — abundant with /false/ Breccia — Calcareous? why /illeg/??
40 — False Breccia — 41 piece of included Breccia
42 — Stratum immediately over Granite
43 44 Siliceous rocks in Gneiss /abounds/

920 Chilean miners.

45 specimen of what I call the carious Porphyry — base slate-coloured not much
Feldspar & part of this glassy — which is not universal
Copper mines of Panuncillo

[67] chiefly Pyrites often associated with white Carb of Lime & Large ~~scales~~ crystal of
greenish mica, latter chiefly on sides of veins — the principal veins about NNW some
transverse — lie in a ridge of varying chiefly quartzose rocks — very puzzling at first,
principally varieties such as (35.) — but less & less coarsely crystallised of a grey color
mixed with a great deal of white granular <u>soft</u> rock (39) (33 & 34) other varieties, these
rocks are to certain degree laminated or stratified &

[68] nearly vertical: low down in the hills there is much compact rock which perhaps is partly
Feldspathic mingled with granular quartziferous rock (37 & 38) — The white
granular & siliceous granular rock is very remarkable in much of its mass, by the
weathered surface being entirely covered with apparently angular imbedded pieces:
these in places form a fissured structure, appear to me to be concretionary origin. —
Some <u>few</u> of these are more or less rounded (40): all the Breccia is of nearly same

[69] nature such as (41) which came from a large fragment. — I have scarcely ever seen,
excepting perhaps at Falklands such a substance. — In some other parts these fragments
were arranged in /waving/ lines parallel to the cleavage or stratification & were
composed of Porphyritic stone, in which crystals of glassy feldspar were visible — there
were few & quite $\angle^r$ their origin must remain

[70] unknown. = The <u>structure</u> of these rocks remind me of white Brecciated soft rock of
Chiloe: Beside the white & more or less compact arenaceous rocks, there was some
/little/ laminated siliceo clay slate (36). — These rocks are traversed by ~~dykes~~ paps of
different porphyries –

 W (A) SSE E Z NNW (granite)

A consists of high hill of Porphyry base greenish or grey — compact sonorous — a spur
bends

[71] & forms part of ridge where mines are, at other extremity there is a pap of a blackish
Porphyry round the (E) side of which the strata fold; in middle there is a small pap
of close grained green rock absolutely beneath the above strata —
These strata completely puzzled me. = To the South of hill (A) judging from fragments
A or road, the whole country consists of unstratified Porphyries

[72] (of which 45 is another specimen) to the NE of the mines, there are ranges of hills,
which run about NW ~~to~~ by NNW, & are composed of granitic or syenitic rocks,

containing little quartz, much Hornblende & black mica: — (In then /these/ hills few mines of Copper). — to the East of these in the distance are escarpement of Porph. Breccia. — I ~~traversed~~ found the granite also right North of

[73] the Panuncillo mines, & following it westerly came in a line with the ridge, to a slaty rock which dipped SW about 60° & almost reposed on the granite, chiefly consisted of large crystals of Feldspar & Hornblende ~~planes L~~ former crystals in planes, hence slaty structure (42) a little further one we had a soft true gneiss — minute black scales of Mica, some associated with grey siliceo-araneceous

[74] rocks such as 43–44 precisely like those in ridge, only containing patches of granitic nature with large crystals of Hornblende = Hence we have superposition of rocks explained = (NB not age) — the association of Porphyry /melted/ hills seems particularly common South of this district to Punitaque; of which doubtless this is same granite in each

[75] case to East tilting the Porph.-Breccia = one of the Copper veins is cut off by stone dyke — /Layers/ of true Porphyry

Tuesday 12th [May 1835] staid all day at Panuncillo — young ½ Chilean & English — look as former & considered himself such

Wednesday 13th [May 1835] Noise of miners — expelling breath, articulating ay-y, ending, in fife whistle — revolting although voluntary = Anecdote of Don Joaquin[921] — when a boy holiday to see Englishmen heresy. —

[76] & contamination sticking to him — Mr Caldcleugh old lady in Coquimbo — Passed today mines of Tambillos, 4 Englishmen, cast iron work — going to drain a large mine —

on road from Panuncillo merely crystalline unstratified rocks. of varying character many Porphyries some purple & some little Quartzose, & granitic rocks —

[77] mine of Tambillos — main vein — a very grand one, with oxides of copper (46) — runs E & W is in a soft decomposing granite, with little quartz neighbouring hills various. Porphyries passing XX into the nature of greenstones & some curious Porphyries — Slept near Punta curious (a rock near this mine 47) /Porph/ Point= Plain xx — white Feldspathic cryst rock /something/ /but/ these crystals & some amorphous: —

[78] with comminuted shells origin ??? —

Thursday 14th [May 1835] — Rode from the Punta to Herradura bay South of Coquimbo — the few rocks which I saw out of plain appeared to be altered slates & some greenstones & porphyries (48) I know not what name to give = This great flat valley expands into the plains

[79] which I suppose extends to /Bañada/, where we come to the Tertiary strata there appears shells on surface. These Tert. strata with marine shells (slight dip W) from a part of the flat valley —

---

921 Joaquin Edwards, Major-domo of copper mines at Panuncillo. For this and the following anecdote see *Beagle diary*, p. 330.

Friday 15th– [May 1835] 49 — 50 .. 51 52 Shells upper Tosca rock (road to
Herradura) 53 — Superior, stratified Stalactiform

[80] Tosca, 54 — Friable Tosca — (55) resting on & blending with latter agglutinated
sand & bits of shells — 56 — 57 — 58 — Syenitic or Hornblendic rocks of point of
Herradura Bay. —
On the road from Panuncillo to the Port, passed over a ridge with slight dip westerly,
the upper layer were hard cream-coloured stalactiform, with lines (53) & beneath which
came an irregular

[81] quantity of white hardish friable Tosca — or Lime (54): the whole /nearly/ resembling
Bahia Blanca of St. Jago Burnt for Lime. — Beneath this came a great mass of coarse
sand bits of shells vacuities. very poorly cemented together & worked as a Freestone
(55). rock hard rather brittle: its upper parts contained several Murex[922] Venuses,
/Odanton/ &c &c 49–52. — Which I imagine are of a the "recent" date = Near to
the beach we came to cliffs forming a

[82] a little plain of white sand, containing infinite shells such as now are actually lying on
the beach — chiefly circular Calyptrea: Turbo — Concholepas — a Venus, the /Quinter/

Donax — Pecten — Abutted against older plain: young old
yet this more modern, than Tosca rock. — Perhaps it is only that the species which now
inhabit the Bay are

[83] different: —
Herradura Bay, the line of coast formed by hills of Syenitic or rather Hornblendic
rocks 56 — — 58 — abundant crystals of Hornblende — generally large & Feldspar.
These hills must have formed a sort of Basin, where places are between them & interior
hills: There is a double system of plains — I only examined the lower which however
I feel nearly sure is part of these only lessened by degradation

[84] consists of a fine grained, whitish sand, agglutinated into a very soft sandstone: this is
/irregular/ & much ferruginous, & with patches of yellow sandstone — few scattered
pebbles, some as large as two fists of crystalline rocks, ~~many~~ several horizontal layers
of irregular formed, thin concretions, almost stratum of very hard brown, fine grained
Sandstone (just as every where else) these contain

[85] very many shells chiefly enormous Mytilus & great Ostrea! of Patagonia
many Coccles & other shells. Turitella &c &c — Venuses &c &c an enormous
quantity of partially silicified bones — I believe, all, certainly many of Cetaceous
animals. — they occur with shells in the concretions: = The ~~plai~~ upper plain is thickly
coated by pebbles in some places cemented by Tosca — I found a layer, filled almost
with large oysters. I do not think they were washed out of

922 A genus of marine snail.

[86] other cliffs, if so existed in Tosca period. = The lower plain, was ~~in places~~ generally covered by Tosca = <u>gravel</u>; the lower *[illeg]* calcareous matter */bedding/* by vein like masses with the Sandstone, yet close above containing the common purple Trochus with tinge of its color Murex & other now living shells — I suspect this apparent <u>gradual</u> junction

[87] in this <u>lower</u> plain */must/* be accidental: (as gravel */at/* <u>St. Julian</u> with white matter). on the syenitic point of land, found patch of small comminuted shells — Concholepas &c &c, part of which were partially cemented with calcareous Tosca — so certainly not brought by man, at an elevation

Upper station — 30.120   } Attached Th. 63.
Lower do   30.376
242 ft   —

Yet on sides of hills scarcely any shells — There occurred a flat patch: Turritella on the coast & high up: —

[88] 16th [May 1835] Took up my abode at Coquimbo.

Monday (18th). [May 1835] Examined plain behind town, side of valley <u>extremely flat</u>, <u>composed</u> of pebbles, apparently when seen, yet certainly forms part of plain of same height with the Tosca rock described yesterday:

[map showing the Coquimbo terraces, with ships anchored off the coast][923]

[89] The main plain */it/* is which runs all up the valleys on its very edge there would appear a small step it is on the step a mile or two South of the town where quarries are, (or still lower ones) & filled with shells & covered by some loose ones: in

923 See the sketch in the geological diary DAR 37.662A.

the mouth of valley besides this there is a ~~small~~ narrow fringe of /double/ two plains (chiefly on north side) which sweep out to seaward & Coquimbo itself appears built on such entrance of town one step & further Easterly another — the bottom flat valley a 4th — To the South

[90] about Herradura a strip of land which divides two ports, seem all a second plain — I do not know to which it corresponds:

In the interior right behind Coquimbo Hills hills of very pale green feldspathic rock where crystals of white Feldspar are (102) not always just discernable

& NB near the Puerta a hill of Porphyry where ingredients were of granite Syenite are imbedded in a do or Euritic reddish base. = I have described the Tosca well — Here beside the shells already seen are very many of a circular Calyptrae — <u>Donax</u> — Turritella

[91] Venus — (Mactra?) Trochus with tinge of color — all of which show close alliance with shells on <u>beach</u>. there was only one Concholepas & no Pecten — & /not/ the straited large Calyptrea. = So that species same, as now, but ~~arrangement~~ grouping different from present beach, or sandy cliffs: so that we have this arrangement

 [section of cliffs]

on Beach shells B Loose recent Tosca recent Ancient A Tosca A Beach C Shells plain of course ~~Beach shells~~ Tosca of A more modern, than Tosca (B) which

[92] cover in succession the old oyster — & I think there will /be/ be found Pecten more Concholepas & Trochus with color — Little sand plain (C) more modern than either & shells identical in proportion with present beach =

I no where saw fragment of big oyster & grand /aureous/ Mytilus — silicified bone or concretion: Above the Tosca there was some calcareous sand, which contained more perfect & more numerous shells of /some/ sort, I carefully avoided picking up these: In the freestone there are in places a good many chiefly Porph pebbles, an

[93] inch or two long & some small flat slaty ones — ~~fragme~~ much entirely composed of bits of shells —

Hence we have upper plain decidedly marine — (How could B Hall[924] fail to see the plains curling out to seaward?) (NB The Tertiary formation does not seem to extend far South of Herradura) at /illeg/ Before Pichidanque &c & we had plenty of proof of lower plains, thus formed — —

The mass of cemented pebbles, with band of broken great oysters

924 Hall 1824.

[94] lying over old Tertiary form, I believe to be true deposition & properly would be covered by the Tosca rock. —

Dip of Tosca beds, to W doubtful whether too much for original deposition: ~~on the~~ First light shower in Coquimbo night of May 14th — effect break ground 2^{d} shower put in corn ~~summer~~ spring ripens, effect on Mule hire for carrying ore. Some Apires can carry 300 lbs up mines — "Honest man — he is <u>old Spaniard</u>" — Coquimbo small very quiet

[95] military, shops all shut. —

NB I hear of Petrified wood in plenty up by Illapel, & petrified Echinus of large oysters & cannon balls at Guasta

| 20th Wednesday | [May 1835]

A here   B 30.018   T 65°

does not represent plain

A	29.980		
	Up the valley		
E	30.110	Therm 70°	
B	29.917 station		6 feet above plain
			represented
A	29.88		
A	29.944 (+ Hill)	Therm 68° —	
B	30.002		plain represented
level	30.308 of sea station		6 feet above

[96]

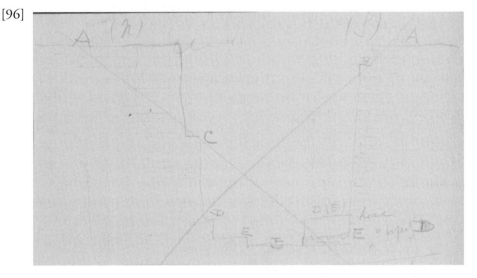

[Coquimbo terraces] A (N) C D E F D (E) B (S) A here E is properly D

(B) not seen on north side: (C) just visible, to West more developed. D & E swallowed up to West in one plain, to East largely expanded

[97] On South side (C not D) E not visible Measurement on South side
In my room 30.240 in Town.

The height of great plan	30.30.8
	29.944
Slope of plain in about 2 miles Easterly	29.944
	−.880

I am not certain whether plain E on South side of section corresponds to E or D. — The measurement in my room above sea, is the height of the lower plain, escarpement taken in middle —

[98] perhaps 30 ft + a broad nearly level sandy space, which corresponds to plain of river: on this plain the principal part of city lies — the next escarpement of about equal

height as last is chiefly subord we then have

Hill, which counting the broard sandy plain (with Chaerus) makes 4th plain (I suspect a little South there is an intermediate one) above which the 5th or grand plain — I believe quarries by Mr Lambert, the above intermediate. — In the section

[99] taking both sides of valley & plain at bottom there are six plains — but many very obscure, some uniting — others /mainly/ shown, by indistinct line of /knobs/. — observed on both sides where point & ravine entered; that three sets of plain were fan shaped, sections; hence great obscurity, but what would be expected at bottom of inland strait —

The difference at (   ) Hill of plain A, B not to be trusted

Earthquake, screams of women great roar — little motion — called severe — appalled the men —

[100] Thursday 21st [May 1835] I notice plain E corresponds to (D); after crossing the plains we come to low hills of the Syenitic greenstones & fragments of Granites, dark greenish & brown porphyritic greenstones & the syenitic Porphyry (red crystals small of Feldspar) mentioned near Coquimbo — These hills were /superceded/ by hills of dull red color of the Porph-Breccia — remarkably plain the latter structure, lines &c, different coloured pebbles — very coarse & little blended & what is curious, the stratification very nearly horizontal afterwards bending up & dipping

[101] small ∠ to SSE. — several dykes & paps of fine green Porph-greenstone, one grand broard dyke, running NNW, SSE of pale Porph rock (59) near which silver mine. — NB amongst the coarse Porph Brecc — Some of the fine crystalline rocks with acicular crystal of Feldspar (60). — Our course about ENE. —

Reached a mountainous grand alpine country — young Chilean José Maria.[925] —

925 José Maria Edwards, Anglo-Chilean, son of the owner of silver mines at Arqueros, with whom CD was travelling.

[102] (22^d) [May 1835] (61) S. Rosa mines; lowest fine grained red bed: (62–63) the green beds. (64) in one locality were near red beds (65) green bed with /Silver/ near vein (66) white Porphyry — minas Don Francisco de Aristoa

66 (67) Brecciated red rock

67 (68) Specimen of poor one Arqueros Mineral & Quillai

(69) The limestone with bits of Corall

(70) do part of Coral reef

(71) Upper blackish hard bed Conch: fracture. —

[103] At the mines of S. Rosa, the rock is very bright green, generally crystalline, with no Brecci structure 62–63, in these is the main vein; this rest on a fine grained crystalline red Sandstone which passes into true Porph-Brecca — These green beds, not often stratified but clearly seen in some places: passes rarely into a slate coloured, true Porph with crystals of white Feldspar: the junction of the red & green, tolerably sudden; in other place rather gradual (64). All these beds certainly belong to P. Brecc

[104] formation: At General Pintos, the grand rich vein lies in a fine grained, but true brecciated rock, little Porph. — In whole mineral silver, (in most rare instance little copper Pyrite) veins run NW & SE — very highly inclined; silver generally pure in Sulph of Barytes; in this latter mine, vein suddenly becomes poor as soon as green bed is reached at S Rosa, as soon (or rather just before), coming to the red — Bed — generally if the red bed is cut through & other is

[105] succeeded by green vein becomes again rich — these mines not a league apart. — At S. Rosa the green rock to the distance of half yard, in one spot contains much silver (65). Considered most rare instance: — Country of Mineral high, undulating, not very uneven, stratification dipping in all ways, having broken up, I think, a general SW dip. — very many dykes (& masses) of a /closeish/ grained, green-greenstone, these chiefly run NW & SW; hence

[106] frequently cut veins. — evidently much troubled district. — To the South of S Rosa — a mile. Cerro Blanco, consists either of compact conch. fracture (66). or same as purple Porph: but white, partly stratified lower part of hill not so — Thin hard beneath purple Porph — may be seen far over country by strata nearly horizontal — at this point a N & S line of violence. has formed anticlinal band, which seems to have expanded, & caused

[107] SW dip & country of innumerable mines. only discovered 9 — years = Leaving the mines (To the East innumerable alternation of red & white bed, all small dip to Cordilleras) We came to grand mass of lmpure Limestone, a good deal of which was formed of nearly parallel tubes like a Corall — the rest more or less pure contained immense quantities of **large oysters** & Gryphites; also some /corticiform/ corall, & shell like Pecten. — Covered by thinly stratified blackish

[108] compact — conch /fracture/ — calcareous? beds. & there by immense thickness of a coarse conglom — Porph — I believe these beds rested also on Porphyry — penetrated by paps of curious plate. greenish Porph = This puts age of all these rocks (& Hornillos)

out of question = such a striking similarity — I believe the white Porphy is inferior to Limestone but am not sure — impure Limestone grand formation — Clouds like plain — different climate here &

[109] Coquimbo — quiet /mines/ in desert busy day. — The strata at mines must be high in series: Near Cerro Blanca — in pale purple Porph — much agate & large nests of quartz crystals rare occurrence. =
~~Friday 22~~ Saturday 23[d] [May 1835] — From the Hacienda of the Marqueros travelling east, various much altered Porphyry — dip where visible to the West till we came to the anticlinal band remarkable for enormous faults, not seen in external form of hill, central parts nearly horizontal, on each side opposite dips, seen on both side of valley — These beds are brownish

[110] x mine easterly
(z) still more beyond which true anticline East dip

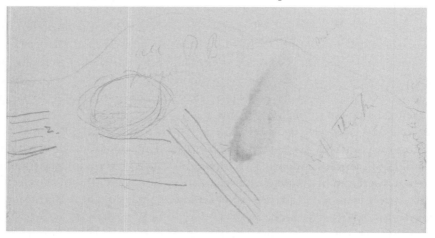

all PB purple z x mere line 150 ft thick W repetition[926]

[111] (72–73–74) The /Horn/ stones like altered clay-slates of Hornillos
75 — Red Sandstone from very high up 76–77 common Red Sandstone beds
(78) Vein with in above with Pitchstone (79). with bed of arenaceous Sandstone Limestone (80). Veins with Gypsum ~~80~~ 81 — Hardened bright green — concent structure Clay bed (82) do with ribby fracture & pebbles —
(83) upper white /rubbly/ bed (84) in parts fine hard compact white Feldspathic
(85) The upper conglomerate
(86) Green stratum of crust Feldspath rock
(87). Part of dyke — 88 Upper Lava
In the P.B. from pebble of diff coloured Porph cemented by white cryst beautiful Carb of Lime & /clay/ cavities with Epidote

926 See the sketch in geological diary DAR 44.21.

[112] from decomposition consist of the Limestone with organic bits; well seen covered & reposing on the Brecc: Porph: on entering on the East dip on North side of valley turned up to the North at Puclaro — We rode obliquely through the Porph-Brecc. greens & purples, above the Limestone till we came to an open spot — with rounded coloured hills very like the Hornillos — the ~~Lowe~~ /first/ or lowest beds were nearly white or pale brown or greenish soft decomposing Porphyries; then some fine grained red crystalline Sandstone: then the first grand mass of Hornstone strata, more or less laminated more or less compact, conch fracture, more

[113] or less black. some quite yellow 72 — 74 always decomposing externally into bright yellow earth. The Horn stones are generally marked dendritic manganese — many parts excessively like the yellower varieties of the laminated rock of the Hornos. = This grand main bed several hundred feet was covered by much of the red — Very little crystalline, none brecciated Sandstone 76–77 — I must except one spot there was Breccia but not Porph: This red sandstone contained some whitish & lilac bands. = was covered by other grand bed of the laminated Hornstones, much of the yellow earthly

[114] sorts — all this formation — traversed by much dykes & paps of bright green cryst rock (87) which in places have white spots of earthy Feldspar: yet it is remarkable there appeared comformable beds of a green crystalline rock, such as (86). In this respect I was forcibly called to mind of the Gypsum of the Portillo — NB in Lower parts of these 3 grand beds, there is an extensive mine of Gypsum; mineralogical nature same as ever, did not visit it but such must have been its position: In the Intermediate Red

[115] Sandstone, there was much black Agate, or rather quartz passing into Pitchstone, I found one small parallel vein (78). Some of the red sandstone — so compact as to become jaspery, others rubbly & soft: There was also a parallel bed of carious brown arenaceous Limestone ~~80~~ 79 (NB in the Porph-Brecc saw a siliceous red Sandstone) near which were intricate veins of Carb of Lime with Gypsum ?? (80) (is the Pitchstone of the Tertiary formation. = An interval of beds were hidden we then had, a much stratified, bright green concretionary structure hardened clay. (81): This contained. — layer of very hard rock

[116] of nearly same nature, with brecciated structure, above this we had a layer of two of red-sandstone, & the clayey Hornstones, & then much of the rubbly green-hardened clay, not so /pure/, with partially brecciated structure, & remarkably for containing in upper parts ~~much~~ several well rounded, pebbles of Porphyry. — This was covered by grand thickness of a fine grained very compact red sandstone (75), & this was banded with white Jaspery rock, which as at /Combarbala/, showed on being wetted, an obscure minute

[117] brecciated structure — (NB forgot injected paps of purplish Porphyry). — Above this red sandstone, we have some hundred feet of white, or occasionally slight greenish, rubbly soft Feldspathic rock (altered white beds) (83) which in part passes into hard, compact conch: fract do rock. (84). = This was covered by 300–400 ft thick of coarse conglomerate, pebbles almost all Porphy — base arenaceous. never Porph: pale

brown as (85) or more general red, lower parts firmly cemented, upper pebbles can be picked out. — certainly altered gravel, just as in Portillo: The upper foot or two is arenaceous

[118]   capped by 200-150 ft of Lava — Slate coloured, or tinge of green, rather fewer plate

crystals of Feldspar than in curious Porph — & some  (88). Seems certainly to have come from lofty mountain to East. = is important in history of curious Poprhyry, to which this Lava certainly belongs. — Riding afterwards round the base of <u>this</u> hill this sort succeeded by Porph. greenstone. Feldspar partially earthy very abundant & this again passed into the syenitic greenstone with true fracture of granite: — All the above great beds dip about 20–30° to East & are all (Conglom & Lava) conformable to P. Breccia = Hills of strange colours, with dykes & paps of green rocks. —

[119]   I can feel no doubt, this upper red Sandstone; Hornstone laminated, green clay with pebbles & Conglomerate corresponds to Tertiary Formation. = These <u>pebbles</u> show that this outside chain subsequent to first upheaval —

Syenite P. B. Limestone P. B Gypsum Horn St R Sandstone Hornst *[illeg]* Clay R. S. Feld White Conglom Lava Syenite[927]

~~Beds all dip~~ — The thickness of the Beds above the much Porph: Breccia must be very many thousand feet: a horizontal section of at least 2 miles of 30°

[120]   Sunday 24th [May 1835] — Stayed whole day at Hacienda of Gualliguaca — Signorita not 17, mere child mother of two children & will soon add another young one to the family of Salzeras. — very pretty — dressed like lady — Some second rate farm house. — Walked up the Valley first rate <u>Chiliean</u> scenery — like scenery of Alps in the Annuals.[928] —

927 See the fair copy of this sketch in DAR 44.21.
928 The Annuals were inexpensive publications introduced in the 1820s, illustrated with high quality steel engravings.

Near the Hacienda there is a small chain, about N & S of Syenite-greenstone (88) good characteristic specimen which seems partly to have up-burst through the superior beds & to have formed another escarpment of P Breccia perhaps covered by the soft beds — These beds do not dip much

[121] Monday 25th [May 1835] 89. Siliceous white sandstone rock beneath Limestone, (90) Layer of yellow, siliceous Limestone (91) superior fine grained red sandstone (92) green dyke (93) Red Sandstone 6 ft from D (94) close to — (95) White Sandstone *[distal]* (96) do close: (97 (white quartz pebbels in) (99) Red Porphyry low down (100) do — high up

We came to another somewhere about N & S chain of granitic rock — real granite & Syenite G. & much dark slate coloured sonorous base with few crystals of Feldspar. Beyond this we again have much Porph Breccia

[122] some much altered some less. —

beds up to about NW by W, which seems owing not to the chain we have passed but to an irregular one which forms Southern W side of Valley of R. Claro. — A section of strata on North Side, well known for petrifications, was examined the strata dipped about North, owing to end of the above chain, exceedingly sheltered & contorted & *[illeg]* traversed by Fault dykes, some of the planes inclined at about 70° or more. = (1) Beginning at bottom, we have the purple, hard sandstone of P. Brecc formation (2) white siliceo Sandstone (89) more or less hard, sometimes passing into the purple varieties

[123] has layers of slaty Sandstone & some hard yellowish calcareous? Sandstone (90): This also contains very great quantities of small quite white quartz pebbles in the white sandstone (97): remarkable quantity, first time I ever saw such a thing, ground strewed with Beans — common Belief. Where did the quartz come from. —

Above the white Sandstones, we have some strata 50–60 ft of dark perhaps semicalcareous rock, (with partially minuted Brecciated structure) almost composed of a large, Terebratula, many twice the size of specimens. — 2 other species, outline

[124] of an ammonite, pieces of Gryphite & perhaps of large oyster: this rock contains a few, round quartz pebbles, is covered by very compact, hard red Sandstone (91). — (All the white Sandstone are brittle & clink) This gradually passes into a (N Red Sandstone) red conglomerate, pebbles both round & angular, several 100 ft thick, *[many]* of the white quartz, in lower parts, & porphyry, pale, chiefly purple & red & whiteish & the red-sandstone & Quartzose rock — in short just such as we know inferior beds to be —

[125] These conformably = covered by another thick mass of similar conglomerate matrix without any porph-structure This is capped by thick bed of crystalline purple Porphyry, in some places traces of stratification in some not = the lower part is ~~not~~ (99) — higher parts assume acicular cryst of Felds. (100) — Covers the Conglomerate, which

appears rather altered at junction. — is capped by the curious Porphyry (in perfection) change sudden — & this again by pale green Porph with numerous cream colored Cryst of Fe & green specks

[126]   are these 3 beds Lavas? or altered rock

9 8 green Porph 7 curious porph
6 Porph? 5 Conglom 4 4 Conglom 3 Shells 2 sand
(9) is a hill composed of central grand mass of curious Porphyry, Syenite & dark slate
Porphyry — The connection of the 3 porphy beds which dip towards this mountain, I
do not know — <u>upper</u> parts of mountains appears stratified. — is the heat of central
parts, altered the upper parts more than the lower; or are they not Lavas which have
flowed, before upheaval = — The age of Limestone is certain. Disturbance consequent
on upheavals, prevented gypsum, & formed green conglomerate. — But some similar
rocks must have been upheaved to have given pebbles = The Porph having in one spot a
stratified aspect is

[127]   most puzzling: — never before /me/ with more crystalline beds above less ones — 6 & 7
certainly Lavas. — (NB the number of Terebratulae quite wonderful) — It is curious as
showing a very early ~~upburst~~ cessation of strata. = Either before Gypsum or latter not
produced. = The Grand bed of Conglomerate (4) & ~~near~~ near the ravine is traversed by
grand dyke of very green greenstone (92) 8 ft wide, high up /dimms/ — passes through
red sandstone with the pebbles at 6 ft distance soft ~~(94)~~ (93) 2 or 3 inches almost white
much harder ~~(95)~~ (94) for two or 3 inches tinged green with matter of dyke — from
volatization: same /foot/ in bed of white sandstone — 95 — 96. = (NB slight spherical
structure in upper surface of Porph (6)) = The Redness of Sandstone does it show no
heat in mass ?? =

[Rio Claro and Gualliquica] Gualliquica (Z) R Claro
The variation in P.B. forms remarkable
(Z) 3 chains form a part /distant/ from it on all sides?

[128]   The question is whether is this gravel same age as Puclaro: Lava appears analogous
nature of gravel do. = The gravel lower part with quartz pebble, as Limestone with
Terebratulae Ammonites — Age nearly certain contemp with P. Brecc. formation. —
I can hardly however imagine all the beds above gypsum so old as this. = Tuesday 26th
[May 1835] returned to Hacienda.

Wednesday 27th [May 1835] returned to Coquimbo: as in going to mines, hills of Syenite: I believe the Limestone occurs with P. Breccia ~~almost~~ to ~~mouth of valley~~ above where the Granite & Porph hills occur: — About 10 miles up valley perfect plain sinks owing to valley rising (the subordinate plains still less) & gradually dies away on irregular sides — I could see nothing of the plain with water a lip — above at 25 miles up no plains valley narrower about 3/0] miles from sea, steps best seen; where they have been measured: 3 most visible which correspond to highest & two on which town stands A: D: E:

[129] Gualliguaica Puclaro Galliguanca

[illeg] Puclaro /Darguia]

Mem: Hypothesis of Salina depression

True Syenite behind Port Coquimbo

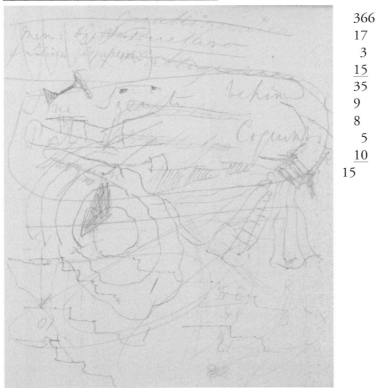

$$
\begin{array}{r}
366 \\
17 \\
3 \\
\underline{15} \\
35 \\
9 \\
8 \\
5 \\
\underline{10} \\
15
\end{array}
$$

[Coquimbo terraces, the step terraces on the lower left appear not to be in CD's handwriting]

[130] Petrified wood in valley of Elque: The terraces dip to centre of valley, as well as Seaward: I have been assured that some Apires have ascended deep mines with 300 Lbs on their back: Mem: silicified rings on the Terebratula of R- ~~Elqu~~ Claro & on the 27th of May, rain having been 13 days before, patches of hills tinged green, thin scattered delicate grass. one inch long like hairs. — soil apparently dry: —

25
Douglass —
Corfield —
Alison petrified wood
Caldcleugh value of Hornillos & other mines —
D^n Eugenio Matta[929] or D^n Diego Carballo — Copiapo. has petrified shells.

[131]   The argentiferous lead mines are very rare, they appear to occur in same formation
one I heard of very near the Granite.
There is silver mine in R. Claro
70 ½

owe C.Fitz Roy 2 dollars & Rial

2.1
17
7 ½
17
11
28
16
 3
13 ½

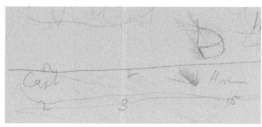

R Capt 2 /Horses/ 2 3 15

[132]   Spunge — Olla — < >/cho/ — Blacking — Milton[930] — Clothes
Washed — Shoes blacking Gertrudio = Griffin Lambert Chiffles
write Letters —
Gypsum — Valley of Coquimbo Ill Almaraz
Tongoy. mixture of white Feldspar & green Chloritic mineral; the Tertiary strata are
found there with <u>Bones</u>: —
Tortoralillo dark Syenitic Greenstone

[133–6 excised]

---

929   Eugenio Matta Spanish co-owner of the San Antonio mine.

930   John Milton. As CD wrote in his *Autobiography*, p. 85, 'Formerly Milton's *Paradise Lost* had been my
chief favourite, and in my excursions during the voyage of the *Beagle*, when I could take only a single
small volume, I always chose Milton.'

[INSIDE BACK COVER]

/person/ in Cordillera noise right beneath
(5 more were felt up in the country same night
Earthquake at Coquimbo 8 PM to 8.15
(20th) [May 1835] — Wednesday
What money has been got out of Arkeross Arqueros
Mar G[931] is paid up to the 15th of May + 20 Rials in Quillota for spurs — Quillay —
/paid/ /illeg/? 98 101

$$14 / 300 (22$$
$$\underline{28}$$
$$20$$
$$\underline{14}$$
$$.6$$
$$28$$
$$\underline{9}$$

[BACK COVER]

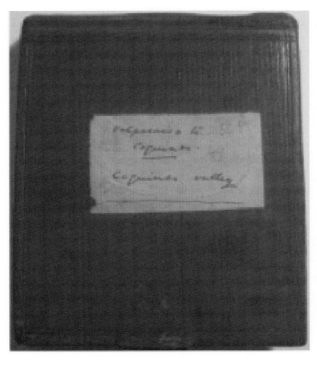

Valparaiso to Coquimbo.
Coquimbo valley

931 Mariano Gonzales.

Textual notes to the *Coquimbo notebook*

[IFC]   1.16.] *Down House number, not transcribed.*
        88202336] *English Heritage number, not transcribed.*
        10] *added by Nora Barlow, pencil, not transcribed.*
 [27]   *page written perpendicular to the spine.*
 [75]   *[Layer] ... Porphyry] added in heavy pencil.*
[96–9] *pages written perpendicular to the spine.*
[110]   *page written perpendicular to the spine. there is a watercolour stain
        on the page.*
[129]   Gualliguaica ... *[Darguia]] in FitzRoy's handwriting?*
        366 ... 15] *written upside down from other entries on page.*
[130]   Douglass ... shells.] *written upside down from other entries on page.*
[131]   70 ... 15] *written upside down from other entries on page.*
[132]   Chiffles] *ink.*
[IBC]   Arqueros] *not in CD's handwriting.*
        Mar G ... ~~Jose?~~] *written perpendicular to the spine.*
        14 ... 6] *ink.*
 [BC]   50P 44] *pencil, not transcribed.*

THE *COPIAPÒ NOTEBOOK*

❧

The *Copiapò notebook* takes its name from the city of Copiapó on the coast of Chile. The notebook is bound in red leather with the border blind embossed: the brass clasp is intact. The front cover has a label of cream-coloured paper (60 × 17 mm) with 'Coquimbo to Copiapò' written in ink. The notebook is similar in shape to the other Type 5 field notebooks begun in 1835 and 1836. It has 51 leaves or 102 yellow-edged pages, 39 of which are blank. It was written in one sequence in June 1835. Darwin used the *Copiapò notebook* to continue recording expedition no. 8, from Valparaiso to Copiapò, and when he arrived at his destination he put it aside, even though it was by no means full. The last dated note is 25 June and the first dated note in the next notebook, *Despoblado*, is 26 June. In this way the *Coquimbo*, *Copiapò* and *Despoblado notebooks* are a trilogy to be considered together.[932]

The *Copiapò notebook* contains only about a dozen sketches, but these are unusually complex and important, several of them forming the basis for Darwin's published 'sketch-section no. 3' on plate 1 of *South America*. He called this an 'eye-section' which, like the other two sections on plate 1, is visually 'stitched together' from a number of the diagrams in the *Copiapò notebook*. As these sections had to 'straighten out' what he actually saw on his zig-zag trek along the river valleys, they are remarkably skilful pieces of graphic integration.

Coquimbo to Copiapò, June 1835

Entries in the *Copiapò notebook* begin on the inside front cover with the note, dated 31 May, reminding Darwin that he owed FitzRoy some money and also that he owed his guide, Mariano Gonzales, '20 Riales in Quillota'. Judging from the entry at the end of the *Coquimbo notebook* this was for some spurs. There is a reference to a 'Don Pedro Jose Barrio Potrero Grande Hills with shells'. The gentleman concerned has not been traced, although there is a famous district in Coquimbo dating from this period called the Barrio Inglés. Also on 31 May Darwin wrote to his sister Catherine describing his mode of travelling and lamenting that: 'every month, my wardrobe becomes less & less bulky — By the time we reach England, I shall scarcely

---

932 Even more so than with the *Coquimbo notebook*, Barlow 1945 gave the *Copiapò notebook* barely one page out of 120 pages devoted to the field notebooks.

have a coat on my back'.[933] The next day, 1 June, was recorded on p. 1. Darwin cited Charles Lambert's opinions concerning veins at the Arqueros silver mine, opinions which Darwin published in *South America*, pp. 211–7. There is now a town called Lambert about 30 km (19 miles) northwest of Coquimbo.

Darwin noted in the *Beagle diary* that he said farewell to the *Beagle* and set out northwards for Ballenar [Vallenar] in the Guasco valley on 2 June. There is a remarkable sketch map, hand drawn by Darwin and now in DAR 44.28 (reproduced below), which shows the coast from Coquimbo to Copiapò with most of the places mentioned by him in the *Copiapò notebook* and the 1835 part of the *Despoblado notebook*.

The notebook records the geological observations from this journey, during which his party met 'occasional troops of mules'. That night he reached a house called Yerbabuena. In one of the most wistful entries in the notebooks, pp. 3–4, which was copied almost verbatim into the *Beagle diary* (p. 334), Darwin noted that: 'the road with a tinge of green, just sufficient to remind one of the freshness of turf & budding flowers in the Spring — travelling in this country produces a constant longing after such scenes, a feeling like a prisoner would have'. On 3 June they pressed on across the 'mountainous rocky desert' to Carizal and Darwin noted that there was 'very little water & that bitter saline' with 'succulent plants' providing the only botanical interest. In the *Beagle diary* Darwin recorded that the only abundant living animal was a species of snail, but in the notebook he recorded 'Many P. St Julian finch & Dinca Turco &… Tapaculo', p. 5, thus painting an ornithologically more interesting picture than any recorded in the previous month. Today some of the world's greatest astronomical observatories are located in the Cordillera in this region because of the extremely dry atmosphere and clear skies.

On Thursday 4 June they 'continued to ride over desert plain with many Guanaco' in the direction of Sauce, p. 7. At Chaneral [Chañaral] they found a 'narrow green valley' with a geological section showing uplifted sea shells 'perhaps at higher level than at Coquimbo'. Further north they 'entered a grand mass of fine true Granite hills', p. 8, then 'a grand mica slate district — mica slate, very much contorted like Chiloe', p. 11, both described briefly in *South America*, p. 217. At Sauce the 'Poor horses [had] nothing but straw to eat, after travelling whole day'.

The next day there was a 'magnificent spectacle of clouds, horizon perfectly true' and the coast seemed to Darwin 'like most broken parts of Chonos Archipelago', p. 12. Darwin compared the clouds filling the vallies with what an ancient ocean would have looked like before that part of the Andes had been elevated to its present height. The mountains were 'covered with tiny bushes encrusted with a gree[n] filamentous Lichen, even the large candlestick cactus of Chile is succeeded by these species'. This description appeared almost verbatim in the *Beagle diary*. In the distance were the snowy Cordillera and when they reached Freirina they found the Guasco valley 'certainly pretty' and 'well wooded with willows'. Darwin felt 'Capt Halls description neutral' but he expected a luxuriant 'valley as at C. Verd Isd.', p. 12. In

933 CCD1: 450.

the *Beagle diary* this became 'all Capt. Hall's beautiful descriptions require a little washing with a neutral tint'.[934] The locals expressed envy that Coquimbo had had rain when Guasco had only seen clouds. Darwin noted that in such a climate the occasional flood did more harm than drought because it covered the valley with sand and stones.

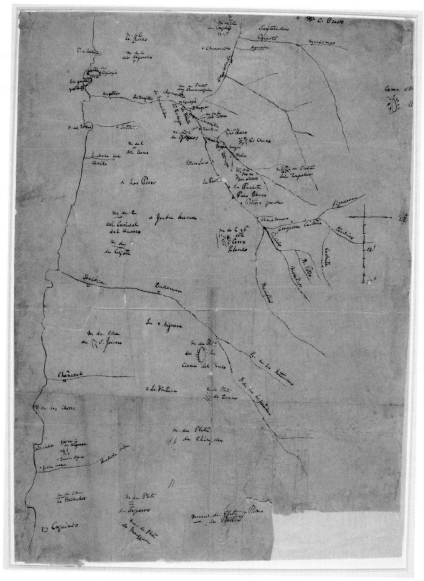

Darwin's hand-drawn map of the coast of Chile between Coquimbo and Copiapò (DAR 44.28). The ink is brown.

934 Keynes, in the *Beagle diary*, indicates that Hensleigh Wedgwood apparently wrote 'A very happy expression' in the margin against this phrase.

On the 6th Darwin 'Rode down to the Port' of Guasco, where he wanted to rest his horses. There an encounter occurred which Darwin recorded on p. 14 but in which, in the *Beagle diary* version, he accused himself of exaggeration in a marginal note. Darwin stayed with a mine owner named Hardy, and called in the evening at the Governor's house. There the Signora was 'the most learned Limenian lady'.[935] He explained in the *Beagle diary*, p. 336, that the lady 'affected blue-stockingism & superiority over her neighbours. Yet this learned lady never could have seen a Map. Mr Hardy told me that one day a coloured Atlas was lying on a Pianoforte & this lady seeing it exclaimed, "Esta es contradanca". This is a country dance! "que bonita" how pretty!' In the notebook Darwin added 'King of Londres'[936] which seems to indicate that the lady thought there was a King of London. Darwin elsewhere recorded meeting people who thought England was 'a large town in London'.[937]

The account continues with an almost painterly 'View up the valley very striking on a clear day infinity of crossing lines blending together in a beautiful haze, distant snow mountains <u>clear</u> outline — formal foreground', pp. 14–5. Again this is repeated nearly word for word in the *Beagle diary*. This is followed by more geology, then a charming list of birds on p. 17: 'Carrancha — T[h]enca — Lozca (Black & Gold finch) Dinca Chingola — Furnarius — Avecassina — Little Grey Bird of mountains — Blue finch with white dot in tail — no Chingola — C[aracara]. Raucanca[938] white tail Callandra'. Several of these species are discussed as observed in northern Chile in the 'Ornithological notes'. On the 7th Darwin 'Ascended hill behind town' to see the terraces which seemed to be more numerous than in Coquimbo. The next day he 'Road [sic] up to Ballenar', p. 19, the 'considerable town' established in 1789 by Ambrosio O'Higgins, father of the more famous Bernardo (1778–1842) who played a key role in the liberation of Chile, in memory of his home town of Ballinagh in Ireland.[939] The view from Ballenar reminded Darwin of the Santa Cruz valley. He stayed there on the 9th and on that day 'Found Terebratulida in the cherty rock as at R. Claro (V[ide] Specimen)', p. 20. Whilst Darwin thought 'Patagonia a garden compared to these plains' there were 'yet dormant seeds' to 'wait for wet year', p. 21. In the *Beagle diary* at this point in his narrative Darwin compared the country to the 'absolute deserts' he saw in Peru, demonstrating that the *Beagle diary* account for northern Chile is retrospective.

935 Barlow 1933, 1945 and Keynes's *Beagle diary* give the word as 'Limerian', although Keynes 2004 notes that the lady was from Lima.
936 Barlow 1945 transcribed this as 'Kiss of Londres'. Londres is Spanish for London.
937 See *Beagle diary* for 9 May 1833, p. 154.
938 Referring to the entry on p. 133, Barlow 1945, p. 242, read this word as 'Raucaria' but gave 'Raucanca?' in the 'Ornithological notes', p. 238. Keynes gives 'Rancanea' in *Zoology notes*, p. 230. The spelling 'Raucanca' used by CD is derived from a typographical error in Drapiez 1830, p. 518, a work of reference aboard the *Beagle*.
939 Harvey 2000.

On 10 June Darwin 'Started for Copiapò'. The ride was 'very desolate but not quite uninteresting'. They encountered a few Indians and 'Many donkey — eat wood' referring to the fact that the donkeys had nothing but stumps of bushes to eat. There was also nothing for Darwin's horses to eat. Darwin was dismayed by the 'contrast of splendid weather & utterly useless weather' by which, judging from the *Beagle diary* entry, he meant 'useless country'.

On the 11th they 'Travelled for 12 hours, never stopping' but the poor horses were 'wonderfully fresh'. The country was 'much prettier than a forest' but the 'Geology (not being able to stop)' was unintelligible, p. 25. Before noon on the 12th they arrived at the Hacienda of Potrero Seco in the Copiapò valley. This was presumably where the town of Potrero Seco is today. Here, in the most northern river valley before the Atacama Desert, Darwin saw a 'Little wren', p. 26.

In the *Beagle diary* Darwin recalled how he spent the 13th and 14th 'geologizing the huge surrounding mountains'. His section was recorded in great detail in the notebook on pp. 27–43, and was published in *South America*, pp. 218–33. Darwin drew part of a spectacular section on p. 37 which is the prototype of the western part of the worked up section now in DAR 44.19.[940] This in turn is the basis of Darwin's published sketch-section no. 3 (*South America*, plate 1), showing at least seven 'axes of elevation'. The Cordillera above Copiapò are named after Darwin, and their northern end is the Ojos del Salado, at 6,893 meters the highest active volcano in the world.

Darwin's northwest–southeast 'sketch-section' up the Copiapò Valley to the base of the main Cordillera of the Andes, published as sketch 3 of plate 1 in *South America*. Copiapò is shown at sea-level on the left, Los Amolanas in the centre and the western base of the Cordillera on the right. Darwin shows seven 'axes of elevation' from west to east, with rock types ranging from granites and porphyries to the Gypseous Formation and porphyritic breccias.

On the 15th Darwin 'Rode up to Los Amolanas', named after the grindstones made from the quartzite there, and the geologizing continued with a second section diagram on p. 44. Darwin noted that the river there was 'size of large muddy brook' which 'for 30 years never reached sea', presumably because not enough snow fell in the Cordillera. There is mention of Don Eugenio Matta, a 'hospitable old Spaniard' with whom Darwin dined and whose name also occurs on p. 130 of the *Coquimbo notebook*.

On 16 June Darwin continued up the valley where he found 'an enormous quantity of Gryphites', p. 50, and other fossils allowing him to correlate the section

940 Partly figured by Browne 1995, p. 271. See also p. xxiv.

with the one he saw on the Rio Claro. The 'Gryphites' were probably the new species of *Gryphaea* described by Forbes in his appendix to *South America*. Darwin drew yet more of the section across pp. 62–3 and on the 18th as the valley became more fertile he collected his 150th specimen. He entered the Jolquera branch of the valley, but at midday Darwin could see no way to penetrate the Cordillera more deeply so he turned back towards the Pacific. He bivouaced and 'experienced a trifling shock of an Earthquake' which is mentioned in the *Beagle diary* but not the notebook. In *Journal of researches*, p. 431, Darwin took the fact that that evening there seemed to be a storm gathering as the starting point for a lengthy discussion, with quotations from Humboldt and others, of the possible link between earthquakes and weather.

On the 19th Darwin 'Returned down the solitary ravine to Los Amolanas' but there was 'difficulty in descending steep mountains', p. 73. The geology was complex but he found it 'very interesting finding grand Volcanic Lava formation of age of Ammonites &c &c separated by what must have been true conglomerates & Brongniarts name for Volcanic sandstone'. This is immediately followed by a reminder to himself to reconsider some much more ancient fossils he found over two years previously '(NB fossils of Falklands of hot country??)', p. 72.[941]

The entry for 20 June was apparently written up at the end of the day as Darwin 'Staid whole day at Hacienda', p. 81. He noted on p. 75 'Examined to day the supergypseous an immense thickness 2000–3000 ft thick almost entirely red Sandstones & Conglomerate'. His specimen tally reached 160 and the geology was difficult: 'No one can imagine such glorious confusion', p. 80.

It was here that Darwin made one of his most significant palaeontological discoveries. He found 'thousand of great blocks of petrified Dycot wood' including 'several in situ', p. 77. One of the petrified trees was 'nearly 6 ft in diameter — What an extraordinary prop.!!' and there were various marine fossils, clearly demonstrating a cycle of subsidence and uplift comparable to the one Darwin had unravelled in the Uspallata range in early April, but here 'age from fossils greater than what I supposed at Uspallata = Now there is good comparison with Humboldt', p. 81.[942] This reference to Humboldt, which followed on a few pages after Darwin described, on p. 76, how the fossil wood was strewn over the surface of the ground, makes sense when compared to the *Santiago notebook*, p. 111. On that page Darwin made explicit reference to the citation by Humboldt in his *Personal narrative*, vol. 6, p. 626, of silicified wood, due to its great hardness, tending to form a residue after erosion of the sediment in which it is embedded. In the *Beagle diary* Darwin combined these two finds into a single short retrospective essay dated 5 April in which he seemed to quote from the *Copiapò notebook*. The notebook entry on p. 80 reads 'The red Conglom &c.... were formed at period of great volcanic agency amongst luxuriant

941 See the introduction to the *Falkland notebook*.
942 See the introduction to the *St. Fe notebook*.

islands' became in the *Beagle diary*, p. 321: 'I can show that this grand chain consisted of Volcanic Islands, covered with luxurious forests; some of the trees, one of 15 ft in circumference, I have seen silicified & imbedded in marine strata.'

In the *Beagle diary* he went on to make the extraordinary supposition that the uplift after the marine sediments were laid down may have in large part 'taken place since S. America was peopled'. Darwin supported this with evidence he gleaned in July 1835 for Indian settlement in Peru at far higher altitudes than he would have expected people to settle, thus suggesting dramatic rates of uplift. By the time he published *South America* in 1846, however, Darwin wisely retreated from such claims, citing instead known uplift at Lima of 'at least eighty-feet since Indian man inhabited that district', p. 246.

On Sunday 21 June Darwin 'Returned to Hacienda of Potrero Seco' where he stayed the night of 12 June with Mr Bingley, an English copper merchant to whom Darwin had a letter of introduction. The next day he went for a long ride: 'Descended the valley to town of Copiapò', pp. 83–4, where he stayed with Mr Bingley for three days. Darwin noted on the 23rd that the 'Town of Copiapò miserable, so often shaken down by earthquakes', p. 92.

Darwin continued to geologize, collecting another nine specimens, and closing the day with an unusually theoretical entry: 'When reading Lyell, often said there ought to be <u>Tertiary</u> strata in other parts of world of the Secondary period, although in Europe from his hypothesis there could not be', p. 91. Lyell 1830, especially chapter 8, suggested, to bolster a 'steady state', anti-progessionist history of the Earth, that it was the shifting distribution of the land and sea which drove climate change, and that there were no 'global' warmer or cooler periods. The apparently warmer conditions recorded in the European Secondary rocks, Lyell argued, were local phenomena and reflected a higher sea to land ratio in Europe during that period. Lyell even went so far as to suggest that giant fossil reptiles, like those known from the European Secondary rocks and which he believed to be better adapted to hot conditions than mammals, might return to Europe if the climate there became tropical once more. Darwin seems to have inferred from this that geologists should thus find cooler '<u>Tertiary</u> strata' on other continents where, during European 'Secondary' times, there was more land and less sea. The notebook entry suggests that Darwin was testing this idea in South America though doubting its validity because Lyell's own 'hypothesis' did not work in Europe. Unfortunately Darwin's note is too ambiguous to determine precisely to which of Lyell's hypotheses he referred.

The following thirty-six pages of the notebook are blank. On pp. 132–3 there is an undated note about a type of fox he called the Culpeu (*Lycalopex culpaeus*), 'pupil round — destroy immense quantity of poultry — Molina's account of boldness true — Bark exactly like a dog when chased — so that I did not know — very heavy animal'.[943] This note is copied almost verbatim into the

---

943 Listed as *Canis magellanicus* in *Mammalia*, pp. 10–2.

specimen list against number 3187 where Darwin added that this individual '& a bitch fox' had together destroyed 'no less than 200 fowls' at a farmhouse during the previous year.[944]

The Culpeu, *Canis magellanicus*, collected in Chile. Plate 5 from *Mammalia*.

This entry is followed by '? I suspect the young of Caracara Raucanca — is brown all over' (apparently referring to specimen number 2029, now identified as *Phalcoboenus albogularis*). Next there is a question whether the Carrancha (*Polyborus brasiliensis*) exists at Copiapò. Since these notes come after a list of rock specimen numbers 107–15, which were collected on 13 June (see *notebook* pp. 27–31), this provides a clue to their date.

Thus ended expedition no. 8, but Darwin still had time before his rendezvous with the *Beagle* in early July to make more geological observations in Chile, so he immediately started planning a trek up the Despoblado valley. Presumably he commenced the *Despoblado notebook* because the remaining thirty-six pages of the *Copiapò notebook* might not be enough for this four day expedition. The earliest entry in the *Despoblado notebook* follows on immediately from the *Copiapò notebook*, as Darwin headed off from Copiapò with Gonzalez, plus a vaqueano and eight mules. The Despoblado trip was to be Darwin's last close encounter with the Andes before sailing north on 6 July for Iquique in Peru.[945] By the time of sailing Darwin would have been in the saddle almost continuously since leaving the *Beagle* at Valparaiso on 11 March, almost four months before.

944 See *Zoology notes*, p. 411, expanded in *Animal notes*, p. 20.
945 Iquique was ceded to Chile in 1879.

[FRONT COVER]

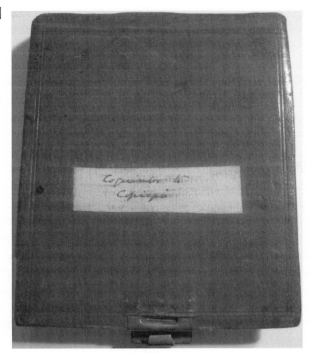

Coquimbo to Copiapò

[INSIDE FRONT COVER]

Charles Darwin

H. M. S *Beagle*

May 31st — Owe Capt FR 6 " 5 riales + 2 " 6 Marianos keep & compass —

20 Riales in Quillota

3 Pesos Coquimbo

Don Pedro Jose Barrio

Potrero Grande Hills with shells

[1] June 1st [1835] Mr Lambert, considers at Arqueros that there are scarcely any true veins; that Silver occurs in wedge-shaped masses with Sulp= of Barytes (certainly M[r] Edward ~~vein~~ mine in vein) — That /there/ — much of S of Barytes without tinge of silver: In short that they are contemporaneous veins (like quartz in mica slate) The disseminated silver in M[r] Edwards mine curious — M[r] L tells me that masses of native copper without grain of intermixture of silver have been found close to others of silver & although the latter

[2] often contains mercury (Men. Gen: Pinto Arsenic — Antimony Sulphur — yet not a grain of copper =

Looking at a chart outside /of/ near coast Copper & Gold — inside Silver & inside again Silver & Lead. — M[r] Lambert. =

Mr Lambert at Arqueros considers Metallic vein better when close to the stone dykes —
Tuesday 2^d [June 1835] Whole road cones & paps of syenitic-greenstone & dark porphyry, some crystalline, greenish, more or less granular altered slates or feldspathic rocks with a

[3]   sort of stratification — a small bay with yellow /ferrugin/ arenaceous; cemented mixture of broken shells — Balanida & Pecten a bit of silicified bone like intermediate; but a Concholepas capped by gravel = North of Coquimbo plain with Tosca rock = Arrived at Yerbabuena — the road with a tinge of green, just sufficient to remind one of the freshness of turf & budding flowers in the Spring — travelling in this country produces a constant longing

[4]   after such scenes, a feeling like a prisoner would have —
Coquimbo has /inn — hor/ our manner of living the universal ones for strangers = met on road occasional troops of mules, only one house after [illeg] leaving Coquimbo —
Wednesday 3^d [June 1835] first part of day over a mountainous rocky desert, like near ~~Quita~~ Conchalee.; — then a long sandy Wretched plain, covered with broken shells, very little water & that bitter saline, streamlets white deposits succulent plants; interior not inhabited — singular mountains of greenstone & syenite having

[5]   saline springs — Many P. St. Julian finch & Dinca Turco & /nest/ Tapacolo = First part Greenstone like 110^{03} & 110^{04}, more or less granular, perhaps mine generally more feldspathic — nature doubtful — Some of the paps syenitic. —
At P. de Chores, green brecciated rock & some purplish do — structure just same as in Cordilleras — fragments, altered slates, greenish & reddish, some more siliceous, some more felspath: traversed by grand & immense N & S dykes of black hornblendic greenstone; the Breccia near

[6]   dyke, very compact, conch: fracture, pieces blending together, like the paps at Quiliman — decidedly altered but not Porph —
Near Carizal, some granites —
Before a great plain of sand & recent broken shells, & a lowstep of plain with very much Tosca rock with shells showing extent of this formation. Many Diff parts of /round/ comminuted shells from 0 to about 200 ft above level of Sea —
Carizal to Sauce

[7]   Thursday 4th [June 1835] — Continued to ride over desert plain with many Guanaco afterwards across remarkably narrow & deep ravines — at Chaneral a narrow green valley —
At Chaneral section, Tosca rock with Donax & Concholepas, perhaps at higher level than at Coquimbo — covering a coarse Losa, with many pebbles beneath which comes a hardish shelly mass, chiefly Balanidae, of an bright ochry color, /smally/ comminuted, cemented & alternating with few layers of a very fine, perhaps aluminous, ochry & sometimes green powder occasional few pebbles — whole ~~bed~~ section very like Coquimbo —

[8]   dip of lower beds, appeared greater than Tosca rock to SE. from an outlying range of hills — I think certainly tilted. = These barren extensive Tosca rock plains very like those near B. Blanca, though there much more grass — here small bushes = it is very

singular how so many shells = At P. de los Leones we /had/ some granite, which by arrangement of black mica & Hornblende is rather a gneiss — Passing /Quebrada/ Onda — we entered a grand mass of fine true Granite hills —

[9] I noticed here an appearance of stratified granite several times regular alternating — the lowest bed was as general mass, with large black cry of mica — then came — broad 18 ft of white-fine grained much quartz feldspar & black /junctures/ above this again common granite, with two or three thin beds. — The ~~lower~~ Main granite had the usual more micaceous black ∠ patches — The lower junction very regular, yet I found few pieces of the white involved; & in upper

[10] junction of white bed, many pieces of the common (& containing the darker ∠ bits) imbedded in = what confusion — Also further on there was a large vein, which bifurcated & thinned away of same black, as ∠ bits. — All this granite had many patches & veins, of /Euritic/ & other granites, many of which died away on each side. — (N B before entering granite, saw some Tosca rock plain at twice elevation as at Coquimbo)
Proceeding NE, we entered

[11] a grand. mica Slate district — mica Slate, much very contorted like Chiloe, rest more even ground, the thin laminae, little micaceous black — dip general high ∠ 70? to about ENE — NNE by E. = Further in country, granite occasionaly appears — In this district some copper mines — The mica slate seems to form a band E of the Granite Hills —
Poor horses nothing but bit of straw to eat, after travelling whole day. — Some of the granite weathers into balls or crusts —

[12] I believe granite near Carizal is a gneiss —
Friday 5th [June 1835] Immense number of /Bulimi/ like B Blanca — magnificent spectacle of clouds, horizon perfectly true — coast like most broken parts of Chonos Archipelago: — Mountains covered with tiny bushes encrusted with a gree[n] filamentous Lichen, even the large candlestick cactus of Chile is succeeded by these species —
Valley of Guasco, from terraces distant tame snowy Cordilleras; exceeding brown hills, narrow flat, well wooded with willows, valley, certainly pretty. Capt Halls[946] description neutral but I expected such valley as at C. Verd Is^d. —
a light shower last May, rain of Coquimbo, a cloudy day here, such envy expressed

[13] three years ago very rainy; flood more injurous than dry year torrents /covering/ flat valley with sand & stones. —
In Cordilleras, near /S./ Yeso I saw a shell like those of R. Claro. = Leaving Sauce again came into some granite with copper mines & then again entered the great mica slate district, which we did not leave till we came into a Trappean ore near Freirina [Freyrina] —

---

946 Hall 1824.

The mica slate has not a very /extreme/ dip; Rock such as at Chiloe is associated with a black micaceous rock, almost like /Amphibolite/ & some brown glossy ones & some even compact, little laminated bluish clay slates. — formation very like Concepcion. — many saline

[14] springs — (is there rain enough for even these few springs?). —
|| Tosca Rock on the Road. — All the mines from description & specimens, appear to lie in the slate formation — ||
In some the vein is associated with Foleaceous rocks & even Soapstone with thin plates of metallic copper = Soapstone good sign — — much Carb of Copper & some beautiful blue vitriol. —
Saturday 6th [June 1835] Rode down to the Port[947] — miserable rocky desert little hole. = Contradanca map /and/ atlas — the most learned Limenian lady[948] — King of Londres — View up the valley very striking on a clear day infinity of crossing lines blending together

[15] in a beautiful haze, distant snow mountains clear outline — formal foreground. = At Port there are a variety of curious varying slaty rocks, often very ferruginous either /jaspery/ or feldspathic (pounded up & used for furnace bricks). sometimes quite Porph. with white crystals certainly seeming to belong to mica slate, — for such is found in valley /perfectly/ characterised, this rock is much traversed by dyke & masses of Trappean rocks apparently belonging to mass of hills mentioned yesterday. — dip to about SW by S — but in other locality to ESE. = Plains of valley more strongly marked than at Guasco

[16] highest one, much greater elevation, some have gravel so firmly cemented into a rock & layers of white clay with obscure vegetable impressions = pebbles white — washed on upper plain = near Port in one of the lower plains a very soft sandstone — on the lowest plains some comminuted shells = I could trace on northern side, including plain of river (which in floods is all washed) 6 plains, & in other place (5)
— the upper ones are extensive & one of the lower ones also rest mere lines on the escarpment

[17] one plain is developed in one spot dies away leaving line of knobs till no evidence remains —
I saw Mem. yesterday Tosca rock in my journey — Saltpetre produced in mouth of furnace. — All these hills highly auriferous —

---

947 Guasco (Huasco).
948 'I called in the evening at the house of the "Governador"; the Signora was a Lime[n]ian & affected blue-stockingism & superiority over her neighbours. Yet this learned lady never could have seen a Map. Mr Hardy told me that one day a coloured Atlas was lying on a Pianoforte & this lady seeing it exclaimed, "Esta es contradanca". This is a country dance! "que bonita" how pretty!' *Beagle diary*, p. 336.

Carrancha — T[h]enca — Loyca[949] (Black & Gold finch) Dinca[950] Chingolo[951] — Furnarius — Avecassina — little Grey Bird of mountains — Blue finch with white dot in tail — no Chingolo — C[aracara]. Raucanca[952] white tail Callandra

Sunday 7th [June 1835] Staid to rest animals. Ascended hill behind town. Beside plain of river there were fine other broad distinct terraces, the

[18] lower ones on the side 2 or 300 yards wide & in places much narrower, (river is removing some) on that escarpment which faces, there is between 3d & 4th from top, another which can be seen for about a mile. — making 7 with bed of river

The two upper plains where I ascended are very broard, the upper one runs far up valley very extensive — lowers quite parallel — where 7 plains there is in one spot one escarpment — where I ascended, there was pretty clear trace of another terrace but not extending far is not counted.

[19] On the Hill where I ascended an imperfect slate (with mines) laminae irregular general N & S directions — 5 steps (beside bed) may be here considered general /as/ 3 in Coquimbo —

Freyrina small village of white houses —

Monday 8th [June 1835] Road up to Ballenar; view when high mountains obscured by clouds very like S. Cruz. Some of the lower plains expand much near Ballenar

Judging by color of hills, the P. Breccia formation begins a little to the E of Ballenar — called 10 leagues Some /Salitrales/ on these plains

[20] Tuesday 9th [June 1835] Staid in Ballenar considerable town just sprung up, owing to mines — food brought from South. — Hear of fossil wood on other side of Cordilleras. = Found Terebratulida in the cherty rock as at R. Claro — (V Specimen) doubtless same as all the specimens from Guashco Alto — & the shells which I saw at Freyrina — I hear of part of ammonite = shells must be excessively abundant at Guasko Alto — Silver mines in all parts there — I can see the P. Brec about 2 leagues E of this place. — At

[21] Ballenar 5 plains magnificently developed besides valley — thickness of highest I should think 600 ft —

Each plain in part very broard — Above highest plain, there seems another with no distinct escarpment — but outlines — extending grand horizon plain to north = Most interesting view from steps north of Ballenar — grand escarpment. East — Hills of Granitic rocks — with these terraces (like Chonos) narrow green line of valley = Patagonia a garden compared to these plains — many absolutely barren — yet dormant seeds — wait for wet year. =

---

949 Molina's name for the starling *Sturnus ruber*, see 'Ornithological notes', p. 214.
950 Chilian name for *Fringilla* see 'Ornithological notes', p. 251.
951 Specimen not in spirits 1615 in *Zoology notes*, p. 394; listed as *Zonotrichia matutina* in *Birds*, p. 91.
952 Specimen 2029 in *Zoology notes*, p. 230, and 'Ornithological notes', p. 238; listed as *Milvago albogularis* in *Birds*, pp. 18–21.

[22] |Wednesday 10th| [June 1835] Started for Copiapò — The highest plain of yesterday north of Town is backed by escarpment, must have formed a bay — there is no higher plain — Very extensive stretching far to mouth = We entered obliquely the P. B. form — apparent extensive escarpment with East dip — but found on margin

of plain W E a slope to W — ~~not~~ scarcely visible — much contortion & irregular dip — each escarpement of a *[illeg]* seems independently stratified = rock chiefly finegrained red Sandstone — Crystalline specks of Carb of lime & some little

[23] *[coarse]* associated with a <u>great</u> <u>deal</u> of the calcareo — cherty bed — with bits of shells — some little purer limestone & softer — is evidently grand formation — seems high in the system — We rode for some time on these irregular *[skutey]* hills — We then came to some low hills of a slatey rock with <u>dip to E</u> 60°–70° — in places silicious clay slate, others Talcaceous — with occasional nodules & crystals — varying character pale color — This is succeeded

[24] to East by extensive range of true granite Hills; two sorts of granite are in <u>distinct</u> mountains close together, common & a ferruginous quartzose little mica, <u>closer</u> grained kind — Beyond this escarpement of true Cordilleras of P.B.
A very desolate but not quite uninteresting ride, followed damp course of *[bank]* about ½ league up nice little stream fine wood — two or three in road *[Ranchitos]* Indians — Many donkey — eat wood — nothing for poor horses = contrast of

[25] splendid weather & utterly useless weather —
|Thursday 11th| [June 1835] Travelled for 12 hours, never stopping — Poor horses wonderfully fresh — nothing to eat at night: — uneven country — Various shades — much prettier than a forest. = Met silver stealers. —
Geology (not being able to stop) unintelligible — The true P.B. seems very comparatively rare — there is a singularly long & regular ridge with E dip — composed of arenaceous Limestone (105) ? Very slaty structure with Terebratulata & fragments of shells — *[others]* with compact siliceo rocks (106)

[26] such as before. — its connection with P.B. I do not know — a separate slaty ridge with Terebratulata, it occurs at considerable absolute elevation. =
|Friday 12th| [June 1835] plains of gravel — 1000^{ft} above the valley of Copiapo = Arrived before noon at Hacienda of Potrero Seco = History of association prevent speculation of Pastures Mules — |Little wren|
|Saturday 13th| [June 1835] Geology in this situation of Los Hornitos very complicated — there is a

[27] Band (Quebrada of Chanuncillo) of the Gypseous strata & its superior strata with an Easterly dip, in parts however there is an *[interline]* dip & (variation in direction E) on borders of valley, in about the band of Gypsum — The strata dip at 45° & upwards I obtained following section but not very perfect. —

(1) Purple sandstone, or rather conglomerate large rounded blocks — cement not Porphyritic, but slightly crystalline. structure large rounded pebbles, even boulders if various porphyries, ¾ of them

[28]   a very pale purple porph with small ~~Porph~~ Cry of F., conch: fracture — just like neighbouring hills of injected rock — veins of Calc spar. — considerable thickness — This grand bed contains *[illeg]* 3 thin strata of Black coarsely laminated Calc? Clay Slate (107):

$\boxed{2^d}$ bed of light green (108) altered argilla bed, slightly nodular structure.

$\boxed{3^d}$. Purple sandstone, slightly crystalline structure, also nodular in small degree (109) —

$\boxed{4^{th}}$ compact <u>dusky</u> green, **finely** brecciated

[29]   feldspathic. altered. Arg. rock (110)

$\boxed{5^{th}}$ Gypsum, very impure, thin layers, sometimes little contorted, (lowest ones alternate with a laminated green rock as 4th) no selenite — rock porous, layer of earthy matter — 20 ft thick —

$\boxed{6^{th}}$ as 4th dusky green

$\boxed{7^{th}}$ gypsum —

$\boxed{8^{th}}$ — a laminated green rock with some layers of siliceous rock (111) veined & specks with Carb of Iron? There is some white rock (113). coarsely veined with this ferruginous rock. — Above this there is a great mass of rather brightly greener rock, with a remarkable nodular or concretionary rock, of all sizes

[30]   strata composed of balls —

$\boxed{9^{th}}$ a grand thickness of <u>Ferruginous</u> coloured (112) soft very much laminated crumbling ~~dusty~~ [dirty/ pale green or brown ~~ven~~ Argillaceous schist — laminae sometimes contorted — crumbling into angular bits = ~~sometimes like~~ dendritic markings manganese veins of Calc. Spar: masses of white patches, few of white rock (113) & ~~some~~ much layers of the Carbonate of Iron & silicious matter (114) — These beds remind me of those of Hornitos & Puclaro. ~~Above~~

$\boxed{10^{th}}$ Above a mountain of dull

[31]   green & sometimes purplish (115) sandstone very compact, with very many Calc-spar veins — & some metallic (<u>distinct</u> of silver gold & copper) — owing to compression of dips could not trace any more = All these rocks, which I call Sandstone have a smooth compact fracture, & are only called sandstone because they show fine brecciated structure =

These formations dip Easterly towards a number of grand hills which appear to have burst through & uplifted them — they consist very pale purple or

[32]   <u>pale clay stone</u> porph. (clearly melted) small crystals of Feldspar. —

$\boxed{\text{Sunday } 14^{th}}$ [June 1835]

116: 117: 118 — 3 varieties of lowest Lava or Porphyry.

119: 120 The great field of ~~pink~~ Lilac lava generally more porphyrytic

121 — White rock above do lava

122 Concretion in this white rock

123 Gypsum impure

124 — Rock which alternates with do

125: Commonest rock of great upper bed

126 — Part of do —

127 — Thin bed or layer in do

[33]   128 — Veins in above strata of Carb of Iron? —

129 — The conglomerate above the Lilac lava & all the X pebbles in do

I examined ~~the~~ a Quebrada of Chanuncillo, del Hornito. The bed of last section wh is lowest or the conglomerate I find 2–300 ft thick; the basis in places certainly is rather porphyritic. (129) — The pebbles from size of fist to that of head very numerous — examined very many all porphyries. the greater part like the injected of Hill of yesterday specimens with /crys/ exemplifying these (NB one of these, by mistake, is part of base of matrix)

[34]   it is manifest, the rock of pebbles was originally a porphyry = difficulty explained by an underlying bed of porphyry which is a Lava — color pale lilac — conch — fracture — structure rather laminated, few crystals — some highly sonorous (119. 120) & on grand scale conformable to upper bed of conglom — with some irregularities — lower edge also parallel — Between this & the Conglom. there is a varying thickness of quite white rock (121), which seems much

[35]   altered hard, /harsh/ — nature doubtful — /smally/ brecciated with little purple spot — remarkable by containing dark rust color. many concretions, I judge from form, globular pear-shaped (like Calc: Sandstone) of some Lilac Porph as Lava stream (122). — This Lilac Lava /is perhaps/ 150–200 ft thick, rests on what in places appears like a bed, in others seems to blend with lofty Porph: Hill, which will be seen is axis of anticlinal band. — The Lilac Lava varies slightly, but this lower bed is always of darker color

[36]   & consists of very many & curious kinds of crystalline rocks (116: 17: 18). — The central parts of Porphyry /probably/ contains /paps/ & part of curious green & brown Porphyry. —

(Mem. pebble of white rock with other pebbles in conglomerate bed). —

With respect to other beds above the conglomerate I find them subject to much variation, up the ravine. The black laminated calc clay Slate increases much in quantity, is very ferruginous in external fracture (like rock of Portillo) — large plates — some brown all way through. — Alternate repeatedly with the red Sandstone — & green beds which are less

[37]

E A B Porphyry Lilic N Sandstone congl Argillaceous beds & gypsum Gravel
Conglomerate Lilac lava Dark Lava Porph B A Porph X Lilac Lava
Conglomerate &c&c G W

abundant — Also ~~these~~ such strata alternate & form part of the /hardened/ clay &
carbonate of Iron beds — Those which show bright yellow in the mountains cannot be
short of 800 ft above exclusive of lower part of Red Conglomerate. (NB) generally some
slate is also below the conglom: & lava = All these laminated ferruginous beds soft, are
proper to Gypsum — in this ravine the Gypsum rather dies away — & becomes very
thin — The

[38] Gypsum, on North side of valley (where there is a mine) almost all the yellow beds abound
in thin layers of Gypsum — rather a Gypseous Tufa — some layers are inch or more
thick, generally thin & impure, with calcareous & these argillaceous Laminae (124. 123)
Above these yellow Argillaceous beds there is at least 1500 ft thick of the rock which
I improperly call Sandstone — its color & structure vary, but smally, a little finer or
coarser, a little more or less green, purplish or brown. — The first & last most
common. — Structure rather nodular

[39] is remarkable by a thin stratum about 80ft apart, each about 18 inches thick of a hard,
black, ferruginous rock — which may [be] seen in middle of these beds running for
miles, & projecting upwards (127) — I believe the great mass — capped by a yellowish
mass & this by a purplish — but the exact limit or superior bed is difficult to determine
on account of great confusion produced by the /form./ shown at (z) in section —
All those beds dip from 30–40 to ESE — the band of upheaval is very great, may be
seen for

[40] for about 15 miles in a N by E to NNE & SSW direction. — its anticlinal dip is seen
on other side of Hill (B). not dipping to WNW but to WSW. Not so highly elevated,
I saw no more of this dip = The strata are thrusted as shown at (Z), apparently by the
posterior elevation of the still higher line of Porph ~~Hills~~ Mountains (A). = I should
think (A) the origin of Lava — it seems that old Volcano point of subsequent
upheaval. —

[41] About 2 miles to North of Hill (B) an important observation. — The Lilac Lava is
found here as everywhere conformable & beneath grand mass of conglom — (covered
by the great gypseous form already alluded to) instead of resting on a perfect porphyry.
it rests on a conglomerate such as that which covers it — but the pebbles are of a darker
tint & not of Lilac Porph, but of various kinds — Now in Hill (B) there is no trace of
such structure, only small trace of being a bed parallel to Lilac, so that

[42] I thought it Lava which blended with its origin amidst curious Porphyries — now it
seems clear, that the hill (B) perhaps forms /part/ of hidden granite, has so altered
this bed, that its true character cannot be recognised. I suppose these beds upper part of
P.B Formation, or those masses which support the Gypseous Formation =
I urge this as a very important instance, as all the beds can be traced over whole
country & there can be no mistakes. —

[43] Hill (B) is /marbled/ with strata, with high dip — bare where facing the valley — dip smaller where the lower beds are seen to be conglomerate — In all these conglomerate pebbles can be easily separated no blending of parts = Can the white Bed above Lilac Lava baked volcanic ashes ? =

Monday 15th [June 1835] Rode up to Los Amolanas valley or rather Quebrada very narrow — very much Alfalfa — river size of large muddy brook — for 30 years never reached sea — now does — send up guards — plenty of snow now secure —

[44]

E Gypseous Sandstone P. B. South side of valley Dark Porph 131 Syenite Green Porph Lilac /Porph/ Posterior elevated W Eugenio Matta Potrero La Punta
After passing for some time between hills of Lilac Porph — (Mem. valley formed by removal of fractured strata between great irregular cones —) we came to some green & dusky /base/ with few crystals of feldspar & specks of Hornblende — Then there are paps of a finely granular syenitic greenstone

[45] & then a great central formation of whitish syenitic greenstone (130) specimen smally crystallised, generally much coarser– This rock is capped & succeeded by some black fine grained micaceous (131) rock which it has penetrated with an extraordinary number of beautiful dykes of a white color consisting of well cryst feldspar with few specks of Hornblende — These dykes on large scale may be seen traversing the succeeding hills to the distance of the two or three

[46] miles & proceeding directly from the syenitic central grand mass —
with the syenite in one spot there is a true granite, white as usual. — The syenite mountains are succeeded by grand dusky coloured Porphyry Hills, which on eastern part, show signs, which gradually become more certain of highly inclined stratification to SE or more southerly. — These consist of the true P.B, although much altered & are of great thickness perhaps from 1500–2000 ft —

[47] such beds, doubtless come below the the conglomerate of yesterday. — To the East the slope of mountain was not much interrupted by a remarkable fault, which brought a small irregular dip to W of the upper gypseous in contact with the SE dip of interior P.B. = (at La Punta the syenite is well developed) It appears looking at

structure on each boundary grand mass of mountains have /fault/ up, posterior to formation of /Gypsum/

[48] Formation. = There is a grand mass of mountains, composed of bright red standstone (a rock which deserves its name) 132, threaded with veins of Gypsum associated with some green & some black laminated rock — but general tint of hills bright red as at Potrero Seco. bright yellow; <u>clearly</u> same formation = Subsequent to the irregular dip towards the P.B. we have

[49] the dips as in section of those upper formations subject to <u>much</u> irregularity traversed by many <u>dykes faults</u> & metallic veins — This reaches opposite to Eugenio Matta — further to Amolanas — am not sure — This is on S side of valley — on North side the first dip with fault is clear, /beyond/ which from faults & compression, there is no /undisturbed/ structure — I see the two side of valley seem in dip in /the/ parts to correspond —

[50] abundant silver mines in all this district. = Gypseous formation like Incas bridge, except that veins instead of grand masses=

Tuesday 16ᵗʰ [June 1835] Ascended to where the famous shells are found — Found there is a grand range with a WSW dip 30°–40° = Hence in the valley which here runs N by W & S by E is a trough between two range & the consequent difficulty is explained. — The shells excessively numerous — an enormous quantity of Gryphites in lower band & above thousands of the univalve — very numerous Terebratulæ

[51] & bits of oysters & some Pectens in all parts = They occur in a jet black thinly stratified /shattey/ thin intermediate strata) calcareous, argillaceous rock (133) conch fracture; although on fracture so black, surface weathered /half/ brown, hence whole mountain precisely like that of Portillo. — Where bits of shells very numerous there are strata of a greyer calcareous rock (134) — These beds are of great thickness, contain very many strata of a yellowish fine grained siliceous sandstone like (135 & 136) — There was one stratum of purple conglomerate

[52] Besides grand, thick, conchiferous layers found two others = Some of the black rock is a little paler & bluer, some with tint of green. — All weather white — Some upper strata were lined with /number/ contemp large veins of Gypsum — Specimen of black rock two crystalline. — Whole mountains very remarkable by the unparallel number of parallel, zizag dykes of all sizes & forms: of very various rock almost all Porphyries some approaching to syenitic greenstone. —

[53] But rock (137) is most common — many crystals many an inch big — base varies from black green to white — 138 is other common variety. — This must have been close to volcanic forces — as many lines for dykes as strata lines = Examined same great chain — on South side of valley the lowest rock is a purplish black coarse fracture, few crystals Lava Porphyry — conformably covered — & suddenly by coarse conglomerate, quite as coarse (no trace of cryst struc in matrix) as at Pluclaro & R Claro Not work of dykes. more cracks filled up, no effect on strata

[54]   colour pale, <u>sudden</u> transition shows lower is true lava — bed some hundred feet thick
       lower half whitish, upper pink — much /marly/ harsh agglutinated coarse sand with
       current cleavage not parallel to strata — This is covered by grand mass of yellowish
       siliceous sandstone (135–136). Softer pieces used for sharpen hence name Los
       Amolanas — This rock contains some band closely cemented with it of white
       conglomerate some

[55]   of purple — Also some of the black Calc or Clay slate — I believe however the grand
       mass of latter rocks comes over this, then Red Sandstone with its Gypsum? —
       Parts of white sandstone hard small white quartz pebbles & seen one piece of ~~quartz~~
       mica &c hence origin — like /Porotas/. —
       The yellowish sandstone in parts great part was greenish & approached character of the
       green rock of Potrero Seco in compactness. I imagine it represents, as this black

[56]   Calc rock with Conch fracture does the /more/ laminated black rock of do seen in both
       in cases connected with Gypsum — The collection of fossils very interesting in itself, &
       as connecting all the localities; all appear to be about age or little older of lower Gypsum
       = The <u>Sandstone</u> & Terebratula & Gryphites show R. Claro same age & the coarse
       conglomerate being here inferior removes doubts about modern age of conglom
       there & Pluclaro — At R. Claro either great streams of lava prevented Gypsum, or
       removed

[57]   or upheaved previously = at the period where these shells great similarity down the
       coast — great lateral volcanic action — As Uspallata a little later =
       (No terrestrial lavas in Chile — except perhaps ? /Canqueros??/ & ~~Portillo~~ ??) I suspect
       there is great replacement & dying away of the different strata of this period = on some
       of the large univalve — oyster & Serpula adhering — a bit of small Ammonite = In the
       conglomerate Porphyres & some jaspery & red sandstones. — where shells conglom's
       non-porph =

[58]   Wednesday 17th [June 1835] (139) altered Sandstone or Lava? (140) matrix of
       conglomerate (141) Lava beneath do — (142) Lava above Sandstone. (143). Syenite
       greenstone — (144) variety of mica slate. remarkable uniformity of syenite greenstone
       in whole range — S Fernando — Copiapò =
       It appears to me that all the mountains about Amolanas might be called one range with
       westerly dip: irregular that is often double or even treble connected by

       faults ▓▓▓▓▓▓▓ [three escarpments separated by faults], sometimes so
       sometimes one single = We have seen yesterday that the bed beneath

[59]   the Conglomerate is a Lava Porphyry — The rest of beds consist of various purple &
       some green more or less Porphyry rock — beneath these came a stone, nature of which
       I am not certain blackish purple fine grained perhaps altered Sandstone (139) —
       beneath which came a stratum of coarse conglomerate not the least blended — matrix
       crystalline 140? — Besides Porphyry there was one pebble of Granite. — a very rare

occurrence — beneath this we had crystalline fine grained porphyrys all of dark colors (141). & after some

[60] few more such strata true Granite. I do not think all strata are here seen, but certainly a great thickness, only very partially examined: contrasted with S. (I found some fragments of pale-coloured true ∠ᵣ Conglom blended Porph Brec) very little true P.B. much more Lava (for from coarse strata of the Conglom so near Granite clearly all the Porphyries which I examined for great thickness are true Lavas)

[61] This section is where Quebrado of Jolquera commences — The Granite traversed by numerous blackish purple dykes — is connected with Syenitic Granite — much (white Feldspar much Quartz & specks of iron rust) — This is the axis of the grand range — The valley (Z) turn up N. with little Easting; on West side the great mountain of Protogine (white) capped by Purple rocks — at its base dipping right on it there are some thinly stratified Sandstone (precisely as before with some of the black rock with the Univalve — Gryphite Pecten Ostrea Terebratulæ

[62]

Mica slate Syenite Much thicker Syenite Porphyry D (Z) A little way to N this is not seen (C)

& veins of Gypsum = This is capped by streams of heavy nearly black Lava (142) remarkable by containing balls rounded or oval of still heavier kind — very hard, egg to cricket ball. — The sandstone varies from 30 to 60 or 80 ft rest on Purple Porphyry full of crystals of white Feld

[63]

Porphyry W Granite Lava Sandstone with shells Porph Lava Granite Porph Conglom Porphyry Conglom Siliceous sandstone Shells Blackstone Gypsum ?

The Amolanas range double in places

This lower mass is clearly divided into several streams by bands of parallel harsh Porphyry highly amygdaloid — This ascends on west side & lies on a range (parallel to Granite) of true coarse grained syenitic greenstone crossing this obliquely in NE — very broard

[64]   good deal of syenite has also mica (143) true Granite character ∠ʳ spots, traversed by paps of diff: kinds some true Granite — (NB some of the Porphˢ with green & yellow spots owing to Epidote): On the East side of the Syenite, a considerable breadth of micaceous slate, with large Quartz veins. (& Granite dykes), in central part contains much green slate (144) containing Feldspar Mica — Chlorite ? with Quartz parallel veins — also some irregular masses of a ferruginous quartose slates

[65]   cleavage about N & S — dip not very regular chiefly E — bent or curved on large scale. = (great fault in this country in /Panilas/) I should think stratified rocks at Amolanas attained thickness 5–7000 ft / at least. about 3 grand divisions —

Thursday 18.ᵗʰ [June 1835] (145) Lowest Grey Porph: (146) Lava compact with little red crystals (147) Lilac Poprhyry — (148) Amygdaloid Poprhyry. (149) Variety of compact Porphyry above latter (150) Red intermediate sediment strata (151) White coarse sandstone volcanic particles

[66]   The mica slate is succeeded by Porphyries, the line of junction appears parallel to stratification. Then a narrow slip which buts up against hills of Syenite with occasional grains of quartz — But further North, these purple strata, little inclined to E extend a good deal into central Cordilleras, but owing to /Northerly/ direction of Quebrada did not follow it = Between mica slate & syenite highly inclined — The purple strata consists of Syenite at least 2000 ft thick.

[67]

??? Porphyry Syenite Porphyry Mica Sl[ate]

clearly stratified: one of the lowest beds is a grey rather harsh Porphyry with a good many crystals of Feldspar (145). — Above this which we have a great thickness of a compact green — mottled Lava (146) which upper part becomes purple waving red water line separated by bright red sedimentary bed like (149 150) from Lilac Poprh above — This (146) Lava /assumes/

[68] often many red little crystals like (149) — Above this grand thickness of Lilac Porph — harsh few crystals of Feldspar. often this substance arranged in short waving lines (147). Above this many kinds — In other section found quantities of thin beds 12–20 ft of dull purple Lava & Lava (149) — separated by parallel bands of Amygdaloid rock (148) mass of crystalline amygdaloid

[69] stone. — these are separated from each other narrow bands 6 inches to 12 inches of baked bright red sedimentary substance (150) — One bed about 20 ft (covered by 8 feet of latter substance & then the purple again the purple Lavas) of white sandstone or rather bits of volcanic rocks firmly cemented. Found some many fragments of true green & purple Porph Breccia — so one stratum does occur in these Purple rocks —

[70] Friday 19th [June 1835] The strata are broken up rather by hills, whose longest axis N & S rather than a chain — if the line of any of these fracture are traced they will not be found far continuous. = The purple rocks seems to extend far in the country Andes: I examined them in many places & found them nearly all Lavas with very thin separation of sedimentary rocks I found one cliff at least 300 ft. of hard rock crumbling fracture, entirely composed of such — chiefly a mottled sort, specks of green

[71] white & red (152). it contained numerous thin layers such as (151) one about 20 ft as (150), another equal thick of 153 — This mottled rock contained some large round pebbles as big as mans head of various Porphyry purple, & its tint varied in its shades — lower part alternated with purple ferruginous Lava = The In all the Porph^y much amygdaloid, some little concentric structure, obscure traces of columnar: It is very interesting

[72] finding grand Volcanic Lava formation of age of Ammonites &c &c separated by what must have been true conglomerates & Brongniarts[953] name for Volcanic sandstone: = (NB fossils of Falklands of hot country??) I took much trouble with formation of Mica Slate — (the green slates external) the other beds certainly are of less crystall structure & ferruginous but seem upheaved bodily through the Breccia formation — The commonest sorts of Porph. Lava have the little red crystals

[73] All this part of valley of Copiapò highly metalliferous, no where have I seen such confused stratification, or so many dykes yet rocks not so much metamorpho = Returned down the solitary ravine to Los Amolanas, difficulty in descending steep mountains. Valley runs up so much to North Valley all formed by connection of low /points/ in lines of upheaval (as seen by inclined dip on each side) enlardeg[954] by water

( ) /here/ generally of this form =

---

953 Alexandre Brongniart (1770–1847), Professor of Mineralogy at the Natural History Museum in Paris and co-author with George Cuvier in 1811 of the classic geological description of the Paris Basin. The specific reference may be to Brongniart 1833 which was in the *Beagle* library (CCD1: 558) and is referred to in the geological diary (DAR 35.396).

954 = Enlarged. An unusual transposition of consonants.

Saturday 20th [June 1835] 154: 55. 56 — Silicified wood from Conglomerate — 157: 58 Sandstone of do Conglomerate: 159 Shell above do — 160 Green dyke in do —

[74] Replacement & causes

It appears to me general that one point of injection is more subject to a quite different class of do. than any other point = Like same volcano different lavas, upheaval on same part of strata: Here where much Gypseous & super Gypseous slate gold — silver —

copper & micaceous iron — with respect to Paniza — very many faults, where so similar beds, hard to discover, no external sign being present: The R. Claro Terebratula is found also here: — Noticed that some of the black rock becomes so thinly stratified & argillaceous, that it is a complete clay – slate. — & some

[75] of the silicious sandstone a coarse one with fragments of shells — The black rock & sandstone replace each other without order, perhaps the ~~latter~~ former rock superior — In same manner the black rock with Gyspum & the Red Sandstone with do replace each other. — Examined to day the supergypseous an immense thickness 2000–3000 ft thick almost entirely Red Sandstones & Conglomerate (difficulty owing to faults as before mentioned except by hammering every stratum — Above the gypseous rock there is an immense

[76] mass of conglomerate, pebbles small — Size of egg cricket ball, few rather larger thickly packed in lines interstices, & occasional seams, of red sandstone (157–158), whole beds of this colour — (some parts purplish) pebbles are easily separated, almost all Porphyries, or crystalline rocks, & some reddish siliceous rock = This bed many hundred feet thick (all westerly dip varying about 30°) remarkable by thousand of great blocks of petrified Dycot wood, whole surface strewed with pieces — (hill due west of house of

[77] Hacienda found several in situ, either in the Sandstone seams or amongst the gravel — horizontal, one ~~had~~ about 8 ft in length apparent with strips of branches apparent; another had a diameter of 5ft. perfect silicified to centre, excepting in few

concentric structures of wood not visible   centre generally marked by /tissue/ of Quartz veins — Most like wood of Hornillos — very black or white siliceous, patches & veins like variety of mica Slate, mistook it even after examination, little of the structure

[78] remains; harsh & brittle, tree easily falls to pieces — enormous rock, one must have belonged to tree nearly 6 ft in diameter — What an extraordinary prop.!! Above the Conglomerate, with alternations, rather a finer grained red Sandstone, 2–300 ft thick upper seams impressions of many shells, amongst which I clearly recognized Gryphaea, & the univalve & many of Bivalve (159) above, grand mass of Lilac Porph, am not quite sure, believe if Lava may be injected

[79]   although nearly parallel across whole mountain, — yet do not feel quite sure, R Conglomerate & /B/ Sandstone (near shells one seam of black rock) up to top of mountain; I believe chiefly latter could not reach it — conformable; these grand mass of strata appear in country, /now/ /commonly/ removed, only saw the cream-coloured mountain capped in one place by Red lead Sandstones — Whole mountain traversed by black Porph — dykes white Feldspathic (all of which I believe like from Syenites)

[80]   & green laminated dyke (160). remarkable as identical with one of R. Claro, & on grand scale, having a sort of of water line or stratified structure. — No one can imagine such glorious confusion —

These beds correspond to compact-misnamed sandstones of Potrero-Seco; The red Conglom &c of Pluclaro (hence surprise) — Hornillos — Uspallata & Conglom. (upon sandstone) of Portillo were formed at period of great volcanic agency amongst luxuriant islands — siliceous springs, after great gypsum deposit, Some from Islands great diversity of

[81]   mineralogical nature — age from fossils greater than what I supposed at Uspallata = Now there is good comparison with Humboldt = I strongly suspect the whole & inferior Gypseous Formation not very ancient:

Staid whole day at Hacienda—

~~Dom~~[ingo] Sunday 21 [June 1835] Earthquakes more frequent on this the hilly side of Andes than at Mendoza — S Juan —

Returned to Hacienda of Potrero Seco

Valley of Copiapo a mere island 12000 inhabitants, three times have removed (owing to mines)

[82]   The section of the country from Los Amolanas to Potrero Seco presents I think only one line of Escarpement besides the grand one — & it is the upper or gypseous parts of this, near the Southern border of the mass of Syenite & Porphyry Hills, therefore that part must not be so much represented in the section of the country —

Monday 22^d [June 1835] (161) Laminated, calcareous, arenaceous rock — (162) Calcareous more so associated with do (163) a sort of Hone

[83]   stone (164) a soft purplish argillaceous stone (165). Stone feldspathic associated with (166) more siliceous, Piedra /Carlo/ of Capperosa — (167.) – (168) Gypseous earth (169) Ochreous do — chief part of vein —

I believe dykes in Chile run in every form & direction = If 30 parellel lines were drawn in this district from E to W I do not believe line of parallel fractures would extend in few parts to above three of them: = Descended the valley to town of

[84]   Copiapò. I see the anticlinal or East dip on E side of Hill B of ~~Lilae~~ Porphyry — is a mere Local Phenomenon The ~~dip~~ strike run into a point on south side of hill, the valley

being on the north: The form is such as that described on a grander scale ~~than~~ at Uspallata — & such would be the form produced in hard layer upheaved by rounded hill — The Lilac Lava & under bed of Conglom or

[85]

B Local P Breccia D Gypseous E W
D Porphyry very obscure, W dip, (E) Gypseous rocks, dip to E — but various —
succeeded by either W. dip or nearly horizontal
Crystalline rock (? whether equivalent?) is succeed by a broad belt of the PB Formation,
there is a little true blended rock, but much pale purple Conglom & the compact
Porph. Lavas — I think greater much

[86]   thickness than in Cordilleras. There /to/ W. blend into hills of Porph — where
stratification dies away, & perhaps succeeded by very obscure dip to W. — Then comes
the same line of hills which was so remarkable in journey of the yellowish Calc
sandstone (161.) associated with other varieties (162 double) These are seen to overlie
an immense mass of beds some absolutely verticaly — others highly inclined, general
grand dip to East of pale cream colour, white, dark

[87]   & pink beds, very thinly stratified, from much equal strata, varying stratification —
excised valleys, & slopes, abrupt of hills give a marked arc = these strata consist of the
black rock, a sort of Hone stone (163), layers of /alternate/ /Yeso/ & the Argillaceous
soapstone (164). purplish or whitish — clearly these are Gypseous formation —
whether the overlying laminated Sandstone & limestone? with corresponds to red

[88]   Conglomerate is doubtful — I think however from remarkable geographical line &
frequent valley it is the upper bed of country = (NB the grand plain of Gravel 2–400 ft
thick which contains is broken formation at Potrero Seco very remarkable) —
(NB in the P Brecc formation there was one, hill of Lilac Porph injected.) =

[89]   The grand band of this formation is succeeded by an W dip in some few spots, more
generally by a broad mass which reaches nearly to Copiapò, which is either nearly
horizontal, or irregularly contorted by various hills & masses of injected rocks. —
fine-grained, greenstones &c &c (NB. true greenstone very rare I recollect no such hill
excepting at Campana & few dykes excepting some fine grained sort & the Porphyritic
sorts) In one of these hills, which has been contorted

[90]   (NB Minerales of all sorts in these hills gold of gypsum age silver & copper) the upper
strata composed of a greenish Feldspathic (165) associated with more siliceous
(166. — in the mines of Sulph of Iron — The vein is very broad & composed chiefly of
yellow ocherous /Cault/. (hence name Tierra Amarilla) (169) associated with
ferruginous Gypseous (168) is not much inclined & parallel to certain marks in hills —
most obscurely stratified (so not quite certain) much sulphate

[91] of iron in Atacama — This Sulp of Iron associated with Sulph of Copper & gypsum & other sandstones (167) which has not so much of the Capperosa for which it is worked When reading Lyell, often said there ought to be <u>Tertiary</u> strata in other parts of world of the Secondary period, although in Europe from his hypothesis there could not be Tuesday 23^{d}) [June 1835] Copiapo seated in broard plain like the valley of Aconcagua (for 25 years only, 2 Governors who have reigned

[92] their proper time) — Town of Copiapò miserable, so often shaken down by earthquakes

I suspect the four grand line[s] of upheaval of tolerably regular — the mass of syenite & porph Hills of Potrero Seco have burst up in Synclinal spot —

Wednesday 24th [June 1835] ⎫
Thursday 25 [June 1835] ⎬ Copiapò

[93] Chaneral Bay, Very fine grained Greenstones marine shells scattered on surface. Also brown quartz rock

Immense quantities of Oysters in strata in English Harbor N of Copiapò

Compact & slaty siliceo—Feldspath lead — coloured rock? at Lavatan: Mem:

[94–131 blank]

[132] culpeu[955] — pupil round — destroy immense quantity of poultry — Molina's account[956]

[133] of Boldness true — Bark exactly like a dog when chased — so that I did not know — very heavy animal —

ʕ I suspect the young of Caracara Raucanca — is brown all over. Carranch infrequent at Guasco — ʕ Exists at Copiapò —

[134] Sandstone

115 Red compact

114 ferruginous /cast of iron/

113 /whole/

112 <u>ferrugi soft laminated</u> clay bed (Sp), <u>nodular strata</u> of /SB/ Laminated G. bed

111 with some siliceous ferruginous veins layers

G

<u>Dusky green sp</u>

/Gr/

110 <u>Dusky</u>[Gr/

109 <u>Purple sp</u>

108 <u>Light green Sp</u>

107 (sh) Purple conglom with 2 layers of (give) thickness. Black laminated Cale. Slate

~~103~~

[135–6 blank]

955 Specimen not in spirits 3187 in *Zoology notes*, p. 411; listed as *Canis magellanicus* in *Mammalia*, pp. 10–2.

956 Molina 1794–5, 1: 330–2.

[INSIDE BACK COVER]

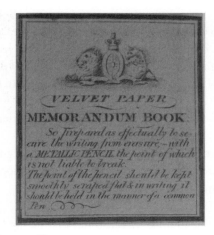

Pills Cigars Maps Letters Pen
Beechey[957] says that Humming Birds stay all winter in
N. California —
/Cerro Bermida/ 3 leagues from Copiapò
/Eribase/ — Pueblo de los Indes 3 Hermanos

[BACK COVER]

957 Beechey 1831, 2: 82. This reference was copied into 'Ornithological notes', p. 252.

Textual notes to the *Copiapò notebook*

[IFC]   1.7.] *Down House number, not transcribed.*
         88202327] *English Heritage number, not transcribed.*
         11] *added pencil by Nora Barlow, not transcribed.*

[37]   *sketch perpendicular to the spine in right margin.*
         'A' and 'B' *deleted transposed in ink.*

[44–93] *pages written perpendicular to the spine.*

[53]   Not work … on strata] *added heavy pencil.*

[62]   D] *ink.*
         (C)] *ink.*

[74]   Replacement & causes] *written over other entries on page.*

[93]   *page written perpendicular to the spine, in pencil.*

[131–2] *top of leaf excised.*

[135–6] *leaf partly excised.*

THE *DESPOBLADO NOTEBOOK*

The *Despoblado notebook* takes its name from the Despoblado Valley (meaning 'depopulated zone') between Copiapò and Chanaral, Atacama Desert, Chile. The notebook is bound in red leather with an embossed border: the brass clasp is intact. The front cover has a (300 × 700 mm) cream-coloured paper label with 'Copiapò — Despoblado Isle of France Cape of Good Hope excursion St Helena.' written in ink. There are 68 leaves or 136 yellow-edged pages. Sadly an excised leaf (43a–44a) has been lost since the notebook was microfilmed. In terms of its square shape and red cover, the *Despoblado notebook* is almost identical to its two predecessors in the expedition no. 8 trilogy, all Type 5 notebooks. The notebook covers late June to September 1835 and May to September 1836.

It is one of the most interesting of all the *Beagle* notebooks. Darwin used it for the last year and three months of the voyage and it was therefore the notebook that virtually circled the world while in use. Darwin first used the *Despoblado notebook* for detailed geological descriptions on the west coast of South America in June 1835, then in the coral islands of the Indian Ocean in the southern winter of 1836. From there he used it in the mountains of South Africa, on St Helena, one of the world's most remote islands, and then on the east coast of South America. Finally the notebook was used back into the northern hemisphere and the Cape de Verds, which had been Darwin's first landfall four and a half years before. Then he was a young man embarking on a great adventure. Now he was a seasoned geologist and naturalist whose reputation at the high table of science was already preceding him.

The *Despoblado notebook* contains the last of Darwin's field notes from Chile but overlaps with the remaining two notebooks for the rest of the voyage. The earliest part of the *Despoblado notebook* was used in late June and through July 1835, with scattered notes until the *Beagle* left mainland South America in September 1835. There is then an eight month gap in use until April or May 1836 when the *Despoblado notebook* came back into occasional use for perhaps five months, as summarised below, up to September 1836. Thus the *Despoblado notebook* was used by Darwin almost to the end of the voyage and as such it also overlaps with the undated pages of the *Santiago notebook* and with the early part of the *Red notebook*.

Curiously Darwin started to fill the *Despoblado notebook* from the back for his exploration of the Despoblado valley at Copiapò in June 1835. There are a few entries at the front from 1835, but most of the front dates from 1836. In order to provide a chronological introduction, it is necessary to start by examining the back pages but to keep referring to front pages where they fit chronologically. The first forty-two of the back pages cover the period from 26 June 1835 to the *Beagle*'s departure on 6 July for Iquique, which was then in Peru but is today in Chile. The notebook was not used for the sea passage but records arrival at Iquique on the same page on 12 July, then departure for the port of Callao on 15 July.

After another four days at sea the *Beagle* reached Callao on 20 July 1835 (p. 45b) whence she departed for the Galapagos on 7 September. There are scarcely any more 1835 entries in the *Despoblado notebook*, with the *Galapagos notebook* filling the gaps for Peru, the Galapagos and Tahiti (July–November 1835), while the *Sydney notebook* covers Australia (January 1836) and Mauritius (late April–May). There is, however, a series of notes in the *Despoblado notebook* around pp. 47b–51b on corals, apparently based on boat soundings when the *Beagle* was in the Indian Ocean. Armstrong 1991b, p. 7, believed that these notes were made at the Cocos (Keeling) Islands (early April), but there is reason to conclude that they relate to Mauritius.[958]

The next dated page in the *Despoblado notebook* is 52b in Cape Town on 31 May 1836 and the following twenty pages or so are all from Africa. By p. 74b Darwin was on St Helena starting on 8 July, then at Pernambuco [Recife] in Brazil in August on p. 87b, ending on p. 91b, which is the reverse of p. 46a.

It is difficult to date the twenty-seven or so front pages, partly because some of them are excised and many are blank. They start in Iquique in 1835 then seem to jump to Mauritius in late April–May 1836, St Helena in July, Pernambuco in August and finally the Cape de Verds in September. As such they overlap with the back pages and seem to include the final field notes of the voyage.[959] If there were any notes from Ascension (July) or Bahia (August) they may have been on the excised pages. There is also a considerable number of blank pages which we have numbered as part of the front sequence but which could equally be labelled as back

---

958 The Cocos (Keeling) Islands are here referred to, following CD, as the Keeling Islands, although they are today almost always called the Cocos Islands. Similarly Mauritius is used even though CD sometimes called it Isle of France, for example on the cover of the *Despoblado notebook*.

959 The September Cape de Verds notes are the last field notes, but there are some jottings from the *Beagle*'s departure from the Azores on the inside back cover of the *Red notebook*, which in our view read as follows: 'Sailed 24th [September 1836] ~~Friday, gale 29th Friday~~ Thursday 29th gale'. That there was a storm on the 29th is confirmed by the meteorological journal in the *Appendix* to FitzRoy's *Narrative*, p. 60. CD disembarked from the *Beagle* for the last time, in Falmouth, on 2 October 1836.

pages. An excised leaf with text (which we found to be now missing from the notebook) (43a–44a) in this sequence relates to St Helena which seems out of the correct date position whether read from the front or the back. We have numbered the excised page by comparing the edge of the leaf visible on the microfilm with the stub of 43a–44a.

Despoblado Valley, June–July 1835

The first notes are on the inside back cover. Although they clearly relate to Chilean geology they are hard to read and are not easily intelligible. Field notes proper, dated 26 June, begin on p. 1b with 'started from Copiapò valley to Despoblado', followed immediately by a record of a finch, '3 sorts of Caracara' and vicuna. Some of the places mentioned are shown on Darwin's hand-drawn map in DAR 44.28 (see p. 485).

Darwin seems not to have enjoyed that first night; the water was so 'Amarga' (putrid) that he 'could not drink Tea' and there was a 'very severe frost'. It is interesting to note that Darwin drew a small sketch in the *Beagle diary* for this date showing how there were small rows of stones apparently aligned with the tributary valleys.[960] When asked what these were all his travelling companions could manage was 'Quien sabe?' (who knows).

The next day Darwin 'Travelled far up ravine of Paypote [Paipote], came to water, & trees of Algarroba', p. 2b, that is mimosa. There was an 'old man watching old smelting furnace', but Darwin felt 'Too much tired to enjoy these situations'. The notes on these pages are rather hard to understand, like the geology they describe which Darwin calls 'exceedingly complicated' in *South America*, p. 229. On p. 3b he recorded finding the all important 'Gryphites & Terebratula', allowing correlation by age with the sediments he saw further south, and on p. 5b he concluded that 'From the quantity of Lava there must have been many volcanoes'.

On p. 7b Darwin referred to a diagram he drew on the inside front cover of the notebook showing 'some very remarkable forms of stratification' of 'thinly laminated red sandstone' enclosed within porphyritic breccia. This diagram was eventually published in *South America*, p. 231 (reproduced below) and seems to be closely related to the one on p. 26b although the latter shows a yoke consisting of breccia rather than sandstone. A small diagram on p. 8b shows what Darwin described as the only case he saw of two 'horizontal Pushes', presumably shear zones, then there is a larger diagram dramatically showing how the strata on the left have been tipped upside-down on the right. This case of inversion through 135° (i.e. $3 \times 45°$) is described in *South America*, pp. 231–2.

960 Compare with the geological diary DAR 36.453v.

Diagram showing complex structures in the Despoblado Valley. A mass of sandstone and conglomerate is shown at the centre; it has been severely down-folded into a 'yolk or urn-formed trough', with the underlying porphyritic conglomerate strata more or less vertical on both sides of the yolk. On the left these strata are severely faulted while on the right their dip radiates 'like the spokes of a wheel'. Fig. 24 from *South America*, p. 231.

The rest of the entry for the 27th is unusually reflective, suggesting perhaps that these pages were written once Darwin had set up camp in the evening. Darwin continued the dialogue with himself concerning the extraordinarily complex geological history of the valley, and on p. 11b he asked himself two difficult questions: 'Are the Pampas solely owing to being leeward of currents has the weight prevented subsequent upheavals'. Darwin was apparently asking himself whether the vast spreads of porphyritic breccia pebbles and gravels which underlie much of the Pampas but not the west coast are there because the Pacific currents swept them away more on that side. His second question was whether the sheer weight of the Pampas gravels had prevented the kind of post-eruption upheavals of which there is clear evidence on the west coast ranges.

On 28 June Darwin 'travelled up to the foot of the Linea — that is a line of mountains which sends water to the East to great lake of Salt many leagues long', p. 12b. The landscape was very harsh with 'very much Puna — snow. Scanty vegetation — I should think — 8000–10000 feet high', p. 15b. There was, however, some zoological interest with 'Several Foxes live on mice — tracks of Vicunna' and partridges which 'in Beveys rise like Grouse & make great chattering', p. 12b, probably reminding Darwin of his pre-*Beagle* sporting days. He referred to the behaviour of the 'partridges' in his 'Ornithological notes', p. 260, under the description of specimen 2823: 'I saw a covey of five rise together. On the wing the[y] ~~made~~ uttered much noise & flew like grouse: were wild: are said never to descent to the lower Cordillera. — Coquimbo'

In *Birds*, p. 117, Darwin's words were printed in only slightly expanded form under the heading *Attagis gayii* (Rufous-bellied Seedsnipe), but there he noted that the Copiapò 'grouse' takes the place of the ptarmigan of the northern hemisphere. They are, therefore, examples of representative species: ones related and filling similar ecological niches in different places, in this case in opposite hemispheres.

That night Darwin had the terrible experience of a 'gale of freezing wind — body benumbed', p. 14b, and he recounted how his vaqueano had as a boy nearly died in the same mountains. The vaqueano's fingers were permanently damaged on that occasion and his brother had frozen to death, along with a large number of mules and cattle. The discursive style of these entries suggests that Darwin was comfortable when he wrote them, rather than in the saddle.

The geology on the 28th continued to be complex. Darwin found a *Terebratula* which provided valuable correlation with those from the Rio Claro, p. 15b. These occurred 'with Gryphite' which could be among the specimens from Copiapò figured on plate 5 of *South America*. Darwin's specimen from the 'grand Tufaceous formation' brought the expedition tally to 173. This rock, he was convinced, was an aqueous deposit as was, in his view, 'the whole top of the Cordilleras' giving 'proof of horizontal upheaval to that amount', p. 19b. At the foot of the mountain was a salt lake 'many leagues long' which dried out in summer.

The entry for 29 June is much shorter but includes a full-page sketch apparently showing how the 'Tufa plain was divided by arms of the sea', pp. 20b–21b. Darwin also noted '(I believe the Carrancha & Chimango are found here) — Certainly the former.' By p. 26b Darwin's opinion on this point was more confident: 'NB. Have not seen the Chimango since leaving Coquimbo'.

By this point the party was turning back to Copiapò and camped that night where they had camped on the 27th. On Tuesday 30 June Darwin collected more specimens taking the tally to 183 and he noted that 'on the rocks no Lichens'. There is an intriguing geological section on p. 26b which is discussed on the previous page. This diagram appears to be closely related to the one on the inside front cover which Darwin referred to on p. 7b and was eventually published in *South America*, p. 231, as fig. 24 (see p. 514 above). Darwin wondered about the enormous thickness of the porphyritic breccia: 'I think above P.B. at least 7000 ft thickness?'

He began his 1 July entry thoughtfully: 'I am surprised on reflection how very few cases have I seen where I can imagine the volcano for all these subaqueous Lavas to have existed', p. 28b. The next two pages are excised, but the geology continues on p. 31b with specimens listed to 185. By p. 32b Darwin 'returned to Valley of Copiapo — smell delicious of the clover fields after the desert journey'. He described the small mud huts at Punta Gorda which was 'at least 7–9 miles from the nearest water' and was obviously baffled by stories of Indians living in the desert mountains. This led to a conjecture, pp. 34b–35b, about climate change and a musing on the smugness of the Christians: 'Only one idea to feed Alpacas — but failing water — I cannot understand — & excessive cold — Has there been a change of climate. I believe stone walls remain for ever: Mem. Druidical mounds —

Houses top of Cerro Bravo — People say they are animals — not being Christians explains anything'.

On 2–3 July Darwin 'Staid in Copiapò', p. 35b. According to the *Beagle diary* this was with Mr Bingley. On the 4th Darwin 'Set out for the Port', p. 36b, and observed that the soil was unproductive due to its high salt content. Darwin seems to refer to an analysis of this salt by an apothecary named Hull.

Darwin reached the port of Caldera on the 5th and found the *Beagle* had arrived on the 3rd, but with Wickham in command, temporarily covering for FitzRoy. That evening Darwin 'gave my "adios" with a hearty goodwill to my companion, Mariano Gonzales, with whom I had ridden so many leagues in Chili.'[961] Gonzales, unlike Syms Covington, was later referred to by name in *Journal of researches*. Presumably Darwin considered Gonzales a local authority and Covington as only a paid assistant.

On the 6th Darwin 'Examined cliffs near the Port' he found masses of shells and 'some large bones'. There were various rocks including 'enormous blocks of granite', p. 41b. On these pages Darwin retrospectively inserted geological specimen numbers 2846–9. No doubt he was relieved at last not to have to carry his specimens too far from the *Beagle*. They sailed north at midday, leaving Chile for the last time and arriving at Iquique, Peru on the evening of 12 July. Darwin may have spent the week at sea arranging his specimens and writing up his notes.

Iquique, July 1835

The *Beagle* stayed at Iquique for three days, but Peru was in political turmoil and Darwin could not do much apart from a two-day visit to a saltpetre works. On p. 44b he noted various pieces of intelligence, including presumably an unspecified person's opinion that 'Absence of Volcanos in Copiapo & Guasco account for frequent Earthquakes'. There are some entries at the front of the notebook which refer to Iquique and on p. 2a Darwin mentioned the 'revolution', the high price of water and the fact that firewood had to be imported. He also referred to the two small mining villages of S. Rosa and Guantajaya which he saw on his way to the saltpetre works.

Callao, July 1835

The *Beagle* sailed north again on the evening of 14 July 1835. This is perhaps the only case of the same date for two different years occurring in the same notebook, as there is an entry for 14 July 1836 on p. 85b. The next date recorded in the notebook is 19 July 1835 'at night outside of Callao', the port of Lima. The passage was short and steady, owing to the south-easterly trade winds which would carry the little ship up to the Galapagos in September.

Darwin recorded on 20 July that they 'Swept in — miserable Callao, Soldiers — green country', p. 44b. He was not impressed, but the next day he went for a walk on the Island of San Lorenzo. His cryptic notebook descriptions 'cold. Drizzle = Callao — flat roofs — Heap of corn — fruits — splendid Castle' are greatly

---

961 *Beagle diary*, p. 344.

expanded in the *Beagle diary* in entries that were obviously written up weeks later. Around this time Darwin put the *Despoblado notebook* aside and entered his field notes for Lima in the *Galapagos notebook*.

There is a mention of the churning of sediment on the beach at Callao in an entry made on 14 July 1836 on p. 86b. This links to two references to Callao in the *Red notebook*, the first referring to the erosion of the cliffs on San Lorenzo, p. 40e, and the second being the 'most manifest example of degradation I ever saw on beach near Callao', p. 95. These entries were, therefore, probably made about a year after Darwin's sojourn in Peru.

Mauritius, April–May 1836

On p. 46b Darwin made a few notes about earthquakes and tsunamis, then on the next page started to record corals and other material taken during depth soundings, which was a major interest to him, as discussed in the introduction to the *Santiago notebook*. These entries, which extend for five pages, are not dated. Immediately following, on p. 52b, Darwin wrote 'Madagascar — African coast ships' recording his first glimpse of the African continent. It is clear from Covington's *Journal* that this sighting was on 13 May 1836.[962]

The sounding notes were made at Mauritius. In *Coral reefs*, p. 80, Darwin recounted 'I sounded with the wide bell-shaped lead which Capt. FitzRoy used at Keeling Island, but my examination of the bottom was confined to a few miles of coast (between Port Louis and Tomb Bay) on the leeward side of the island.' There is a strong correlation between the notebook lists and the published account for Mauritius. There are two particularly telling notes on pp. 48b–49b. Here Darwin recorded, firstly, at 15 fathoms 'splendid Astrea', and in *Coral reefs* off Mauritius 'twice at the depth of 15 fathoms, the arming was marked with a clean impression of an Astraea', and secondly the '*[immense]* Caryophyllia' at 30 fathoms is almost certainly the 'large Caryophyllia' at the same depth off Mauritius.[963] On p. 50b he referred to the 'harbour mouth', which Keeling lacks. Darwin was making his sounding notes on Mauritius about the same time he expressed his fascination with coral reefs to his sister Caroline in his letter dated 29 April 1836.[964] Two of the mountains on Mauritius which Darwin climbed, La Puce and (if the identification of 'P.B.' is correct) Peter Botts, are mentioned on the stub of p. 7a. Since Darwin's other observations on the Island are in the *Sydney notebook*, there seems to be no notebook coverage for Keeling. Intriguingly, this might be linked to the lack of any extant notebooks for Hobart Town, Tasmania or King George's Sound, whereas there are field notes of various descriptions preserved at Cambridge University

962 Young 1995.
963 *Coral reefs*, pp. 81–2.
964 CCD1: 494.

Library for all three of these consecutive locations from January to mid April 1836.[965]

Cape of Good Hope, June 1836

The next entry marks the sighting around 13 May of Madagascar on p. 52b and arrival in Simon's Town, Cape of Good Hope, on 31 May 1836.[966] Darwin's first note was probably made afloat in the Bay: '2 or 300 white houses scattered along beach backed by barren wall of rocks'. In the *Red notebook* he reminded himself to check if the depth was '70 fathoms 20 miles from the shore?'[967]

The next entry is 'Wednesday morning' (i.e. 1 June 1836) but actually records Darwin's trip by gig to Cape Town in the afternoon. The *Beagle diary* account expands on the trees and scenery, but lacks the delightful notebook simile of the white Wijnberg [Wynberg] houses 'as from a town dropped in country'. Darwin 'arrived after dark', p. 53b, and it was difficult finding quarters because most beds were taken by passengers from some ships which arrived that day from India. On 2 June he 'wandered & rode about town' and was impressed by the extraordinary bullock wagons and by the 'really very splendid wall of the well known table mountain'.

Darwin then commenced his geological account of the Cape which was eventually included in *Volcanic islands*. On notebook pp. 54b–55b he recorded briefly the slates and sandstones of 'Lions hill'. On Saturday 4 June he 'started on a short ride of four days first to the Paarl', p. 55b. Paarl is about 60 km (37 miles) northeast of Cape Town and Darwin arrived that evening with time to ascend the granite hills behind the town which, as he wrote in the *Beagle diary*, reminded him of northern Chile. En route, however, the houses and flowers reminded him of eastern South America and a ferruginous sandstone reminded him of King George's Sound in Australia, pp. 55b, 57b.[968]

The next day Darwin went 'across pass of French Hoeck [Franschhoek]', a 'considerable work', where the Dutch were hospitable, 'but [did] not like the English', p. 58b. Darwin likened the country to Wales and noted 'Emancipation not popular to any people, yet will answer'. To judge from a series of sketches, he was intrigued by the bare white sandstone, by lines of faults dipping *en echelon* south-eastwards and by 'very remarkable stratification'. He arrived at the toll bar at one o'clock in the afternoon and took lodgings at 'Comfortable Mr Holms house'.

On 6 June Darwin crossed over the irregular hills to join Sir Lowry Cole's pass, on the east side of False Bay. The entry on pp. 63b–64b is probably one of the most

---

965 Armstrong 1985, Armstrong 1991b; Banks and Leaman 1999.

966 See Armstrong 1991a for a superb account of CD's three weeks at the Cape, based partly on entries in the *Despoblado notebook*.

967 *Red notebook*, p. 15e.

968 CD's observations on King George's Sound are analysed in great detail in Armstrong 1985.

powerful descriptive passages in any of the *Beagle* notebooks and is far more vivid than the *Beagle diary* account:

> perfect chaos = country very desolate solitary mountainous, few animals, farm houses in valleys — no trees, wild deer large white vultures like Condors — Band of mountains. When we arrived in evening at Mr Gadney's found party of five men Boer — runaway rascals spirited fellows — long guns. leathern breeches, poor horses = grey, rocky tame mountains, most monotonous ride

On 7 June there was a 'hot wind, gale from N[orth]'. Darwin described the sandstones sitting above granite, with slates at the Lions Rump which he compared to those he saw previously at Lions Hill, p. 65b. He described how the slate 'becomes compact & homogenous' as it approaches the granite 'then in parts white spots appear Feldspathic which the weathering makes rock cellular, then granular & black micaceous fine grained mica in spots'.[969]

On p. 68b Darwin noted 'Dr Smyth — Karroo Clay Slate' with a small sketch and on the next few pages refers to Dr Andrew Smith who showed Darwin a series of exposures of the granite/slate junction. These revealed features such as 'ghosts' of the slate within the granite which convinced Darwin that the granite was injected in a liquid state, a view not universally accepted at the time. He also traced the transition from slate into gneiss at the Lions Rump and decided that the foliation of the gneiss was of metamorphic origin and did not necessarily indicate the original sedimentary layering. Darwin declared this view forcefully in *South America*, p. 165, and specifically mentioned Sedgwick and Lyell as geologists from whom he differed on this point. Darwin's analysis of the junction between the deeply weathered slates and the granite at this important site is discussed with reference to his specimens in Herbert 2005, pp. 126–8.

On p. 72b Darwin seems to have estimated the thickness of the sandstones as a staggering 12,000 feet but this is clearly a mistake for 1,200, 2,100 or 2,000 feet. This great formation, which is part of the mainly Permian and Triassic Karroo System, is astonishingly similar to rocks of the same age across the Atlantic in Brazil. These formations were continuous before Africa and South America were split apart by the creation of the Atlantic Ocean starting some 150 million years ago. Darwin made the trans-Atlantic comparison himself, in the last paragraph of *Volcanic islands*, p. 151: 'Mr. Schomburgk has described a great sandstone formation in Northern Brazil, resting on granite, and resembling to a remarkable degree, in composition and in external form of the land, this formation of the Cape of Good Hope.'

The dates 8–10 June barely figure in the notebook, presumably because Darwin was 'not very well' on the 9th, p. 72b, although it is known that Darwin visited Sir

---

969 This description appeared in clearer form in a footnote to *Volcanic islands*, pp. 148–9.

Thomas Maclear (1794–1879), the Astronomer Royal, on 10 June.[970] There is no record of the 11th but on the 12th there is the entry 'Museum long walk' and on the 13th 'evening Colonel Bell'. This last event was significant because Darwin cited it in his *Autobiography*, p. 107, where he recalled dining with 'Lady Caroline Bell' who, while being a great admirer of Sir John Herschel, the astronomer at that time residing at the Cape, commented 'that he always came into a room as if he knew that his hands were dirty, and that he knew that his wife knew that they were dirty.'[971]

The 14th warrants no entry but on the 15th Darwin noted his visit to Sir John Herschel whose clear exposition of the methods of scientific investigation had so influenced Darwin while he was at Cambridge.[972] It is well known that Herschel's philosophy of science was of profound importance to the way Darwin developed his theoretical positions, firstly his coral reef theory but most importantly his theory of evolution by natural selection. It was also Herschel who coined the phrase 'mystery of mysteries', in a letter from the Cape to Lyell dated 20 February 1836,[973] for the origin of new species, a phrase Darwin used in the opening paragraph of the *Origin of species*. In the *Beagle diary* Darwin stated that his encounter with Herschel 'was the most memorable event which, for a long period, I have had the good fortune to enjoy.' The notebook entry seems to be the only known Darwin manuscript record of the precise date of this event. It is clear from Darwin's letter to Henslow of 9 July that in addition to dining with Sir John he also 'saw him a few times besides'.[974]

On 16 June Darwin 'Returned to ship' and on the 17th went for a 'Walk with Sulivan', p. 73b. On the 18th they 'Sailed — Evening' and commenced the long passage across to England, bound initially for St Helena. There was also a longer term value to Darwin of his stay at the Cape. His lengthy discussions with Andrew Smith, continued back in England, concerning the large mammals of Africa were of great importance in convincing Darwin that sparse vegetation was not the explanation for the extinction of the large South American mammals he discovered. Darwin discussed this at length in his *Journal of researches*.[975]

St Helena, July 1836

The notebook is silent for ten days but records crossing the Tropic of Capricorn on 29 June and the arrival at St Helena on 8 July, p. 74b. At some point a few days after crossing the Tropic the *Beagle* would also cross the Greenwich meridian and would

970 Crompton and Singer 1958.
971 *Autobiography*, p. 107. It seems that CD was mistaken about the Lady's name. According to the *DNB* Sir John Bell's (1782–1876) wife was of Russian extraction and was named after Catherine the Great not Caroline.
972 Herschel 1831.
973 See Wilson 1972, pp. 438–9.
974 CCD1: 500.
975 See Thackeray 1982–3 and introduction to the *St. Fe notebook*.

effectively have circumnavigated the globe. From that time onwards Darwin and his shipmates may have had their thoughts firmly set on returning to England. It was to be 'a sore discomfiture' for Darwin when he realized on 23 July that the *Beagle*, instead of continuing north from Ascension, would make a major detour back to South America.

On p. 74b Darwin described the general aspect of the rock fortress of St Helena and noted how the 'small town' of James Town on the volcanic island's northwest side stretched up the 'little flat valley' behind. He 'Walked up ladder hill — curious' then 'to High Knoll castle or Telegraph' which reminded him of an 'old Welsh castle'. Darwin was obviously impressed by the scenery, with 'deep valleys or naked pinnacles' and the 'little white houses placed in most marvellous position', p. 75b.

At this point it is important to consider the notes on St Helena Darwin made in the front of the notebook. These are on pp. 12a–14a and on the excised and currently missing pp. 43a–44a. Although secondary to the notes on the back pages they are interesting in referring to compass bearings of landscape features. Unfortunately the eight pages following p. 14a are excised. These may well have included field notes from Ascension.

Darwin also made some notes about St Helena in the *Sydney notebook* (pp. 59a–62a) but because those are followed by entries relating to Mauritius they presumably derive from something Darwin was reading about St Helena. Darwin's *Sydney notebook* notes are geological and in particular they refer to subfossil *Bulimus* land snails and the bones and eggs of albatrosses. It seems most likely that the book Darwin was reading when he wrote those notes was Robert Seale's *Geognosy of the island of St Helena* (1834) as Darwin cited Seale's book in reference to finds of albatross bones in *Volcanic islands*, p. 90. It is also very likely that the two men met and discussed these topics, perhaps in England, as Darwin wrote that Seale 'gave me a large collection' of the snails.[976] Darwin was very interested in the destruction of the island's native trees and land-snails by pigs and goats introduced in 1502 and he discussed this subject in *Natural selection*, p. 192.

Darwin's *Despoblado notebook* notes from St Helena complement his exceptionally disjointed *Beagle diary* account which suffers from having lost two pages of insertions, and from an obviously inserted paragraph on the destruction wrought by species introduced to the Island. The notes also complement Darwin's published description in *Journal of researches*, in which he referred to St Helena as 'this little world within itself', and his highly detailed chapter devoted to St Helena in *Volcanic islands*. Chancellor 1990 summarized Darwin's manuscripts relating to St Helena and craters of elevation, a theory Darwin was working on at almost exactly the same time as he was distilling the essentials of his theory of natural selection, in September 1838.

976 *Volcanic islands*, p. 89.

Darwin was always proud of his geological work on St Helena. In his *Autobiography* Darwin singled out his elucidation of the structure of 'certain islands, for instance, St Helena' as among the scientific results of the *Beagle* voyage which gave him 'high satisfaction'. Hearl 1990 gives an excellent short account of Darwin's stay on the Island which is a useful supplement to Quentin Keynes's more general account (reprinted 2005) and that of Armstrong 2004. Hearl was able to track down the sad fate of Robert Seale's 'gigantic model' of the Island which was of considerable interest to Darwin.[977] Hearl informed us (personal communication) that the model, which must have been c. 2 meters across, went from Addiscombe College to the Royal Artillery Museum at the Rotunda in Greenwich (now closed). According to Gosse 1938 the model was broken up around 1930.[978]

On 9 July Darwin 'Obtained Lodgings in country in centre of Is^d near stones throw of Nap[oleon']s grave', p. 75b. At some point during the day he wrote his last voyage letter to Henslow, in which he asked the older man to propose him for Fellowship of the Geological Society.[979] In this letter he also referred to St Helena as 'this little centre of a distinct creation', a phrase which became 'a little centre of creation' in his geological diary (DAR 38.920). At this stage in the voyage Darwin was acutely interested in island biogeography. He was probably anxious to test Lyell's speculations in the second volume of the *Principles of geology* that St Helena would have acted as 'focus of creative force'. It was also at almost exactly this time that Darwin wrote the famous 'Ornithological notes' passage about the relationships between organisms on oceanic islands which appeared to 'undermine the stability of Species.'[980]

On seeing how close Napoleon's grave was to the road Darwin was struck by all the 'bombast & nonsense' he heard about it, which he dismissed as 'sublime & ridiculous', pp. 75b–76b. Keynes 1979, p. 359, reproduces Syms Covington's sketch of the tomb. The next day 'Things improved' and Darwin 'hired guide 55 years old — feet like iron — mulatto'. Together they 'Walked to Flagstaff [Hill] along race course' and 'Passed long wood', the house in which Napoleon 'really lived & died very poor'. Darwin was appalled by the 'Walls scored with names of Seamen & Merchant captains'. In a barely legible entry he seems to state that 'It appeared degradation, like profaning old castle', p. 77b.

On 11 July Darwin 'Walked Bencoulen Plain down to Prosperous Bay' on the Island's northeast coast. He thought it absurd that there were so many guard houses

977 Chancellor 1990.
978 We are grateful to Paul Evans of the Royal Artillery Library, Woolwich, for checking the Rotunda catalogues and for confirming that the model was there from 1862 to 1930. Sadly all the other records were destroyed by fire bombs in 1940.
979 See CCD1: 500.
980 See the introduction to the *Galapagos notebook*.

and forts, now that 'Prosperous' was no longer an appropriate epithet. The references on pp. 78b and 83b to flocks of terns 193 km (120 miles) out at sea are clearly later insertions, as is clear from the expansion of these references in the 'Ornithological notes', p. 265, that Darwin saw these terns off Bahia sometime in the first week of August. On p. 78b he drew a delightful 'pin' drawing of a man (presumably himself) standing on the windy edge of the cliff. Although the air was still at his head, reaching over the edge of the cliff revealed a strong gush of wind. Darwin remarked in the *Beagle diary* that the cliff was about 300 meters high. Barlow 1945, p. 255, stated that her father, Darwin's son Horace (1851–1928), demonstrated this wind effect to her as a child, suggesting that Darwin had shown his own children what he experienced on St Helena.

There is no entry for 12 July but on the next day Darwin visited 'Lott & Lotts Wife' on the south side of the Island. These are both odd-shaped pinnacles of phonolite which Darwin described in *Volcanic islands*, p. 85. It is clear from the notebook entry that Darwin could see that they were the upstanding remnants of an east-north-east trending dyke which he could trace as far as 'Man of Wars Birds Roost', p. 79b. His notes show that he was trying to understand the geological structure of the Island. Was it a group of volcanoes, or one big volcano? His mature view, sketched out in *Journal of researches* and later detailed in *Volcanic islands*, was that it was an example, comparable to Mauritius and St Jago (Cape de Verds), of an elevated volcanic island where the outer rim was being uplifted more than the centre. St Helena, Darwin thought, is the northern part of a great crater, the southern half having been 'removed by the waves of the sea'.[981]

On p. 81b Darwin drew a section through a trap dyke truncated by erosion at the top, then covered by more recent rock, thus demonstrating the long intervals of time between eruptions, which he thought were 'probably terrestrial'. This sketch was worked up via a version in DAR 38.931 and eventually published as fig. 10 of *Volcanic islands*, p. 82.

On p. 86b he drew another sketch linked to notes about the 'effect of prolonged elevation' on the evolution of cliffs, a subject he developed in the *Red notebook* and in *Notebook A*.[982] He also drew a sketch showing how dykes seem to 'stretch' a volcano, a theme he discussed with reference to St Helena in the *Red notebook* (e.g. p. 59). The diagram was reworked in a draft version in DAR 38.929 to become fig. 8 in Darwin's discussion of the role of dykes in volcanic districts in *Volcanic islands*, p. 76.

981 *Journal of researches*, p. 581.
982 Chancellor 1990.

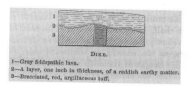

'Dike.' from St Helena. Fig. 10
from *Volcanic islands*, p. 82.

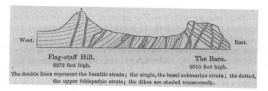

Crater on St Helena. 'The double lines represent the
basaltic strata; the single, the basal submarine strata; the
dotted, the upper feldspathic strata; the dikes are shaded
transversely.' Fig. 8 from *Volcanic islands*, p. 76.

Darwin made two comparisons between what he saw on St Helena with
what he saw earlier in the Galapagos. On p. 85b he drew what might be a
sketch of Lott's Wife 'like at Galapagos' and on p. 86b he seems to indicate
that at both places sand may be derived from shells which over time lost all
visible trace of their structure. He then simply noted 'Sailed 14th', as the *Beagle*
left St Helena, followed by some jottings on crops he saw at Sir William Doveton's
house, perhaps indicating the source of Darwin's interest in the introduced
plants of St Helena. He also noted that English game laws applied because there
were so many introduced game birds; p. 86b ends with mention of the beach at
Callao.

Pernambuco, August 1836

The *Beagle* continued north-westwards to Ascension where she arrived on 19 July
1836, but this was a Tuesday so the next note 'Arrived Friday', p. 87b, must refer
to arrival at Pernambuco on 12 August, as does the next extant note in the front of
the notebook on p. 23a. The *Beagle* was at Ascension until 23 July, had turned
west-south-westwards to Bahia on the Brazilian coast, arriving on 1 August and
then set off for Pernambuco on the 6th. There are, therefore, no field notes in the
Despoblado notebook from Ascension or Bahia (the latter are on loose sheets in
DAR 38.954–5).

The passage to Pernambuco was unpleasant 'heavy rains, winds easterly', p. 87b,
and Darwin was disenchanted by the 'filthy narrow streets' and the 'filthy old Hags
of Nuns'. According to the *Beagle diary*, however, he enjoyed his canoe trip to 'old
Olinda'.

On p. 88b Darwin referred to Lyell's *Principles of geology* 'Vol. I P. 314 Refers
to Hoff about T del Fuego & Patagonia', this was a cross reference to vol. 2,
p. 476 of the massive compilation of geological facts in von Hoff 1822–4. Lyell
intended to define the southern limits of the Andes, and numbered von Hoff
among those who conceived the Andes 'to extend into Tierra del Fuego and
Patagonia'. It is by no means obvious why Darwin made this reference at this
point in the notebook.

The last two back pages include mention of shelly sandstone at Pernambuco indicating 'elevation possible', p. 90b, and 'Mangroves like rank grass in Church Yard', p. 91b. In the *Beagle diary*, p. 435, this became: 'The bright green colour of these [mangrove] bushes always reminds me of the rank grass in a Church-yard: both are nourished by putrid exhalations; the one speaks of death past, the other, too often, of death to come.'

On p. 23a Darwin recorded the various types of organisms at different depths 'Off Pernambuco'. He summarised this information in the *Beagle diary*, noting that the sandy bar through which ships had to pass did not support true corals. His description of the bar was published separately.[983]

Cape de Verds, September 1836

The *Beagle* left Brazil on 17 August and crossed the equator on the 21st. The next note in the *Despoblado notebook* is 'arrived on 31st', p. 24a, which must refer to the *Beagle*'s landfall at the Cape de Verds on 31 August 1836. Darwin noted 'No kingfishers at present. Sparrow building — land rather greener' as a result of recent rain, in contrast to the arid conditions on his first visit to the archipelago in January 1832. Darwin finally noted some bearings and distances. He left St Jago on 4 September, heading for the Azores, arriving on the 20th and leaving on the 24th. His next stop was England.

---

983 Darwin 1841; *Shorter publications*, pp. 137–9, reprinted in *Coral reefs* 2d ed. (1874), p. 265.

[FRONT COVER]

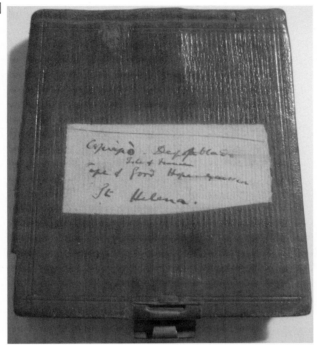

Copiapò — Despoblado
Isle of France
Cape of Good Hope excursion
St Helena.

[INSIDE FRONT COVER]
*[illeg]*y 26–27
Luccocks Notes of R de Janeiro & (M. Video) 4to 1820[984]

984 Luccock 1820.

[1a]

Limit of Basin Thinly stratified beds P. B. P. B.[985]

[2a] *[illeg]* vessel
revolution
Price of water 18 — P.B. 1 B — ½ m. — no wood. —
Sand — detritus. absolute desert S Rosa — Guantajaya state more curious Tufaceous
deposit like lake

[3a–4a excised]

[5a] Mem copy notes in Humboldt Vol V P I[986] —
V. Beechey
[N.B.] The Barking <*bird*>[987]

[6a] Bits of dead coral volcanic pebbles.
Flagstaff & La Puce
15° 50'

[7a] P. B. & La Puce 13.1

[8a–10a largely excised]

[11a] 57 (3)
bright pink granular centre envelope distinct
rather smaller than orifice their orifice of cell
8 or 12 perhaps in a circle in flocculent matter[988]

---

985 This diagram was eventually published in *South America*, p. 231.
986 Humboldt 1814–29, vol. 6.
987 See specimen 2532 in *Zoology notes*, p. 279.
988 Notes on dissecting coralline algae under the microscope, specimen 3686 collected at the Cape of
      Good Hope in June 1836, *Zoology notes*, p. 51, and *Beagle plants*, pp. 199–200.

[12a] Holdfast Tom & Prosperous Bay. between these a great level basin, with track near
coast of ossiferous soil, sea birds, not now inhabitants of beach.
Dianas peak is part of the Basaltic chain.

[13a] Flagstaff & the Barn parts of the Crater.
Lott worth visiting /point/ of basalt though stratified rocks

[14a]

Dianas peak
Sandy Bay
Lotts Pillar &
Wife

Flagstaff Hill &
barn
Bencoolen Plain
Prosperous Bay
℉ Iron at Longwood (near Napoleon)? 7 miles from
Gregories Valley
Turk Caps Bay

[15a–22a excised]

[23a] Off Pernambuco.
17 Fathoms essentially red corallina small pieces in continental dead & living; few
minute fragments of shells & opercula
40 fathoms /20/ fathoms very fine grained
minute yellowish brown particles of shells?
~~10~~ 9 Fathoms Corallina & shells rounded articles of quartz rather coarse for sandstone

[24a] V. Plan for kind of bottom close to anchorage
arrived on 31st [August 1836] —
No kingfishers at present. Sparrow building — land rather greener — Boabab full of
leaves = interesting country — /barren/ stony steps

[25a] Centre of Flat hill   N 36 W
right extreme   n 32 W
N by E Hill   n 13 W
SE extreme of table   n 1 E
mountain with horizontal strata N. From ship

[26a] From cove. West of Quail Isd to extreme SE point 3 & ¼.
~~5~~ 4 miles & ⅓ ½ from extreme SE point to where I last could see. the white /beach/ may
be called 8 miles

[27a–42a blank]

[43a] 2 points of flat hill —   280
extreme N escarpment. of distant west hill —   298
Summit of N by E Hill   305

S.E. edge of table land   ~~337~~

                                                            327

extends at least to   334

Right hand flat hill   285

[44a]   Extreme point 250 to Flagstaff. Bluff 48° Extreme low 62°

NB 4 miles distant

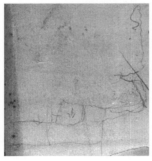

Some other Bearing in Note Book

[45a blank]

[BACK COVER]

[INSIDE BACK COVER]

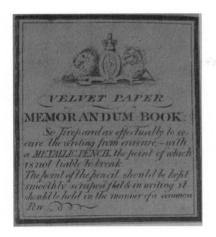

/Mepal/ — Copiapo 405 Nautical miles
/Guantajaya/ /Pasco/ Copiapò */Arqueto/* Mr Caldcleugh
veins NNW & SSE
Carb of copper at top.

[1b]   (Friday 26th.) [June 1835] Started from Copiapo valley to Despoblado (silver spurs —
for burial)
T del Finch 3 sorts of Caracara ||
Vicuna.^{989} as far as near Guasco
Noticed 3 very distinct plains all on grand scale. the lowest ones the grand one of city.
Hills all about Copiapò. the greenstone — Syenite — also some dark slate Porphyry.
Slept at the Agua amarga — putrid. could not drink tea — very severe

[2b]   frost road very desert bring wood three day journey on donkeys. —
(Saturday 27th) [June 1835] Travelled far up ravine of Paypote, came to water, &
trees of Algarroba^{990} one man watching old smelting furnace road joins Pampas
N of Rioca Too much tired to enjoy these situations & perfectly bare hills =
During these two days the geology (our course being about NE — E — SE —
E &c) has been

[3b]   chiefly a repetition of Valle — Leaving the Syenite we chiefly traversed a mass (I have
not attempted a section from extreme complication of stratification) of the upper
strata, reddish fine conglomerate yellowish & greenish Sandstone with bands of

989   The Vicuna (*Lama vicugna*) is the smallest of the llamas and lives at high altitudes in South America.
It was much-prized by the Incas for its soft fur.
990   (=Nimora) a Mimosa plant.

pebbles, some of the black rock, & the greyish limestone full of pieces of Gryphites & Terebratula — The number of green dykes astonishing, viz. standing out—

[4b]

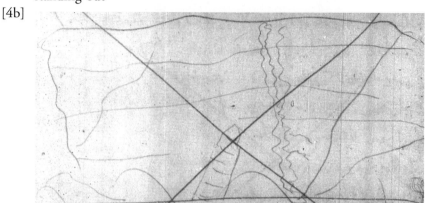

[section]

occasional hills of crystalline rocks one of which was of a curious variety (170) with the sedimentary beds were enormous streams of Lava Lilac & pale &

[5b] very dark Porphyry — also above a white Sandstone the curious Porph in great quantity. = From the quantity of Lava there must have been many volcanoes. surely these isolated cryst hills were such May seem to have little relation as axis with stratification but I think they generally have. — If they are centers of old volcanos, submarine volcano the stream must directly flow out of a hole = these cryst hills frequently stand isolated

[6b] I am surprised the center & stream are not oftener seen, connected — removal & upheaval being on same spot chief cause = Upon some inclined strata, & at corresponding height of 2 diff hills, horizontal mass (like gravel, in height of Potrero Seco) of a very white, very light, Tufa, grains, scales of mica — calcareous — clearly Volcanic —

[7b] interesting as being the 3^d instance of Lava after upheaval of strata, yet beneath probably the sea = from Gravel, nature & obscure horizontal lines = I observed some very remarkable forms of stratification 1st is figured at end of book[991] & is remarkable by the outer & lower rock being a true Porph — Brecc — which I saw no where else — the inner strata curved more than a yoke (U) are of thinly laminated red sandstone & some

[8b] black rock — there are many horizontal Pushes — I never saw this double case but several of both single —

991 The sketch is found on p. 1a.

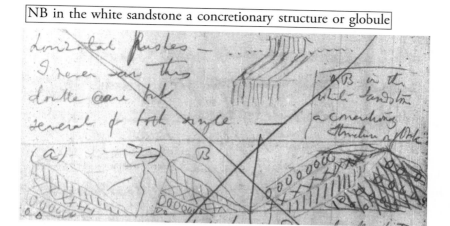

[section]⁹⁹² (a) (z) B

Two little successive hillocks had a double dyke or upheaval — separated by wedge shaped valley the (X) bed a <u>Lilac</u> Porph by a succession of evident faults, is tilted from about

[9b] 45° to vertical & from than till it leans down to an opposite dip of about 45° — which is the head of the little ravine where it rests on continuation of (B) hillock whose strata are also highly inclined, — but I do not know the exact juxtaposition of the strata, when there is an this <u>apparent</u> anticlinal ridge: the x strata being pale & generally denuded shows a curious sinuous band, which may be

[10b] Produced by twisting a slip of paper white sandstone pebbles Porphyries The strata have been twisted 135°. —

The great period of Volcanic activity <u>S of Copiapo</u>. (Mem. Mr Lambert notes) was submarine — also I believe nearly all the hills are likewise so, that the outer ridges are subsequent. — that these upper strata removed so as even to see <u>granite</u> generally

[11b] on coast — then P B — Dykes generally subsequent to metallic veins & alteration of sediments — Valleys — Are the Pampas solely owing to being leeward of currents has the weight prevented subsequent upheavals — Chief of these Lavas before tilt — some subsequent, but subaqueous — others terrestrial to this day —

[12b] Sunday 28th [June 1835] We travelled up to the foot of the Linea — that is a line of mountains which sends water to East to great lake of Salt many leagues long = Partridge⁹⁹³ near snow line in Beveys rise like Grouse & make great chattering. Several foxes live on mice — Tracks of Vicuna. This first Cumbre very tame outline nearly on plain = on the road valleys very flat (in one spot very high up below junction

---

992 See *South America*, p. 231.

993 Probably specimen 2823 in *Zoology notes*, p. 410, and 'Ornithological notes', p. 260; listed as *Attagis gayii* in *Birds*, p. 117.

of /Maricongo/ & /Azufre/ 2 very distinct plains or fringes above the broard flat base = a <u>large</u> side ravine mouth /barred/ up owing to their being

[13b]   no drainage 20 or 30 ft high — water course only just perceptible in soft flat bottom, ~~base~~ sides of soft, immense crumbling mountains not scored by ravines; quite bare — singular aspect of tranquility — All owing to extreme dryness. — Specimen (171) is part of injected mountain; same rock forms immense dykes 10–15 yards wide traversing mountains also a brecciated mass like a Lava, & on upper part of mountain obscure appearance of being like a bed — This hill

[14b]   belongs to system of Porphyry hills (170). — (172) a red Porph <u>Lava</u> = The night dreadfully cold, gale of freezing wind — body benumbed — History of Vaqueano — boy 14 years old = gale of wind — May — clear sky — no snow — stones flying — stoop to ride = out of 200 mules 14 only lived — 30 cows died — Brother tried to return on foot, body by a /led/ mule skin — clothes — hair — 2 years afterwards other brother lost several fingers & toes. this man could scarcely use his, dreadfully

[15b]   swelled. — returned with fresh men to look after Cargo. in return snowed up 3 days with usual clothes No hunger or thirst — yet lost no fingers up to chin = crouched round fire = In evening climbed up to near Cumbre. very much Puna — snow. scanty vegetation — I should think — 8000–10000 feet high = at the very head of the valley on inclined layers of whitish sandstone & limestone Terebratula (with Gryphite — /R Claro/ Terebrat) there is a grand Tufaceous formation many 100 ft thick — this can be traced from one mountain to other

[16b]   forming mere caps to the first mentioned. — It is horizontal & divided into two masses, the lower reddish (173) much harder included fragments smaller — the upper is ~~174~~ quite white, soft crumbling (hence in places are only left groups of pinnacles) contains much mica — grains of quartz, & bits of granite — Mem. Mr Lambert glazed granite, together with bits of Porphyry, but 19/20

[17b]   are ∠^r fragments, chiefly but not all, small grey Trachytic Lava — In upper parts there was some rounded stones — The plain seems extensive, but I am informed on road to the other side, there is an immediate descent, & continual ascent, & descent till we come to other side, the plain dips to W & small ∠ or rather the little bits of Caps still left — The white Tufa in parts seems highly calcareous like that of yester

[18b]   day, before yesterday & brecciated — (is covered by bed ~~several~~ 200 ft thick of some darker substance, not Lava) <u>horizontally</u> junction very regular, cannot doubt aqueous deposit = ~~product~~ of sea subsequent to upheaval & amongst outlying islands —

Having crossed first line proceeding to north about 15 leagues, over dead level plain (with very much Puna), this plain covered with shingle — This & the

[19b]   aqueous Tufa (although $\angle^r$ fragments) NB did not find the source of
        Trachytic Lava but probably the red rounded hills which rose in places, the whole
        top of the Cordilleras are here a recent aqueous deposit, being horizontal, proof of
        horizontal upheaval to that amount In this plain at foot of a mountain is a grand
        Lake; many leagues long, not very deep, but much mud — Salt deposited Snow
        water, dried in summer, deposited in layers very hard — mules enter by a
        Peninsula

[20b]   Monday 29th) [June 1835] Returned to same sleeping place — (I believe the
        Carrancha & Chimango are found here) — Certainly the former. — The Plain of Salina
        is surrounded by mountains, but very broad, & nearly two days journey long — may be
        Lake — (Mem Lake South Pampas of Mendoza) It is manifest that this Tufa plain was
        divided by arms of the sea — because main valley has steps, hence if divided *[surely]*
        deposited —

[21b]

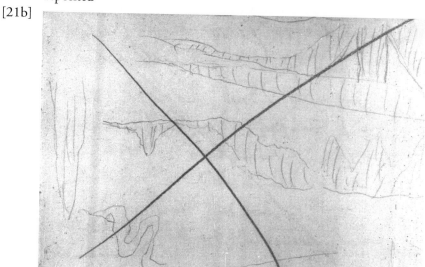

                                                                               [dykes?]

[22b]   ~~Mon~~ (Tuesday 30th) [June 1835] (174) Tufa, from far to ~~East~~ West near junction
        of San Andres. — 175 — 176. 177 — Porph Lava 1st most abundant = 178–79 —
        Porphyritic Brecc Little above San Andrès = 180 green dykes — 181 — laminated red
        Sandstone? 182 — injected mass with do?? (183) Grand Conglomerate resulting from
        Granite rocks = I see the Tufa formation extends both side

[23b]   of great valley = The system of dykes figured (formerly) is the most extensive I have
        any where seen (on the rocks no Lichens) (180). — Traverse a thickness of many
        thousand feet = *[are]* from hillocks of same sort of rock = I found a spot where vertical
        strata allowed them to be measured — Red sandstone above Porphy Brecc Strata
        400 y^{ds} — 300

[24b] of white Sandstone (all thinly laminated) 200 y^{ds} of purple /sonorous/ thinly laminated
rock, the nature of which not certain (180), intimately connected perhaps injected
globular curious rock (181) — Independent of these a little way to W many thousand
feet thick of coarse red conglomerate, in some places

 mon

[25b] almost loose pebbles, in others closely cemented (182). with the dykes fragment of
petrified wood —
The drawing of stratification on adjoining page is singularly complicated it is an oblique
section of one end of a trough-shaped stratification = specimen of Porph Breccia comes
from this spot — not very perfect Porph — fragments blended & ∠^{r} chiefly (NB ∠^{r} n
rounded great difference between P.B. & conglomerate of this country — stratification
wonderfully contorted, not far from here

[26b]

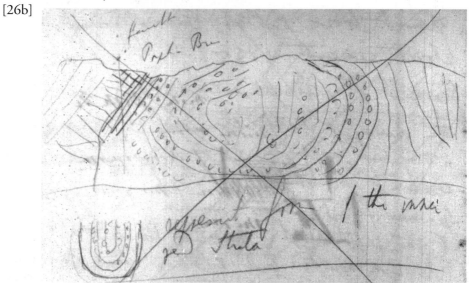

[section]^{994} fault Porph Brecc represent form of the inner red strata
N B. Have not seen the <u>Chimango</u> since leaving Coquimbo

[27b] ~~great irregularity in neighbourhood~~ measurements were made — between the measured
beds & the great coarse red conglomerate, there were 2 or 3000 ft of strata which from

---

994 See the diagram on the inside front cover.

valleys & irregularity in stratification I do not know which system they belong to — I think above P B. at least 7000 ft thickness? I see the conglomerate is stained green for some inches each side of dykes, but not otherwise much altered: — /Bey/

[28b] ~~Monday~~ Wednesday July 1ˢᵗ [1835] I am surprised on reflection how very few cases have I seen where I can imagine the volcano for all these subaqueous Lavas to have existed. — a weathered volcano would probably represent a single cone or various cones of various crystalline rocks separated by an interval from their streams the sides composed of scoriae & ashes being weaker ~~than~~ more exposed than the solid field. —

I saw yesterday (2) amidst a group of Porph hills a mass which sent out on each side a well developed field of Lava which at some distance covered /conformably/

[29b–30b excised]

[31b] a section of about 7000 ft of Lower of P B formation gave. — ⌐1¬ˢᵗ or lowest P Brecc Conglom & Breccia smaller particles blended — larger yet distinct, basis partially crystalline several 100 ft thick — ⌐2¬ Purple porphyry 60 ft quite distinct (184) ⌐3¬ Porph Brecc great thickness ⌐3 4¬ Greenish Porph (185) 80 ft (5) slightly crimson, position with the mechanical rock quite evident. (5). Porph Brecc or conglomerate 30 ft

(6) The same Porph (185) about 30 ft — I believe covered again by the dull blackish slightly purplish Conglom

[32b] I returned to Valley of Copiapo — smell delicious of the clover fields after the desert journey:

At P Gorda at least 8 7–9 miles from the nearest water & that in very small quantity, & Indians had no animals — land quite desert — every two or three years there is a little rain pasture enough for a few weeks (but not enough for Horticulture) = Not a concealed place, now where there are these tiny /Pozos/ are very much concealed — cannot be said to have run away — At this P Gorda 6 or seven little apartments partly constructed of mud

[33b] (& this mud surpasses in hardness present mud /mem/) chiefly stones — figure & some like the huts, in pass of Aconcagua: — Also in front of side ravine, in several places two piles of stones, apparently as a mark = I hear of great numbers of houses in all parts of Cordilleras & not only in the Passes: some between the ranges where there is Puna, desert excepting for a very short time in summer & snow all winter — many of these groups of

[34b] at immense elevation where there is no water — Where at present silver mines could hardly be worked, & that with advantage of having animals — are considered as places of residence Indian corn, bits of woollen threads, Guanaco & I have heard of arrow of Agate precisely like T del Fuego. I hear of these groups of houses in all parts — Valley of Aconcagua /&c/ /Maupas/ &c — /illeg/ Only one idea to feed Alpacas — but failing water — I cannot understand — & excessive cold —

[35b]   Has there been a change of climate. I believe stone walls remain for ever: Mem.
Druidical mounds —
Houses top of Cerro Bravo — People say they are animals — not being Christians
explains anything: —
Thursday 2 — Friday 3^d [July 1835] Staid in Copiapò Mine of Silver — Gold —
Copper — in the Syenite near town — magnetic iron. in the district — Quicksilver
also & Lead =

[36b]   Saturday 4th [July 1835] Set out for Port Valley soon expands large patches friable white
as snow, miserable sort of sterile grass not fit & brownish green — some large
Algarroba — Rocks, which I examined granite — G. syenite — Greenstone & slate —
coloured porph remarkable by the great number of dykes chiefly N & S — & from
hardness forming the /sub/ ridge — so that every

[37b]   part of valley has been highly disturbed & is metalliferous — Valley more level & soft
earth — difference with other Chilean valleys — cause & effect — on side below the
Ramadilla — two plains on the side, the upper one being hard crystalline rock
smoothed down & interstices filled up by numerous pebbles — Perhaps the broad plain

[38b]   of valley is alluvial, at some former period, or from some cause too horizontal for transportal
of gravel — the side plains as before, with gravel. — Salitra — analysis by Apothecary
Hull^{995} — Thick /crumbling/ mass in the lower patches of valley & more or less all over
it — water tasted by it = Mountains remarkable by being covered with sand — like Dunes
even

[39b]   these from 1000 ft to 2000 ft high
Sunday 5th [July 1835] — Saw some Clay <u>Slate</u> or rather siliceous Feldspathic slate &
Hornblendic Slate (2848–49) & again near coast true Granite = small bits of plains with
sandy earth hence perhaps valley is not alluvial = On the plain certainly higher than at
Guasco, the recent Donax, & Venus ( ) & Pecten, in cemented grains, of sand & shells,
like Losa, or a

[40b]   brownish Tosca rock — Beneath this various coloured yellow or greenish sand — seams
of Gypsum — many large pebbles of Granite & Large Oysters = formation very same as
Coquimbo
Monday 6th [July 1835] Examined cliffs near the Port = chiefly a /hardish/ yellow mass
of comminuted shells I believe of the recent period = there were layers of soft white &
even reddish sandstone, & thinly ~~lam~~

[41b]   ~~inated~~ stratified Tripoli (2847) in this very great number, chiefly horizontal veins of
Gypsum = some large bones = The upper part gravel also more decidedly of the
recent period with very many Venus (2846). <u>Pecten</u> Oliva Barnacles & other species
Venus (& whole Barnacle) — In this fine sediment enormous blocks of granite 2 yards
long 1 — wide 4 ft high. From the ship a little way to

995 Possibly James G. Hull, resident of Santiago, Chile.

[42b] north the Strata appeared inclined as if tilted — (Resting the Strata on the worn surface
of the granite — *[234]*
Iquique 20° 13 – S
Sailed from Copiapo 6th [July 1835] — reached Iquique 12th
Sailed 15th [July 1835]

[43b] 13th

Sea Beach	30.382 —	T 64. —
First Plain	28.450 —	T 65 –
Salt Petre	27.026	–
~~22~~		T 44
Iquique	30.360	T 63

Pintado[996] at Iquique
Sailed from Lima 7th of September [1835]

[44b] I hear of the Crater of extinct Volcano — Hot water — Sulphur — Smoke — near
Iquique & two in Atacama. —
Absence of Volcanos in Copiapo & Guasco account for frequent Earthquakes
Probable steps same height in diff valleys

[45b] 19th [July 1835] at night outside of Callao = short — *[cloudy]* passage —
20th [July 1835] Swept in — miserable Callao, Soldiers — green country
21st [July 1835] Isd of St Lorenzo — not so desert in the region *[cloud]* — much
[Amancaes][997] — several other plants —
cold. Drizzle = Callao — flat roofs — Heap of corn — fruits — splendid Castle
not only low down —

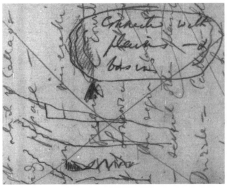

Connected with plains — & basins
Valleys marine — Alluvium
Rise gradual
periods of long repose or greater upheaval

recent period —

---

996 The Cape Petrel, which CD later noted 'often approached close to the stern of the *Beagle* … the
constant attendants on vessels traversing these southern seas.' listed as *Daption capense* in *Birds*,
p. 140.
997 'beautiful yellow lilies', *Beagle diary*, p. 348; see *Beagle plants*, p. 181.

[46b]   Remarkable coincidence in overthrow of Juan Fernandez & Concepcion in the year
35 + 51 — each time great wave. parallel case to wave on coast of Lisbon & Madeira.
Playfair discussion on valleys — compare Sand Hills /Oyster/ English Harbor.
Salt in Guasco or at Copiapo — Extent of Tosca

[47b]   6 7 Coral, few grains of broken sh[ell]
Sand chiefly, with some Coral, many soundings
12 Fathom Lamelliform
C. Madrepore[998] seriatopora no doubt living upright
15 F. dead Seriatopora & a /mark/ of Lamelliform

[48b]   20 ½ Fathom — Madrepore of the stony kind — mark of Lamelliform
When the coral luxuriant arming quite clean

14 Fathom fine grained astrea[999] or   [corals]
15 splendid astrea /5/ /dead/ & frequently dead

[49b]   Seriatopora

27	all sand
30	coarse do
20	living Seriatopora[1000]
18	do.
20	do.
30	coral probablyliving
xx a /immense/ Caryophillia	

[50b]

33	F fine sand
24	F do
20	Seriatopora
18	do & mark of some (III)? Stony coral
18	Seriatopora

    11   fathom Lamelliform on /sound/ to ship & therefore nearer harbour mouth

[51b]   Lead nearly 4 inches in diameter. I struck the ground many times which if sand would
not effect the lead, but would if Coral. Diameter of Lead nearly 4 inches.
30 xx) A large patch arming indented & fluted: clean:
Apparent /in/ tank, /sandy/

998   Reef-building hydrocorals. See *Zoology notes*, p. xiii and p. 310 note, and description of Keeling
specimen 3560, p. 307.

999   'Astrea' (CD's spelling in *Coral reefs* is 'Astraea') has no directly equivalent modern generic name,
although it is probably the stony coral *Acanthastrea*. *Zoology notes*, p. 419 identifies CD's Keeling
specimen not in spirits 3608 as 'possibly a *Favia*' i.e. a brain coral.

1000   This may correspond to specimen not in spirits 3633 'Seriatopora, in 20 Fathom water', *Zoology
notes*, p. 419, collected at Mauritius.

[52b] Madagascar — African coast ships
deep /there/ 20–30 fathoms
delay — mon monoceros
Tuesday 31st [May 1836] arrived in evening Simons Town, 2 or 300 white houses
scattered along beach backed by very barren wall of rocks —
Wednesday morning [1 June 1836] after noon took Gig for Cape Town
pleasant drive, succulent plants, heaths, rather nice about Wijnberg oaks, scotch
firs, shady leaves white houses as /from/ a town dropped in country but always
close to road

[53b] nice houses — country naturally very barren great flat with solitary houses as on the
Pampas, separating mountains from island mountains — arrived after dark
2 Thursday [June 1836] wandered & rode about town — lodging houses — churches
streets with trees square. parade backed by /really/ very splendid wall of the well known
table

[54b] mountain — waggon most extraordinary feature 18 bullocks like a field of cattle all
caught, 6 & 8 in /hand/ horses & mules — quite English town. about 15000 in town
Lions hill, Slate /compact/ blue. semicrystalline (calcareous?) covered irregularly by pale
not hard sandstone, in parts laminated & micaceous

[55b] veins of quartz not very compact — dendritic manganese — a kind of cleavage (NW ??)
common to underlying Slate & Sandstone?
Tertiary ferruginous earthy sandstones, unequal structure like K G Sound —
4 Saturday [June 1836] started on a short ride of four days; first to the Paarl; sandy
flat &

[56b] hills with thin pasture, many small flowers Oxalis like M: Video = solitary white houses
like in Pampas more tidy — Evening arrived at sort of boarding house. — ascended
splendid round granite hills behind village, row of all tidy white houses & gardens,
avenues of small oaks & many vineyards. — general color brownish green no trees
pretty well watered — Table mountain very

[57b] level over flat — Western mountains reddish & grey, not fine forms, 3–4000 ft. —
Extraordinary fissures & boulders of Granite, caverns beneath — On road, very
quartzose white sandstones & some do breccias = A concretionary
ferruginous stone very common with cavities, & yellowish. sandy clay — Mem
coincidence of ferruginous stone & casts of trees here & K G Sound

[58b] Sunday 5 [June 1836] Across pass of French Hoeck, considerable work, grumble
at toll bar. the Dutch hospitable but not like the English. Emancipation not popular
to any people, yet will answer — valley something like Wales on E side of pass
wild valley white quartz green grass no trees — solitary — Comfortable Mr Holms
house Toll

[59b] Bar — 1° oclock —
East of pass all mountains ranges short declining towards S End. with faults all dipping
to SSE or SE compass 20°–30° degrees — but on pass, there is a spur sent

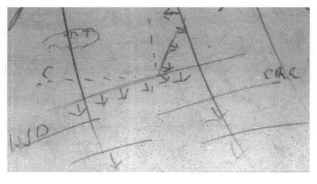

[map]

[60b] off to the North, with dip to E & W this dip bends in a circle at summit of pass till it joins the E & W ranges — very remarkable stratification — intersection of two lines — In the E & W lines strata dipping towards range 60° summit arched & flattened

[sketch of hill] other part

[61b] curvilinear — lower part of sandstones, /siliceous/ slightly ferruginous, sometimes very much so, sometimes white & laminated —
[sketch of strata]
patches of white quartz veins
Above this about 80 ft thick of <u>softer</u> ferruginous
containing /numerous/ angular & semirounded small
pebbles of white quartz & again above rather more compact

[62b] [cliff section]

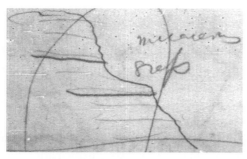

micaceous gneiss

Remarkable Table Mountain horizontal stratification
Monday 6 [June 1836] I saw the E & W ranges South of Caledon, the mountains in the curved road to the Palmiet River is a
& Zonder end do

[63b] perfect chaos = country very desolate solitary mountainous, few animals, farm houses in valleys — no trees, wild deer large white vultures like Condors — Band of mountains.

When we arrived in evening at Mr Gadney's[1001] found party of five men Boer —
runaway rascals spirited fellows —

[64b]　long guns. leathern breeches, poor horses = grey, rocky tame mountains, most
monotonous ride

　　　7th [June 1836] Sir Lowry Coles Pass, fine cut, hot wind, gale from N Sand dunes,
mountainous Sandstone resting on fine grained granite, strata inclined

[65b]　X Clay Slates weathering into sandstone.
Clay Slates Lions Rump dipping pretty regularly to E by N compass at about 45°. —
approaching Granite (same which underlies Lions Head) becomes compact &
/homogenous/ — then in parts

[66b]　white spots appear Feldspathic which the weathering makes rock cellular, then
granular & black micaceous fine grained mica in spots — slate laminated same direction
mean line of junction parellel, running cross country pale siliceous granite (by /chance/)
near junction dykes of do

[67b]　Junction very /extraord/ some hundred yards granite appears veins with /this/ kind of
mica Slate, streaks appear dissolved /long/ gradually increasing with here & there patch
of large grained (as commonly) granite — Many dykes in granite. E & W? /country/

[68b]　horizontal fault, some dykes at ⊥° — (no true Boulders)
Dr Smyth[1002] — Karroo Clay Slate

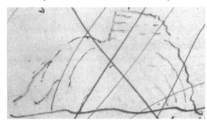 [Karroo Hill? South Africa]

Granite remarkable from weathering. hollow. cavities
Near Green Point patch of sand. coated with Tosca rock

[69b]　impressions of trees, carts like K.G Sound.
On plains of /true/ Tosca rock V Specimen On /old/ Isthmus new Ferruginous
sandstone Smith
Junction of clay Slates & white /siliceous/ Granite
N B. many quartz veins in granite near junction

[70b]　(Z1) nearest
(Z2) a little further
(Z3) considerably further
(Z4) some hundred yards distan/ce/ Slate not altered

1001　William Gadney, a British merchant resident in Cape Town who turned his Sea Point home into a
　　　lodging-house after a business failure.
1002　Andrew Smith (1797–1872), naturalist and explorer stationed in South Africa, 1821–37.

White siliceous granite within junction near to Clay Slate

Walked to Lions Head

[71b] & so to Rump — The Slate is affected to the distance of ¼ of mile, micaceous & compact & granular or dotted, — quite /gneiss/ close to number (Z X) direction invariable, dip to opposite point or vertical, white granite veins — a

[72b] A not very hard, slightly siliceous ferruginous sandstone immediately on granite, & red shale —

Sandstone 12000 ft —

Story of the /Cuentas/

8 Wednesday [June 1836] Coast to ship

9 Thursday [June 1836] not very well

10 Friday [June 1836] long walk

[73b] Sunday [12 June 1836] Museum long walk

13 Monday [June 1836] evening Colonel Bell[1003]

15 [June 1836] — Sir J Herschel[1004]

16 [June 1836] Returned to ship

17 [June 1836] Walk with Sulivan[1005]

18 [June 1836] Sailed. — Evening

Mem at C of Hope SE bearing corrected wrongly

[74b] 29th [June 1836] Crossed Tropic: Light northerly winds arrived on Friday 8th [July 1836] in morning: rock wall round Isd — Volcanic ∴ harbourless, thinly stratified — Fortresses mingled with rocks —

small town. little flat valley, — not magnificent forts — Walked up ladder hill — curious = Walked to High Knoll castle or Telegraph: very picturesque at a distance. like old Welsh castle. = Wonderful contrast decomposed humid rocks, green vegetation. from

[75b] mem rocks of coast; fine scenery, fir grows pretty because not confined — every patch of ground cultivated little white houses placed in most marvellous position

deep valleys or naked pinnacles = All talk English — very poor the poor appear — Saturday. [9 July 1836] Obtained Lodgings in country in centre of Isd near stones throw of Nap[oleon']s grave — no romance cottage & road close by & bombast & nonsense, sublime &

[76b] ridiculous = Mist & wretched cold nice Cottage = Sunday [10 July 1836]

Things improved. — hired guide 55 years old — feet like iron — mulatto.

So many times crossed, that has not disagreeable look — quiet very civil old man, was slave, has 40£ to pay for freedom. How is this? Walked to Flagstaff along race course, elevated plain — Passed long wood, cultivated fields — rather but no very bleak — gentlemans houses — hovel where he really lived & died very poor — I took shelter during heavy rain,

1003 John Bell (1782–1876), army officer.

1004 John Frederick William Herschel (1792–1871), astronomer, mathematician, chemist and philosopher of science.

1005 Bartholemew James Sulivan (1810–90), second Lieutenant on the *Beagle*.

[77b] shutter rattling about Walls scored with names of Seamen & Merchant captains. It appeared degradation, like profaning old castle — Green plain short thin grass — few Syngenesia trees = great red & white Hills called Flagstaff & great Black barn: ~~Lime~~ Uninteresting view: every Govt. path blockaded — such wild little spots, old picket houses. — Wild arid villages — Even near sea Cactus & 2 or 3 plants in very small quantities

[78b] Monday [11 July 1836] Walked Bencoulen Plain down to Prosperous Bay — passed Guard House & fort — 2 soldiers, 2 invalids at old Telegraph House, where army cause appears to me absurd ~~& in~~ Wonderfully protected & guarded. —
Returned across very pretty valley, fir woods — yellow flowering gorge, willow trees & little brook — cottages & small white houses & green rocky hills —

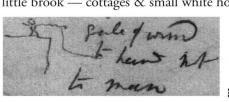 gale of wind to hand not to man

[figure of a man, possibly Darwin, on cliff, St Helena] Mem Tern 120 to sea. — in flock.

[79b] Wednesday [13 July 1836]
Lott & Lotts wife. I believe parts of same ENE dike: = Lott Lower part laminated /curvilinear/ globular — angular — /columno/ — radiated has effected strata on both sides not quite Δ base rather broader 2–300 ft high = linear train of volcanos = dike of nearly ⊥° to this & great isolated peaks. I believe even Man of Wars Birds Roost — /if/ /this constitution/. — Greenstone pale a green /compact/ was reminded of peaks in

[80b] Greenstone countries = Sandy valley chaos of reds yellows /& hollow consist/ chiefly of decayed red scoriae, yellow /wacke/ full of /augite/ crystals cellular lavas & other decomposed kinds many dikes lower parts amygdaloid = great amphitheatre incompletely to S.W. compass = Barn black strata end above, the colored strata dipping to SE by E (Comp) & in front by sandy /bay/ beach, obscure

[81b] seaward dip — to West nothing to be seen = Wall of modern Crater — chiefly decomposed red scoriae & as before mentioned with lower ledge & dipping outwards mem wall — /certainly/ Crater. further west than High peak which /in part/ —

 [dyke on St Helena][1006]

In part on /outside/ /Hence/ probably terrestrial —
Looking round Is^d colored beds

1006 See the fair copy in the geological diary DAR 38.931v; a later version of this sketch appeared in *Volcanic islands*, p. 82.

[82b] in very small proportion on West side; beneath Horse pasture, black bed. quaqua-versal
       dip, whole way to High Knoll Hill, & even to extreme WSW (comp) end of Island —
       Crater both ancient (if such existed) & modern quite vanished to SSW & S (true) —
       High Hill appears mass of greenstone. — great cone —

[83b] also perhaps mass beneath Man & Horse — (N.B. name is Long Range & Not
       Mountain) & ~~other~~ knolls as in Crater —

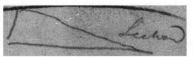 Leeward

Tern 120 miles

action of sea — greatest where high tidal. mem effect on prolonged elevation modern
conglomerate —

Calcareous sand (in parts agglutinated into hard sand from having /vein/ — in patch 12
tion must have contained some calcareous matter. — Blasted because so soft !!

[85b] ENE & WSW Longer axis?

Lotts wife   [Lot's Wife?] like at Galapagos — modern conglomerates
Sailed 14th [July 1836] ≠ Soda plant[1007] Game laws ≠ English vegetation shading into
rocks instead of leafless woods —
Sir W. Doveton[1008] — /bread/, — Fruit, number of plants, Tobacco — /found/ on top
of High Rock proportionate numbers

[86b] Sand at Galapagos —
       Structure of shell sand, sometimes no trace of shells
       Effect of lower mass being so stretched with dikes?? when some of the dike 3 &

400 ft   [perhaps the crater between Flagstaff Hill
and the Barn, St. Helena, showing stretching due to multiple injections][1009]

1007 'Partridges & pheasant are tolerably abundant; the Is[d] is far too English not to be subject to strict
      game laws. I was told of a more unjust sacrifice to such ordinances, than I ever heard of even in
      England: the poor people formerly used to burn a plant which grows on the coast rocks, & export
      soda; — a peremptory order came out to prohibit this practice, giving as a reason, that the Partridges
      would have no where to build!' *Beagle diary*, pp. 429–30.

1008 William Doveton (1753–1843), magistrate and judge. Napoleon famously died after a picnic at
      Doveton's house in 1820.

1009 See the sketch in the geological diary DAR 38.929 and *Volcanic islands*, p. 76 fig. 8.

Anomalous action of sea Southern rocks all buoyed with Kelp yet the Il Defenso rocks like those in the /Hebrides/ bear most clear token of degradation: Water so muddy Callao beach /ratling/ perfect mill. —

[87b] Arrived Friday [12 August 1836] — bad passage, heavy rains, winds Easterly — Pernambuco low green, land, rising a little toward old Olinda Pilot ran in very curious breakwater — Light-house Walked through town, filthy narrow streets, tall houses, prisons — built on mere sand banks /gained from/ sea — Connected by broad

[88b] wooden bridge — population of black & brown. flithy old Hags of Nuns — White men appear as foreigners — A good many horses wild. young Portuguese — Structure of reef, width length

Lyell Vol. I P. 314 Refers to Hoff about T del Fuego & Patagonia[1010]

[89b] Granitic sand. coarse composed of /rounded/ pebbles of quartz distinct horizontal strata — /analog/ with Bahia, sand spotted with red & yellow clayey matter —

layers & lumps of greasy lithomarge upper parts, red alluvium — Water worn

[90b] highest land in country part of plain — round great big yellow limestone becoming arenaceous & aluminous yellow clay No shells — But conglomerate & sandstone with shells in quantity, /from/ neighbourhood recent elevation possible

[91b] Mangroves like rank grass in Church Yard — Swamps /unhealthy/ Olinda Old Town Ill-natured Brazilians

Did the metamorphosed gneiss at Bahia contain dike before metamorphic action?

Textual notes to the *Despoblado notebook*

[IFC] 1.6] *Down House number, not transcribed.*
88202326] *English Heritage number, not transcribed.*

[1a] *page written perpendicular to the spine.*

[5a–10a] *pages partly excised.*

[11a] *page written perpendicular to the spine.*

[IBC] *written perpendicular to the spine.*
12] *added by Nora Barlow, pencil, not transcribed.*

[2b–42b] *pages written perpendicular to the spine.*

[24b] *'Mon'] (apparently a nib test) and sketch in ink.*

[25b] *ten scribbles like 'MM' written in ink and strong scratching, suggesting that CD used the page to test his pen.*

[27b] *there is pink and mauve watercolour paint on this page.*

[31b] *cut near the spine as a result of the excision of 29b–30b by a sharp knife.*

1010 Lyell 1830–3, 1: 314: 'Of these great regions, that of the Andes is one of the best defined. Respecting its southern extremity, we are still in need of more accurate information, some conceiving it to extend into Terra del Fuego and Patagonia.' Lyell referred to Hoff 1822–4, 2:476.

[43b] Sailed … September] *added pencil.*
[44b] *page written perpendicular to the spine.*
[45b] not only … period] *in ink written perpendicular to the spine.*
[46b] *page written perpendicular to spine.*
[51b] |Diameter … inches] *added pencil.*
[52b] mon monoceros] *added ink.*
[67b] Many … [country]] *added pencil.*
[68b] *added heavy pencil.*

THE *SYDNEY NOTEBOOK*

⌐

The *Sydney notebook* is the last of Darwin's *Beagle* field notebooks. It takes its name from the city of Sydney, Australia. The notebook is bound in black leather with the border blind embossed: the hinge of the brass clasp is missing. The front of the notebook has a (60 × 30 mm) label of cream-coloured paper with 'Sydney Mauritius' written in ink. There are 50 leaves or 100 pages in the notebook. The entries were written in two sequences; one covering 92 pages from the front cover, the other eight pages from the back cover. The notebook covers 12 January to 30 April 1836. It is the only notebook Darwin never used before 1836. Darwin put it aside on 27 January for perhaps three months, until May 1836, when the *Beagle* was at Mauritius. The *Sydney notebook* is also unique in being the only one never used in South America, so that the geological observations recorded in it are to be found published only in *Coral reefs* and *Volcanic islands*.

There is no other known notebook which covers any place between Sydney and Mauritius. There seems, therefore, to be a gap in notebook coverage for Hobart (5–16 February), where Darwin turned twenty-seven years of age, King George's Sound (6–13 March) and Keeling (1–11 April). Darwin's field notes for these locations were written on loose sheets now preserved at Cambridge University Library. Darwin used the *Sydney notebook* in Mauritius (29 April–8 May), then the *Despoblado notebook* for the Cape of Good Hope, St Helena, Brazil and the Cape de Verds.

In order to place the *Sydney notebook* in context, it is helpful to reconsider Darwin's activities in the preceding months. The *Beagle* left Tahiti bound for New Zealand on 26 November 1835, giving Darwin the chance to draft his 'Coral Islands' paper.[1011] Gruber and Gruber 1962 have shown how, by seeing coral islands with 'the eye of reason', Darwin created a theory in that paper with strong logical parallels to the theory of natural selection which he formulated in England three years later. Mauritius was to provide him with additional observations of fringing reefs, and he linked these together as a historical series of increasing degrees of subsidence via barrier reefs to atolls. This methodological approach, of

---

1011 See the introduction to the *Santiago notebook*.

integrating a series of observable natural phenomena and then inferring a historical sequence from that series was a keystone of Darwin's argument for descent with modification in the *Origin of species*.

Darwin may have at least started his coral paper between the Galapagos and Tahiti, as his opening line, 'Although I have personally scarcely seen anything of the Coral Islands in the Pacifick Ocean' is unlikely to have been written after just spending a month looking at Pacific coral reefs. It is obvious, however, from statements such as 'Wytootacke (seen by the *Beagle*)' (DAR 41.6), that the bulk of the coral paper was written after sailing through the Cook Islands on 3 December 1835.

The *Beagle* arrived at the Bay of Islands, New Zealand, on 21 December and stayed for nine days.[1012] On 30 December she continued on much the same bearing on the twelve-day voyage to Sydney, where she arrived at Port Jackson on 11 January 1836. The *Sydney notebook* was first used on 16 January, three days before the much-discussed 'antlion' *Beagle diary* passage. Darwin used the notebook for his rather gruelling excursion to Bathurst, whence he returned, putting the notebook aside, on 27 January.

The *Sydney notebook* contains few scientific notes apart from geology and questions about coral growth. There are only a few mentions of the famous Australian mammals and no mention at all of the antipodian 'antlion'.[1013] Of course Darwin knew that the closer the ship took him back to England the less chance there would be for him to collect species new to science, at least on land, so his scientific notes began to thin out. To obtain a rounded picture of Darwin's response to Australia it is, therefore, essential to read the *Beagle diary* alongside the notebook. It is crucial to appreciate, however, that in the remaining months of the voyage Darwin had more and more time to start sifting out what he saw that would break new scientific ground.

While in Australia, Darwin seems to have become rather despondent at the prospect of so many more months before returning to England. In his second from last voyage letter to Henslow, written 28 January, he lapsed into introspection:

> Certainly I never was intended for a traveller; my thoughts are always rambling over past and future scenes; I cannot enjoy the present happiness, for anticipating the future; which is about as foolish as the dog who dropt the real bone for its' shadow.[1014]

The *Beagle* left Sydney on 30 January. As FitzRoy wrote to the Hydrographer on 3 February, poor Darwin was always 'a martyr to confinement and sea-sickness when under way'[1015] and the passage down to Tasmania was no exception.

1012 CD's activities there are very well described by Laurent and Campbell 1987 and by Armstrong 2004.
1013 See Armstrong 2004, p. 170.
1014 CCD1: 484.
1015 Quoted by Francis Darwin 1912.

An overview of the *Sydney notebook* shows that the first two back pages were excised and the remaining six back pages are undated lists of equipment and people's names which were probably written in Sydney and Hobart Town. Field notes begin on p. 1a in Sydney on 16 January 1836 and there are more or less continuous, mainly geological entries up to 27 January on about p. 57a. There follow perhaps eight or nine pages of less coherent, mainly geological entries which are difficult to date but which were probably written on Mauritius. Page 65a seems to date to 5 May and these apparently Mascarene entries continue up to 'sailed 9th [May]' on p. 79a. In the following account the Sydney and Mauritius sections are treated separately, including everything after p. 57a under Mauritius.

The *Sydney notebook* has been well served by scholars.[1016] As discussed in the introduction to the *Santiago notebook*, Darwin may have been using that notebook for theoretical jottings in parallel with the *Sydney notebook* and in addition seems to have opened the *Red notebook* around the time the *Beagle* left Mauritius.[1017] There are some cross references to Sydney and Mauritius in the *Red notebook*, as for example p. 126, on droughts in Sydney, p. 17, on 'Mrs Power at Port Louis', who is mentioned in the *Sydney notebook* entry for 8 May 1836, and pp. 71–2 and pp. 118–20 on the lavas of Mauritius.

Sydney, January 1836

The back pages of the notebook are impossible to date precisely as they consist of lists of names and places, apparently in Australia but also including Valparaiso, together with one of the most detailed lists of equipment in any of the notebooks. They appear to relate to preparations for Darwin's trip to Bathurst. 'Dr Jennerett', on the inside front cover ('Dr Jennerat' on p. 6b) was Henry Jeanneret (1802–86), surgeon, dentist and amateur botanist in Hobart Town, Tasmania. 'Quamby' is a bluff in Tasmania.

On p. 1a the entry begins '16th [January] Saturday Left Sydney'; Darwin commenced his inland expedition to Bathurst, some 190 km (118 miles) away. He noted that there were 'fine trees' but that they were 'all peculiar'. He ate lunch at a 'nice little Public House' which Nicholas and Nicholas 1989 locate at Parramatta, then 'rode on to Emu ferry on the Nepean: a broard river still as a pool', p. 2a. He was impressed by the 'escarpment of Blue mountains' and by the 'beautiful precision'

1016 The Australian section of Chancellor's transcription was published in part by Nicholas and Nicholas 1989 in the first three chapters of their magnificent treatment of CD's thirty-eight days in Australia. Nicholas and Nicholas 1989, p. 23, provide numerous quotations from the notebook and photographs of the front cover and first page. The covers and pp. 1a, 3a, 5a, 7a, 9a and 11a are reproduced as a colour facsimile in van Wyhe 2008a, p. 31. Laurent and Campbell 1987 add much valuable background information about CD's Australian journeys and Armstrong 2004 gives excellent summaries of CD's fieldwork in both Australia and Mauritius.

1017 CD presumably had the *Red notebook* since February 1833 when he began using its twin, the *St. Fe notebook*.

of the 'black men' with their 'throwing darts', p. 3a, and he drew a comparison with the 'Fuegians going to fight some other people', p. 4a. In the next entry Darwin returned to technical descriptions of the clays, sandstones and 'Granitic Trappean rock', p. 5a, then noted 'Black men. See marks of Oppossum's feet. — chief food; no home', p. 6a.

The next page is dated 'Sunday 17th Started 6 o'clock — ferry'. Darwin commenced his ascent of the Blue Mountains, noting the 'Singularly uniform tint', p. 8a. There were 'pretty birds, magnificent parrots' and after lunch at the Weatherboard Inn he walked to see the 'most magnificent. astounding & unique view'.[1018]

The 'stupendous cliffs' of 600 meters (2,000 feet) of white quartzite were so vertical that Darwin could 'pitch a stone over perhaps 800 ft', pp. 9a–10a. His published descriptions are to be found in the last chapter of *Volcanic islands*. As Armstrong 2004 points out, Darwin's interest in elevation distorted his interpretation of the Blue Mountains which he concluded, incorrectly, had formed under the sea. That night he stayed at Black Heath, which was 'comfortable as Welsh Inn', p. 11a, an opinion expressed in many travellers' accounts of the time. Descriptions of the strata continue and on p. 14a Darwin was reminded by a 'patch of shale' of 'Mica Slate in Gneiss at Rio' which he saw almost four years previously.

On 'Monday 18th' Darwin continued westward up the 'Vale of Clwyd [Clwydd]' to Govett's Leap where the 'grand valley' was 'full of blue mist from rising sun' arising from the eucalyptus trees and giving the mountains their name, pp. 16a–17a. He recorded being told a 'bad account of men' who were 'quite impossible to reform'. There were 'white Coccatoos & Crows' and 'wild dogs tamed: copulate freely', p. 18a a theme Darwin returned to in the *Santiago notebook*, p. 130. He noted his somewhat harsh view of Sydney, meaning Australia, as a 'poor country; to be improved but limited' and 'not comparable to N. America', p. 19a.

Darwin continued describing the sandstones and granites of the valley. On p. 23a he predicted that 'The Chart will give correct idea of peninsula & Islands of the grand plain' which were in his view formed not by 'present causes' but he was clearly unsure, as 'Sea could not excavate?', p. 24a. Perhaps the explanation was 'Elevation acted upon by sea?', p. 25a.

On 'Tuesday 19th' Darwin 'Staid at Mr [Andrew] Browne's' sheep farm and went 'Kangeroo hunting'. Although he probably enjoyed the gallop, due to his 'usual ill luck' Darwin 'did not see one'. He did, however, kill a 'Kangeroo Rat' and 'Saw several ornithorynch: like water rats, in movements & habits:' and he 'Shot one', p. 26a. According to the *Beagle diary*, it was on this day that Darwin observed an antlion in its little 'conical pitfall' on a 'sunny bank'.

1018 Nicholas and Nicholas 1989, p. 35 provide a photograph of the 'oak' tree planted at Wentworth Falls on the site of the Inn in 1936 to mark the centenary of CD's visit. They also provide, p. 48, beautiful photographs of John Gould's illustrations of some of the 'magnificent parrots'.

In the *Beagle diary*, pp. 402–3, Darwin considered what a 'Disbeliever' might make of this case of 'double creation', with one antlion in the northern hemisphere and one extremely similar one in the southern:

> Without doubt this predacious Larva belongs to the same genus, but to a different species from the Europæan one. — Now what would the Disbeliever say to this? Would any two workmen ever hit on so beautiful, so simple & yet so artificial a contrivance? I cannot think so. — The one hand has worked over the whole world. A Geologist perhaps would suggest, that the periods of Creation have been distinct & remote, the one from the other; That the Creator rested in his labor.

This passage has been discussed by many scholars as it strongly suggests that by early 1836 Darwin was already thinking that whatever process 'created' species would have to explain representative species in widely separated parts of the world. Darwin had already noted the Copiapò grouse apparently 'representing' the English grouse of his sporting youth.[1019] And there were several other examples recorded in his *Beagle* notebooks.

Darwin was no doubt greatly struck by the peculiarity of the marsupial 'Kangeroo Rat' (*Potorous tridactylus*) and especially by the 'ornithorych', the duck-billed platypus (*Ornithorhynchus anatinus*), an animal so strange that when first sent to London in 1798 it was thought to be a taxidermist's hoax. At the time of Darwin's visit few Europeans credited the Aborigines' belief that the platypus laid eggs and in fact this was not demonstrated in Darwin's lifetime. Darwin might have wondered why such completely different mammals were created in Australia, yet the little antlion in its sandpit was virtually identical to the ones in Europe.

As a student at Christ's College, Cambridge, Darwin was captivated by the logic of William Paley's *Evidences of Christianity* which taught that species were perfectly designed for their environments and that this design was evidence of the Creator.[1020] The possibility, as Lyell might have argued, that two creations had taken place at different epochs, 'that the Creator rested in his labor', would be beside the point for Darwin as it would merely substitute two logically inconsistent events for one.

Perhaps Darwin realized that a single, natural 'creation' of the antlion, followed by some sort of migration, would make more sense to the 'Disbeliever', since there would be no need for logical consistency in a natural creation. From today's perspective we can see the enormous irony of what Darwin subsequently did to Paley's logic. In the *Origin of species* Darwin accepted the adaptation of organisms to their 'conditions of life' but substituted the merciless scrutiny of natural selection for Paley's Creator as the explanation for this adaptation.

1019 See the introduction to the *Despoblado notebook*. See for example Armstrong 2004.
1020 There is a tradition that CD's rooms at Christ's College, Cambridge, were once William Paley's. No College records have been found to substantiate this.

Sadly there is no mention of the antlion in the notebook, which continues with a list of rocks including 'One layer of coal nearly a foot thick', p. 29a, and a 'Blue Slate with impressions of leaves', p. 30a. He assumed the valley was 'modelled by water' but was puzzled by the valley exit being 'a narrow crack a few hundred yards wide, with stupendous vertical sides', p. 32a, so how had the water 'removed the whole mass of rock'?, p. 34a. 'After sea, the lake saw an exit', p. 35a, is clearly Darwin's visualization of the sequence of events, and there was '[countless] time to form so much coal & sandstone', p. 36a.

The next day (20 January) Darwin noted the 'Squatters Huts' and 'Crawlers' at Bathurst. Darwin explained in the *Beagle diary* the distinction between the 'squatters' and the 'crawlers', both varieties of ex-convict. There was a 'hot wind [and] clouds of dust', p. 36a, and 'here & there a good house'. Darwin noted the 'gentleman houses' and the 'Soldiers &c &c', p. 38a, and the prevailing drought 'R. Macquarie just flowing', p. 37a. He 'Was told not to form too high an opinion of Australia' on the basis of what he saw at Bathurst. From this he wrote in the notebook 'my opinion [is] stamped', p. 38a, and in the *Beagle diary* he declared that he felt no danger of forming an overly high opinion on that basis!

Darwin's rock collection reached specimen 12 'Hornblendic Greenstone', p. 39a and 'Mem. Capt King Mica Slate', p. 41a presumably referred to a specimen collected by Philip Parker King (1793–1856), commander of the *Beagle*'s first voyage, perhaps from South America, which Darwin had seen. The two men met on 26 January at King's house at Dunheved. Darwin discerned 'two grand formations' at Bathurst, p. 45a, a 'Primitive' one of granite 'smoothed over with shingle & Diluvial (as would be called) matter', p. 44a.

On 21 January Darwin 'Rode about Bathurst — saw nothing — pleasant mess party', p. 46a. On 22 January he was no doubt anxious to leave the baked town of Bathurst. He started on his way back to Sydney across the 'O'Connel Plains' and 'baited at Midday', that is stopped for lunch at a farmhouse, p. 48a. His hosts were '2 years from England'. The somewhat laconic 'pretty daughter' entry may conceal an impatience to return to the company of English ladies. That evening he arrived amid 'great fires' at another farmhouse where there was 'general civility' but during the night Darwin endured 'horrid filth'.

The next day the fires were still 'raging' and Darwin set out for the Weatherboard, arriving there soon enough and before dark walking again to the Cascade. On p. 50a he noted 'I do not perceive any difference in manners at the Inns from England'. On the 24th Darwin was 'Ill in bed'. As Nicholas and Nicholas 1989, p. 62, suggest, he may have been exhausted from a tortuous journey in temperatures that were in the 40°s every day. On the 25th, however, it was 'cold — great contrast with former weather' and there was 'Quiet drizzly rain: all still dripping from eaves'. Darwin wondered if this change would be 'Perhaps good for me', p. 50a.

On the 26th Darwin went to see 'Capt King' who presented him with a copy of his paper on barnacles and molluscs. The next few notebook entries can certainly be read as a record of some of King's opinions on the geology Darwin saw, for example 'much Quartz rock South of Bathurst — King', p. 52a. The 'pieces of shale' Darwin sketched, p. 53a seem likely to be the 'patches of shale' he compared in *Volcanic islands*, p. 132, to the similar fragment of gneiss from Rio de Janeiro.[1021] He noted the 'current cleavage' of the sandstones, p. 54a which is today called cross-bedding to avoid confusion with metamorphic cleavage. Again, this became a published discussion in *Volcanic islands*.

The next day is the last recorded in the notebook from Sydney. Darwin 'Returned Mac Arthurs', that is rode with King to his brother-in-law Hannibal Macarthur's big house for lunch in Parramatta, p. 54a. There were some more 'nice looking young ladies' there, two of whom subsequently married shipmates of Darwin's. After lunch he rode by himself into Sydney, where he spent the next two days writing to his sister Susan and to Henslow, visiting Conrad Martens and perhaps collecting insects. From Martens he bought two watercolours for three guineas each: 'River Santa Cruz' and 'The *Beagle* in Murray Narrow, *Beagle* Channel'.[1022] On 30 January the *Beagle* weighed anchor and made all sail from Port Jackson for Hobart Town.

Mauritius, May 1836

As explained above, the notebook entries from p. 56a to p. 64a are almost impossible to date and do not relate to Australia. The first reference is to Huafo Island off Chiloe, then the rest of p. 57a and p. 58a are concerned with coral reefs, and seem to indicate that Darwin had reached Mauritius and picked up the notebook again after the long gap for the rest of Australia and Keeling. The *Beagle diary* records that Darwin and Stokes met the surveyor-general Captain Lloyd on Mauritius on 3 May. Since Lloyd was an expert on Panama, the note 'Corals on Panama Coast = Rodriguez' on p. 58a suggests that the meeting, also referred to in the introduction to the *Santiago notebook*, had occurred. That Darwin was keenly interested in the corals of Panama is confirmed by the reference to their absence there, on p. 72a.

Slates and sandstones are mentioned, pp. 58a–59a, then a note about 'Mr Seales — Museum' which is a reference to Robert Seale's Museum on St Helena. As explained in the introduction to the *Despoblado notebook* this note and all the following entries on p. 59a to the start of p. 62a are a response to reading Seale 1834.

The next entry on p. 62a relates to Ascension and Fernando Noronha and Darwin seems to have thought that what he was reading was 'precious nonsense'. Obviously Darwin was familiar with the geology of Fernando Noronha and declared the

---

1021 See the introduction to the *Rio notebook*.

1022 The watercolours are reproduced in Keynes 1979: No. 150, 'The *Beagle* in Murray Narrow, *Beagle* Channel' (pp. 116, 395), and No. 193, 'River Santa Cruz' (pp. 201, 397).

account 'trash!!!', p. 63a.[1023] The entries then switch to 'granite in the [South] Shetlands', with 'vast blocks of granite' at Kemp Bay, p. 64a, the 'Lion's Head' and 'Turk's Bay', p. 65a.

Whether or not these pages were written on Mauritius, where the *Beagle* arrived on 29 April 1836, the first and only dated entry 'Thursday Rode on Elephant' is unambiguously 5 May on the island. Captain Lloyd was taking Darwin to see 'Flat plain covered with Coral' on the only elephant on Mauritius, p. 66a. The elevated coral was on the southwest side of the island and was eventually described in *Volcanic islands*, where Darwin also discussed the volcanic structure of the island: 'Grand quaqua versal dip on all the west side; certainly distinct craters on the Isl^{d.}', p. 70a, later gave him the evidence for including Mauritius as an example of a 'crater of elevation'.

Darwin noted that the 'Elephant [was] noiseless' and he may have written this while actually sitting on the animal, p. 65a. He described the landscape as 'charming country mango avenues, nice gardens' and the 'Mimosa hedges, Sugar cane, prosperity', p. 72a, clearly left a favourable impression.

Darwin was by now confident that his coral reef theory was a breakthrough and he started to collect information on reef distribution whenever it presented itself: 'Coast of Guinea? Coral?', p. 72a, and 'coral grows highest & most solid & apparently more abundantly on windward side', p. 73a. He drew two sketches, the first a section through the reef 'off Grand Port', the second a map showing the 'bites' or indentations along the reef where fresh water streams came into the sea, but he declared 'I do not understand this', p. 75a. His observations on the fringing reefs of Mauritius were eventually published in chapter three of *Coral reefs*.

On p. 77a Darwin mentioned Mrs Power and 'Miserable quarrels between French & English'. On the next page he noted 'Hindoo convicts, most extraordinary white beards black as negros plenty of intellect' and he judged the English parts of the island better than the French island of Bourbon.

There are no more entries from Mauritius and finally the *Beagle* 'sailed 9th [May 1836]', p. 79a. In the *Beagle diary* entry for that day he confessed that: 'Since leaving England I have not spent so idle & dissipated a time. I dined almost every day in the week: all would have been very delightful, if it had been possible to have banished the remembrance of England.' The little *Beagle* still had many thousands of miles to sail before Darwin would see the green fields of Shropshire again.

The next time Darwin selected a field notebook it was for the last time and that notebook saw him all the way back to England. It was the *Despoblado notebook*, as Darwin prepared, three weeks after leaving Mauritius, to disembark at the Cape of Good Hope.

1023 See the introduction to the *Cape de Verds notebook*.

[FRONT COVER]

Sydney
Mauritius

[INSIDE FRONT COVER]
Bathurst 2232
Waimate
Dr Jennerett[1024] Rtt St
Quamby [Bluff]

[1a] 16.th = Saturday [January 1836]
Left Sydney — soon entered country — excellent roads, turnpike: Pothouses
too much wood land — some fine trees, all peculiar — rails instead of Hedges: many
Carts Gigs — Phaetons & Horses /from/ garrison soldiers

[2a] Lunched. nice little Public House. — pleasant distinction of Ranks — Napoleon &
Dan -O'connel[1025]
Rode on to Emu ferry on the Nepean: a broard river still as a pool, small body of
running water. —
Poor pasture land; but

1024 Henry Jeanneret (1802–86), surgeon, dentist and amateur botanist in Hobart Town, Tasmania.
1025 Daniel O'Connell (1775–1847), Irish statesman.

[3a] here & there Farm Houses: on the Nepean pretty scene, escarpment of Blue mountains, contemptible edge of great plain. cultivated land:
Party of black men, beautiful precision in throwing darts: speak English — Merry

[4a] fellows: distance about 100 y^{ds} Caps at about 30 — curious throwing stick: painted white (like Fuegians) going to fight some other people: not nearly so degraded a set, as I expected, all clothed: Prisoners constable & soldiers
Sandstone: frequently ~~many~~ Laminated Shale beds at top of Clays with

[5a] much <u>Clay</u> Iron Stone
Slight irregularities in the stratification:
Sandstone generally moderately hard; thinly stratified, on coast dip inwards — near the Nepean first meet pebbles of *[illeg]* Granitic Trappean rock & *[siliceous]* Sandstones

[6a] Appears to descend to valley by steps: plain at base composed of fine Alluvial *[strongly]* sandy soil stratified, lying on a coarse conglomerate of above pebbles; great Escarpement of Blue Mountains
Black men, see marks of Oppossum's feet. —
chief food: no home: —

[7a] Sunday 17th — [January 1836] Started 6 oclock — ferry. ascent of Blue Mountains. ~~great~~ fine woods. —
Then plain, uneven, many valleys; gradually rises great deception, when the elevation is considered of nearly 3000 ft —
[Barren] woods: poles — pale & peculiar green

[8a] Singularly uniform tint, in bushes the vertical leaves singular effect: pretty birds, magnificent parrots: baited at the Weatherboard;[1026] walked mile & ½ to see Cascade: most magnificent. astounding & unique view, small valley not lead to expect such scene: rill of water,

[9a] *[Rainbow]* fluttering: semicircular; cliff white or reddish, so vertical that pitch a stone over perhaps 800 ft, quartz wall, about 2000 ft, grand valley, sea of forest; necessary to go 16 miles to reach base of fall

[10a] certainly most stupendous cliffs I have ever seen, some stand in middle like Island. —
Stratum 3000 ft thick cracked, cracks enlarged —
Rode on to Gardners;[1027] occasionally glimpse of part of same valley

[11a] called Clwyd from the wooded plain, could not see the bottom; Reached Black Heath, comfortable as Welsh Inn, old Soldier — 15 beds, in a barren mountain ~~70~~ 60 miles from Sydney:[1028] Waggon with Wool. Bullocks —

1026 'In the middle of the day we baited our horses at a little Inn, called the Weather-board.' *Beagle diary*, p. 309.

1027 Andrew Gardner, a Scottish ex-soldier and ex-convict who established the Scotch Thistle Inn (1831), later Blackheath Inn.

1028 'The Blackheath is a very comfortable inn, kept by an old Soldier; it reminded me of the small inns in North Wales.' *Beagle diary*, p. 401.

[12a] Edge of Blue Mountains Strata ~~appears~~ to dip to plain; I believe owing to deposition

from thickness of strata: Thin structure excessively common, all the
planes well developed, dip to various points:

[13a] On a grand scale as seen from Cascade Horizontal strata & Sandstone
grains of quartz & much ferruginous matter, plates, hollow balls, & Claystone
Iron — line of small Quartz pebbles; rare at edge of Blue Mountains, common

[14a] at Cascade & even Inch long *[as before]*

Blackheath: At edge of B. Mountains in Sandstone, patch of Shale   like Mica
Slate in Gneiss at Rio —
Grand Sandstone planes slopes to East,

[15a] valleys, most extraordinary

heads like arm of sea, appears quite out of proportion to little streams in great
fissures cleared by the sea — It is an immense Sandstone formation: —

[16a] Monday 18th [January 1836] — Early in morning went to East side about 2 & ½ miles
to Govetts Leap — the same absolute ⊥° precipices in form of cones not affected by
streams; horizontal strata; grand valley full of blue mist from rising sun: Descended Mt
Victory; an extraordinary undertaking. —
Old soldier of Blackheath who had travelled[1029]
account of the women:

[17a] M[r] Browne[1030] men, very sensitive Scotchman, bad account of men, not reformation,
or punishment, not happy but do not quarrel, excepting when drunk, — quite
impossible to reform. — In Vale of Clwyd — Granite 2500 ft country improves, trees
park like scattered, then covering of pasture, green

[18a] Sheep-down; whole country one aspect: white Coccatoos & Crows: wild dogs tamed:
copulate _freely_: M[r] Brownes farm, fine wheat fields harvest time just concluded 7000
sheep shearing — harvest, possess 15000 — 80 convicts; no women — miserable
although from scene ought to be happy. —

1029 Possibly Andrew Gardner.
1030 Andrew Brown(e), farmer and superintendent of Wallerawang. See Nicholas and Nicholas 1989,
      pp. 44–5, and *Beagle diary*, p. 401. CD carried a letter of introduction from the owner, James
      Walker.

[19a]   Sydney — poor country; to be improved but limited: no Canals or rivers — not comparable to N. America:
M. Victoria Sandstone, with thin lines of quartz pebbles; some even inch long — ferruginous plates Quartz (Feldspar? & Mica decomposed?) Some shaly red marl, <u>white</u> soft aluminous

[20a]   fine grained Sandstone: In whole great plain no pebbles of Granite; hence no debacles, any small ones would be decomposed:
Vale of Clwyd Granite porph. with large Cysts: of Feldspar. concentric structure (scenery hills surmounted by balls) 3 or 4 large dykes

[21a]   of decomposing greenstone spec (1) = Veins of Quartzose Granite = & paps of reddish Granite. with large Crystals of Hornblende: — This is superimposed by the Sandstone cliffs, as is one of the great valleys — Perhaps at Weatherboard Sandstone 2000 ft thick

[22a]   towards Mr Brownes. /Siliceous/ & Erratic rock containing grains of Quartz & scales of Mica, at distance cliffs of Sandstone & in neighbourhood Sandstone conglomerate with the pebbles of Siliceous rocks & Eurites — Granite of Vale of Clwyd about

[23a]   2500 ft above sea. — Valleys most extraordinary expand with great depression surrounded by absolutely ⊥° cliffs — The Chart will give correct idea of peninsula & Islands of the grand plain: in Crust or points of plains

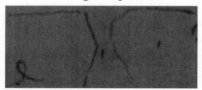 notches at head

[24a]   of opposite valleys: not present causes: forms that of marine bays.
Sea could not excavate? Sand banks, mem current cleavage & stratification edge of Blue Mountains? Cracks too large? Elevation

[25a]   acted upon by the sea?
Tuesday 19th [January 1836] Staid at Mr Browne's: Kangeroo hunting on the barren slopes country can be galloped over in all parts — thin turf — peculiar character
shadeless open forest park: — did not see one — My usual ill luck: killed Kangeroo

[26a]   Rat:[1031] — In the evening walked up Cox's river. chain of pools, so dry a country. — Saw several Ornithorynch:[1032] like water rats, in movements & habits: — Shot one. — Wild dog trap. —
Already pushing sheep /nearly/ into remote interior:

[27a]   decidedly a sterile country: Forgery or gambling — /always/ attempt a sporting character.

1031 The Long-nosed Potoroo (*Potorous tridactylus*).
1032 The Duck-billed platypus (*Ornithorhynchus anatinus*).

(1) Greenstone. Vale of Clwyd

(2) & 3 White Sandstone white cement.

(4) Bluish (Calc?) clay slate

(5) Black Carbonaceous do

(6) Coal

(7) Coal other locality

(8) Blue Calc Clay S. *[illeg]*

[28a]   Close to house, coarse Granite — Mica imperfect: the Sandstone Conglomerate reposes on this —

At Wolgan: grand valley surrounded by cliffs of Sandstone: with ~~much~~ a good deal of shale & Clay Slate: Section of one part gave a whitish or ferruginous Sandstone

[29a]   soft /brittle/, grain of quartz cemented by white aluminous matter (2 3) covering a Bluish (Calc?) Clay Slate. (4) this alternates with black (5) carbonaceous slate, coal (6) & Sandstone & /muddy/ shale. /Many lines/ One layer of coal nearly a foot thick. —

In other parts of country

[30a]   there was the Blue Slate with impressions of leaves (8) & in another good stratum of coals as before (7): — Also Conglomerates Sandst. & the ordinary sandstones. — Strata all nearly = tal:

Valley of Wolgan: large

[31a]   ordinary form like a bay with arms: so precipitous, then with great /labor/ in one spot a cattle track has been /cut/ down, generally vertical walls, many hundred feet high: sides perpendicular reaching to general

[32a]   level of country, about 7 miles long, & ½ one miles broad — appears to have all been removed & form modelled by water: yet exit of the valley, is by a narrow crack a few hundred yards wide, with stupendous vertical sides. no cattle

[33a]   can pass /out/ & twice the Surveyors have attempted to pass down the bed of the small river but have failed

Capertee is said to have same structure, but hollow /on/ valley far more extensive

In Wolgan cattle can

[34a]   never be lost: It is impossible water could have removed the whole mass of rock, & /nearly/ hollow a crevice at its exit — Spaces of the crust have not been elevated: being detached by fissures from surrounding country: sea occupying

[35a]   them hollows would /model/ & give vertical sides: (Interesting to reflect on forms of lower hypogene rocks) ~~rivers~~ after sea /tides/ then rivers, the lake saw the exit: — Shale far more abundant than near the coast. — Granite centre supports

[36a]   luxuriant vegetation & hard sandy soil: /countless/ time to form so much coal & Sandstone —

Wednesday 20th [January 1836] To Bathurst Squatters Huts. = Crawlers = The flat-bottomed valley: path through bushes: found high road — hot wind clouds of dust. — Downs of Bathurst: undulating

[37a]   ☐110☐ 2560–3000 ft. Willow trees brown. — Scattered hovels in groups & here & there a good house, whole plain divided by rails into fields: thin crops of grain & still thinner pastures river chain of pools — R. Macquarie just flowing Was told not to form too high an opinion of Australia

[38a]   from this & not too low from Road. — my opinion stamped —
Dividing ridge — nothing: waters into vast interior
Several gentlemans houses about 5 miles apart from one from other — chapel Soldiers &c &c

[39a]   (9) Limestone
(10) Primitive Greenstone
(11) Glossy Clay Slate
(12) Hornblendic Greenstone
Dividing range low, 20 ft covered with Sandstone:
In close neighbourhood & /doubtless/ foundation rock is a reddish compact Quartz rock, or siliceous

[40a]   Sandstone & a white Cryst Limestone (9) which I am assured is the same as at Mudgee & Wellington (doubtless bones to be found) On West slope — coarse granitic Porph with Quartz & Feldspar, & very little mica. further on compact Clay Slates & much Trappean rocks, siliceous rocks, which have the character of those

[41a]   called Primitive granites. — (9)
Beyond the Green Man there is a considerable formation of a glossy Clay Slate, cleavage nearly vertical (East dip) running N & S (Mem. Capt King[1033] Mica Slate) Slate has /variety/ in its beds in direction

[42a]   of laminae (11) is one of the more glossy kind
In many parts of road, when cleavage was not developed, could not tell this slate from that in the Sandstone Carboniferous[1034] series: —
Very many large snow white Quartz veins hence pebbles of do. & form

[43a]   Mica slate —
Crossing the slate, came to Granite, where only Quartz & Feldspar are present. —
Again beyond this a Hornblendic Greenstone (12) — There are on the borders of Bathurst downs which are elevated

[44a]   2 & 3000 ft. Consist of Granite & Primitive rocks smoothed over with shingle & Diluvial (as would be called) matter — I imagine land to West which is higher from form is similarly constituted —
Hence two grand

1033 Philip Parker King (1793–1856), commander of the *Beagle*'s first voyage, hydrographer and company manager in Australia.
1034 Geological period now known to be *c*. 345–280 million years ago, named after the extensive coal deposits formed at that time. Early Secondary period. More generally, meaning bearing carbon.

[45a] formations are seen: The Primitive one, although to appearance is low where crossed
must be high because much more elevated than Bathurst
Flag Post which being 2560 this land is probably higher than the Sandstone plains
[46a] which front the sea of the older formation —
21st [January 1836] — Rode about Bathurst — saw nothing — pleasant mess party —

(22nd) [January 1836] O Connel Plains —   gravel terrace.
same class of country to Cox's river. — Open woodland. almost all
[47a] Granite & /there/ Quartz & Feldspar — which passes into a reddish Porphyry & a
white Euritic one, in several parts some darker colored Trappean rocks — ~~Red~~ Much
Quartz <u>rock</u>. — ~~essentially~~ at spot where we
[48a] baited at Midday.
2 years from England: pretty daughter: at evening great fires: — 43 miles, hilly road —
general civility —
23^d^ [January 1836] — to Weatherboard last night horrid filth, in morning
[49a] raging fire. — Arrived soon walked to Cascade,
Sandstone almost composed of Quartz-pebbles — extraordinary figures: the soft
sandstone may almost be described as forming irregular balls in the harder kinds
Granite of yesterday with angular dark fragments
[50a] I do not perceive any difference in manners at the Inns from England
24th [January 1836] Ill in Bed
25th [January 1836] Quiet drizzly rain: all still dripping from eaves, undulating
woodland horizon of lost in thin mist — cold — great contrast
[51a] with former weather: Perhaps good for me, Jobs — comfort nice /girl/ rain for three weeks
26th [January 1836] Capt King —
Edge of Blue Mountain dip to sea; from irregular thinning. I think current deposition on
[52a] banks, not elevation: above plain to Paramatta which is inconsiderable —
Sandstone plain higher than granitic country — No = & granite (much Quartz rock
South of Bathurst — King) Sandstone of Blue Mountain 4000 ft Blackheath /either/
grains of quartz, fine cemented by

[53a] ferruginous matter or white powder: included pieces of shale ∠^r^:
Diluvium 200–300 ft above Nepean R. — (Gorge of Nepean river). latter only 9 miles
from Tide: pebbly siliceous & many Trappean rocks, one
[54a] curious kind — Traps hence Siliceous Sandstones:
I think most current cleavage toward Sea Side of great Sandstone plain —
27th [January 1836] Returned Mac Arthurs[1035]

1035 Hannibal Hawkins Macarthur (1788–1861), Australian colonist, politician, businessman, wool
pioneer and brother-in-law of Philip Parker King.

[55a]   Greater elevation of the coast has thrown drainage into the interior
Escarpement of B Mountain not formed by Nepean

[56a]

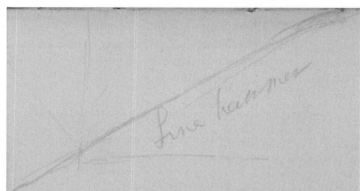

Line Hammer

[57a]   Huons Is$^{d.}$ named by D'Entrecasteaux[1036]
Rota de la Beche with cliffs of raised coral[1037]
Dillons account of Perouse — Mannicolo perfect encircling reef[1038]
SE of Radack

[58a]   Many reefs

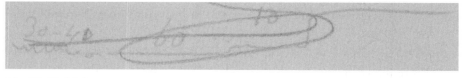

30–40 60 10 S

~~P. Journal~~

Craters. Corals. on Panama coast = Rodriguez[1039] =

Much quartz rock — *[Graywacke]* & Slate Staten Land

[59a]   Sandstone & S Shetland
Mr. Seales — Museum[1040]
Jasper & quartzose rocks vertical strata of gypsum
Limestone from Sandy Bay

1036   Antoine Raymond Joseph de Bruni d'Entrecasteaux (1739–93), French navigator who explored the Australian coast in 1792.

1037   De la Beche 1831, p. 142: 'the isle of Rota; where corals, resembling those now found in the neighbouring seas, occur in cliffs.' CD referred to the same page on the IBC.

1038   Dillon 1829.

1039   A volcanic island 550 km north-east of Mauritius with an encircling coral reef. Cited in *Coral reefs*, p. 242.

1040   Robert F. Seale, geologist, who had a museum on St Helena. These notes and all the following entries on p. 59a to the start of p. 62a are a response to reading Seale 1834.

Prosperous Bay Gregories Valley — /Voluta & crystallised/ Limestones

[60a] Buccinum vel Helix dextra near summit of Flagstaff Hill 1900 ft. high — Bencoolen plain ~~1500~~ 1576 ft —

Bone of Diomedea exulans?[1041] 3–100 ft below surface Prosperous Bay, in a /Quarry/ 14 ft deep egg

[61a] Most of hill quartz veins!

Phonolites & porphyries

Layers of Salt with Sulph of Lime in Turk's Cap Bay

Limestone principally from Sandy Bay. — Oolitic[1042] Limestone Flagstaff Hill

[62a] Limestone hills, from Lots wife to the sea

Account of Ascension

Salt & Gypsum incrustations

Limestone — Fernando Noronha — geology /precious nonsense/,

[63a] not truly volcanic because no scoriae!! (Yet we have seen amygdaloid)

The peak he calls a greenstone composed of Feldspar Quartz & Hornblende!!

is it so? trash!!! the quartz the great /compose/

[64a] embedded siliceous schist !!!!!

/Binary/ granite in the /South/ Shetlands

There are vast blocks of granite in Kemp Bay

appears to be NW & SE traversing from

[65a] the Lion's Head appearing abundantly in Turk's Bay

Thursday [5 May 1836] Rode on Elephant to between R. Rempant & house — ~~(N.B. more than 18ft water between reef & land at Flack)~~ — Elephant noiseless — charming country mango avenues, nice

[66a] gardens. — Flat plain covered with Coral, two lumps — which one must have formed islands on reefs, & such a one exists — (1) about 98 ft high, circular, 20 above sea, strata, inclined at ∠ 8 lowest ½ coarse sand, largely stratified, upper half

[67a] coarser, softer & containing great rounded rocks of Basalt, much rounded coral. part of which partly growing; 2 astrea & the /fine/ Keeling kind: & the 2^d species of do Isl^d ? branched Madrepore[1043]

From /illeg/ /partly/

[68a–69a blank]

[70a] /nearly/ surface of reef

Here we have proof of foundation of reef

agreement of blocks & growing coral —

Shores Basaltic much /belong/ Corallina but not coral ?? —

1041 The wandering albatross.

1042 Composed of tiny balls of lime.

1043 Specimen not in spirits 3634 'Branching Millepora, part of it encrusting a tubiform shell.' *Zoology notes*, p. 419, and note on p. 308.

[71a] Reef in parts dry — Here more astrea because more sea: — other /patch/ dipping
nearly 18° — water worn, same constitution 200 yards inland

[72a] N B Grand quaqua versal dip on all the West side; certainly distinct craters on the Isl[d.] —
M[r] Lloyd[1044] on the reef 5 miles from the shore; all deep

[73a] water between? —
Patch of rather coarse Tosca rock on Lava. very cellular subaerial — no trace of organic
remains — ?Spring? 200 ft above the sea —

[74a] Very little coral on coast of Panama
Capital roads — Mimosa hedges, Sugar cane, prosperity — Indian population snow
white beards
Coast of Guinea? Coral? —

[75a] Both, Dyne & Radcliff say that the coral grows highest & most solid & apparently more
abundantly on windward side
Reef off grand Port, so high that record to seaward —

[76a] Coral islets

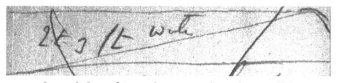

2 to 3 ft water deep

water /outside/, surface rather smooth with Corallina —
Excepting where main channel for fresh water 2–4 ft of depth. —

[77a] Reef <u>very seldom</u> attached to shore — Frequently in bites
small reef —

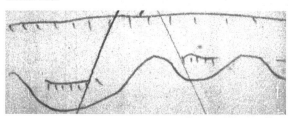

I do not understand this —

[78a] Corals very different out & inside reef —
Channel seldom more than a few feet deep can wade outside
At Grand Port, at Low water spring tides several ft above high

[79a] water perhaps whole reef elevated
M[rs] Power[1045] thinks that ~~reef~~ that /cratering/ /opens/ towards the grand hill?
Miserable /quarrels/ between French & English

1044 John Augustus Lloyd (1800–54), civil engineer and surveyor. Surveyor-General, Mauritius, 1831–49.
1045 Mrs Power was a resident of Port Louis, Mauritius. Possibly the wife of Colonel James Power, Royal
Artillery. She is also mentioned in the *Red notebook*, p. 17: 'M[rs] Power at Port Louis talked of the
<u>extraordinary</u> freshness of the streams of Lava in Ascencion known to be inactive 300 years?' See also
a reference to a Dr Power of Mauritius in *Descent* 1: 335.

[80a] Hindoo convicts, most extraordinary white beards black as negros plenty of intellect negroes state, poor people

Prosperity of English /government/ roads, contrast to Bourbon

[81a] Sailed 9th [May 1836] —

[82a–92a blank]

[BACK COVER]

[INSIDE BACK COVER]

12 of the little *[Quires] from the Captain*

Blackheath 3411 N^r 1

Wolgan
Capita
[1b–2b excised]
[3b]   Paper — Chaffers[1046]
      Mr Dring[1047]
      Capt King
      Bathurst
      Mica Slate
      Trappean Rocks
      Mr Blaxland[1048]
      Van Diemens Land
[4b]   [M^r] Dring — Tobacco = B Soap
      Dollars changed
      Inkstand
      Pencils — Blotting Paper — Writing Paper — Portfolio? Bramah Pens Common do
      German Books: Spelling Dict:
      Taylor — Boot-maker
      Dentist —
[5b]   Peppermint — Hops Carb of Soda & Magnesia = Laudanum Lozenges
      ½ oz Tinct Term. *[Muriaticæ]*
      Lavander water —
      Truss —
      Second Penknife
[6b]   M^r Th Walker[1049] James Mr Clay —
      Mr Bennet
      Mr Dring — Tobacco & Soap
      [Tooth] Dr Jennerat
      Valparaiso Letters
[7b]   ~~Inkstand~~ Taylor
      Fools cap — Paper Take size
      Tooth-Brushes
      Pill Boxes
      Bramah pens
[8b]   2 Lb of common soap & Tobacco

1046 Edward Main Chaffer, Master on the *Beagle*.
1047 John Edward Dring acting Purser on the *Beagle*.
1048 Probably Gregory Blaxland (1778–1853) or John Blaxland (1769–1845), wealthy landowners and
      merchants in Australia.
1049 James Walker (1784–1856), former army officer and owner of Wallerawang.

Textual notes to the *Sydney notebook*

[FC]   LT *[V]* 15], *not in CD's handwriting, not transcribed.*

[IFC]   1.3.] *Down House number, not transcribed.*
88202323] *English Heritage number, not transcribed.*
Bathurst 2232] *written perpendicular to the spine in right margin, not transcribed.*
14] *added by Nora Barlow, pencil, not transcribed.*

[37a]   R. Macquarie just flowing] *added heavy pencil.*

[IBC]   Blackheath ... No 1] *written perpendicular to the spine in left margin.*
Wolgan Capita] *written upside down from other entries on page.*

[3b]   Capt King ... Land] *written upside down from other entries on page.*

CHRONOLOGICAL REGISTER

⟿

By Kees Rookmaaker

H. M. S. *Beagle* left England on 27 December 1831 and returned on 2 October 1836, 1741 days later.[1050] This chronology supersedes all previous itineraries of the voyage. The nights Darwin spent on board or on land have been calculated using the *Beagle diary*, and for the nights on land, which territories he visited. Darwin stayed ashore for some days when the *Beagle* was travelling elsewhere. According to these calculations, Darwin had the following itinerary.

At sea	581 days or 33.3 %
At anchor	566 days or 32.6 %
On land	594 days or 34.1 %

Darwin's nights on board the *Beagle*:[1051] 1145 nights or 65.8 %.

Darwin's nights on land: 596 nights or 34.2 % (955 nights, or 55%, were spent in South America).

Darwin spent his nights in the following territories, either on land or on the *Beagle* at anchor. Nights on board the *Beagle* while traversing coastal waters are not included.

Cape de Verds	27 days	New Zealand	9 days
Brazil	126 days	Australia	38 days
Banda Oriental	156 days	Keeling	11 days
Argentina	222 days	Mauritius	10 days
Falklands	64 days	Cape of Good Hope	18 days
Chile	375 days	Ascension	4 days
Peru	51 days	St Helena	6 days
Galapagos	25 days	Azores	6 days
Tahiti	10 days		

1050 Monday 16 November 1835 did not exist due to crossing the international dateline. Barlow 1967, appendix VI (pp. 241–3) gave a list of sections of the journey, stating the number of days at sea and on land. When the days are added up (with a month taken as 31 days), the totals are: 533 days at sea, 1147 on land, total 1680 days. Hence the total is 61 days less than the correct total. Barlow calculated the period 28 April to 23 July 1833 as 56 days, which should be 87 days. And the period 29 January to 7 March 1834 as 9 days, which should be 32 days. In both cases a month was missed.

1051 When CD spent a day on land but was on board at night, the date is calculated as being on board, with the *Beagle* probably in harbour.

The *Beagle* field notebooks contain references to *c.* 474 days or 26 % of the journey. These are listed chronologically below.

Date	Notebook/page	Locality
18 January 1832	*Cape de Verds notebook*: 15b	Praia, St Jago
20 January 1832	*Cape de Verds notebook*: 24b	Praia, St Jago
23 January 1832	*Cape de Verds notebook*: 31b	Signal Post Hill, St Jago
26 January 1832	*Cape de Verds notebook*: 34b	Ribera Grande, St Jago
02 February 1832	*Cape de Verds notebook*: 36b	St Domingo, St Jago
03 February 1832	*Cape de Verds notebook*: 38b	Praia, St Jago
20 February 1832	*Cape de Verds notebook*: 44b	Fernando Noronha
25 February 1832	*Cape de Verds notebook*: 46b	Quail Island, St Jago
29 February 1832	*Cape de Verds notebook*: 56b	Bahia, Brazil
03 March 1832	*Cape de Verds notebook*: 62b	Bahia, Brazil
05 March 1832	*Cape de Verds notebook*: 63b	Bahia, Brazil
13 March 1832	*Cape de Verds notebook*: 68b	Bahia, Brazil
14 March 1832	*Cape de Verds notebook*: 70b	Bahia, Brazil
15 March 1832	*Cape de Verds notebook*: 72b	Bahia, Brazil
21 March 1832	*Cape de Verds notebook*: 73b	Bahia–Rio de Janeiro
23 March 1832	*Cape de Verds notebook*: 73b	Bahia–Rio de Janeiro
27 March 1832	*Cape de Verds notebook*: 13a,14a,15a	Abrolhos Islets
28 March 1832	*Cape de Verds notebook*: 16a	Abrolhos Islets
29 March 1832	*Cape de Verds notebook*: 74b	Abrolhos Islets
08 April 1832	*Rio notebook*: 1b	Excursion to Rio Macae
09 April 1832	*Rio notebook*: 4b	Mandetiba Ingetado
10 April 1832	*Rio notebook*: 9b	Campos Novos
11 April 1832	*Rio notebook*: 12b	Venda da Matto
12 April 1832	*Rio notebook*: 14b	Socego
13 April 1832	*Rio notebook*: 20b	Socego
14 April 1832	*Rio notebook*: 21b	Lennon's estate
15 April 1832	*Rio notebook*: 21b	Lennon's estate
16 April 1832	*Rio notebook*: 26b	Socego
17 April 1832	*Rio notebook*: 27b	Socego
18 April 1832	*Rio notebook*: 30b	Socego
19 April 1832	*Rio notebook*: 31b	Venda de Matto
20 April 1832	*Rio notebook*: 32b	Campos Novos
21 April 1832	*Rio notebook*: 32b	Rio Combota
22 April 1832	*Rio notebook*: 34b	Fregueria de Tabarai
23 April 1832	*Rio notebook*: 38b	Praia Grande
24 April 1832	*Rio notebook*: 38b	Rio de Janeiro
26 April 1832	*Rio notebook*: 39b	Rio de Janeiro
27 April 1832	*Rio notebook*: 39b	Rio de Janeiro
28 April 1832	*Rio notebook*: 40b	Rio de Janeiro
29 April 1832	*Rio notebook*: 40b	Rio de Janeiro

Date	Notebook/page	Locality
30 April 1832	*Rio notebook*: 41b	Rio de Janeiro
01 May 1832	*Rio notebook*: 41b	Rio de Janeiro
09 May 1832	*Cape de Verds notebook*: 77b	Rio de Janeiro
15 May 1832	*Cape de Verds notebook*: 78b	Rio de Janeiro
19 May 1832	*Cape de Verds notebook*: 78b	Rio de Janeiro
27 May 1832	*Cape de Verds notebook*: 18a	Rio de Janeiro
30 May 1832	*Cape de Verds notebook*: 19a	Rio de Janeiro (Corcovado)
02 June 1832	*Cape de Verds notebook*: 81b	Rio de Janeiro
05 June 1832	*Cape de Verds notebook*: 24a	Rio de Janeiro
07 June 1832	*Cape de Verds notebook*: 26a	Rio de Janeiro
10 June 1832	*Cape de Verds notebook*: 84b	Rio de Janeiro
16 June 1832	*Rio notebook*: 41b	Rio de Janeiro
05 July 1832	*Rio notebook*: 47a	Rio de Janeiro–Monte Video
06 July 1832	*Rio notebook*: 47a	Rio de Janeiro–Monte Video
07 July 1832	*Rio notebook*: 47a	Rio de Janeiro–Monte Video
08 July 1832	*Rio notebook*: 47a	Rio de Janeiro–Monte Video
09 July 1832	*Rio notebook*: 47a	Rio de Janeiro–Monte Video
14 July 1832	*Rio notebook*: 48a	Rio de Janeiro–Monte Video
15 July 1832	*Rio notebook*: 49a	Rio de Janeiro–Monte Video
16 July 1832	*Rio notebook*: 49a	Rio de Janeiro–Monte Video
18 July 1832	*Rio notebook*: 50b	Rio de Janeiro–Monte Video
15 August 1832	*Rio notebook*: 59a	Monte Video
22 September 1832	*Rio notebook*: 62a	Bahia Blanca
02 October 1832	*Rio notebook*: 67a	Bahia Blanca
06 October 1832	*Rio notebook*: 70a	Bahia Blanca
01 November 1832	*Buenos Ayres notebook*: 4b	Monte Video–Buenos Ayres
02 November 1832	*Buenos Ayres notebook*: 4b	Buenos Ayres, Argentina
03 November 1832	*Buenos Ayres notebook*: 5b	Buenos Ayres
04 November 1832	*Buenos Ayres notebook*: 6a	Buenos Ayres
05 November 1832	*Buenos Ayres notebook*: 6a	Buenos Ayres
06 November 1832	*Buenos Ayres notebook*: 7a	Buenos Ayres
07 November 1832	*Buenos Ayres notebook*: 7a	Buenos Ayres
19 December 1832	*Rio notebook*: 20a	Good Success Bay, Tierra del Fuego
20 December 1832	*Rio notebook*: 19a	Good Success Bay
19 January 1833	*Buenos Ayres notebook*: 14a	Navarin Island
20 January 1833	*Buenos Ayres notebook*: 18a	Navarin Island
21 January 1833	*Buenos Ayres notebook*: 21a	Navarin Island
22 January 1833	*Buenos Ayres notebook*: 25a	Ponsonby Sound
23 January 1833	*Buenos Ayres notebook*: 27a	Ponsonby Sound
24 January 1833	*Buenos Ayres notebook*: 32a	Ponsonby Sound
25 January 1833	*Buenos Ayres notebook*: 34a	Ponsonby Sound
26 January 1833	*Buenos Ayres notebook*: 38a	Ponsonby Sound
27 January 1833	*Buenos Ayres notebook*: 40a	Ponsonby Sound

Date	Notebook/page	Locality
28 January 1833	*Buenos Ayres notebook*: 46a	Ponsonby Sound
29 January 1833	*Buenos Ayres notebook*: 51a	Tierra del Fuego
30 January 1833	*Buenos Ayres notebook*: 55a	Darwin Sound
31 January 1833	*Buenos Ayres notebook*: 59a	Tierra del Fuego
01 February 1833	*Buenos Ayres notebook*: 62a	Tierra del Fuego
02 February 1833	*Buenos Ayres notebook*: 64a	Whaleboat Sound
03 February 1833	*Buenos Ayres notebook*: 66a	Tierra del Fuego
04 February 1833	*Buenos Ayres notebook*: 68a	Tierra del Fuego
05 February 1833	*Buenos Ayres notebook*: 72a	Tierra del Fuego
06 February 1833	*Buenos Ayres notebook*: 73a	Ponsonby Sound
07 February 1833	*Buenos Ayres notebook*: 78a	Goree Sound (Navarin Island)
16 February 1833	*Falkland notebook*: 5a	Hardy Peninsula
02 March 1833	*Falkland notebook*: 8a	Port Louis, East Falkland
06 March 1833	*Falkland notebook*: 9a	Port Louis, East Falkland
09 March 1833	*Falkland notebook*: 10a	Port Louis, East Falkland
12 March 1833	*Falkland notebook*: 12a	Port Louis, East Falkland
19 March 1833	*Falkland notebook*: 13a	Port Louis, East Falkland
20 March 1833	*Falkland notebook*: 16a	Port Louis, East Falkland
21 March 1833	*Falkland notebook*: 20a	Port Louis, East Falkland
22 March 1833	*Falkland notebook*: 22a	Port Louis, East Falkland
25 March 1833	*Falkland notebook*: 25a	Port Louis, East Falkland
28 March 1833	*Falkland notebook*: 26a	Port Louis, East Falkland
06 April 1833	*Falkland notebook*: 27a	Patagonian coast, Argentina
09 April 1833	*Falkland notebook*: 27a	Patagonian coast
17 April 1833	*Falkland notebook*: 28a	St Josephs Bay
10 May 1833	*Falkland notebook*: 35a	To Las Minas
11 May 1833	*Falkland notebook*: 38a	House of Don Juan Fuentes
12 May 1833	*Falkland notebook*: 43a	Las Tapas on Rio Marmaraga
13 May 1833	*Falkland notebook*: 45a	North of the Rio Polanco
14 May 1833	*Falkland notebook*: 48a, 3b	Las Tapas on Rio Marmaraga
15 May 1833	*Falkland notebook*: 54a, IBC	Las Tapas on Rio Marmaraga
16 May 1833	*Falkland notebook*: 56a	Las Minas
17 May 1833	*Falkland notebook*: 59a	Near Las Minas
18 May 1833	*Falkland notebook*: 64a	House of Sebastian de Pimiento
19 May 1833	*Falkland notebook*: 68a	House of Sebastian de Pimiento
20 May 1833	*Falkland notebook*: 73a	Maldonado
31 May 1833	*Falkland notebook*: 76a	Maldonado
02 August 1833	*Falkland notebook*: 88a	Maldonado to Rio Negro
03 August 1833	*Falkland notebook*: 88a	Rio Negro
04 August 1833	*Falkland notebook*: 87a, 89a	Rio Negro
05 August 1833	*Falkland notebook*: 87a	Carmen de Patagones
06 August 1833	*Falkland notebook*: 94a	Rio Negro
07 August 1833	*Falkland notebook*: 94a	Rio Negro
08 August 1833	*Falkland notebook*: 96a	Rio Negro

Date	Notebook/page	Locality
09 August 1833	*Falkland notebook*: 105a	Rio Negro
10 August 1833	*Falkland notebook*: 106a	Rio Negro
11 August 1833	*Falkland notebook*: 109a	Rio Negro–Bahia Blanca
12 August 1833	*Falkland notebook*: 111a	Rio Negro–Bahia Blanca
13 August 1833	*Falkland notebook*: 112a	Rio Colorado
14 August 1833	*Falkland notebook*: 114a	Rio Colorado
15 August 1833	*Falkland notebook*: 115a	Rio Colorado
16 August 1833	*Falkland notebook*: 116a	Rio Colorado
17 August 1833	*Falkland notebook*: 120a	Bahia Blanca
18 August 1833	*Falkland notebook*: 124a	Bahia Blanca
19 August 1833	*Falkland notebook*: 125a	Bahia Blanca
20 August 1833	*Falkland notebook*: 125a	Bahia Blanca
21 August 1833	*Falkland notebook*: 126a	Bahia Blanca
22 August 1833	*Falkland notebook*: 129a	Bahia Blanca
23 August 1833	*Falkland notebook*: 130a	Bahia Blanca
24 August 1833	*Falkland notebook*: 139a	Bahia Blanca
25 August 1833	*Falkland notebook*: 139a	Bahia Blanca
26 August 1833	*Falkland notebook*: 141a	Bahia Blanca
27 August 1833	*Falkland notebook*: 142a	Bahia Blanca
28 August 1833	*Falkland notebook*: 143a	Bahia Blanca
29 August 1833	*B. Blanca notebook*: 1a, 3a	Bahia Blanca (Punta Alta)
30 August 1833	*B. Blanca notebook*: 4a	Bahia Blanca
31 August 1833	*B. Blanca notebook*: 4a	Bahia Blanca (Punta Alta)
01 September 1833	*B. Blanca notebook*: 7a	Bahia Blanca
02 September 1833	*B. Blanca notebook*: 11a	Bahia Blanca
03 September 1833	*B. Blanca notebook*: 12a	Bahia Blanca
04 September 1833	*B. Blanca notebook*: 12a	Bahia Blanca
05 September 1833	*B. Blanca notebook*: 15a	Bahia Blanca
06 September 1833	*B. Blanca notebook*: 26a	Bahia Blanca
07 September 1833	*B. Blanca notebook*: 27a	Bahia Blanca
08 September 1833	*B. Blanca notebook*: 29a	Foot of Sierra
09 September 1833	*B. Blanca notebook*: 38a	Foot of Sierra
10 September 1833	*B. Blanca notebook*: 44a	Sauce Posta
11 September 1833	*B. Blanca notebook*: 45a	Third Posta
12 September 1833	*B. Blanca notebook*: 48a	Third Posta
13 September 1833	*B. Blanca notebook*: 53a	Third Posta
14 September 1833	*B. Blanca notebook*: 54a	Between 3rd and 4th Posta
15 September 1833	*B. Blanca notebook*: 55a	5th Posta
16 September 1833	*B. Blanca notebook*: 57a	8th Posta
17 September 1833	*B. Blanca notebook*: 60a	10th Posta
18 September 1833	*B. Blanca notebook*: 63a	12th Posta, south of Rio Salada
19 September 1833	*B. Blanca notebook*: 65a	Guardia del Monte
20 September 1833	*B. Blanca notebook*: 67a	Buenos Ayres (Mr Lumb)
21 September 1833	*B. Blanca notebook*: 68a	Buenos Ayres (Mr Lumb)

Date	Notebook/page	Locality
27 September 1833	*St. Fe notebook*: 9a	Near Luxan
28 September 1833	*St. Fe notebook*: 9a	Arrecife River
29 September 1833	*St. Fe notebook*: 11a	St Nicholas
30 September 1833	*St. Fe notebook*: 12a	Colegio de St Carlos
01 October 1833	*St. Fe notebook*: 16a	Near Monge River
02 October 1833	*St. Fe notebook*: 19a	Santa Fe
03 October 1833	*St. Fe notebook*: 22a	Santa Fe
04 October 1833	*St. Fe notebook*: 23a	Santa Fe
05 October 1833	*St. Fe notebook*: 23a	Rio Parana
06 October 1833	*St. Fe notebook*: 25a	Rio Parana
07 October 1833	*St. Fe notebook*: 28a	Rio Parana
08 October 1833	*St. Fe notebook*: 31a	Rio Parana
09 October 1833	*St. Fe notebook*: 32a	Rio Parana
10 October 1833	*St. Fe notebook*: 32a	Rio Parana
11 October 1833	*St. Fe notebook*: 36a	Rio Parana
12 October 1833	*St. Fe notebook*: 37a	On boat to Buenos Ayres
13 October 1833	*St. Fe notebook*: 38a	On boat to Buenos Ayres
14 October 1833	*St. Fe notebook*: 39a	On boat to Buenos Ayres
15 October 1833	*St. Fe notebook*: 40a	On boat to Buenos Ayres
16 October 1833	*St. Fe notebook*: 42a	On boat to Buenos Ayres
17 October 1833	*St. Fe notebook*: 44a	On boat to Buenos Ayres
18 October 1833	*St. Fe notebook*: 45a	On boat to Buenos Ayres
19 October 1833	*St. Fe notebook*: 47a	On boat to Buenos Ayres
20 October 1833	*St. Fe notebook*: 49a	Near Punta de St Fernando
21 October 1833	*St. Fe notebook*: 50a	Buenos Ayres
02 November 1833	*St. Fe notebook*: 55a	Boat (Packet) to Monte Video
03 November 1833	*St. Fe notebook*: 55a	Boat (Packet) to Monte Video
04 November 1833	*St. Fe notebook*: 55a	Monte Video
05 November 1833	*St. Fe notebook*: 55a	Monte Video
06 November 1833	*St. Fe notebook*: 56a	Monte Video
14 November 1833	*Banda Oriental notebook*: 5	Canelones
15 November 1833	*Banda Oriental notebook*: 5	Post house of Cufrè
16 November 1833	*Banda Oriental notebook*: 6	Post house of Cufrè
17 November 1833	*Banda Oriental notebook*: 6	Colonia del Sacramiento
18 November 1833	*Banda Oriental notebook*: 9	Arroyo de St Juan
19 November 1833	*Banda Oriental notebook*: 15	Arroyo de las Vivoras
20 November 1833	*Banda Oriental notebook*: 17	At a Rancho
21 November 1833	*Banda Oriental notebook*: 23	At a Rancho
22 November 1833	*Banda Oriental notebook*: 26	Estancia of the Berquelo
23 November 1833	*Banda Oriental notebook*: 27	Mercedes, Capella Nueva
24 November 1833	*Banda Oriental notebook*: 30	Mercedes, Capella Nueva
25 November 1833	*Banda Oriental notebook*: 32	Mercedes, Capella Nueva
26 November 1833	*Banda Oriental notebook*: 33	Return to Monte Video
27 November 1833	*Banda Oriental notebook*: 35	San Jose de Mayo

Date	Notebook/page	Locality
28 November 1833	*Banda Oriental notebook*: 35	Monte Video
27 December 1833	*Buenos Ayres notebook*: 87a	Port Desire
29 December 1833	*Buenos Ayres notebook*: 88a	Creek near Port Desire
02 January 1834	*Port Desire notebook*: 3	Port Desire
03 January 1834	*Port Desire notebook*: 4	Port Desire
09 January 1834	*Port Desire notebook*: 13	Port St Julian
10 January 1834	*Port Desire notebook*: 13	Port St Julian
16 January 1834	*Port Desire notebook*: 13	Port St Julian
17 January 1834	*Port Desire notebook*: 14	Port St Julian
20 January 1834	*Port Desire notebook*: 14	Port Desire
21 January 1834	*Port Desire notebook*: 18	Port Desire
03 February 1834	*Port Desire notebook*: 19	Port Famine
06 February 1834	*Port Desire notebook*: 22	Port Famine Mount Tarn
25 February 1834	*Port Desire notebook*: 26	Wollaston Island, Tierra del Fuego
27 February 1834	*Port Desire notebook*: 29	Navarin Island
16 March 1834	*Port Desire notebook*: 33	Berkeley Sound, East Falkland
17 March 1834	*Port Desire notebook*: 35	Valley at Rincon del Toro
18 March 1834	*Port Desire notebook*: 37	Valley at Rincon del Toro
19 March 1834	*Port Desire notebook*: 38	Berkeley Sound
23 March 1834	*Port Desire notebook*: 39	Berkeley Sound
31 March 1834	*Port Desire notebook*: 41	Berkeley Sound
14 April 1834	*B. Blanca notebook*: 69a	Mouth of Rio Santa Cruz
18 April 1834	*Banda Oriental notebook*: 38	Along Rio Santa Cruz
19 April 1834	*Banda Oriental notebook*: 38	Along Rio Santa Cruz
20 April 1834	*Banda Oriental notebook*: 39	Along Rio Santa Cruz
21 April 1834	*Banda Oriental notebook*: 41	Along Rio Santa Cruz
22 April 1834	*Banda Oriental notebook*: 42	Along Rio Santa Cruz
23 April 1834	*Banda Oriental notebook*: 45	Along Rio Santa Cruz
24 April 1834	*Banda Oriental notebook*: 47	Along Rio Santa Cruz
25 April 1834	*Banda Oriental notebook*: 49	Along Rio Santa Cruz
27 April 1834	*Banda Oriental notebook*: 63	Along Rio Santa Cruz
28 April 1834	*Banda Oriental notebook*: 67	Along Rio Santa Cruz
29 April 1834	*Banda Oriental notebook*: 75	Along Rio Santa Cruz
30 April 1834	*Banda Oriental notebook*: 79	Along Rio Santa Cruz
01 May 1834	*Banda Oriental notebook*: 81, 87, 90	Along Rio Santa Cruz
02 May 1834	*Banda Oriental notebook*: 94	Along Rio Santa Cruz
03 May 1834	*Banda Oriental notebook*: 98	Along Rio Santa Cruz
04 May 1834	*Banda Oriental notebook*: 100	Along Rio Santa Cruz
05 May 1834	*Banda Oriental notebook*: 101	Along Rio Santa Cruz
06 May 1834	*Banda Oriental notebook*: 102	Along Rio Santa Cruz
07 May 1834	*Banda Oriental notebook*: 103	Along Rio Santa Cruz
08 May 1834	*Banda Oriental notebook*: 104	Mouth of Rio Santa Cruz

Date	Notebook/page	Locality
09 May 1834	*Banda Oriental notebook*: 107	Mouth of Rio Santa Cruz
12 May 1834	*Banda Oriental notebook*: 106, 107	Santa Cruz–Port Famine
25 May 1834	*Banda Oriental notebook*: 107	Santa Cruz–Port Famine
02 June 1834	*B. Blanca notebook*: 76a, 80a	Port Famine
09 June 1834	*B. Blanca notebook*: 81a	Magdalen Channel
14 August 1834	*Valparaiso notebook*: 1a	Valparaiso, Chile
15 August 1834	*Valparaiso notebook*: 2a	Hacienda de San Isidro
16 August 1834	*Valparaiso notebook*: 9a	Campana or Bell Mountain
17 August 1834	*Valparaiso notebook*: 17a	Campana or Bell Mountain
18 August 1834	*Valparaiso notebook*: 30a	Hacienda de San Isidro
19 August 1834	*Valparaiso notebook*: 41a	Quillota
20 August 1834	*Valparaiso notebook*: 43a	Mines of Jajuel
21 August 1834	*Valparaiso notebook*: 46a	Mines of Jajuel
22 August 1834	*Valparaiso notebook*: 53a	Mines of Jajuel
23 August 1834	*Valparaiso notebook*: 64a	Mines of Jajuel
24 August 1834	*Valparaiso notebook*: 74a	Mines of Jajuel
25 August 1834	*Valparaiso notebook*: 75a	Mines of Jajuel
26 August 1834	*Valparaiso notebook*: 78a	Cerro del Talguen
27 August 1834	*Valparaiso notebook*: 81a	Santiago
28 August 1834	*Valparaiso notebook*: 86a	Santiago
29 August 1834	*Valparaiso notebook*: 86a	Santiago
30 August 1834	*Valparaiso notebook*: 96a	Santiago
02 September 1834	*Santiago notebook*: 2	Santiago
05 September 1834	*Santiago notebook*: 3	Hacienda outside Santiago
06 September 1834	*Santiago notebook*: 4	Rancagua
07 September 1834	*Santiago notebook*: 7	Cauquenes
08 September 1834	*Santiago notebook*: 14	Cauquenes
09 September 1834	*Santiago notebook*: 19	Cauquenes
10 September 1834	*Santiago notebook*: 26	Cauquenes
11 September 1834	*Santiago notebook*: 30	Cauquenes
12 September 1834	*Santiago notebook*: 30	Cauquenes
13 September 1834	*Santiago notebook*: 33	River Claro
14 September 1834	*Santiago notebook*: 33	San Fernando
18 September 1834	*Santiago notebook*: 37	Yaquil near Rancagua
19 September 1834	*Santiago notebook*: 53	River Tinderidica
20 September 1834	*Santiago notebook*: 54	West of Rancagua
21 September 1834	*Santiago notebook*: 59	West of Rancagua
22 September 1834	*Santiago notebook*: 59	Navedad
23 September 1834	*Santiago notebook*: 60	Navedad
24 September 1834	*Santiago notebook*: 67	Towards Valparaiso
25 September 1834	*Santiago notebook*: 67	Casa Blanca
26 September 1834	*Santiago notebook*: 67	Casa Blanca
27 September 1834	*Santiago notebook*: 67	Valparaiso

Date	Notebook/page	Locality
24 November 1834	*Port Desire notebook*: 56	Chacao, Chiloe
25 November 1834	*Port Desire notebook*: 59	Huapilenou, Chiloe
26 November 1834	*Port Desire notebook*: 60	Caucahue Island, Chiloe
27 November 1834	*Port Desire notebook*: 63	Quinchao Island, Chiloe
28 November 1834	*Port Desire notebook*: 65	Quinchao Island, Chiloe
29 November 1834	*Port Desire notebook*: 67	Castro, Chiloe
30 November 1834	*Port Desire notebook*: 71	Castro, Chiloe
01 December 1834	*Port Desire notebook*: 79	Lemuy Island, Chiloe
02 December 1834	*Port Desire notebook*: 84	Lemuy Island, Chiloe
03 December 1834	*Port Desire notebook*: 91	Lemuy Island, Chiloe
04 December 1834	*Port Desire notebook*: 94	P. Chagua, Chiloe
05 December 1834	*Port Desire notebook*: 96	P. Chagua, Chiloe
06 December 1834	*Port Desire notebook*: 99	Caylen, Chiloe
07 December 1834	*Port Desire notebook*: 103	San Pedro Island, Chiloe
19 January 1835	*Galapagos notebook*: 3a	Chiloe
10 February 1835	*Santiago notebook*: 73	Valdivia
11 February 1835	*Santiago notebook*: 79	Excursion from Valdivia
22 February 1835	*Santiago notebook*: 87	Valdivia to Concepcion
24 February 1835	*Santiago notebook*: 87	Mocha Island
12 March 1835	*St. Fe notebook*: 227a	Valparaiso
14 March 1835	*Galapagos notebook*: 4a	Valparaiso to Santiago
15 March 1835	*Galapagos notebook*: 7a	Santiago
18 March 1835	*St. Fe notebook*: 66a	To Portillo Pass
19 March 1835	*St. Fe notebook*: 69a	To Portillo Pass
20 March 1835	*St. Fe notebook*: 75a, 79a	Portillo Pass
21 March 1835	*St. Fe notebook*: 93a,127a	Far side of Portillo Pass
22 March 1835	*St. Fe notebook*: 131a	Far side of Portillo Pass
23 March 1835	*St. Fe notebook*: 132a	Los Arenales
24 March 1835	*St. Fe notebook*: 133a	Estancia of Chaquaio
25 March 1835	*St. Fe notebook*: 145a	Estacado
26 March 1835	*St. Fe notebook*: 148a	Luxan
27 March 1835	*St. Fe notebook*: 152a	Mendoza
28 March 1835	*St. Fe notebook*: 153a	Mendoza
29 March 1835	*St. Fe notebook*: 154a	Villa Vicencio
30 March 1835	*St. Fe notebook*: 157a	Hornillos
31 March 1835	*St. Fe notebook*: 164a	Hornillos
01 April 1835	*St. Fe notebook*: 177a	Uspallata
02 April 1835	*St. Fe notebook*: 183a	Pulvadera
03 April 1835	*St. Fe notebook*: 189a	Rio de las Vacas
04 April 1835	*St. Fe notebook*: 196a	Puente del Inca
05 April 1835	*St. Fe notebook*: 201a	Ojos del Agua
06 April 1835	*St. Fe notebook*: 211a	Ojos del Agua, Guard House
07 April 1835	*St. Fe notebook*: 214a	Ojos del Agua, Guard House
08 April 1835	*St. Fe notebook*: 214a	Villa de St Rosa

Date	Notebook/page	Locality
09 April 1835	*St. Fe notebook*: 218a	Cuesta of Chacabuco Colina
10 April 1835	*St. Fe notebook*: 218a	Santiago
15 April 1835	*St. Fe notebook*: 218a	Santiago to Valparaiso
20 April 1835	*St. Fe notebook*: 224a	Valparaiso
23 April 1835	*Galapagos notebook*: 12a	Valparaiso
27 April 1835	*Coquimbo notebook*: 1	Vino del March
28 April 1835	*Coquimbo notebook*: 1	Quillota
29 April 1835	*Coquimbo notebook*: 3	Quillota
30 April 1835	*Coquimbo notebook*: 4	Catapilco
01 May 1835	*Coquimbo notebook*: 6	Longotomo
02 May 1835	*Coquimbo notebook*: 8	Quilimar
03 May 1835	*Coquimbo notebook*: 12	Conchalee
04 May 1835	*Coquimbo notebook*: 16	Illapel
05 May 1835	*Coquimbo notebook*: 19	Illapel
06 May 1835	*Coquimbo notebook*: 20	Los Hornos
07 May 1835	*Coquimbo notebook*: 25	Los Hornos
08 May 1835	*Coquimbo notebook*: 48	Combarbala
09 May 1835	*Coquimbo notebook*: 57	Mineral of Punitague
10 May 1835	*Coquimbo notebook*: 59	Ovalle
11 May 1835	*Coquimbo notebook*: 64	Panuncillo
12 May 1835	*Coquimbo notebook*: 75	Panuncillo
13 May 1835	*Coquimbo notebook*: 75	The Punta
14 May 1835	*Coquimbo notebook*: 78	Coquimbo
16 May 1835	*Coquimbo notebook*: 88	Coquimbo
18 May 1835	*Coquimbo notebook*: 88	Coquimbo
20 May 1835	*Coquimbo notebook*: 95, BC	Coquimbo
21 May 1835	*Coquimbo notebook*: 100	Mine of Edwards
22 May 1835	*Coquimbo notebook*: 102	Mine of Edwards
23 May 1835	*Coquimbo notebook*: 109	Hacienda of Don Jose
24 May 1835	*Coquimbo notebook*: 120	Hacienda of Don Jose
25 May 1835	*Coquimbo notebook*: 121	Rio Claro
26 May 1835	*Coquimbo notebook*: 128	Hacienda of Don Jose
31 May 1835	*Coquimbo notebook*: FC	Coquimbo
01 June 1835	*Copiapò notebook*: 1	Coquimbo
02 June 1835	*Copiapò notebook*: 2	Yerba Buena
03 June 1835	*Copiapò notebook*: 4	Carizal
04 June 1835	*Copiapò notebook*: 7	Sauce
05 June 1835	*Copiapò notebook*: 12	Guasco
06 June 1835	*Copiapò notebook*: 14	Guasco
07 June 1835	*Copiapò notebook*: 17	Guasco
08 June 1835	*Copiapò notebook*: 19	Ballenar
09 June 1835	*Copiapò notebook*: 20	Ballenar
10 June 1835	*Copiapò notebook*: 22	Valley above Ballenar
11 June 1835	*Copiapò notebook*: 25	Valley above Ballenar

Date	Notebook/page	Locality
12 June 1835	*Copiapò notebook*: 26	Hacienda of Potrero Seco
13 June 1835	*Copiapò notebook*: 26	Hacienda of Potrero Seco
14 June 1835	*Copiapò notebook*: 32	Hacienda of Potrero Seco
15 June 1835	*Copiapò notebook*: 43	Hacienda of las Amolanas
16 June 1835	*Copiapò notebook*: 50	Hacienda of las Amolanas
17 June 1835	*Copiapò notebook*: 58	Jolquera valley
18 June 1835	*Copiapò notebook*: 65	Jolquera valley
19 June 1835	*Copiapò notebook*: 70	Hacienda of las Amolanas
20 June 1835	*Copiapò notebook*: 73	Hacienda of las Amolanas
21 June 1835	*Copiapò notebook*: 81	Hacienda of Potrero Seco
22 June 1835	*Copiapò notebook*: 82	Copiapò (Mr Bingley)
23 June 1835	*Copiapò notebook*: 91	Copiapò
24 June 1835	*Copiapò notebook*: 92	Copiapò
25 June 1835	*Copiapò notebook*: 92	Copiapò
26 June 1835	*Despoblado notebook*: 1b	Copiapò
27 June 1835	*Despoblado notebook*: 2b	Ravine of Paypote
28 June 1835	*Despoblado notebook*: 12b	Cordilleras Maricongo
29 June 1835	*Despoblado notebook*: 20b	Ravine of Paypote
30 June 1835	*Despoblado notebook*: 22b	Copiapò
01 July 1835	*Despoblado notebook*: 28b	Copiapò
02 July 1835	*Despoblado notebook*: 35b	Copiapò
03 July 1835	*Despoblado notebook*: 35b	Copiapò
04 July 1835	*Despoblado notebook*: 36b	Copiapò port
05 July 1835	*Despoblado notebook*: 39b	Copiapò port
06 July 1835	*Despoblado notebook*: 40b	Copiapò–Iquique, Peru
12 July 1835	*Despoblado notebook*: 42b	Iquique
15 July 1835	*Despoblado notebook*: 42b	Iquique-Lima
19 July 1835	*Despoblado notebook*: 45b	Lima
20 July 1835	*Despoblado notebook*: 45b	Lima
21 July 1835	*Despoblado notebook*: 45b	Lima
22 July 1835	*Galapagos notebook*: 8a	Lima
29 July 1835	*Galapagos notebook*: 17a, 25a	Lima
03 August 1835	*Galapagos notebook*: 25a	Lima
31 August 1835	*Despoblado notebook*: 24a	Lima
07 September 1835	*Despoblado notebook*: 43b	Lima–Galapagos
19 September 1835	*Galapagos notebook*: 18b	Chatham Island, Galapagos
20 September 1835	*Galapagos notebook*: 18b	Chatham Island, Galapagos
30 September 1835	*Galapagos notebook*: 29a, BC	Albemarle Island, Galapagos
12 October 1835	*Galapagos notebook*: 36b	James Island, Galapagos
13 October 1835	*Galapagos notebook*: 40b	James Island, Galapagos
14 October 1835	*Galapagos notebook*: 42b	James Island, Galapagos
16 October 1835	*Galapagos notebook*: 45b	James Island, Galapagos
17 October 1835	*Galapagos notebook*: 50b	James Island, Galapagos

Date	Notebook/page	Locality
18 October 1835	*Galapagos notebook*: 50b	Albemarle Island, Galapagos
18 November 1835	*Galapagos notebook*: 52b	Tahiti
19 November 1835	*Galapagos notebook*: 61b	Tahiti
20 November 1835	*Galapagos notebook*: 64b	Matavai, Tahiti
21 November 1835	*Galapagos notebook*: 65b	Matavai, Tahiti
22 November 1835	*Galapagos notebook*: 65b	Matavai, Tahiti
23 November 1835	*Galapagos notebook*: 66b	Matavai, Tahiti
25 November 1835	*Galapagos notebook*: 66b	Papiete, Tahiti
26 November 1835	*Galapagos notebook*: 66b	Tahiti-New Zealand
16 January 1836	*Sydney notebook*: 1a	Emu ferry (Penrith), Australia
17 January 1836	*Sydney notebook*: 7a	Blackheath Inn
18 January 1836	*Sydney notebook*: 16a	Walarawang
19 January 1836	*Sydney notebook*: 25a	Walarawang
20 January 1836	*Sydney notebook*: 36a	Bathurst
21 January 1836	*Sydney notebook*: 46a	Bathurst
22 January 1836	*Sydney notebook*: 46a	Tarana
23 January 1836	*Sydney notebook*: 48a	Weatherboard Inn
24 January 1836	*Sydney notebook*: 50a	Weatherboard Inn
25 January 1836	*Sydney notebook*: 50a	Weatherboard Inn
26 January 1836	*Sydney notebook*: 51a	Dunheved
27 January 1836	*Sydney notebook*: 54a	Sydney
05 May 1836	*Sydney notebook*: 65a	Port Louis, Mauritius
31 May 1836	*Despoblado notebook*: 52b	Simon's Town, Cape of Good Hope
01 June 1836	*Despoblado notebook*: 52b	Cape Town
02 June 1836	*Despoblado notebook*: 53b	Cape Town
04 June 1836	*Despoblado notebook*: 55b	Paarl
05 June 1836	*Despoblado notebook*: 58b	Franschhoek
06 June 1836	*Despoblado notebook*: 62b	Franschhoek
07 June 1836	*Despoblado notebook*: 64b	Cape Town
08 June 1836	*Despoblado notebook*: 72b	Cape Town
09 June 1836	*Despoblado notebook*: 72b	Cape Town
10 June 1836	*Despoblado notebook*: 72b	Cape Town
12 June 1836	*Despoblado notebook*: 73b	Cape Town
13 June 1836	*Despoblado notebook*: 73b	Cape Town
15 June 1836	*Despoblado notebook*: 73b	Cape Town
16 June 1836	*Despoblado notebook*: 73b	Simon's Town
17 June 1836	*Despoblado notebook*: 73b	Simon's Town
18 June 1836	*Despoblado notebook*: 73b	Cape of Good Hope–St Helena
29 June 1836	*Despoblado notebook*: 74b	St Helena
08 July 1836	*Despoblado notebook*: 74b	St Helena
09 July 1836	*Despoblado notebook*: 75b	St Helena
10 July 1836	*Despoblado notebook*: 76b	St Helena

Date	Notebook/page	Locality
11 July 1836	*Despoblado notebook*: 78b	St Helena
13 July 1836	*Despoblado notebook*: 79b	St Helena
14 July 1836	*Despoblado notebook*: 85b	St Helena–Ascension
12 August 1836	*Despoblado notebook*: 87b	Pernambuco
31 August 1836	*Despoblado notebook*: 34b	Cape de Verds

EXPEDITION EQUIPMENT

⌒

This list is an attempt to collect all known references to the clothing and equipment Darwin carried on his inland expeditions, when the notebooks were used. Only items for which there is documentary evidence are listed. The main sources are the *Beagle* notebooks and *Beagle diary*. Of course it must be borne in mind that listing an item in a notebook does not necessarily mean that Darwin had the item with him. This is only unambiguously the case if he actually referred to using the item in his notes, for example pushing a bird off a tree with his gun on the Galapagos. He sometimes carried a rifle but there are few references for this, other than 12 September 1832 (*Beagle diary*, p. 104). In the *Autobiography* Darwin explained that in the early years of the voyage he did his own shooting, but gradually handed this over to his servant Syms Covington. The instruction 'compare with blowpipe' (*Santiago notebook*, p. 13) indicates that Darwin had left this item of mineralogical equipment on the *Beagle*.[1052] Cathy Power kindly provided the English Heritage accession numbers (EH) for items believed to be in the Down House Collection.

Clothing

'Belt' (also used for measuring) *Port Desire*: 128, CCD1: 248.
'Boots' *Galapagos*: 5b, 'thick boots' *Beagle diary*: 363, 'Strong Boots' *St. Fe*: 232a.
'cape of the Indian-rubber cloth' *Beagle diary*: 184.
'Gloves' *Falkland*: 4b.
'Handkerchief' (used for measuring length) *Buenos Ayres*: 80a, (used to wrap specimens) Falkland 52a, 'Blank silk handkerchief *St. Fe*: 5a, 'silk pocket-handkerchief *Mammalia*: 31.
Hat: '[Big Hat]' *St. Fe*: 4a. (Panama hat EH 88202307)
'Mackintosh' *St. Fe*: 5a.
'Nightcap' *Falkland*: 4b, 'Night caps' *Galapagos*: 11b.
'Poncho' *Falkland*: 3a.
'shirt' *Buenos Ayres*: 18b, *Galapagos*: 5b.

[1052] See Herbert 2005 for photographs.

Shoes: 'shoes' *St. Fe*: 6b, *Galapagos*: 30a, 'Shoes blacking' *Coquimbo*: 132, '4 pair of very strong walking shoes' CCD1: 314–5.
'Shooting Jacket' *St. Fe*: 5a, 237a.
Stockings: 'Woollen Stockings' *St. Fe*: 5a, *Galapagos*: 11b, *Falkland*: 4b 'Worsted stockings' *St. Fe*: 189a.
'Trousers' *St. Fe*: 5a, 237a, 'Drawers' *Buenos Ayres*: 18b.
Waistcoat: 'Flannel waistcoat' *Beagle diary*: 311.

Food, drink and medicine

'Bread' *Galapagos*: 5b, *St. Fe*: 5a, 2b, *B. Blanca*: 2a.
Canteens: 'Two Metal Canteens' Shipley and Simpson 1909: 8.
'Cheese' [traps baited with] *Birds*: 53 *et al.*
'Chocolate' *Galapagos*: 5b.
'flask with water' *Falkland*: 82a.
'flask of spirits' *Beagle diary*: 373 (in the original manuscript: '~~bottle~~ flask of spirits').
'Laudanum Lozenges' *Sydney*: 5b.
'Mattee' *Galapagos*: 2b, 3b.
'Meat' *Galapagos*: 5b.
'Medicin' *Galapagos*: 8b,11b.
'Peppermint' *Sydney*: 5b.
'Carb of Soda & Magnesia' *Sydney*: 5b.
'Pills' *St. Fe*: 5a, *Copiapò*: IBC.
'Tea' *Galapagos*: 5b.
'Yerba' *B. Blanca*: 2b.
'Sugar' *Galapagos*: 7b, *St. Fe*: 5a, 2b, *B. Blanca*: 2a.

Personal

Books: 'a single small volume' of Milton's *Paradise Lost*, *Autobiography*: 85, 'Milton' *Despoblado*: 132. Apart from Milton, no evidence has been found that Darwin carried any books with him and there is some evidence that he did not, e.g. 'no books' *Falkland*: 124a. The only exception is Desquiron de Saint-Agnan ed. 1821, Darwin recorded reading in *B. Blanca*: p. 12a, but presumably this was present in the house where he was staying. He may have taken Lyell's *Principles of geology* and other books which were his personal property, but the *Beagle* library rules expressly prohibited taking the library books off the ship (see CCD1: 554). Darwin does, however, list numerous books that he wanted to read or re-read, e.g. Humboldt 'of course' *Falkland*: 74a and it is certainly possible that he used *Werner's nomenclature of colours* by Patrick Syme (1821) to capture the colours of specimens before they started to fade, e.g. condor iris 'scarlet red' *Banda Oriental*: 66 and hawk iris 'honey yellow' *Port Desire*: 53.
'bracers' *Galapagos*: 7b.
'Candles' *Galapagos*: 5b.

'Cigaritos' *Galapagos*: 3b.

'Cigars' *Buenos Ayres*: 1a, *St. Fe*: 5a–2b, *Galapagos*: 5b, *B. Blanca*: 2a.

'Comb' *Falkland*: 4b.

'knapsack' *Falkland*: 52a.

'Lavander water' *Sydney*: 5b.

'Letter paper' *Galapagos*: 7b.

Matches: 'Prometheans' *Falkland*: 4b, 'Promethians' *Beagle diary*: 155.

'Money' *St. Fe*: 4a, *Galapagos*: 9b.

'Mule-shoe' *Galapagos*: 6b.

Oil: 'Sweet smelling oil' *Galapagos*: 11b.

'Passport' *Falkland*: 4b, *Galapagos*: 5b, *St. Fe*: 6a.

'Razors' *Galapagos*: 11b.

'Saddle for mule' *Galapagos*: 3b.

'Snuff' *St. Fe*: 5a, *Galapagos*: 6b, 'Wood Snuff Box' *St. Fe*: 5a.

'Soap' *Sydney*: 6b.

'Stirrups' *Buenos Ayres*: 12b, *Galapagos*: 7b.

'Tobacco' *Sydney*: 8b.

'Tooth-Brushes' *Sydney*: 7b, *Galapagos*: 7b.

Watch: *Buenos Ayres*: 12b, 'Watch key & glass' *Buenos Ayres*: 3.

Tools and instruments

Barometer: 'Mountain barometer' *Beagle diary*: 68, 'Barometer' *Falkland*, 6a, CCD1:399, *South America*, many references but also 'not having a barometer with me' p. 45. (EH 88202303)

'bed & a Kettle, & a pot, a plate & basin' CCD1: 449.

'Bees wax' *St. Fe*: 5a.

'Big bag' *St. Fe*: 5a.

'Bladders' *St. Fe*: 53a.

Bottles: 'Big Bottles' *B. Blanca*: 3a, *St. Fe*: 5a, 'Bottle small with large mouth' *St. Fe*: 5a, 'Spider Bottle', *Zoology notes*: 38, 'collecting bottles covered with leather' *Journal of researches*: 601.

Boxes: 'Pill Boxes' *Galapagos*: 11b, *Buenos Ayres*: 16b, 'pill-boxes', 'chip pill-boxes', 'tin boxes' *Journal*: 601.

'Chissel' *Galapagos*: 6b, 'chisels … for fossils' Darwin 1849: 160.

'clinometer' Herschel: 160. (EH 88202343)

Compass: 'Kater compass' *B. Blanca*: 47a, 'Katers Compass' *St. Fe*: 4a, 2b, *Beagle diary*: 186.

'Corks for Jars' *St. Fe*: 5a.

'Fish hooks' *St. Fe*: 2b.

Firearms: Pistols: 'small pistol' *Beagle diary*: 46, 'Pistol, balls, powder' *Falkland*: 4b, *St. Fe*: 5a, *Galapagos*: 4b. (Pocket pistol EH 88202355), Rifle: 'our rifles' *Beagle diary*: 104, 'my gun'353, *B. Blanca*: 7a, 'Gunpowder & shot' *St. Fe*: 6a, *Falkland*: 4b, *St. Fe*: 5a, *Galapagos*: 4b. 'wadding' *Falkland*: 75a.

'fly-nippers' *Beagle diary*: 22.

'forceps' *Journal of researches*: 38.

'Gold Leaf' *St. Fe*: 5a.

Hammers: '25 pound hammer' *Coquimbo*: 65, Herschel: 229, 'small hammer' *Falkland*: 4b, 'Hammer & [tools]', 'hammer' Buenos Ayres: 15b, 'a heavy hammer, with its two ends wedge-formed and truncated, a light hammer for trimming specimens' Darwin 1849: 160. (EH 88202681)

'Hygrometer' *Beagle diary*: 31, 70. (EH 88202304)

Knives: 'I wear a large clasp-knife, in the manner of sailors fastened by a string round my neck' *Beagle diary*: 158, *St. Fe*: 5a, 'Big knife' *Falkland*: 4b, 'Pen knife' *St. Fe*: 5a, 'Second Penknife' *Sydney*: 4b.

Lenses: 'a pocket-lens with three glasses' Darwin 1849: 160. A lens with two is implied on *St. Fe*: 126a. (EH 88202694/ 2693)

Nets: 'a plain strong sweeping-net', *Journal*: 601, 'Fly net' *Beagle diary*: 46, 'insect net, with scissor handles' Harmer and Ridewood 1910: 24, 'casting net' [for fish] *Narrative* 2: 343.

'Note book' *Buenos Ayres*: 1a, 12b, *Falkland*: 4b, *B. Blanca*: 3b, *St. Fe*: 6a, *Valparaiso*: 98a

'Paper for Plants' *St. Fe*: 5a, 7a, 2b.

Pencils: 'silver pencil case' *Beagle diary*: 26, '[Sharp] Pencil' *Galapagos*: 7b, *Sydney*: 4b 'Spare pencil' *Falkland*: 4b.

Pens: 'Bramah pens', *Buenos Ayres*: 1a, *Falkland*: 4b, *Sydney*: 4b, 7b, 'Common [pens]', *Sydney*: 4b.

'[pincer] for insects & bottle' *Rio*: 50b

'Pix axe' *Rio*: 50b, *St. Fe*: 2b, 'pickaxe for fossils' Herschel: 160.

'Protractor' *St Fe*: 5a

'Rosin' *St. Fe*: 5a.

'Salt petre' *Falkland*: 4b.

'Scissors' *Buenos Ayres*: 1a.

'Sextant & artificial horizon' *Beagle diary*: 336.

'small Vasculums' *Falkland*: 75a.

'Spirit bottle' *Falkland*: 4, 'Jar with spirits' *St. Fe*: 2b, 'Keg of Spirits' *Galapagos*: IBC.

'tape measure', CCD1: 314. (EH 88202759)

'telescope' *Beagle diary*: 184. (EH 88202348/ pocket telescope 88202342)

'Thermometer' Galapagos: 8a, 'therm: in pocket' *Rio*: 5b

'Tow' *St. Fe*: 5a, *Falkland*: 83a.

Traps: 'Mice & Rat traps' *St. Fe*: 5a, 'Mice traps' *Falkland*: 75a, 'mouse traps' *Beagle diary*: 160, ('baited either with cheese or meat' *Mammalia*: 53, 'baited with a piece of bird': 49) 'Many contrivances for catching animals' *Beagle diary*: 74.

BIBLIOGRAPHY

Items known to be in the *Beagle*'s library (see CCD1 Appendix IV) are marked [*Beagle*]. Some works now in the Darwin Library in Cambridge University Library contain inscriptions written during the voyage, these are included in parentheses. Works cited in the notebooks, which are not otherwise known to be in the *Beagle* library, are marked with *.

Items currently available on *Darwin Online* (http://darwin-online.org.uk/) are marked [DO].

Adler, Saul 1959. Darwin's illness. *Nature* 4693: 1102–3.

Alban, John. 1990. The wider world. In Griffiths, Ralph A. ed. *The city of Swansea: challenges and change*. Gloucester: Sutton, pp. 114–29.

Allan, Mea. 1977. *Darwin and his flowers. The key to natural selection*. London: Faber and Faber.

Animal notes: Keynes, Richard ed. 2005. Charles Darwin's *Beagle* animal notes (1832–33). [DAR 29.1.A1–A49] [DO]

Anon. 1833. *The nautical almanac and astronomical ephemeris for the year 1834. Published by order of The Lords Commissioners of the Admiralty*. London: John Murray.

Armstrong, Patrick H. 1985. *Charles Darwin in Western Australia: a young scientist's perception of an environment*. Nedlands, W.A.: University of Western Australia Press.

—. 1991a. Three weeks at the Cape of Good Hope 1836: Charles Darwin's African Interlude. *Indian Ocean Review* 4(2): 8–13,19.

—. 1991b. *Under the Blue Vault of Heaven: A study of Charles Darwin's Sojourn in the Cocos (Keeling) Islands*. Nedlands, Western Australia: Indian Ocean Centre for Peace Studies.

—. 1992. *Darwin's desolate islands: a naturalist in the Falklands, 1833 and 1834*. Chippenham: Picton.

—. 2002. Antlions: A link between Charles Darwin and an early Suffolk naturalist. *Transactions of the Suffolk Natural History Society* 38: 81–6.

—. 2004. *Darwin's other islands*. London and New York: Continuum.

Aubuisson de Voisins, Jean François d'. 1819. *Traité de géognosie, ou exposé des connaissances actuelles sur la constitution physique et minérale du globe terrestre*. 2 vols. Strasbourg: F. G. Levrault. [*Beagle*: (Inscription in vol. 1: 'C. Darwin HMS *Beagle*')]

Autobiography: Barlow, Nora. 1958. *The autobiography of Charles Darwin 1809–1882. With the original omissions restored. Edited and with appendix and notes by his grand-daughter Nora Barlow*. London: Collins. [DO]

Azara, Félix d'. 1801. *Essais sur l'histoire naturelle des quadrupèdes de la Province du Paraguay écrits depuis 1783, jusqu'en 1796, avec une appendice sur quelques reptiles; et formant suite nécessaire aux Oeuvres de Buffon. Traduits sur le manuscrit inédit de l'Auteur, par M. L. E. Moreau-Saint-Méry.* Paris: Imprimerie de C. Pougens, Librairie de Madame Huzard.

—. 1802–5. *Apuntamientos para la historia natural de los páxaros del Paraguay y Rio de la Plata.* 3 vols. Madrid: Imprenta de la Viuda de Ibarra.*

—. 1809. *Voyages dans l'Amérique Méridionale par F. de Azara depuis 1781 jusqu'en 1801, contenant la description géographique, politique et civile du Paraguay et de la rivière de La Plata, l'histoire de la découverte et de la conquête de ces contrées; des détails nombreux sur leur histoire naturelle, et sur les peuples sauvages qui les habitent. Publiés d'après les manuscrits de l'auteur, avec une notice sur sa vie et ses écrits, par C. A. Walckenaer; enrichis de notes par G. Cuvier ... suivis de l'histoire naturelle des Oiseaux du Paraguay et de La Plata par le même auteur, traduite, d'après l'original espagnol, et augmentée d'un grand nombre de notes par M. Sonnini. Accompagnés d'un Atlas de vingt-cinq planches.* Paris: Dentu.*

Banks, Maxwell R. and David Leaman eds. 1999. Charles Darwin's field notes on the geology of Hobart Town – A modern appraisal. *Papers and Proceedings of the Royal Society of Tasmania* 133(1): 29–50. [DO]

Barlow, Nora ed. 1933. *Charles Darwin's diary of the voyage of H. M. S. Beagle.* Cambridge: University Press.

—. ed. 1945. *Charles Darwin and the voyage of the Beagle.* London: Pilot Press. [DO]

—. ed. 1967. *Darwin and Henslow. The growth of an idea.* London: Bentham-Moxon Trust, John Murray. [DO]

Barrett, Paul H., Gautrey, Peter J., Herbert, Sandra, Kohn, David, Smith, Sydney eds. 1987. *Charles Darwin's notebooks, 1836–1844: Geology, transmutation of species, metaphysical enquiries.* London: British Museum (Natural History); Cambridge: Cambridge University Press.

Bartolomé, Gerardo and Glickman, Barry 2008. *Patagonia con los ojos de Darwin.* Ushuaia: Zagier & Urruty.

Beagle diary: Keynes, Richard ed. 1988. *Charles Darwin's Beagle diary.* Cambridge: Cambridge University Press. [DO]

Beagle plants: Porter, Duncan M. 1987. Darwin's notes on *Beagle* plants. *Bulletin of the British Museum (Natural History) Historical Series* 14, No. 2: 145–233. [DO]

Beechey, Frederick William. 1831. *Narrative of a voyage to the Pacific and Beering's Strait, to co-operate with the polar expeditions performed in His Majesty's ship Blossom, in the years 1825, 26, 27, 28.* 2 vols. London: Henry Colburn and Richard Bentley. [*Beagle*]

Benton, Michael J. 1990. *Vertebrate palaeontology.* London: Unwin Hyman.

Bernardin de Saint Pierre, Jacques Henri. 1773. *Voyage a l'Isle de France, a l'Isle de Bourbon, au Cap de Bonne-Espérance, &c.: avec des observations nouvelles sur la nature & sur les hommes.* vol. 1. Amsterdam. [*Beagle*]

Birds: Gould, John. 1838–41. Part 3 of *The zoology of the voyage of H. M. S. Beagle. Birds. By John Gould. Edited and superintended by Charles Darwin.* London: Smith Elder and Co. [DO]

Blainville, Henri Marie Ducrotay de. *et al.* 1834. Rapport sur les resultants scientifiques du voyage de M. Alcide d'Orbigny ... par MM. De Blainville, Brongniart, Savary, Cordier. *Nouvelles Annales du Muséum National d'Histoirs Naturelle* 3: 84–115. [*Beagle*]

Boué, Ami. 1830. [Formations secondaires autour du monde.] *Journal de Géologie* 2 (6): 205–7.*

Bougainville, Louis Antoine de. 1772. *A voyage round the world: performed by order of His Most Christian Majesty, in the years 1766, 1767, 1768, and 1769 by Lewis de Bougainville, colonel of foot, and commodore of the expedition, in the frigate La Boudeuse, and the store-ship L'Etoile; translated from the French by John Reinhold Forster*. London: Printed for J. Nourse and T. Davies. [*Beagle*]

Brongniart, Alexandre. 1833. Rapport fait à l'Academie Royale des Sciences, sur les travaux de M. Gay. *Annales des Sciences Naturelles* 28: 26–35. [*Beagle*]

Browne, Janet. 1995. *Charles Darwin: voyaging*. London: Jonathan Cape.

Buch, Christian Leopold von. 1813. *Travels through Norway and Lapland, during the years 1806, 1807, and 1808. Translated from the original German by John Black. With notes and illustrations, chiefly mineralogical, and some account of the author, by Robert Jameson ... Illustrated with maps and physical sections*. London: Henry Colburn. [*Beagle*]

Burchell, William John. 1822–4. *Travels in the interior of Southern Africa*. 2 vols. London: Longman, Hurst, Rees, Orme, and Brown. [*Beagle*]

Byron, George Anson. 1826. *Voyage of H. M. S. Blonde to the Sandwich Islands, in the years 1824–25*. London: John Murray. [*Beagle*]

Caldcleugh, Alexander. 1825. *Travels in South America, during the years 1819–20–21: containing an account of the present state of Brazil, Buenos Ayres, and Chile*. 2 vols. London: John Murray. [*Beagle*] [DO]

Candolle, Alphonse de. 1855. *Géographie botanique raisonnée ou exposition des faits principaux et des lois concernant la distribution géographique des plantes de l'époque actuelle*. 2 vols. Paris: Victor Mason and Geneva: J. Kessmann. [DO] [*Beagle*]

Candolle, Augustin Pyramus de. 1820. Géographie botanique. In Cuvier, G. ed. *Dictionnaire des Sciences Naturelles*, Paris, 18: 359–422.

—. 1839–40. *Vegetable organography; or an analytical description of the organs of plants*. Translated by Boughton Kingdon. 2 vols. London: Houlston & Stoneman.

Catling, David C. 2009. Revisiting Darwin's voyage. In Kelly, A. *et al.* eds. *Darwin: For the love of science*. Bristol: Cultural Development Partnership.

CCD: Burkhardt Frederick H. *et al.* eds. 1985–. *The correspondence of Charles Darwin*. 16 vols. Cambridge: Cambridge University Press.

Chancellor, Gordon. 1990. Charles Darwin's St Helena Model notebook. *Bulletin of the British Museum (Natural History) Historical Series* 18(2): 203–28. [DO]

—. Rookmaaker, Kees and van Wyhe, John eds. 2007. 'Chiloe Jan^r. 1835'. [DAR 35.328, 328a–328j] [DO]

—. 2008. The *Beagle* paintings of John Chancellor (1925–84). *The Linnean Special Issue* No 9: 49–60.

Chancellor, Gordon, diMauro, Angelo, Ingle, Ray and King, Gillian. 1988. Charles Darwin's *Beagle* collections in the Oxford University Museum. *Archives of Natural History* 15 (2): 197–231.

Cleaveland, Parker. 1816. *An elementary treatise on mineralogy and geology: being an introduction to the study of these sciences, and designed for the use of pupils, – for persons attending lectures on these subjects, – and as a companion for travellers in the United States of America*. Boston: Cummings and Hilliard; Cambridge: Cambridge University Press.*

Colp, Ralph. 1977. *To be an invalid: the illness of Charles Darwin*. Chicago and London: Chicago University Press.

Coral reefs: Darwin, Charles Robert. 1842. *The structure and distribution of coral reefs. Being the first part of the geology of the voyage of the* Beagle, *under the command of Capt. Fitzroy, R. N. during the years 1832 to 1836.* London: Smith Elder and Co. [DO]

Coral reefs 2d ed.: Darwin, Charles Robert. 1874. *The structure and distribution of coral reefs.* 2d ed. London: Smith Elder and Co. [DO]

Crompton, A. W. and Singer, R. 1958. Darwin's visit to the Cape. *Quarterly Bulletin of the South African Library* (September): 9–11.

Cuvier, Georges. 1830. *The animal kingdom arranged in conformity with its organization ... with additional descriptions of all the species hitherto named, and of many not before noticed.* By Edward Griffith and others. Supplementary volume on the fossils. London: Whittaker, Treacher and Co. [*Beagle*, number of volumes on board unknown.]

Darwin, Charles Robert. [1835]. [Extracts from letters addressed to Professor Henslow]. Cambridge: [privately printed]. [DO]

—. 1837a. On certain areas of elevation and subsidence in the Pacific and Indian oceans, as deduced from the study of coral formations. [Read 31 May] *Proceedings of the Geological Society of London* 2: 552–4. [DO]

—. 1837b. A sketch of the deposits containing extinct Mammalia in the neighbourhood of the Plata. [Read 3 May] *Proceedings of the Geological Society of London* 2: 542–4. [DO]

—. 1838. On the connexion of certain volcanic phænomena, and on the formation of mountain-chains and volcanos, as the effects of continental elevations. [Read 7 March] *Proceedings of the Geological Society of London* 2: 654–60. [DO]

—. 1839. Observations on the parallel roads of Glen Roy, and of other parts of Lochaber in Scotland, with an attempt to prove that they are of marine origin. [Read 7 February] *Philosophical Transactions of the Royal Society* 129: 39–81. [DO]

—. 1840. On the connexion of certain volcanic phenomena in South America; and on the formation of mountain chains and volcanos, as the effect of the same powers by which continents are elevated. [Read 7 March] *Transactions of the Geological Society of London* (Ser. 2) 5 (3): 601–31, pl. 49, fig. 1. [DO]

—. 1841. On a remarkable bar of sandstone off Pernambuco, on the coast of Brazil. *The London, Edinburgh and Dublin Philosophical Magazine* (Ser. 3) 19 (October): 257–60, 1 text figure. [DO]

—. 1842a. On the distribution of the erratic boulders and on the contemporaneous unstratified deposits of South America. [Read 14 April 1841] *Transactions of the Geological Society* Part 2, 3 (78): 415–31. [DO]

—. 1842b. Notes on the effects produced by the ancient glaciers of Caernarvonshire, and on the boulders transported by floating ice. *The London, Edinburgh and Dublin Philosophical Magazine* 21 (September): 180–8. [DO]

—. 1844. Observations on the structure and propagation of the genus *Sagitta*. *Annals and Magazine of Natural History* 13 (January): 1–6, 1 plate. [DO]

—. 1846. On the geology of the Falkland Islands. [Read 25 March] *Quarterly Journal of the Geological Society of London* 2: 267–79. [DO]

—. 1849. Section VI: Geology. In John Frederick William Herschel ed., *A manual of scientific enquiry; prepared for the use of Her Majesty's Navy: and adapted for travellers in general.* London: John Murray, pp. 156–95. [DO]

—. 1870. Notes on the habits of the pampas woodpecker (*Colaptes campestris*). [Read 1 November] *Proceedings of the Zoological Society of London* no. 47: 705–6. [DO]

—. 1880. Darwin's reply to a vegetarian. *Herald of Health and Journal of Physical Culture* n.s. 31: 180. [DO]

—. 1881. The parasitic habits of *Molothrus*. *Nature* 25 (17 November): 51–2. [DO]

Darwin's insects: Smith, Kenneth G. V. 1987. Darwin's insects: Charles Darwin's entomological notes, with an introduction and comments by Kenneth G. V. Smith. *Bulletin of the British Museum (Natural History) Historical Series* 14 (1): 1–143. [DO]

Darwin Online: Wyhe, John van ed. 2002–. *The complete work of Charles Darwin Online*. [DO]

Daubeny, Charles Giles Bridle. 1826. *A description of active and extinct volcanoes; with remarks on their origin, their chemical phaenomena, and the character of their products, as determined by the condition of the earth during the period of their formation. Being the substance of some lectures delivered before the University of Oxford, with much additional matter*. London: W. Phillips; and Oxford: Joseph Parker. [*Beagle*]

Davy, Humphry. 1830. *Consolations in travel, or the last days of a philosopher*. Edited by John Davy. London: John Murray.*

De la Beche, Henry Thomas. 1831. *A geological manual*. London: Treuttel & Wurtz, Treuttel Jun. & Richter. [*Beagle*]

Desjardin, Julien. 1832. Extract from 'Analyse des travaux de la Société d'Histoire Naturelle de l'Ile Maurice, pendant la 2de année.' *Proceedings of the Zoological Society of London* 2: 111–12.

Desquiron de Saint-Agnan ed. 1821. *Historia del proceso de la reine de Inglaterra escrita en francés con presencia de documentos fidedignos recogidos en Londres por A. T. Desquiron de St. Agnan y traducida al castellano por el ciudadano don Juan Valle y Codés*. Barcelona: Imprenta nacional de la Viuda Roca.

Dillon, Peter. 1829. *Narrative and successful result of a voyage in the South Seas: performed by order of the government of British India, to ascertain the actual fate of La Peyrouse's expedition*. 2 vols. London: Hurst, Chance.*

DNB: Oxford dictionary of national biography. http://www.oxforddnb.com/

Dobson, Jessie. 1959. *Charles Darwin and Down House*. Edinburgh and London: Churchill Livingstone (reprinted 1971).

Drapiez, P. A. J. 1830. Vautorins. *Dictionnaire classique d'histoire naturelle*. Paris: Rey et Gravier, Amable Gobin et Cie, vol. 16, p. 518. [*Beagle*]

Duperrey, Louis Isidore. 1826–30. *Voyage autour du monde … sur la Corvette de sa Majesté, La Coquille, pendant les années 1822, 1823, 1824 et 1825. Zoologie, par MM. Lesson et Garnot*. 3 vols. Paris: Arthus Bertrand. [*Beagle*]

Earthworms: Darwin, Charles Robert. 1881. *The formation of vegetable mould, through the action of worms, with observations on their habits*. London: John Murray. [DO]

Ellis, William. 1829. *Polynesian researches, during a residence of nearly six years in the South Sea Islands, including descriptions of the natural history and scenery of the Islands, with remarks on the history, mythology, traditions, government, arts, manners, and customs of the inhabitants*. 2 vols. London: Fisher, Son, & Jackson. [*Beagle*]

Encyclopedia Britannica. 20 vols., 1 vol. *Supplement*. 6th ed. Edinburgh, 1823. [*Beagle*]

Estes, Gregory, Grant, K. Thalia and Grant, Peter R. 2000. Darwin in Galapagos. His footsteps through the archipelago. *Notes and Records of the Royal Society* 54: 343–68.

Eyton, Thomas Campbell. 1838. *A monograph on the Anatidæ or Duck Tribe*. London: Longman, Orme, Brown, Green, & Longman; Shrewsbury: Eddowes.

Falconer, John Downie. 1937. Darwin in Uruguay. *Nature* (24 July): 138–9.

Falkner, Thomas. 1774. *A description of Patagonia and the adjoining parts of South America, containing an account of the soil, produce, animals, vales, mountains, rivers, lakes, &c. of those countries; the religion, government, policy, customs, dress, arms, and language of the Indian inhabitants; and some particulars relating to Falkland Islands.* Hereford: printed by C. Pugh. [*Beagle*]

Febres, Andrés. 1765. *Arte de la lengua general del reyno de Chile, Calle de la Encarnación.* Lima.*

Fish: Jenyns, Leonard. 1840–2. *Fish.* Part 4 of *The zoology of the voyage of HMS* Beagle. *Edited and superintended by Charles Darwin.* London: Smith Elder and Co. [DO]

Fitzinger, Leopold Joseph Franz Johann. 1826. *Neue Classification der Reptilien nach ihren natürlichen Verwandtschaften: Nebst einer verwandtschafts-Tafel und einem Verzeichnisse der Reptilien-Sammlung des K. K. zoologischen Museum's zu Wien.* Vienna: J. G. Heubner.*

FitzRoy, Robert. 1836. Sketch of the surveying voyages of his Majesty's ships Adventure and Beagle, 1825–1836. Commanded by Captains P. P. King, P. Stokes, and R. Fitz-Roy, Royal Navy. *Journal of the Geological Society of London* 6: 311–43. [DO]

—. 1837a. Extracts from the diary of an attempt to ascend the River Santa Cruz, in Patagonia, with the boats of his Majesty's sloop Beagle. By Captain Robert Fitz Roy, R. N. *Journal of the Royal Geographical Society of London* 7: 114–26. [DO]

—. 1837b Notice of the Mountain Aconcagua in Chile. *Journal of the Royal Geographical Society of London* 7: 143–4

—. and Darwin, Charles Robert. 1836. A letter, containing remarks on the moral state of Tahiti, New Zealand, &c. *South African Christian Recorder* 2 (4) (September): 221–38. [DO]

Fleming, John. 1822. *The philosophy of zoology; or, A general view of the structure, functions and classification of animals.* 2 vols. Edinburgh: A. Constable. [*Beagle*]

Forster, John Reinhold. 1778. *Observations made during a voyage round the world, on physical geography, natural history, and ethic philosophy.* London: G. Robinson. [*Beagle*]

Fossil mammalia: Owen, Richard. 1838–40. Part 1 of *The zoology of the voyage of H. M. S.* Beagle. *Fossil mammalia. Edited and superintended by Charles Darwin.* London: Smith Elder and Co. [DO]

Frézier, Amédée-François. 1717. *A voyage to the South-Sea, and along the coasts of Chili and Peru, in the years 1712, 1713, and 1714. Particularly describing the genius and constitution of the inhabitants, as well Indians as Spaniards: their customs and manners; their natural history, mines, commodities, traffick with Europe.* London: Printed for J. Bowyer. [*Beagle*]

Funes, Gregorio. 1816–7. *Ensayo de la historia civil del Paraguay, Buenos Aires y Tucuman.* Buenos Ayres: M. J. Gandarillas.*

Gay, Claude. 1833. Aperçu sur les recherches d'histoire naturelle faites dans l'Amérique du Sud, et principalement dans le Chili, pendant les années 1830 et 1831. *Annales des Sciences Naturelles* 28: 369–93. [*Beagle*]

Gosse, Philip. 1938. *St Helena 1502–1938.* Cassell: London.

Grant, Peter R. 1984. Recent research on the evolution of land birds on the Galápagos. *Biological Journal of the Linnean Society* 21: 113–36.

Green, Toby. 1999. *Saddled with Darwin.* London: Wiedenfeld and Nicolson.

Grove, Richard. 1985. Charles Darwin and the Falkland Islands. *Polar Record* 23: 413–20.

Gruber, Howard E. and Gruber, Valmai. 1962. The eye of reason: Darwin's development during the *Beagle* voyage. *Isis* 53: 186–200.

Hacq, J. M. 1826. *Mapa fisico y politico del alto y bajo Peru: nota, esta mapa esta corejido con presencia de las observaciones e itinerarios de los oficiales facultatives que han accompañado a los ejercitos en sus diferentes operaciones, cujos trabajos se reunieron a la cosecuencia de los ultimos acontecimientos por orden del gobierno.* Paris: J. M. Darmet.

Hall, Basil. 1824. *Extracts from a Journal written on the coast of Chili, Peru, and Mexico, in the years 1820, 1821, 1822.* 2 vols. Edinburgh: Printed for Archibald Constable; London: Hurst, Robinson & Co. [*Beagle*] [DO]

Harmer, Sidney F. and Ridewood, W. G. eds. 1910. *Memorials of Charles Darwin: a collection of manuscripts portraits medals books and natural history specimens to commemorate the centenary of his birth and the fiftieth anniversary of the publication of "The origin of species."* 2d edition. London: British Museum (Natural History). Special guide No. 4. [DO]

Harvey, Robert. 2000. *Liberators: South America's savage wars of freedom 1810–30.* London: John Murray.

Head, Francis Bond 1826. *Rough notes taken during some rapid journeys across the Pampas and among the Andes.* London: John Murray. [*Beagle*]

Hearl, Trevor W. 1990. *Darwin's island.* Cheltenham: Privately printed.

Helms, Anton Zacharias. 1807. *Travels from Buenos Ayres, by Potosi, to Lima.* London: Printed for R. Phillips.*

Henslow, John Stevens. 1844. Rust in wheat. *Gardeners' Chronicle* (28 September): 659.

Herbert, Sandra. 1982. Les divergences entre Darwin et Lyell sur quelques questions géologiques. In Conry, Yvette ed. *De Darwin au Darwinisme: science et idéologie.* Paris: J. Vrin, pp. 69–76.

—. 1991. Charles Darwin as a prospective geological author. *British Journal for the History of Science* 24: 159–92. [DO]

—. 1995. From Charles Darwin's portfolio: An early essay on South American geology and species. *Earth Sciences History* 14, no. 1, pp. 23–36. [DO]

—. 1999. An 1830s view from outside Switzerland: Charles Darwin on the "Beryl Blue" glaciers of Tierra del Fuego. *Eclogae Geologicae Helvetiae* 92: 339–46. [DO]

—. 2005. *Charles Darwin, geologist.* Ithaca, NY; London: Cornell University Press.

—. 2007. Doing and knowing: Charles Darwin and other travellers. In Wyse Jackson, P. N. ed. *Four centuries of geological travel.* London: Geological Society Special Publications 287, pp. 311–23.

Herschel, John Frederick William. 1831. *A preliminary discourse on the study of natural philosophy.* Part of Dionysius Lardner's *Cabinet cyclopædia.* London: Longman, Rees, Orme, Brown & Green; John Taylor.

—. 1833. *A treatise on astronomy.* In Lardner, D. ed. *The cabinet cyclopaedia.* London: Longman, Orme, Brown, Green … Longmans, and John Taylor.

Hodge M. J. S. 1983. Darwin and the laws of the animate part of the terrestrial system (1835–1837); on the Lyellian origins of his zoonomical explanatory program. *Studies in History of Biology* 6: 1–106.

—. 1985. Darwin as a lifelong generation theorist. In Kohn, David ed., *The Darwinian heritage.* Princeton: Princeton University Press, in association with Nova Pacifica.

Hoff, Karl Ernst Adolf von. 1822–4. *Geschichte der durch Ueberlieferung nachgeweisen natürlichen Veränderungen der Erdoberfläche.* 3 vols. Gotha: J. Perthes.

Hopkins, R. S. 1969. *Darwin's South America.* New York: The John Day Company.

Hudson, William Henry. 1870. A third letter on the ornithology of Buenos Ayres. [Read 24 March] *Proceedings of the Zoological Society of London* No. 38: 158–60.

Humboldt, Alexander von. 1811. *Political essay on the kingdom of New Spain. Containing researches relative to the geography of Mexico, the extent of its surface and its political division into intendancies, the physical aspect of the country, the population, the state of agriculture and manufacturing and commercial industry, the canals projected between the south sea and Atlantic ocean, the Crown revenues, the quantity of the precious metals which have flowed from Mexico into Europe and Asia, since the discovery of the new continent, and the military defence of New Spain. Translated by John Black.* 4 vols. London: Longman, Hurst, Rees, Orme, Brown. [*Beagle*: (Inscription, both volumes: 'Chas Darwin Buenos Ayres')]

—. 1814–29. *Personal narrative of travels to the equinoctial regions of the New Continent, during the years 1799–1804, by Alexander de Humboldt, and Aimé Bonpland ; with maps, plans, &c. written in French by Alexander de Humboldt, and trans. into English by Helen Maria Williams.* 7 vols. London: Longman, Hurst, Rees, Orme and Brown. [*Beagle*: (Vols. 1 and 2, in one, 3d edition, inscribed from J. S. Henslow to CD 'on his departure', September 1831)]

Huxley, Julian and Kettlewell, H. B. D. 1965. *Charles Darwin and his world.* London: Thames and Hudson.

Journal: Wyhe, John van ed. Darwin's personal 'Journal' (1809–1881) [DAR 158] [DO]

Journal of researches: Darwin, Charles Robert. 1839. *Narrative of the surveying voyages of His Majesty's Ships Adventure and* Beagle *between the years 1826 and 1836, describing their examination of the southern shores of South America, and the* Beagle*'s circumnavigation of the globe. Journal and remarks. 1832–1836.* London: Henry Colburn. [DO]

Journal of researches 2d ed.: Darwin, Charles Robert. 1845. *Journal of researches into the natural history and geology of the countries visited during the voyage of H. M. S.* Beagle *round the world, under the Command of Capt. Fitz Roy, R.N.* 2d edition. London: John Murray. [DO]

Judd, John Wesley. 1909. Darwin and geology. In Seward, Albert Charles ed. 1909. *Darwin and modern science. Essays in commemoration of the centenary of the birth of Charles Darwin and of the fiftieth anniversary of the publication of The origin of species.* Cambridge: Cambridge University Press. [DO]

Keynes, Quentin. 2005. St. Helena: the forgotten island. Reprinted from *The National Geographic Magazine* August 1950 with an introduction by Simon Keynes. Jamestown: Museum of St. Helena and The Society of Friends of St Helena.

Keynes, Richard ed. 1979. *The* Beagle *record. Selections from the original pictorial records and written accounts of the voyage of HMS* Beagle. Cambridge: Cambridge University Press.

Keynes, Richard. 2003. *Fossils, finches and Fuegians: Charles Darwin's adventures and discoveries on the* Beagle*, 1832–1836.* London: Harper Collins.

Keynes, Simon ed. 2004. *Quentin Keynes, explorer, film-maker, lecturer and book-collector 1921–2003.* Cambridge: privately printed.

King, Phillip Parker. 1827. *Narrative of a survey of the intertropical and western coasts of Australia performed between the years 1818 and 1822.* London: John Murray. [*Beagle*]

Kohn, David. 1980. Theories to work by: rejected theories, reproduction, and Darwin's path to natural selection. *Studies in History of Biology* 4: 67–170.

Kotzebue, Otto, von. 1821. *Voyage of discovery in the South Sea, and to Behring's Straits, in search of a north-east passage: undertaken in the years 1815, 16, 17, and 18, in the ship Rurick.* 3 vols. London: Printed for Sir Richard Phillips and Co. [*Beagle*]

Laurent, J. and Campbell, M. 1987. *The Eye of Reason: Charles Darwin in Australasia*. North Wollongong: University of Wollongong Press.

Living Cirripedia: Darwin, Charles Robert. 1851 [= 1852]. *A monograph of the sub-class Cirripedia, with figures of all the species. The Lepadidæ; or, pedunculated cirripedes*. London: The Ray Society. [DO]

Lowe, Percy R. 1936. The finches of the Galapagos in relation to Darwin's conception of species. *Ibis* 13th ser. 6: 310–21.

Luccock, John. 1820. *Notes on Rio de Janeiro, and the southern parts of Brazil: taken during a residence of ten years in that country, from 1808 to 1818*. London: S. Leigh.*

Lyell, Charles. 1830–3. *Principles of geology, being an attempt to explain the former changes of the earth's surface, by reference to causes now in operation*. 3 vols. London: John Murray. [*Beagle*: (Inscriptions: vol. 1 (1830), 'Given me by Capt. F. R C. Darwin'; vol. 2 (1832), 'Charles Darwin M: Video. Novem^r 1832'; vol. 3 (1833): 'C. Darwin')] [DO]

McBirney, A. R. and Williams, Howel. 1969. Geology and petrology of the Galapagos Islands. *Geological Society of America Memoir* 118: 1–197.

MacCulloch, John. 1820. On peat. *Edinburgh Philosophical Journal* 2: 40–59.*

Mammalia: Waterhouse, George R. 1838–9. Part 2 of *The zoology of the voyage of H. M. S. Beagle. Mammalia. Edited and superintended by Charles Darwin*. London: Smith Elder and Co. [DO]

Mariner, William. 1817. *An account of the natives of the Tonga Islands in the South Pacific Ocean*. 2 vols. London: printed for the author.

Mawe, John. 1825. *Travels in the gold and diamond districts of Brazil; describing the methods of working the mines, the natural productions, agriculture, and commerce, and the customs and manners of the inhabitants: to which is added a brief account of the process of amalgamation practised in Peru and Chili. A new edition, illustrated with colored plates*. London: Printed for Longman, Hurst, Rees, Orme, Brown, and Green. [*Beagle*: (Inscription: 'Chas. Darwin Octob: 1832 Buenos Ayres')]

Miers, John. 1826. *Travels in Chile and La Plata, including accounts respecting the geography, geology, statistics, government, finances, agriculture, manners and customs, and the mining operations in Chile*. London: Printed for Baldwin, Cradock, and Joy. [*Beagle*] [DO]

Milton, John. 1667. *Paradise lost. A poem written in ten books*. London: Peter Parker, Robert Butler, and Mathias Walker. [*Beagle*]

More letters: Darwin, Francis & Seward, Albert Charles. eds. 1903. *More letters of Charles Darwin. A record of his work in a series of hitherto unpublished letters*. 2 vols. London: John Murray. [DO]

Molina, Giovanni Ignazio. 1794–5. *Compendio de la historia geografica natural y civil del Reyno de Chile*. Part 1. Madrid: Don Antonio de Sancha, 1794; (Compendio de la historia civil del Reyno de Chile). Part 2. Madrid: 1795. [*Beagle*: (Inscription: 'Charles Darwin Valparaiso 1834')]

Morris, John and Sharpe, Daniel. 1846. Description of eight species of brachiopodous shells from the Palæozoic rocks of the Falkland Islands. *Quarterly Journal of the Geological Society of London* 2: 274–8. [DO]

Morton, Nicol. 1995. In the footsteps of Charles Darwin in South America. *Geology Today* (September–October): 190–5.

Murray, John. 1826. *Experimental researches on the light and luminous matter of the glow-worm, the luminosity of the sea, the phenomena of the chameleon, the ascent of the spider into the atmosphere, and the torpidity of the tortoise, etc.* Glasgow: W. R. M'Phun.*

Murray, Lindley. 1824. *An English grammar, comprehending the principles and rules of the language, illustrated by appropriate exercises and a key to the exercises.* 5th ed. 2 vols. York: Longman, Hurst, Rees, Orme, Brown, and Green. [*Beagle*: A copy in Darwin Library – CUL of the two-volume edition of 1824 has the signature 'Robert FitzRoy 1831']

Narrative 1: King, Philip Parker. 1839. *Narrative of the surveying voyages of His Majesty's Ships Adventure and* Beagle *between the years 1826 and 1836, describing their examination of the southern shores of South America, and the* Beagle's *circumnavigation of the globe. Proceedings of the first expedition, 1826–30, under the command of Captain P. Parker King, R. N., F. R. S.* London: Henry Colburn. [DO]

Narrative 2: FitzRoy, Robert. 1839. *The narrative of the voyages of H. M. Ships Adventure and* Beagle. *Proceedings of the second expedition, 1831–36, under the command of Captain Robert Fitz-Roy, R.N.* London: Henry Colburn. [DO]

Narrative Appendix: FitzRoy, Robert. 1839. *Narrative of the surveying voyages of His Majesty's Ships Adventure and* Beagle *between the years 1826 and 1836, describing their examination of the southern shores of South America, and the Beagle's circumnavigation of the globe. Appendix to Volume II.* London: Henry Colburn. [DO]

Natural selection: Stauffer, Robert C. ed. 1975. *Charles Darwin's Natural Selection; being the second part of his big species book written from 1836 to 1858.* Cambridge: Cambridge University Press. [DO]

Nicholas, Frank W. and Nicholas, Jan M. 1989. *Charles Darwin in Australia.* Cambridge: Cambridge University Press.

OED: *The Oxford English dictionary.* (http://www.oed.com/)

Orbigny, Alcide Charles Victor Dessalines, d'. [1834]–47. *Voyage dans l'Amérique méridionale: le Brésil, la République orientale de l'Uruguay, la République argentine, la Patagonie, la République du Chili, la République de Bolivia, la République du Pérou-executé pendant les années 1826, 1827, 1828, 1829, 1830, 1831, 1832, et 1833.* Paris: Bertrand; Strasbourg: Levrault.

Origin of species: Darwin, Charles Robert. 1859. *On the origin of species by means of natural selection, or the preservation of favoured races in the struggle for life.* London: John Murray. [DO]

Origin of species 6th ed: Darwin, Charles Robert. 1872. *On the origin of species by means of natural selection, or the preservation of favoured races in the struggle for life.* London: John Murray. [DO]

'Ornithological notes': Barlow, Nora. ed. 1963. Darwin's ornithological notes. With introduction, notes & appendix by the editor. *Bulletin of the British Museum (Natural History) Historical Series* 2 (7): 201–78.

Parodiz, Juan José. 1981. *Darwin in the New World.* Leiden: E. J. Brill.

Pauly, Daniel. 2004. *Darwin's fishes: an encyclopedia of ichthyology, ecology, and evolution.* Cambridge: Cambridge University Press.

Pearson, Paul N. 1996. Charles Darwin on the origin and diversity of igneous rocks. *Earth Sciences History* 15: 49–67. [DO]

—. and Nicholas, C. J. 2007. 'Marks of extreme violence': Charles Darwin's geological observations at St Jago (São Tiago), Cape Verde Islands. In Wyse Jackson, P. N. ed. *Four centuries of geological travel*. London: Geological Society Special Publications 287, pp. 239–53.

Pennant, Thomas. 1771. *Synopsis of quadrupeds*. Chester: by J. Monk.*

Péron, François. 1807. *Voyage aux Terres Australes*. Vol. 1. Paris: Imprimerie Impériale; A. Bertrand.*

Phillips, William. 1816. *An elementary introduction to mineralogy: including some account of mineral elements and constituents; explanations of terms in common use; brief accounts of minerals, and of the places and circumstances in which they are found: Designed for the use of the student*. London: Printed, and sold by William Phillips. [*Beagle*]

Playfair, John. 1802. *Illustrations of the Huttonian theory of the earth*. Edinburgh, London: William Creech, Printed for Cadell and Davies. [*Beagle*]

Porter, Duncan M. 1985. The *Beagle* collector and his collections. In Kohn, David ed. *The Darwinian heritage*. Princeton, N.J.: Princeton University Press.

Puigmal, Patrick. 2006. *¡Diablos, no pensaba en Chile hace tres años!: cartas inéditas sobre la independencia de Chile, Argentina y Perú (1817–1825) Joseph Albert Bacler D'Albe estudio biográfico y prosopográfico*. Osorno: PEDCH, Universidad de los Lagos.

Rachootin, Stan Philip. 1985. Owen and Darwin reading a fossil: *Macrauchenia* in a boney light. In Kohn, David ed. *The Darwinian heritage*. Princeton: University Press, pp. 155–84.

Red notebook: Herbert, Sandra ed. 1980. The red notebook of Charles Darwin. *Bulletin of the British Museum (Natural History) Historical Series* 7: 1–164. [DO]

Rhodes, Frank H. T. 1991. Darwin's search for a theory of the Earth: symmetry, simplicity and speculation. *British Journal for the History of Science* 24: 193–229.

Richardson, Constance. 1933. Petrology of the Galapagos Islands. In Lawrence John Chubb, Geology of Galapagos, Cocos, and Easter Islands. *Bernice B. Bishop Museum-Bulletin* 110: 45–67.

Richardson, Samuel. 1781. *The history of Sir Charles Grandison, in a series of letters publ. by the editor of Pamela. To which is added A brief history of the treatment which the editor has met with from certain booksellers and printers in Dublin*. 7 vols. London: W. Strahan. [*Beagle*]

Rosen, Brian Roy. 1982. Darwin, coral reefs, and global geology. *Bioscience* 32: 519–25.

Rookmaaker, Kees ed. 2006. Darwin's *Beagle* diary [the first unabridged transcription]. [DO]

—. ed. 2008. Darwin's Edinburgh notebook. [DAR 118] [DO]

Rudwick, Martin J. S. 1974. Darwin and Glen Roy: a 'great failure' in scientific method? *Studies in the History and Philosophy of Science* 5: 97–185.

—. 1982. Charles Darwin in London: the integration of public and private science. *Isis* 73: 186–206.

—. 1985. *The great Devonian controversy: The shaping of scientific knowledge among gentlemanly specialists*. Chicago, London: University of Chicago Press.

—. 2005. *Lyell and Darwin, geologists: Studies in the earth sciences in the age of reform*. Aldershot: Ashgate.

—. 2008. *Worlds before Adam: The reconstruction of geohistory in the age of reform*. Chicago: University of Chicago Press.

Schweber, Silvan S. 1985. The wider British context in Darwin's theorizing. In Kohn, David ed., *The Darwinian heritage*. Princeton: Princeton University Press, pp. 35–71.

Scoresby, William. 1820. *An account of the Arctic regions, with a history and description of the northern whale-fishery*. Edinburgh: A. Constable and Co.*

Scrope, George Julius Poulett. 1825. *Considerations on volcanos, the probable causes of their phenomena, the laws which determine their march, the disposition of their products, and their connexion with the present state and past history of the globe; leading to the establishment of a new theory of the earth*. London: W. Phillips. [*Beagle*]

Seale, Robert F. 1834. *The geognosy of the Island of St Helena, illustrated in a series of views, plans and sections; accompanied with explanatory remarks and observations*. London: Ackermann and Co.*

Secord, James A. 1991. The discovery of a vocation: Darwin's early geology. *British Journal for the History of Science* 24: 133–57.

[Shipley, Arthur Everett and Simpson, James Crawford eds.] 1909. *Darwin centenary: the portraits, prints and writings of Charles Robert Darwin, exhibited at Christ's College, Cambridge 1909*. [Cambridge: Cambridge University Press]. [DO]

Shorter publications: Wyhe, John van. 2009. *Charles Darwin's shorter publications 1829–1883*. Cambridge: Cambridge University Press.

Simkin, Tom. 1984. Geology of Galapagos. *Biological Journal of the Linnean Society* 21: 61–75.

Sloan, Phillip. 1985. Darwin's invertebrate program, 1826–1836. In Kohn, David ed. *The Darwinian heritage*. Princeton, N.J.: Princeton University Press.

South America: Darwin, Charles Robert. 1846. *Geological observations on South America. Being the third part of the geology of the voyage of the* Beagle, *under the command of Capt. Fitzroy, R. N. during the years 1832 to 1836*. London: Smith Elder and Co. [DO]

Spix, Johann Baptist von and Martius, Carl Frederick Philip von. 1824. *Travels in Brazil in the years 1817–1820, undertaken by command of His Majesty the King of Bavaria*. 2 vols. London: Printed for Longman, Hurst, Rees, Orme, Brown, and Green. [*Beagle*: (Inscription in vol. 2: 'Chas. Darwin Octob: 1832 Buenos Ayres')]

Stanbury, David. 1977. *A narrative of the voyage of HMS* Beagle *being passages from the 'Narrative' written by Captain FitzRoy selected and edited by David Stanbury*. London: The Folio Society.

Steinheimer, Frank D. 2004. Charles Darwin's bird collection and ornithological knowledge during the voyage of H. M. S. *Beagle*, 1831–1836. *Journal of Ornithology* 145 (4): 300–20, (appendix [pp. 1–40]). [DO]

Stevenson, William Bennet. 1825. *A historical and descriptive narrative of twenty years' residence in South America: containing the travels in Arauco, Chile, Peru, and Colombia; with an account of the revolution, its rise, progress, and results*. 3 vols. London: Hurst, Robinson & Co.*

Stoddart, David R. ed. 1962. Coral islands. *Atoll Research Bulletin* no. 88: 1–20. [DO]

—. 1976. Darwin, Lyell, and the geological significance of coral reefs. *British Journal for the History of Science* 9: 199–218.

—. 1995. Darwin and the seeing eye: iconography and meaning in the *Beagle* years. *Earth Sciences History* 14 (November): 3–22.

Sulloway, Frank J. 1982a. The *Beagle* collections of Darwin's finches (*Geospizinae*). *Bulletin of the British Museum (Natural History) Zoology Series* vol. 43 no. 2: 49–94.

—. 1982b. Darwin and his finches: The evolution of a legend. *Journal of the History of Biology* 15: 1–53.

—. 1982c. Darwin's conversion: The *Beagle* voyage and its aftermath. *Journal of the History of Biology* 15: 325–96.

—. 1983. Further remarks on Darwin's spelling habits and the dating of *Beagle* voyage manuscripts. *Journal of the History of Biology* 16: 316–90.

—. 1984. Darwin and the Galapagos. *Biological Journal of the Linnean Society*, 21: 29–59.

—. 1985: Darwin's early intellectual development: an overview of the *Beagle* voyage (1831–1836). In Kohn, David ed., *The Darwinian heritage*. Princeton, N.J.: Princeton University Press, pp. 121–54.

Syme, Patrick. 1821. *Werner's nomenclature of colours: with additions, arranged so as to render it highly useful to the arts and sciences, particularly zoology, botany, chemistry, mineralogy, and morbid anatomy. Annexed to which are examples selected from well-known objects in the animal, vegetable, and mineral kingdoms.* 2d ed. Edinburgh: Blackwood. [*Beagle*]

Taylor, James. 2008. *The voyage of the* Beagle*: Darwin's extraordinary adventure aboard Fitzroy's famous survey ship.* London: Anova Books.

Thackeray, J. F. 1982–3. On Darwin, extinctions and South African fauna. *Discovery* 16 (2): 2–11.

Thomson, Keith. 1995. *HMS* Beagle*: The story of Darwin's ship.* London: W. W. Norton.

Thunberg, Carl Peter. 1795–6. *Travels in Europe, Africa and Asia, performed between the years 1770 and 1779.* London: W. Richardson and J. Egerton.*

Ulloa, Antonio de. 1806. *A voyage to South America: describing at large the Spanish cities, towns, provinces, &c. on that extensive continent: undertaken, by command of the king of Spain, by Don George Juan and Don Antonio de Ulloa. Tr. from the original Spanish; with notes and observations; and an account of the Brazils. By John Adams.* 2 vols. 4th ed. London: Printed for J. Stockdale. [*Beagle*] [DO]

Vargas y Ponce, Josef de. 1788. *Relacion del ultimo viage al Estrecho de magellanes de la Fragata de S. M. Santa Maria de la Cabeza en los anos de 1785 y 1786.* Madrid: Por la viuda de Ibarra, hijos y compañia.*

Variation: Darwin, Charles Robert. 1868. *The variation of animals and plants under domestication.* 2 vols. London: John Murray. [DO]

Volcanic islands: Darwin, Charles Robert. 1844. *Geological observations on the volcanic islands visited during the voyage of H. M. S. Beagle, together with some brief notices of the geology of Australia and the Cape of Good Hope. Being the second part of the geology of the voyage of the* Beagle*, under the command of Capt. Fitzroy, R. N. during the years 1832 to 1836.* London: Smith Elder and Co. [DO]

Wafer, Lionel. 1699. *A new voyage and description of the Isthmus of America, giving an account of the author's abode there.* London: James Knapton.*

Waterton, Charles. 1833. The habits of the carrion crow. *Magazine of Natural History, and Journal of Zoology, Botany, Mineralogy, Geology, and Meteorology* 6 (No. 32): 208–14.*

Wilson, Leonard G. 1972. *Charles Lyell: the years to 1841: the revolution in geology.* New Haven and London.

Winkler Hans, Christie, David A. and Nurney, David. 1995. *Woodpeckers: a guide to the woodpeckers, piculets and wrynecks of the world.* Robertsbridge: Pica Press.

Winslow, John H. 1975. Mr Lumb and Masters Megatherium: an unpublished letter by Charles Darwin from the Falklands. *Journal of Historical Geography* 1: 347–60.

Wyhe, John van. 2004a. *Phrenology and the origins of Victorian scientific naturalism.* Aldershot: Ashgate.

—. 2004b. Was phrenology a reform science? Towards a new generalization for phrenology. *History of Science* 42: 313–31.

—. 2007a. Mind the gap: Did Darwin avoid publishing his theory for many years? *Notes and Records of the Royal Society* 61: 177–205. [DO]

—. 2007b. 'The position of the bones of Mastodon (?) at Port St Julian is of interest' [DAR 42.97–9] [DO]

—. 2008a. *Darwin*. London: Andre Deutsch.

—. 2008b. [Darwin's notes on reading Sumner's *Evidence of Christianity*]. [DAR 91.114–8] [DO]

Young, Victoria. ed. 1995. *The journal of Syms Covington*. http://www.asap.unimelb.edu.au/bsparcs/covingto/chap_1.htm (Accessed 9 June 2007.)

Yudilevich Levy, David and Le-Fort, Eduardo Castro eds. 1995. *Darwin en Chile (1832–1835): viaje de un naturalista alrededor del mundo por Charles Darwin*. Editorial Universitaria. Santiago de Chile.

Zoology: Darwin, Charles Robert. ed. 1838–43. *The zoology of the voyage of H. M. S. Beagle*. 5 vols. London: Smith Elder and Co. [DO]

Zoology notes: Keynes, Richard ed. 2000. *Charles Darwin's zoology notes & specimen lists from H. M. S. Beagle*. Cambridge: Cambridge University Press. [DO]

INDEX

⏝